Methods in Protein Biochemistry
Edited by Harald Tschesche

Methods in Protein Biochemistry

Edited by Harald Tschesche

DE GRUYTER

Editor

Prof. Dr. em. Harald Tschesche
Universität Bielefeld
Lehrstuhl für Biochemie I
Universitätsstraße 25
33615 Bielefeld, Germany
harald.tschesche@uni-bielefeld.de

This book has 89 figures and 17 tables.

Front cover image kindly provided from Prof. David A. Lomas.

ISBN 978-3-11-025233-0
e-ISBN 978-3-11-025236-1

Library of Congress Cataloging-in-Publication Data

> Methods in protein biochemistry / edited by Harald Tschesche.
>   p. ; cm.
> Includes bibliographical references and index.
> ISBN 978-3-11-025233-0 (alk. paper)
> I. Tschesche, Harald.
> [DNLM: 1. Proteins—analysis. 2. Proteins—chemistry.
> 3. Proteins—isolation & purification. 4. Proteomics—methods.   QU 55]
> LC classification not assigned
> 572'.67—dc23   2011029558

Bibliografic information published by the Deutsche Nationalbibliothek

The Deutsche Nationalbibliothek lists this publication in the Deutsche Nationalbibliografie; detailed bibliographic data are available in the Internet at http://dnb.d-nb.de.

© Copyright 2012 by Walter de Gruyter GmbH & Co. KG, Berlin/Boston

Printing and Binding: Hubert & Co. GmbH & Co. KG, Göttingen
Printed in Germany
www.degruyter.com

Preface

The enormous progress in protein biochemistry within the past five decades is the result of the very rapid development and progress in analytical methods. Therefore, it is the intention of this book to provide the continuously growing family of researchers involved in the study of proteins with a compendium that reviews the latest developments in the very special fields of protein studies in vitro and in vivo. It is not the aim of this book to provide a complete listing of all efforts made in a particular field but, rather, to give experts the opportunity to evaluate particular developments and to offer the inexperienced investigator an opportunity to orient himself or herself among the literature and to evaluate the chosen method in accordance with his or her special needs.

Of course, this book can only provide a highly selected assortment from the multiplicity of analytical methods, which I selected from authors that I met at research meetings, such as the International Association of Protein Structure Analysis and Proteomics (IAPSAP) and the Max-Bergman-Kreis (MBK), or that I contacted on recommendation or because of personal acquaintance. Hence, I would like to thank Ettore Appella, Carl Anderson, Jan Johansson, and Roland Schauer for support and valuable suggestions.

It is hoped that this book will help to extend methodological knowledge, assist the reader in the evaluation and application of these methods, facilitate the approach to actual scientific problems, and help avoid time-consuming and unnecessary errors.

Münster, August 2011
Harald Tschesche

Editor

Prof. Dr. Harald Tschesche (em.), Professor of Biochemistry at the University of Bielefeld studied Chemistry at the University Bonn and Chemistry and Microbiology at the University Heidelberg, where he received his doctorate with the Nobel Laureate Prof Wittig. After a year as research associate at the Massachusetts Institute of Technology (MIT) he worked with Prof. Weygand at the Technical University in Munich (TUM) and habilitated in 1970. He got nomination for chairs at the Universities of Braunschweig, Essen, and Bielefeld. He published 366 peer reviewed scientific articles and edited 9 books including the German translation of the Lehninger, Nelson, Cox: Principles in Biochemistry, 1994. He received the Max-Bergmann Medal 1994 for Peptide and Protein Research; and is a member of the board of directors for the International Association for Protein Structure Analysis and Proteomics (IAPSAP) and an executive board member of the association for Research and Technology Transfer (GFT) at the University Bielefeld.

List of contributing authors

Sarka Beranova-Giorgianni, Ph.D.
Associate Professor
University of Tennessee Health
Science Center
Department of Pharmaceutical Sciences
Pharmacy Building, Room 445
881 Madison Avenue
Memphis, TN 38163
sberanova@uthsc.edu
Chapter 13

Prisca Boisguerin
Institute for Medical Immunology
Charité - Universitätsmedizin Berlin
Berlin, Germany
prisca.boisguerin@charite.de
Chapter 15

David M. Byers
Department of Biochemistry and
Molecular Biology
Department of Pediatrics
Atlantic Research Centre
Dalhousie University
Halifax, Canada
Chapter 16

Julio J. Caramelo
Institute of Biophysics
and Biochemistry
School of Pharmacy
and Biochemistry
University of Buenos Aires
Argentina
Chapter 4

List of contributing authors

Li Chen
National Life Science BU Manager
Millipore
Billerica, Massachusetts
United States
Chapter 13

Ivan R. Corrêa Jr.
New England Biolabs
Ipswich, United States
Chapter 12

John E. Cronan, Jr.
Departments of Microbiology and
Biochemistry
University of Illinois
Urbana, Illinois
United States
Chapter 16

Charles M. Deber
Department of Biochemistry
University of Toronto
Toronto, Canada
Chapter 3

Clive Dennison
School of Biochemistry, Genetics and
Microbiology
University of KwaZulu-Natal
Scottsville, Pietermaritzburg
South Africa
Chapter 1

Dominic M. Desiderio
Department of Neurology
The University of Tennessee
Health Science Center
ddesiderio@uthsc.edu
Chapter 13

Michael Dreiseidler
Institute for Cell Biology
University of Bonn
Germany
dreiseidler@uni-bonn.de
Chapter 2

Beat Fierz
Laboratory of Synthetic Protein Chemistry
Rockefeller University
New York, United States
Chapter 5

Robert F. Fischetti
Argonne National Laboratory
Biosciences Division
rfischetti@anl.gov
Chapter 10

Rudolf Glockshuber
Institute of Molecular Biology and
Biophysics
ETH Zurich
Switzerland
Chapter 11

Francesco Giorgianni, Ph.D.
Assistant Professor
University of Tennessee Health Science
Center
Department of Pharmaceutical Sciences
Pharmacy Building, Room 456
881 Madison Avenue
Memphis, TN 38163
Chapter 13

Bibek Gooptu
Institute of Structural and Molecular
Biology
Department of Biological Sciences
University of London
United Kingdom
Chapter 18

Ariele Hanek
New England Biolabs
Ipswich, United States
Chapter 12

Johannes M. Herrmann
Cell Biology
University of Kaiserslautern
Kaiserslautern, Germany
Chapter 6

Hisashi Hirano
Yokohama City University
Advanced Medical Research
Center
School of Nanobioscience
Yokohama, Japan

hirano@yokohama-cu.ac.jp
Chapter 7, 14

Jörg Höhfeld
Institut für Zellbiologie
Universität Bonn
Bonn, Germany
hoehfeld@uni-bonn.de
Chapter 2

Yoko Ino
Graduate School of Nanobioscience
Yokohama City University
Suehiro 1-7-29, Tsurumi
Yokohama 230-0045, Japan
Chapter 7

Roza Maria Kamp
Beuth Hochschule für Technik Berlin
University of Applied Sciences
Berlin, Germany
Chapter 14

Reiko Kazamaki
Higashicho 4-28-13-208
Koganei, Tokyo
184-0011, Japan
Chapter 7

Thomas Kiefhaber
Technische Universität München
Chemistry Department & Munich Center
for Integrated Protein Science
Garching, Germany
Chapter 5

Alexander Leitner
Institute of Molecular Systems Biology
ETH Zurich
Switzerland
Chapter 8

David A. Lomas
Department of Medicine
University of Cambridge
Cambridge Institute
for Medical Research
Wellcome Trust/MRC Building
Cambridge, United Kingdom
Chapter 18

Dean R. Madden
Dartmouth Biochemistry
Darthmouth Medical School
Hanover, New Hampshire
United States
Chapter 15

Oleg A. Makarov
National Institute of General Medical
Sciences and National Cancer Institute
Collaborative Access Team
Biosciences Division
Argonne National Laboratory
Argonne, IL 60439
Chapter 10

Lee Makowski
Dana Research Center
Chemistry/Chemical Biology
Northeastern University
Boston, United States
Chapter 10

Valentina Mileo
Chiesi Farmaceutici S.P.A.
Parma, Italy
v.mileo@chiesigroup.com
Chapter 13

Kerstin Moehle
Department of Chemistry
University of Zurich
Switzerland
Chapter 17

Peter W. Murphy
Department of Biochemistry and
Molecular Biology
Department of Pediatrics
Atlantic Research Centre
Dalhousie University
Halifax, Canada
Chapter 16

Allen M. Orville
Department of Biology
Brookhaven National Laboratory
Upton, New York
United States
Chapter 9

Armando J. Parodi
Fundacion Instituto Leloir
Buenos Aires, Argentina
Chapter 4

Andreas Reiner
Department of Molecular
and Cell Biology
University of California
Berkeley
United States
Chapter 5

Jan Riemer
Cell Biology
University of Kaiserslautern
Kaiserslautern, Germany
Chapter 6

John A. Robinson
Department of Chemistry
University of Zurich
Switzerland
robinson@oci.uzh.ch
Chapter 17

Elden E. Rowland
Department of Pediatrics
Atlantic Research Centre
Dalhousie University
Halifax, Canada
Chapter 16

Kunigunde Stephani-Kosin
Beuth Hochschule für Technik Berlin
University of Applied Sciences
13347 Berlin
Germany
Chapter 14

David V. Tulumello
Department of Biochemistry
University of Toronto
Toronto, Ontario
Canada M5S 1A8
Chapter 3

Anna Ulbricht
Institute for Cell Biology
University of Bonn
Germany
Chapter 2

Toni Vagt
Institute of Molecular Biology and
Biophysics
ETH Zurich
Switzerland
Chapter 11

Gerrit Volkmann
Institut für Biochemie
Westfälische Wilhelms-Universität
Münster
Münster, Germany
Chapter 16

Rudolf Volkmer
Institute for Medical Immunology
Charité-Universitätsmedizin
Berlin
Berlin, Germany
rve@charite.de
Chapter 15

Shenglan Xu
National Institute of General Medical
Sciences and National Cancer
Institute
Collaborative Access Team
Biosciences Division
Argonne National Laboratory
Argonne, IL 60439
Chapter 10

Contents

Preface v
Editor vii
List of contributing authors vii
Abbreviations xvii
Acknowledgements xxv

1 Three-phase partitioning 1
    1.1   Method 2
    1.2   The mechanism of TPP 3
    1.3   A practical example – the isolation of cathepsin L from liver tissue 4
    1.4   Other applications 5
              Clive Dennison

2 Folding and degradation functions of molecular chaperones 13
    2.1   Introduction 13
    2.2   The domain structure of Hsc/Hsp70 13
    2.3   The Hsc/Hsp70 reaction cycle 15
    2.4   Cochaperones determine the function of Hsc/Hsp70 16
    2.5   In vitro reconstitution and functional analysis of the Hsc/Hsp70 chaperone system 16
    2.6   Measuring the ATPase activity of Hsc/Hsp70 18
    2.7   Determining chaperone activity 18
    2.8   In vitro reconstitution of chaperone-assisted ubiquitylation 19
    2.9   Concluding remarks 21
              Michael Dreiseidler, Anna Ulbricht, and Jörg Höhfeld

3 Membrane protein folding in detergents 23
    3.1   Introduction 23
    3.2   Interactions of membrane proteins with detergents 24
    3.3   Techniques to characterize TM proteins in detergents 29
    3.4   Applications of TM protein-detergent complexes 35
    3.5   Conclusions 41
              David V. Tulumello and Charles M. Deber

4 Glycoprotein-folding quality control in the endoplasmic reticulum 47
    4.1   Introduction 47
    4.2   Glycoprotein-folding quality control (QC) 47
    4.3   The UGGT 49
    4.4   GII 52
    4.5   CNX and CRT 54

4.6	ERp57	57
4.7	Methods to study glycoprotein folding QC	58

Julio J. Caramelo and Armando J. Parodi

5 Conformational dynamics in peptides and proteins studied by triplet-triplet energy transfer ... 73

5.1	Introduction	73
5.2	Concept of TTET experiments to study intrachain loop formation in polypeptide chains	73
5.3	Diffusion-controlled loop formation in unstructured polypeptide chains	79
5.4	Detection of fast conformational fluctuations in folded peptides and proteins by TTET	85
5.5	Conclusions	91

Andreas Reiner, Beat Fierz, and Thomas Kiefhaber

6 Protein import into the intermembrane space of mitochondria ... 95

6.1	Introduction	95
6.2	The mitochondrial IMS	95
6.3	The mitochondrial disulfide relay	96
6.4	The sulfhydryl oxidase Erv1	96
6.5	The oxidoreductase Mia40	98
6.6	Substrates of the mitochondrial disulfide relay	98
6.7	Methods to study mitochondrial protein translocation	99
6.8	General comments to the analysis of thiol-disulfide redox states	105
6.9	Outlook	107

Jan Riemer and Johannes M. Herrmann

7 On-membrane identification of gel-resolved proteins by matrix-assisted laser desorption ionization mass spectrometry (MALDI-MS) ... 113

7.1	Introduction	113
7.2	Methods for identifying proteins electroblotted onto the PVDF membrane	116
7.3	General comments to the analysis of proteins on membranes	118
7.4	PVDF membranes or diamond-like carbon-coated (DLC) stainless steel plates?	124
7.5	Concluding remarks	124

Yoko Ino, Reiko Kazamaki, and Hisashi Hirano

8 Analysis of protein complexes using chemical cross-linking and mass spectrometry ... 127

8.1	Introduction	127
8.2	Reagents for chemical cross-linking	127
8.3	The chemical cross-linking workflow	131
8.4	MS and data analysis	133
8.5	Practical examples	136

8.6	The use of spatial constraints for modeling	137
8.7	Conclusion and outlook	138
	Alexander Leitner	

9 Single-crystal spectroscopy correlated with X-ray crystallography provides complementary perspectives on macromolecular function ... 143

9.1	Introduction	143
9.2	Ionizing radiation: essential for crystal structures; a problem and a reagent	146
9.3	Cofactors in biology provide spectroscopic access to reaction cycles	147
9.4	Single-crystal spectroscopy correlated with X-ray diffraction	150
9.5	Correlated studies at beamline X26-C of the NSLS	153
9.6	Future prospects	159
	Allen M. Orville	

10 Wide-angle X-ray solution scattering (WAXS) ... 165

10.1	Introduction	165
10.2	Sample preparation	166
10.3	Sample-handling robot	167
10.4	Data collection	168
10.5	Data processing	169
10.6	Structural information	171
10.7	Size and shape	171
10.8	Secondary and tertiary structure	172
10.9	Quaternary structure	173
10.10	Structural changes	175
10.11	Unfolding	177
10.12	Molecular modeling	179
10.13	Modeling of structural fluctuations	180
10.14	Outlook	182
	Lee Makowski, Oleg A. Makarov, Shenglan Xu, and Robert F. Fischetti	

11 Where purity matters: recombinant versus synthetic peptides in beta amyloid formation ... 187

11.1	Amyloid fibrils possess a defined quaternary structure	187
11.2	The importance of purity for reproducible kinetics of amyloid fibril formation in vitro: the Aβ as an example	189
11.3	Future challenges for the characterization of fibrillar structures	192
	Toni Vagt and Rudi Glockshuber	

12 Chemical modification of proteins in living cells ... 197

12.1	Introduction	197
12.2	Site-specific labeling of proteins with chemical probes	198
12.3	Selecting an appropriate labeling technique	206
12.4	Live cell applications	206
12.5	Technical Protocols for SNAP- tag labeling	211
	Ariele Hanek and Ivan R. Corrêa Jr.	

13 Proteomics of human bronchoalveolar lavage fluid: discovery of biomarkers of chronic obstructive pulmonary disease (COPD) with difference gel electrophoresis (DIGE) and mass spectrometry (MS) 219
13.1 Introduction 219
13.2 Application of DIGE platform to COPD biomarker discovery 222
13.3 Outlook 233
Francesco Giorgianni, Valentina Mileo, Li Chen, Dominic M. Desiderio, and Sarka Beranova-Giorgianni

14 Proteomic analysis of Duchenne muscular dystrophy (DMD) 235
14.1 Introduction 235
14.2 Materials 236
14.3 Methods 237
14.4 Results and Discussion 240
14.5 Conclusion 248
Kunigunde Stephani-Kosin, Hisashi Hirano, and Roza Maria Kamp

15 Target-oriented peptide arrays in a palliative approach to cystic fibrosis (CF) 249
15.1 Introduction 249
15.2 PDZ domains 250
15.3 CF 252
15.4 Role of PDZ domains in CFTR trafficking 253
15.5 Target-oriented peptide arrays 254
15.6 An engineered peptide inhibitor of CAL extends the half-life of ΔF508-CFTR 256
15.7 Methods 260
15.8 Outlook 264
Prisca Boisguerin, Dean R. Madden, and Rudolf Volkmer

16 Probing protein dynamics in vivo using backbone cyclization: bacterial acyl carrier protein as a case study 271
16.1 Introduction 271
16.2 In vivo protein cyclization, biophysical analyses and functional assays 275
16.3 Outlook 288
Gerrit Volkmann, Peter W. Murphy, Elden E. Rowland, John E. Cronan Jr., and David M. Byers

17 The protein epitope mimetic approach to protein-protein interaction inhibitors 295
17.1 Introduction 295
17.2 Mechanisms of protein-protein interactions 295
17.3 Small-molecule screening approaches 299
17.4 Protein epitope mimetic approaches 301
John A. Robinson and Kerstin Moehle

18 The structural biology of α_1-antitrypsin deficiency and the serpinopathies 325
- 18.1 Clinical phenotypes of the serpinopathies 325
- 18.2 The serpin mechanism of protease inhibition 327
- 18.3 Folding, misfolding and polymerization 328
- 18.4 Serpin folding 330
- 18.5 Dissecting the pathways of polymerization 330
- 18.6 Cellular processing of polymers 337
- 18.7 Stem cell technology to generate models of disease 340
- 18.8 Conclusions 340
  Bibek Gooptu and David A. Lomas

Index 345

Abbreviations

Aβ	amyloid β peptide
ACP	acyl carrier protein
ACP-L46W	*Vibrio harveyi* ACP with leucine to tryptophan mutation at position 46
ACTH	adrenocorticotropic hormone
AD	Alzheimer's disease
AMP	adenosinmonophosphate
ANS	anilinonaphtalene sulfonic acid
AP	acceptor peptide
APP	amyloid precursor protein
ATF6	activating transcription factor 6
ATP	adenosine triphosphate
AUC	analytical ultracentrifugation
BAG-1	Bcl2-associated athanogen
BAL	bronchoalveolar lavage fluid
BC	O^2-Benzylcytosine
BG	O^6-Benzylguanine
BiP	binding protein
BirA	biotin ligase
BSA	bovine serum albumin
BS^{2}G	bis(sulfosuccinimidyl) glutarate
BS^3	bis(sulfosuccinimidyl) suberate
BTP	bromothenylpteridine
BVA	biological variation analysis
CAL	CFTR-associated ligand
CALI	chromophore-assisted laser inactivation
CAP	chaperone-assisted proteasomal degradation
CASA	chaperone-assisted selective autophagy
CCD	charge coupled device (two-dimensional position sensitive detector)
CD	circular dichroism
CDI	1,1'-carbonyldiimidazole
CDT	1,1'-carbonyldi(1,2,4-triazole)
CF	cystic fibrosis
CFTR	cystic fibrosis transmembrane conductance regulator
CGHC	cysteine-containing active sites
CHAPS	3-[(3-cholamidopropyl)-dimethylammonio]-1-propanesulfonate
CHIP	carboxy terminus of Hsc70 interacting protein
CHO	choline oxidase
CI2	chymotrypsin inhibitor 2
CJD	Creutzfeld-Jakob disease

CMA	chaperone-mediated autophagy
CMC	critical micelle concentration
CNX	calnexin
CoA	coenzyme A
ConA	concanavalin A
COPD	chronic obstructive pulmonary disease
CP	O^6-benzyl-4-chloropyrimidine
CRIPT	cysteine-rich interactor of PDZ three
CRT	calreticulin
CRYAB	crystallin B
CSD	charge state distribution
cycL46W	backbone-cyclized variant of ACP-L46W
(cyc)ORF	(cyclization) open reading frame
CZ	cruzipain
DCM	dichloromethane
DDM	*n*-dodecyl-β-D-maltoside
deoxy HbA	deoxygenated form of human hemoglobin
DhaA	haloalkane dehalogenase
DHR	discs large homology repeat
di-α HbA	carbonmonoxy form of human hemoglobin in which the two α-chains have been linked covalently by a single glycine residue
DIA	differential in-gel analysis
DIGE	difference gel electrophoresis
DIPEA	diisopropylethylamine
DLC plate	diamond-like carbon-coated plate
Dlg	discs large
DM	n-decyl-β-D-maltoside
DMD	Duchenne muscular dystrophy
DMF	dimethylformamide
DMSO	dimethyl sulfoxide
DNA	deoxyribonucleic acid
Dol	dolichol
DPC	*n*-dodecyl phosphocholine
DSG	disuccinimidyl glutarate
DSS	disuccinimidyl suberate
dSTORM	direct stochastic optical reconstruction microscopy
DTT	dithiothreitol
Eco	*Escherichia coli*
EDC	ethyl diisopropyl carbodiimide
eDHFR	*Escherichia coli* dihydrofolate reductase
EDTA	ethylenediaminetetraacetic acid
EEDQ	ethyl 1,2-dihydro-2-ethoxyquinoline-1-carboxylate
EM	electron microscopy
Endo H	endo-β-N-acetylglucosaminidase H
EPR	electron paramagnetic resonance

ER	endoplasmic reticulum
ERAD	ER-associated degradation
ERGIC	ER-Golgi intermediate compartment
ESES	epilepsy of slow-wave sleep
ESI-LIT/TOF MS	electrospray ionization-linear ion trap/time-of-flight mass spectrometry
ESI-Q/TOF MS	electrospray ionization-quadrupole/time-of-flight mass spectrometry
FAD	familial Alzheimer's disease
FENIB	familial encephalopathy with neuroserpin inclusion bodies
FMOC	9-fluorenylmethoxycarbonyl
FP	fluorescent protein
FP	fluorescence polarization
FRET	fluorescence resonance energy transfer
FRET	Förster resonance energy transfer
FTIR	Fourier transform infrared spectroscopy
GABA	γ-Aminobutyric acid
GI	glucosidase I
GII	glucosidase II
GFP	green fluorescent protein
GpA	glycophorin A
GPCR	G protein-coupled receptor
GSH	glutathione
GST	glutathione S-transferase
GSTO1	glutathione S-transferase omega 1
GuK	a guanylate kinase domain
hAGT	O^6-alkylguanine-DNA alkyltransferase
HB	hydrogen-bonding
HbA	adult human hemoglobin
HbCO A	carbonmonoxy form of adult human hemoglobin
HEPES	4-(2-hydroxyethyl)-1-piperazineethanesulfonic acid
hGH	human growth hormone
hGHR	hGH-receptor complex
Hip	Hsc/Hsp70 interacting protein
hIPSCs	human-induced pluripotent stem cells
His_6	hexahistidine sequence
HMPA	4-hydroxymethylphenoxyacetic acid
HPLC	high-performance liquid chromatography
Hsc70	70 kDa heat-shock cognate protein
HSMR	*Halobacterium salinarum* small multidrug resistance transporter
HspBP1	Hsp70 binding protein 1
HumLib	peptide library of human C-terminal protein sequences
I_C	C-terminal split intein fragment
IDP	intrinsically disordered protein
IEF	isoelectric focusing

IM-MS	ion mobility–mass spectrometry
IMS	intermembrane space
$I_N(H)$	N-terminal split intein fragment (with C-terminal His$_6$-tag)
IPG	immobilized pH gradient
IPTG	isopropyl β-D-1-thiogalactopyranoside
IPTG	isopropyl-β,D-thiogalactopyranoside
IRE1	inositol-requiring kinase 1
IR-MALDI-TOF MS	infrared–matrix-assisted laser desorption ionization–time-of-flight mass spectrometry
I_{sc}	short-circuit current
KD	dissociation constant
kDa	kilo Dalton
LAP	LplA acceptor peptide
LC	liquid chromatography
LDAO	lauryl dimethylamine oxide
linL46W	linear variant of ACP-L46W with identical primary sequence as cycL46W
LplA	lipoic acid ligase
LTQ OT	linear ion trap Orbitrap
LTQ OT MS	linear ion trap Orbitrap mass spectrometry
MAGUKs	membrane associated guanylate kinases
MALDI	matrix-assisted laser desorption/ionization
MALDI MS	matrix-assisted laser desorption ionization mass spectrometry
MALDI-TOF MS	matrix-assisted laser desorption ionization–time-of-flight mass spectrometry
MALDI-TOF/TOF MS	matrix-assisted laser desorption ionization–time-of-flight/time-of-flight mass spectrometry
Man 6-P	mannose 6-phosphate
MARS	multiple affinity removal system
MBP	maltose binding protein
MD	molecular dynamics
MEM	minimal essential medium
MES	2-(N-morpholino)ethanesulfonic acid
met HbA	oxidized form of adult human hemoglobin (liganded with a water molecule as ligand)
MHC	major histocompatibility complex
MINT	molecular interaction database
MLFTPP	macroaffinity ligand-facilitated three-phase partitioning
MPDZ	human multiple PDZ domain protein
MRH	mannose 6-P receptor homologous
MS	mass spectrometry
MS/MS	tandem mass spectrometry
M_W	molecular weight
MX	macromolecular crystallography

NADH	nicotinamide adenine dinucleotide
NCL	native chemical ligation
NEM	N-ethyl maleimide
NG	n-nonyl-β-D-glucoside
NHB	non-hydrogen-bonding
NHERF	Na+/H+ exchanger regulatory factor
NMP	N-methylpyrolidone
NMR	nuclear magnetic resonance
Nu	nucleophile
OG	n-octyl-β-D-gluocoside
OM	outer membrane
ORF	open reading frame
OST	oligosaccharyltransferase
PAGE	polyacrylamide gel electrophoresis
PALM	photoactivated localization microscopy
PAS	periodic acid Schiff
PCA	principal components analysis
PCR	polymerase chain reaction
PDB	Protein Data Bank
PDI	protein disulfide isomerase
PDZ	initial letter of PSD-95, Dlg, ZO-1
PEG	polyethylene glycol
PEM	protein epitope mimetic
PERK	PKR-like ER kinase
PES	polyethersulfone
PID	protein interaction domain
PIR	protein interaction reporter
PMF	peptide mass fingerprinting
PMSF	phenylmethylsulphonyl fluoride
PMT	photomultiplier tube
pNPG	p-nitrophenyl α-D-glucopyranoside
POI	protein of interest
PPI	protein-protein interaction
PPTase	4'-phosphopantetheinyl transferase
PRIME	probe incorporation mediated by enzymes
PrP^C	prion protein, cellular form
PrP^{SC}	prion protein, scrapie form
PSD-95	postsynaptic density 95
PSF	point spread function
PTFE	polytetrafluoroethylene
PVDF	polyvinylidene difluoride
QC	quality control
RCL	reactive center loop
Rg	radius of gyration

RG	reactive group
r-HbA	adult human hemoglobin exhibiting the double mutation (αV96W/βN108K)
RhoBo	rhodamine-derived bisboronic acid
RNA	ribonucleic acid
ROS	reactive oxygen species
SAR	structure–activity relationships
SAXS	small-angle X-ray solution scattering
SBA	soybean agglutinin
SCX	strong cation exchange
SDS	sodium dodecyl sulfate
SDS-PAGE	sodium dodecyl sulfate–polyacrylamide gel electrophoresis
SE-AUC	sedimentation equilibrium analytical ultracentrifugation
SEC	size exclusion chromatography
SH2	src-homology 2
SH3	src homology 3
SICLOPPS	split intein-mediated circular ligation of peptides and proteins
SkMCs	human skeletal muscle cells
SM	supplement mix
SMART	simple modular architecture research tool
SMCGM	skeletal muscle cell growth medium
SMR	small multidrug resistance transporter
SOD	superoxide dismutase
SPFO	sodium perfluorooctanoate
SPPS	solid-phase peptide synthesis
SPR	surface plasmon resonance
SrtA	Sortase A
SSA	sulfosalicylic acid
SSE	secondary structure elements
STED	stimulated emission depletion microscopy
STORM	stochastic optical reconstruction microscopy
SubAna	substitutional analysis
TBS	Tris-buffered saline
t-BuOH	tertiary butanol (2-methylpropan-2-ol)
TCA	trichloracetic acid
TCCD	two-color coincidence detection
TDI	thiol-disulfide interchange
TEV	Tobacco etch virus
TFA	trifluoroacetic acid
TIBS	triisobutylsilane
TM	transmembrane
TM4	TM helix four
TMP	trimethoprim
TNS	trypsin neutralization solution
TOF	time-of-flight

TOF/TOF	tandem time-of-flight
TOM	translocase of the outer membrane of mitochondria
TPP	three-phase partitioning
TR-WAXS	time-resolved wide-angle X-ray solution scattering
TS	transition state
TTET	triplet-triplet energy transfer
Ubc	ubiquitin conjugating enzyme
UGGT	UDP-Glc:glycoprotein glucosyltransferase
UPR	unfolded protein response
Vha	vibrio harveyi
WAXS	wide-angle X-ray solution scattering
WT	wild-type
WW	domain with two highly conserved tryptophans
ZO-1	zonula occludens-1
2-DE	two-dimensional gel electrophoresis
2-DE/MS	two-dimensional gel electrophoresis/mass spectrometry
2D-PAGE	two-dimensional polyacrylamide gel electrophoresis

Acknowledgements

Chapter 2

Work in the authors' lab was supported by the Deutsche Forschungsgemeinschaft.

Chapter 3

This work was supported, in part, by grants to C.M.D. from the Canadian Institutes of Health Research (CIHR FRN-5810) and from the Canadian Cystic Fibrosis Foundation. D.V.T. holds a graduate award from the CIHR Strategic Training Program in Protein Folding and Interaction Dynamics: Principles and Diseases. We thank Bradley Poulsen for helpful discussions.

Chapter 8

The author acknowledges the other members of the cross-linking team in the Aebersold lab (Franz Herzog, Abdullah Kahraman and Thomas Walzthoeni) and Prof. Ruedi Aebersold for valuable discussions and support. This research is funded by the European Union 7th Framework program PROSPECTS (Proteomics Specification in Space and Time, grant no. HEALTH-F4–2008–201648).

Chapter 9

Portions of this work were supported by the Office of Biological and Environmental Research, U.S. Department of Energy (FWP BO-70), and the National Center for Research Resources (2-P41-RR012408) of the National Institutes of Health. Some of the data for this review were measured, in part, at beamline X26-C of the National Synchrotron Light Source (NSLS) at the Brookhaven National Laboratory. Use of the NSLS was supported by the U.S. Department of Energy Office of Basic Energy Sciences, under Contract DE-AC02–98CH10886. I am grateful to the many users who collaborate with me and the PXRR group as we to continue to improve our capabilities at beamline X26-C, especially those enumerated in the text above. I also thank the entire PXRR staff and the many beamline scientists at ESRF, SLS, Diamond and Soliel for their kind openness and for sharing their insights from developing similar capabilities in Europe.

Chapter 10

The work reviewed here has received substantial support from our collaborators during the development of WAXS over the past decade. We thank them for many discussions, idea, and good humor during long nights of data collection. Tobin Sosnick provided the ubiquitin WT and mutants used to obtain the results described. This work is supported by a research grant from the NIH (R01GM-085648 to L. Makowski). Development of the sample-handling robot was supported by an SBIR to Shamrock Structures, LLC (R43GM085933 to R. Walter and L. Makowski). We thank Rick Walter and Steve Schiltz of Shamrock Structures for support of this effort. BioCAT is a National Institutes of Health-supported Research Center RR-08630. Use of the Advanced Photon Source, an Office of Science User Facility operated for the U.S. Department of Energy (DOE) Office of Science by Argonne National Laboratory, was supported by the U.S. DOE under Contract No. DE-AC02–06CH11357.

Chapter 14

The research was supported by the European Social Fund (ESF) and the Special Coordination Fund for Promoting Science and Technology, Creation and Innovation Centers and Advanced Interdisciplinary, Japan. We thank Dr. Kimura from Riken Institut in Yokohama for the comparison of the 2D-gels using Progenesis software. We are grateful to Prof. Dr. Speer and Dr. von Moers for providing of DMD-cells, and the European Social Fund for Promoting Science and Technology in Japan for the financial support.

Chapter 18

BG is a Wellcome Trust Intermediate Clinical Fellow. This work is supported by grants from the Medical Research Council (UK), the Engineering and Physical Sciences Research Council (UK), the Biotechnology and Biological Sciences Research Council, the Wellcome Trust, the British Lung Foundation, GlaxoSmithKline and Papworth NHS Trust.

1 Three-phase partitioning

Clive Dennison

Three-phase partitioning (TPP) is a relatively novel method, which uses tertiary butanol (2-methylpropan-2-ol) (t-BuOH) and ammonium sulfate to form a two-phase system in which macromolecules precipitate at the interface (thus comprising a third phase). TPP is proving increasingly useful for the concentration and purification of proteins and other biomacromolecules. It should not be confused with phase-forming methods using hydrophilic polymers.

Rex Lovrien, of the University of Minnesota, serendipitously discovered TPP. He was of the opinion that, within cells, enzymes did not operate in a purely aqueous environment, but that, in addition to water, there were other compounds present, including salts and organic compounds. To establish what effect these might have on enzymes, Lovrien tested enzyme activity in a variety of salt/organic solvent mixtures (Tan and Lovrien 1972).

With the particular combination of ammonium sulfate and t-BuOH, it was found that the solution split into three phases, with protein being precipitated in a phase between the aqueous and t-BuOH phases. Lovrien's group developed the method further, first for the isolation of components of the cellulase complex of *Trichoderma reesei* (Odegaard et al. 1984) and later as a general protein isolation method (Lovrien et al. 1987).

In 1984–1985, I was on sabbatical in Minnesota. As I was about to leave for home, Lovrien pressed some papers in my hand, saying we might be interested in a method he was developing. We tested it on a number of lysosomal proteinases, the focus of our studies, and were amazed at the results. For example, we were able to isolate cathepsin D in a single day (Jacobs et al. 1989), and, in some cases, more than 100% recovery of activity was obtained. In an attempt to understand the method, I returned to Minnesota for discussions with Lovrien, which led to our joint publication (Dennison and Lovrien 1997).

Since then, the method has been adopted universally by increasing numbers of researchers, who have greatly expanded its applications. General reviews of the method have been provided by Pike (1989), Pike and Dennison (1989a, 1989b), Dennison and Lovrien (1997), Dennison et al. (2000a), Dennison (2003) and Ward (2009).

A common experience of researchers trying TPP for the first time is one of amazement at the simplicity and effectiveness of the method, which seems to fly in the face of common "wisdom" about protein isolation, that is, that organic solvents denature proteins at room temperature and that, in any step, one should expect less than 100% recovery of activity. Moreover, it is very surprising to find a "bucket" method that can be applied on a micro- or macroscale and that can achieve results equal to, or better than, sophisticated methods such as chromatography. In the words of Ward (2009:28), "Three-phase partitioning represents an incredibly simple, rapid method for extracting, purifying and concentrating proteins." This initial amazement and surprise matures into curiosity about how the method achieves what it does, and several researchers have addressed this question.

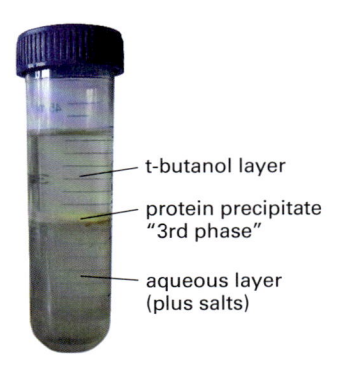

Figure 1.1: The distribution of phases in three-phase partitioning.

1.1 Method

TPP involves the salting out of proteins from a mixture of t-BuOH and water. t-BuOH is normally infinitely miscible with water, but upon addition of sufficient ammonium sulfate, the solution splits into two phases: a lower aqueous phase containing the salt and an alcohol-rich upper phase. If proteins (and cell wall polysaccharides and DNA [Ward 2009]) are additionally present in the original solution, they may precipitate into a third layer, between the aqueous and t-BuOH layers. The upper t-BuOH layer contains dissolved membrane lipids and low-molecular-weight pigments (Ward 2009). What proteins precipitate into the interfacial layer depends on their concentration and specific properties, and on the concentration of ammonium sulfate, and this forms the basis for protein concentration and fractionation.

In practice, TPP is most useful early in a protein isolation, where it serves to remove nonprotein material and concentrate and fractionate the protein constituents. Typically, ca. 30% t-BuOH may be added to an aqueous extract, followed by stepwise increments of ammonium sulfate, the protein precipitate being collected for assay after each addition of ammonium sulfate. In this way an optimal "cut" of ammonium sulfate concentrations may be determined. Variations on this overall approach are possible to separate proteins from DNA and cell wall components (Ward 2009).

TPP has some advantages over conventional salting out with ammonium sulfate. With conventional salting out, the solution becomes increasingly dense with increasing ammonium sulfate concentration, until the stage is reached where it becomes difficult to sediment the precipitated protein. However, in TPP, the precipitated protein floats on the aqueous layer, and so, as this becomes increasingly dense, the precipitate floats more and more easily. Secondly, the precipitate from TPP has a relatively low salt concentration, so TPP can be followed immediately by ion exchange, for example, without prior desalting as is required with conventional salting out.

TPP appears to depend on the unique properties of t-BuOH, which is described as a "differentiating" cosolvent (one that emphasizes the differences between proteins) as opposed to C1 and C2 alcohols, which are described as "leveling" cosolvents (which tend to obscure the differences between proteins). Uniquely among common organic solvents, t-BuOH stabilizes proteins rather than denaturing them (Dennison and Lovrien 1997). A common property of agents that stabilize proteins is that they are excluded from the protein interior (Timasheff et al. 1982). In the case of t-BuOH, this may be due to its "bushy" structure, which makes it too large to enter the protein interior.

> t-BuOH also reversibly inhibits the activity of many enzymes (Dennison et al. 2000a), a phenomenon that may be related to its protein-stabilizing effect.

The activity of many enzymes depends upon the flexibility of their structure, and, perhaps, in stabilizing an enzyme's structure, it may be made too rigid to function.

Many proteins are isolated from homogenates, in which intracellular compartments are destroyed and cellular components may be inappropriately mixed, leading to homogenization artifacts. For example, the lysosomal proteinase, cathepsin L, occurs in endosomal compartments, while its endogenous inhibitor, stefin B, occurs in the cytoplasm. After homogenization of liver tissue, Pike et al. (1992) isolated (artifactual) covalent complexes of these two molecules, joined by a thioester bond. When the homogenization was done in 30% t-BuOH, no complex was found and the yield of free enzyme was increased. It was concluded that an (unidentified) enzyme in the homogenate (without t-BuOH) probably catalyzed formation of the thioester bond between cathepsin L and stefin B, and that t-BuOH inhibited this enzyme.

The ability of t-BuOH to inhibit enzymes may thus minimize homogenization artifacts and suggests that homogenization in 30% t-BuOH may be a profitable avenue to explore in any protein isolation from tissue (Dennison et al. 2000a, 2000b).

TPP must be tested for each specific application, as it doesn't work in all cases. In our laboratory, we have found that TPP causes dimerization of serum albumin, it is not suitable for isolation of IgG antibodies and, for the isolation of bromelain, it is less successful than cold acetone precipitation (unpublished observations).

1.2 The mechanism of TPP

Dennison and Lovrien (1997) explained the mechanism of TPP in terms of the antichaotropic (i.e. kosmotropic) effects of both t-BuOH and ammonium sulfate. Neither the large hydrated sulfate ion nor the "bushy" t-BuOH molecule is able to penetrate the interior of proteins, and thus both serve to stabilize protein structure.

With protein suspended in a mixture of water and t-BuOH, protein is envisaged as equilibrating with the ambient proportions of the two solvents. Upon addition of ammonium sulfate, the sulfate ion becomes hydrated, growing larger in the process and sequestering water, so that less is available to the t-BuOH and protein. As less water becomes available, the protein equilibrates with the new available proportions of t-BuOH and water. Addition of further ammonium sulfate eventually sequesters sufficient water to bring the protein to its solubility limit, at which point it precipitates. However, at this point, the protein will be equilibrated with a preponderance of t-BuOH, which presumably binds to surface hydrophobic patches and lowers its density such that the protein floats between the aqueous and organic phases.

The hydrated sulfate ion is relatively large, having 14 water molecules in its first hydration layer and possibly more in a second layer (Cannon et al. 1994), and can thus additionally serve as a molecular crowding or exclusion agent. This renders even less of the water available to the protein. Molecular exclusion is suggested by the observation of Pike and Dennison (1989b) that larger proteins tend to be precipitated at lower concentrations of ammonium sulfate than smaller proteins.

That ammonium sulfate abstracts and sequesters water is shown by the observation of Pike and Dennison (1989b) that in TPP the density of the t-BuOH–rich phase decreases with increasing ammonium sulfate concentration, suggesting that it is increasingly dehydrated.

Dennison and Lovrien (1997) and Ward (2009) have assumed that the distribution of precipitated proteins into the interface in TPP is due to density differences. This may be the case, but Jauregi et al. (2002) have noted that with water-butanol and water-hexanol two-phase systems, interfacial partitioning of crystals occurs even when their density exceeds that of the individual liquid phases.

Kiss and Borbás (2003) suggest that interfacial tension and hence emulsion stability may play a role in protein separation by TPP. Supporting this, Borbás et al. (2003) have shown that proteins precipitated at the interface exist in the form of spherical globules.

1.3 A practical example – the isolation of cathepsin L from liver tissue

Cathepsin L is a proteolytic enzyme of the endocytic system of cells, where it is thought to be stored in lysosomes in a low-pH environment, below the enzyme's pI, possibly ionically bound to negatively charged LAMP-2 integral membrane proteins (Pillay et al. 2002). The enzyme is active below and at neutral pH but is stable only below neutral pH (Dehrmann et al. 1995).

In previous studies, Pike et al. (1992) used a system of homogenization → pH 4.2 precipitation → TPP → cation exchange on S-Sepharose → molecular exclusion chromatography on Sephadex G-75 and isolated the enzyme in the free form and as a covalent complex with its cytoplasmic inhibitor, stefin B. Later it was discovered that 30% t-BuOH inhibits many enzymes, including cathepsin L, and that homogenization in 30% t-BuOH obviates formation of the covalent complex of cathepsin L and stefin B (Dennison et al. 2000a, 2000b).

A rational approach to the isolation of cathepsin L would thus involve the following steps: (i) homogenization in pH 4.2 buffer (partly because subsequent TPP precipitation is more effective below a protein's pI [Dennison and Lovrien 1997]) containing sufficient t-BuOH to give a final concentration of 30% (to inhibit proteases and obviate complex formation), (ii) TPP fractionation and (iii) cation exchange chromatography at a pH below neutral, where the enzyme is both stable and positively charged (as it is in its intracellular environment).

Method. Thawed rabbit liver was homogenized in a Waring Blendor in a mixture of two parts of homogenization buffer (150 mM Na-acetate buffer, 1 mM EDTA, pH 4.0) and sufficient t-BuOH to give 30% (v/v) t-BuOH in the homogenate. The homogenate was centrifuged (9,000 × g, 20 min, 4°C), the supernatant was adjusted to pH 4.2 with dilute acetic acid and the mixture was again centrifuged (15,100 × g, 20 min, 4°C). Ammonium sulfate was added to the supernatant to constitute 20% (w/v) (as defined by

Pike and Dennison [1989b]), and the precipitate pellet, suspended between the aqueous and t-BuOH phases, was harvested by centrifugation (15,100 × g, 10 min, 4°C) and discarded. The aqueous and t-BuOH phases were recombined and mixed, and a further amount of ammonium sulfate was added to bring the total to 45% (w/v). The pellet was harvested as before and redissolved, in one-tenth the volume of the original pH 4.2 supernatant, in Buffer A (20 mM Na-acetate buffer, 1 mM Na2EDTA, 0.02% NaN3, pH 5.0). The redissolved solution was centrifuged (15,100 × g, 10 min, 4°C) to remove any insoluble material. The clarified solution was applied to a column of S-Sepharose (2.5 × 18.3 cm), and unbound material was eluted with Buffer A. Bound material was eluted with a gradient of 0–300 mM NaCl in Buffer A in ca. six column volumes. Fractions active against Z-Phe-Arg-NHMec were pooled. The purification of cathepsin L was 782.3-fold, with a 10.7% yield. Sodium dodecyl sulfate–polyacrylamide gel electrophoresis (SDS-PAGE) analysis showed no evidence of a covalent complex of cathepsin L and stefin B (cf. Pike et al. 1992).

1.4 Other applications

An ever-increasing number of enzymes, enzyme inhibitors, other proteins, polysaccharides and oils have been isolated by TPP (▶Table 1.1).

1.4.1 Green fluorescent protein

Green fluorescent protein is a useful marker protein for the study of cellular function and was the subject of studies that led to the 2008 Nobel Prize in Chemistry. TPP has revolutionized the isolation of green fluorescent protein (Jain et al. 2004; Penna and Ishii 2002; Thompson and Ward 2002; Ward 2009). Ward (2009) reported that, using TPP, a stage of purification was achieved in an hour or two that previously took several months. This echoes the experience of Jacobs et al. (1989) who, with TPP, were able to isolate cathepsin D in one day as opposed to a week, without TPP.

1.4.2 Membrane proteins

Advances in various fields must often await the development of suitable techniques, and, for the study of membrane proteins, TPP may be just such a technique. Peltier et al. (2004), for example, used a modified TPP method to analyze salt-stripped thylakoid membranes of *Arabidopsis thaliana* chloroplasts and identified 242 proteins, of which at least 40% were integral membrane proteins. These authors used n-butanol, instead of t-BuOH, as they found that this was better at extracting pigments, yet still formed three phases.

1.4.3 Enhanced activity

A curious feature of TPP (unexpected in a protein isolation) is that there is often an increase in activity after TPP (Dennison et al. 2000b; Roy and Gupta 2004; Singh et al. 2001). Singh et al. (2001) determined the crystal structure of TPP-modified proteinase K at 1.5 Å resolution, in an attempt to determine the origin of the enhanced activity (a 210% increase in specific activity, in this case). The overall structure of the TPP-treated

Table 1.1: Proteins and other macromolecules isolated by TPP

Macromolecule isolated	Reference
cellulase complex of *Trichoderma reesei*	Odegaard et al. (1984)
cathepsin D	Jacobs et al. (1989)
cathepsin L	Pike and Dennison (1989a)
human carbonic anhydrase I and II, catalase and superoxide dismutase	Pol et al. (1990)
human manganese superoxide dismutase	Deutsch et al. (1991)
cathepsin L	Pike et al. (1992)
proteases from *Trypanosoma brucei brucei*	Troeberg et al. (1996)
pro-MMP-9 from human sputum	Price et al. (2000)
alkaline phosphatase from chicken intestine	Sharma et al. (2000)
pectinases	Sharma and Gupta (2001a)
wheat germ bifunctional protease/amylase inhibitor	Sharma and Gupta (2001b)
phospholipase D from *Dacus carota*	Sharma and Gupta (2001c)
DNA	Jánosi and Szamos (2001)
cathepsin L from sheep liver	Irving et al. (2002)
cathepsin B from bovine liver	Pillay and Dennison (2002)
soybean trypsin inhibitor	Roy and Gupta (2002)
xylanase	Sharma and Gupta (2002a)
alginates	Sharma and Gupta (2002b)
green fluorescent protein	Thompson and Ward (2002)
chitosan	Sharma et al. (2003)
wheat germ agglutinin	
glucoamylase and pullulanase	Mondal et al. (2003a)
α-amylases	Mondal et al. (2003b)
turnip peroxidase	Singh and Singh (2003)
starch	Mondal et al. (2004)
immunoreactive excreted-secreted proteins of *Corynebacterium pseudotuberculosis*	Paule et al. (2004)
thylakoid membrane proteome of *Arabidopsis thaliana*	Peltier et al. (2004)
pyroglutamyl peptidase type I from *Trypanosoma brucei*	Morty et al. (2005)
carp and corn proteins	Szamos (2006)
cysteine proteinases from potato leaves	Popovic and Brzin (2007)
ragi (*Eleusine coracana*) bifunctional amylase/protease inhibitor	Saxena et al. (2007)

α-galactosidase and invertase from *Aspergillus oryzae*	Dhananjay and Mulimani (2008)
exo-polygalacturonase from Aspergillus *sojae*	Dogan and Tari (2008)
Ipomoea peroxidase	Narayan et al. (2008)
α-galactosidase from *Aspergillus oryzae*	Dhananjay and Mulimani (2009)
trypsin inhibitor from legume seeds	Wati et al. (2009)
DNA	Ujhelyi (2009)
invertase from baker's yeast	Akardere et al. (2010)
invertase from tomato	Özer et al. (2010)
protease from *Calotropis procera* (milkweed)	Rawdkuen et al. (2010)
cysteine peptidases from *Trypanosoma congolense*	Pillay et al. (2010)

enzyme proved similar to the original, and the hydrogen bonding system of the catalytic triad was intact. However, some of the water molecules in the substrate binding site were replaced by acetate ions, and, most strikingly, the TPP-modified structure had a temperature factor (B-factor) more than twice that of the original enzyme. They concluded that the increased enzymatic activity could be attributed to the presence of an acetate ion at the active site and/or to the enzyme's more excited state, as reflected by the higher B-factor.

1.4.4 Refolding

TPP appears to be able to refold reversibly denatured proteins and to extract active proteins from inclusion bodies (Raghava et al. 2008; Roy et al. 2004, 2005; Sardar et al. 2007). Coupled with the ability of t-BuOH to selectively extract proteins on a size basis from microbial cells (Raghava and Gupta 2009), this suggests that TPP could be a useful tool in genetic engineering endeavors.

The ability of TPP to refold proteins suggests that the higher specific activity often observed after TPP may be partly due to renaturation of partially denatured proteins.

1.4.5 Denaturation of hemoglobin

Not all proteins are stable to TPP, and an early observation in studies on the isolation of cathepsin D from spleen was that TPP denatures hemoglobin (Jacobs et al. 1989). In the context of that study, this was an advantage, as hemoglobin is present in spleen in overwhelming excess and its removal by selective denaturation facilitated the simple isolation of cathepsin D. This observation opened the door to the isolation of erythrocyte enzymes (Deutsch et al. 1991; Pol et al. 1990). The erythrocyte enzymes isolated all had quaternary structures, so the denaturation of hemoglobin is not simply a consequence of its quaternary structure.

1.4.6 Affinity TPP

Roy and Gupta (2002) and Sharma and Gupta (2002a) have reported on an extension of the TPP method that improves its selectivity, which they have called macroaffinity ligand-facilitated three-phase partitioning (MLFTPP).

In this method, an appropriately chosen macomolecular affinity ligand is added to a mixture, containing the protein for which it has affinity. The ligand selectively binds to the protein and subsequently separates out in TPP as a ligand-protein complex. Using this approach, soybean trypsin inhibitor was purified 13-fold with 72% recovery, and wheat germ agglutinin and wheat germ lipase were purified with 99% activity recovery (with 40-fold purification) and 94% activity recovery (with 27-fold purification), respectively.

1.4.7 Other uses of TPP

Besides proteins, carbohydrate polymers can also be concentrated and purified by TPP. Those studied to date include alginates (Sharma and Gupta 2002b), chitosan (Sharma et al. 2003) and starch (Mondal et al. 2004).

TPP has also been successfully used to purify DNA (Jánosi and Szamos 2001; Shabani 2009; Ujhelyi 2009) and in the extraction of oils from oilseeds (Saifuddin 2006; Shah et al. 2004; Sharma and Gupta 2004; Sharma et al. 2002).

Adányi et al. (2005) have used TPP as a tool in the construction of biosensors.

References

Adányi, N., Szabó, E.E., Váradi, M., and Szamos, J. (2005). Interfacial enzyme partitioning as a tool for constructing biosensors. Acta Aliment 28, 329–338.

Akardere, E., Özer, B., Çelem E.B., and Önal S. (2010). Three-phase partitioning of invertase from baker's yeast. Sep Purif Technol 72, 335–339.

Borbás, R., Kiss, É., and Murray, B.S. (2003). Interfacial rheology and interfacial gelation partitioning. In: Food colloids, biopolymers and materials, E. Dickenson and T. van Vliet, eds (Cambridge, UK: Royal Society of Chemistry), pp. 368–376.

Cannon, W.R., Petit, B.M., and McCammon, J.A. (1994). Sulfate anion in water: model structural, thermodynamic and dynamic properties. J Phys Chem 98, 6225–6230.

Dehrmann, F.M., Coetzer, T.H.T., Pike, R.N., and Dennison, C. (1995). Mature cathepsin L is substantially active in the ionic milieu of the extracellular medium. Arch Biochem Biophys 324, 93–98.

Dennison, C. (2003). A guide to protein isolation. 2nd edn. Book series: Focus on Structural Biology, vol. 3 (Dordrecht, the Netherlands: Kluwer Academic Publishers B.V.).

Dennison, C., and Lovrien, R.E. (1997). Three-phase partitioning: concentration and purification of proteins. Protein Expr Purif 11, 149–161.

Dennison, C., Moolman, L., Pillay, C.S., and Meinesz, R.E. (2000a). t-Butanol: nature's gift for protein isolation. S Afr J Sci 96, 159–160.

Dennison, C., Moolman, L., Pillay, C.S., and Meinesz, R.E. (2000b). Use of 2-methylpropan-2-ol to inhibit proteolysis and other protein interactions in a tissue homogenate. An illustrative application to the extraction of cathepsins B and L from liver tissue. Anal Biochem 284, 157–159.

Deutsch, H.F., Hoshi, S., Matsuda, Y., Suzuki, K., Kawano, K., Kitagawa, Y., et al. (1991). Preparation of human manganese superoxide dismutase by tri-phase partitioning and preliminary crystallographic data. J Mol Biol 219, 103–108.

Dhananjay, S., and Mulimani, V. (2009). Three-phase partitioning of α-galactosidase from fermented media of *Aspergillus oryzae* and comparison with conventional purification techniques. J Ind Microbiol Biotechnol *36*, 123–128.

Dhananjay, S.K., and Mulimani, V.H. (2008). Purification of α-galactosidase and invertase by three-phase partitioning from crude extract of *Aspergillus oryzae*. Biotechnol Lett *30*, 1565–1569.

Dogan, N., and Tari, C. (2008). Characterization of three-phase partitioned exo-polygalacturonase from *Aspergillus sojae* with unique properties. Biochem Eng J *39*, 43–50.

Irving, J.A., Pike, R.N., Dai, W., Bromme, D., Worrall, D.M., Silverman, G.A., et al. (2002). Evidence that serpin architecture intrinsically supports papain-like cysteine protease inhibition: engineering alpha1-antitrypsin to inhibit cathepsin proteases. Biochemistry *41*, 4998–5004.

Jacobs, G.R., Pike, R.N., and Dennison, C. (1989). Isolation of cathepsin D using three-phase partitioning in t-butanol/water/ammonium sulfate. Anal Biochem *180*, 169–171.

Jain, S., Singh, R., and Gupta, M.N. (2004). Purification of recombinant green fluorescent protein by three-phase partitioning. J Chromatogr A *1035*, 83–86.

Jánosi, A., and Szamos, J. (2001). Comparison of two methods in purification of meat-DNA for PCR. Acta Aliment *30*, 113–118.

Jauregi, P., Hoeben, M.A., van der Lans, R.G.J.M., Kwant, G., and van der Wielen, L.A.M. (2002). Recovery of small bioparticles by interfacial partitioning. Biotechnol Bioeng *78*, 355–364.

Kiss, É., and Borbás, R., (2003). Protein adsorption at liquid/liquid interface with low interfacial tension. Colloids Surfaces B Biointerfaces *31*, 169–176.

Lovrien, R., Goldensoph, C., Anderson, P.C., and Odegaard, B. (1987). Three phase partitioning (TPP) via t-butanol: enzymes separations from crudes. In: Protein purification: micro to macro, R. Burgess, ed. (New York: A.R. Liss), pp. 131–148.

Mondal, K., Sharma, A., and Gupta, M.N. (2003a). Macroaffinity ligand-facilitated three-phase partitioning for purification of glucoamylase and pullulanase using alginate. Protein Expr Purif *28*, 190–195.

Mondal, K., Sharma, A., and Gupta, M.N. (2004). Three phase partitioning of starch and its structural consequences. Carbohydr Polym *56*, 355–359.

Mondal, K., Sharma, A., Lata, R., and Gupta, M.N. (2003b). Macroaffinity ligand-facilitated three-phase partitioning (MLFTPP) of α-amylases using a modified alginate. Biotechnol Prog *19*, 493–494.

Morty, R.E., Bulau, P., Pellé, R., Wilk, S., and Abe, K. (2005). Pyroglutamyl peptidase type I from *Trypanosoma brucei*: a new virulence factor from African trypanosomes that de-blocks regulatory peptides in the plasma of infected hosts. Biochem J Immediate Publication, BJ20051593.

Narayan, A.V., Madhusudhan, M.C., and Raghavarao, K.S.M.S. (2008). Extraction and purification of *Ipomoea* peroxidase employing three-phase partitioning. Appl Biochem Biotechnol *151*, 263–272.

Odegaard, B., Anderson, P.C., and Lovrien, R. (1984). Resolution of the multienzyme cellulase complex of *Trichoderma reesei* QM9414. J Appl Biochem *6*, 158–183.

Özer, B., Akardere, E., Çelem, E.B., and Önal, S. (2010). Three-phase partitioning as a rapid and efficient method for purification of invertase from tomato. Biochem Eng J *50*, 110–115.

Paule, B.J., Meyer, R., Moura-Costa, L.F., Bahia, R.C., Carminati, R., Regis, L.F., et al. (2004). Three-phase partitioning as an efficient method for extraction/concentration of immunoreactive excreted-secreted proteins of *Corynebacterium pseudotuberculosis*. Protein Expr Purif *34*, 311–316.

Peltier, J-B., Ytterberg, J., Sun, Q., and van Wijk, K.J. (2004). New functions of the thylakoid membrane proteome of *Arabidopsis thaliana* revealed by a simple, fast, and versatile fractionation strategy. J Biol Chem *279*, 49367–49283.

Penna, T.C.V., and Ishii, M. (2002). Selective permeation and organic extraction of recombinant green fluorescent protein (*gfpuv*) from *Escherichia coli*. BMC Biotechnol. 2:7. http://www.biomedcentral.com/1472-6750/2/7 (last accessed on 10/21/2011).

Pike, R.N. (1989). Three-phase partitioning in aqueous/t-butanol mixtures: a useful new protein fractionation method. S Afr J Sci *85*, 424–425.

Pike, R.N., Coetzer, T.H.T., and Dennison, C. (1992). Proteolytically active complexes of cathepsin L and a cysteine proteinase inhibitor; purification and demonstration of their formation in vitro. Arch Biochem Biophys *294*, 623–629.

Pike, R.N., and Dennison, C. (1989a). A high-yield method for the isolation of cathepsin L. Prep Biochem *19*, 231–245.

Pike, R.N., and Dennison, C. (1989b). Protein fractionation by three phase partitioning (TPP) in aqueous/t-butanol mixtures. Biotechnol Bioeng *33*, 221–229.

Pillay, C., and Dennison, C. (2002). Cathepsin B stability, but not activity, is affected in cysteine:cystine redox buffers. Biol Chem *383*, 1199–1204.

Pillay, C., Elliott, E., and Dennison, C. (2002). Endolysosomal proteolysis and its regulation. Biochem J *363*, 417–429.

Pillay, D., Boulangé, A.F., and Coetzer, T.H.T. (2010). Expression, purification and characterisation of two variant cysteine peptidases from *Trypanosoma congolense* with active site substitutions. Protein Expr Purif *74*, 264–271.

Pol, M.C., Deutsch, H.F., and Visser, L. (1990). Purification of soluble enzymes from erythrocyte hemolysates by three phase partitioning. Int J Biochem *22*, 179–185.

Popovic, T., and Brzin, J. (2007). Purification and characterization of two cysteine proteinases from potato leaves and the mode of their inhibition with endogenous inhibitors. Croat Chem Acta *80*, 45–52.

Price, B., Dennison, C., Tschesche, H., and Elliott, E. (2000). Neutrophil tissue inhibitor of matrix metalloproteinases-1 occurs in novel vesicles that do not fuse with the phagosome. J Biol Chem *275*, 28308–28315.

Raghava, S., Barua, B., Singh, P.K., Das, M., Madan, L., Bhattacharyya, S., et al. (2008). Refolding and simultaneous purification by three-phase partitioning of recombinant proteins from inclusion bodies. Protein Sci *17*, 1987–1997.

Raghava, S., and Gupta, M.N. (2009). Tuning permeabilization of microbial cells by three-phase partitioning. Anal Biochem *385*, 20–25.

Rawdkuen, S., Benjakul, S., Pintathong, P., and Phanuphong Chaiwut, P. (2010). Three-phase partitioning of protease from *Calotropis procera* latex. International Conference on Agriculture and Agro-Industry (ICAAI2010) Food, Health and Trade, 19–20 November 2010: A Celebration of the 12th Anniversary of Mae Fah Luang University, Chiang Rai, Thailand, p. 1.

Roy, I., and Gupta, M.N. (2002). Three-phase affinity partitioning of proteins. Anal Biochem *300*, 11–14.

Roy, I., and Gupta, M.N. (2004). α-Chymotrypsin shows higher activity in water as well as organic solvents after three phase partitioning. Biocatalysis Biotransformation *22*, 261–268.

Roy, I., Sharma, A., and Gupta, M.N. (2004). Three-phase partitioning for simultaneous renaturation and partial purification of *Aspergillus niger* xylanase. Biochim Biophys Acta Proteins Proteomics *1698*, 107–110.

Roy, I., Sharma, A., and Gupta, M.N. (2005). Recovery of biological activity in reversibly inactivated proteins by three phase partitioning. Enzyme Microb Technol *37*, 113–120.

Saifuddin, N. (2006). Optimization of aqueous enzymatic oil extraction from kernel of oil palm (*Elaeis guineensis*) using three phase partitioning and microwave irradiation. Pak J Anal Environ Chem 7, 31–38.

Sardar, M., Sharma, A., and Gupta, M.N. (2007). Refolding of a denatured α-chymotrypsin and its smart bioconjugate by three-phase partitioning. Biocatalysis Biotransformation 25, 92–97.

Saxena, L., Iyer, B.K., and Ananthanarayan, L. (2007). Three phase partitioning as a novel method for purification of ragi (*Eleusine coracana*) bifunctional amylase/protease inhibitor. Process Biochem 42, 491–495.

Shabani, Z. (2009). Rapid purification of recombinant plasmid DNA from Escherichia coli via aqueous three-phase partitioning. M.A. thesis, Columbia University, New York. http://www.columbia.edu/cu/biology/grad/biotech/theses/2009_theses.htm#zs (last accessed on 10/21/2011).

Shah, S., Sharma, A., and M. N. Gupta (2004). Extraction of oil from *Jatropha curcas* L. seed kernels by enzyme assisted three phase partitioning. Ind Crops Prod 20, 275–279.

Sharma, A., and Gupta, M.N. (2001a). Purification of pectinases by three-phase partitioning. Biotechnol Lett 23, 1625–1627.

Sharma, A., and Gupta, M.N. (2001b). Three phase partitioning as a large-scale separation method for purification of a wheat germ bifunctional protease/amylase inhibitor. Process Biochem 37, 193–196.

Sharma, A., and Gupta, M.N. (2001c). Purification of phospholipase D from *Dacus carota* by three-phase partitioning and its characterization. Protein Expr Purif 21, 310–316.

Sharma, A., and Gupta, M.N. (2002a). Macroaffinity ligand-facilitated three-phase partitioning (MLFTPP) for purification of xylanase. Biotechnol Bioeng 80, 228–232.

Sharma, A., and Gupta, M.N. (2002b). Three phase partitioning of carbohydrate polymers: separation and purification of alginates. Carbohydr Polym 48, 391–395.

Sharma, A., and Gupta, M.N. (2004). Oil extraction from almond, apricot and rice bran by three-phase partitioning after ultrasonication. Eur J Lipid Sci Technol 106, 183–186.

Sharma, A., Khare, S.K., and Gupta, M.N. (2002). Three phase partitioning for extraction of oil from soybean. Bioresour Technol 85, 327–329.

Sharma, A., Mondal, K., and Gupta, M.N. (2003). Some studies on characterization of three phase partitioned chitosan. Carbohydr Polym 52, 433–438.

Sharma, A., Sharma, S., and Gupta, M.N. (2000). Purification of alkaline phosphatase from chicken intestine by three-phase partitioning and use of phenyl-Sepharose 6B in the batch mode. Bioseparation 9, 155–161.

Singh, N., and Singh, J. (2003). A method for large scale purification of turnip peroxidase and its characterization. Prep Biochem Biotechnol 33, 125–135.

Singh, R.K., Gourinath, S., Sharma, S., Roy, I., Gupta, M.N., Betzel, Ch., et al. (2001). Enhancement of enzyme activity through three-phase partitioning: crystal structure of a modified serine proteinase at 1.5 Å resolution. Protein Eng 14, 307–313.

Szamos, J. (2006). Behaviour of carp proteins and corn proteins in three-phase partitioning. Acta Aliment 35, 109–116.

Tan, K.H., and Lovrien, R. (1972). Enzymology in aqueous-organic cosolvent mixtures. J Biol Chem 247, 3278–3285.

Thompson, C.M., and Ward, W.W. (2002). Three-phase partitioning (TPP) a rapid and preparative tool for GFP. In: Bioluminescence & chemiluminescence: progress and current applications, P.E. Stanley and L.J. Kricka, eds (Singapore: World Scientific), pp. 115–118.

Timasheff, S.N., Arakawa, T., Inoue, H., Gekko, K., Gorbunoff, M.J., Lee, J.C., et al. (1982). The role of solvation in protein structure stabilization and unfolding. In: The biophysics of water, F. Franks and S.F. Mathias, eds (New York: Wiley), pp. 48–50.

Troeberg, L., Pike, R.N., Morty, R.E., Berry, R.K., Coetzer, T.H.T., and Lonsdale-Eccles, J.D. (1996). Proteases from *Trypanosoma brucei brucei*: Purification, characterisation and interactions with host regulatory molecules. Eur J Biochem *238*, 728–736.

Ujhelyi, G. (2009). Studying the contamination of the glyphosphate-tolerant soybean from soy containing food products and the risk of its translocation into the organism using PCR method. Ph.D. thesis, Corvinus University of Budapest, Hungary.

Ward, W.W. (2009). Three-phase partitioning for protein purification. Innovations Pharm Technol *28*, 28–34.

Wati, R. K., Theppakorn, T., Benjakul, S., and Rawdkuen, S. (2009). Three-phase partitioning of trypsin inhibitor from legume seeds. Process Biochem *44*, 1307–1314.

2 Folding and degradation functions of molecular chaperones

Michael Dreiseidler, Anna Ulbricht, and Jörg Höhfeld

2.1 Introduction

Molecular chaperones are of central importance for protein homeostasis (proteostasis). They possess the ability to interact with nonnative polypeptides and thereby facilitate protein folding, protein sorting and protein degradation (Arndt et al. 2007; Hartl and Hayer-Hartl 2009). The activity of many chaperones is regulated by cochaperones, which modulate the interaction with nonnative clients or facilitate a cooperation with other proteostasis factors. A particularly versatile chaperone system, which operates in the mammalian cytoplasm and nucleus, comprises constitutively expressed Hsc70 (also known as Hsp73 or HspA8), stress inducible Hsp70 (also known as Hsp72 or HspA1) and a large plethora of Hsc/Hsp70-regulating cochaperones (Arndt et al. 2007). Based on the identification and functional characterization of these cochaperones, performed with increasing depth and comprehensiveness over the past 30 years, we now understand many chaperone-assisted processes at a molecular level (Hartl and Hayer-Hartl 2009; Kettern et al. 2010). In this chapter, we will describe experimental approaches for analyzing the function of molecular chaperones and their regulation by cochaperones with an emphasis on the Hsc/Hsp70 chaperone system. The provided protocols should be helpful for chaperone researchers but also for scientists who stumble across these chaperones as interaction partners of their protein of interest. Such interactions are often considered to be "unspecific," and many proteomic facilities will score these interactions accordingly. Highly relevant functional interactions may therefore remain unexplored, which could hamper the efforts to understand the role of chaperones in protein homeostasis. The methods described here will allow rapid verification of whether a protein interacts with Hsc/Hsp70 as a client or a cochaperone and whether it regulates chaperone activity.

2.2 The domain structure of Hsc/Hsp70

Members of the Hsp70 chaperone family display a conserved domain architecture, comprising an amino-terminal ATPase domain, a central peptide-binding region and a carboxy-terminal domain that forms a lid over the peptide-binding region (▶Figure 2.1). Notably, the ATPase domain and the lid domain function as main cochaperone docking sites, whereas nonnative clients are contacted through the central peptide-binding region. This division of labor between the different domains provides an excellent basis for distinguishing between client and cochaperone binding. Deletion fragments of Hsc/Hsp70 that cover individual chaperone domains were used in binding studies to verify whether observed interactions with the chaperones reflect recognition as a nonnative client or a functional cooperation with a cochaperone (Demand et al. 1998).

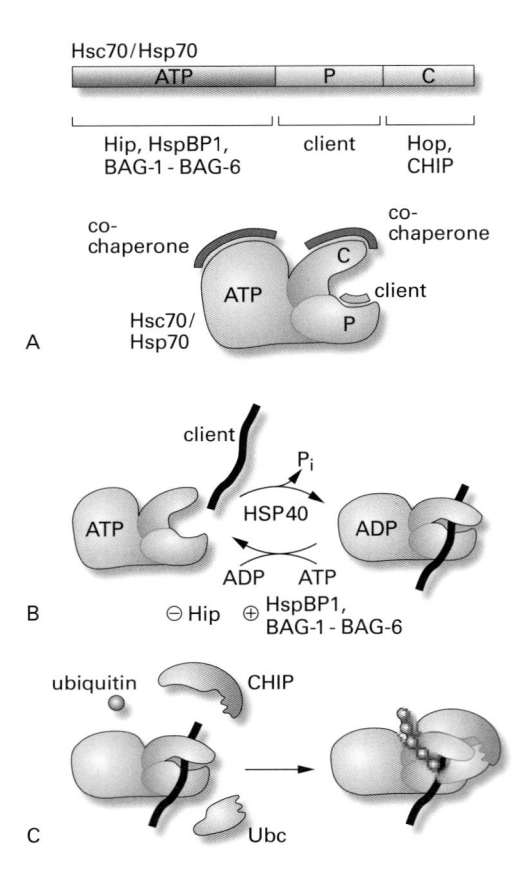

Figure 2.1: Cochaperones regulate the activity of constitutively expressed Hsc70 and stress-inducible Hsp70 in the mammalian cytoplasm and nucleus. (A) Hsc70 and Hsp70 possess an amino-terminal ATPase domain (ATP), followed by the central peptide-binding domain (P) and the carboxy-terminal lid domain (C). Whereas the peptide-binding domain contacts the nonnative chaperone client, the amino- and carboxy-terminal domains provide binding sites for diverse cochaperones. (B) The dynamic interaction of Hsc/Hsp70 with a client protein is regulated by the chaperone's ATPase cycle. The ATP-bound state represents an open conformation with the lid domain raised allowing for rapid binding and release of the client. Hsp40-stimulated ATP hydrolysis drives Hsc/Hsp70 in the ADP-bound state, characterized by stable client binding. Nucleotide exchange facilitates client release and is under regulation by different cochaperones. Hip slows down nucleotide exchange, whereas HspBP1 and the BAG domain cochaperones BAG-1 to BAG-6 act to facilitate this step in the reaction cycle. (C) The cochaperone CHIP switches chaperone activity from protein folding to protein degradation because it acts as a chaperone-associated ubiquitin ligase. CHIP binds to the carboxy terminus of Hsc70 and Hsp70 and then mediates ubiquitylation of the chaperone-bound client protein in cooperation with a ubiquitin-conjugating enzyme (Ubc). Ubiquitylation triggers client degradation by the proteasome or the autophagosome/lysosome system.

2.2.1 Chaperone client or cochaperone? Getting answers from the yeast-2-hybrid system

To distinguish between client and cochaperone binding to Hsc70, fragments of the chaperone that cover or lack certain domains were used in interaction studies. We performed such an analysis using the yeast-2-hybrid system (Demand et al. 1998). The coding regions for full-length rat Hsc70, the amino-terminal ATPase domain and the carboxy-terminal lid domain of the chaperone were subcloned in frame into plasmid systems encoding the DNA-binding domain (BD) of the Gal4 transcription factor (Clonetech), resulting in the expression of BD-Hsc70 fusion proteins in transformed yeast cells. Coding regions for Hsc70-interacting proteins were subcloned in frame into plasmid systems for the transcription-activating domain of Gal4. The yeast strain AH109 was transformed with the resultant constructs in different combinations, and double transformants were selected according to the protocol of the manufacturer (Clonetech). The strain AH109 is unable to synthesize histidine because of a mutation affecting the His3 gene but contains a wild-type copy of the His3 gene under control of the Gal4 promotor. In case of an interaction between Hsc70 (or Hsc70 fragments) and the protein of interest, a functional Gal4 transcription factor is reconstituted, and the corresponding transformants gain the ability to grow on histidine-free medium. A cochaperone will show binding to full-length Hsc70 and the amino- or carboxy-terminus, whereas a client protein will only interact with the full-length chaperone but not with the amino- or carboxy-terminus.

2.3 The Hsc/Hsp70 reaction cycle

Hsc/Hsp70 chaperones interact with nonnative clients in a dynamic manner that is regulated by ATP binding and hydrolysis (Mayer and Bukau 2005). When ATP is bound at the amino terminus of Hsc/Hsp70 the chaperone is in an open conformation. The carboxy-terminal lid domain is raised up, allowing the peptide-binding domain to interact with the client protein (Figure 2.1B). Upon ATP hydrolysis the lid domain closes, leading to a stabilization of the chaperone/client interaction. Finally, client release is achieved through ADP release and ATP rebinding, which opens the lid domain again. Mammalian Hsc70 and Hsp70 display very low ATPase activity on their own (Minami et al. 1996). Going through the ATPase cycle at a physiologically relevant rate therefore requires a cochaperone of the J-domain protein family (also termed Hsp40 cochaperones when referring to mammalian cells). The J-domain cochaperone stimulates the ATP hydrolysis step of the reaction cycle and thereby facilitates client binding (Figure 2.1B) (Minami et al. 1996). Through cooperation with a J-domain protein, a core Hsc/Hsp70 chaperone machinery is formed, which seems to be sufficient for many cellular processes. However, additional cochaperones have been identified that modulate the ATPase cycle at the level of the ADP-bound conformation. The Hsc/Hsp70-interacting protein Hip, for example, stabilizes the ADP-bound state and thereby promotes chaperone/client interactions (Höhfeld et al. 1995). On the other hand, the Hsc/Hsp70-binding protein HspBP1 and also the different members of the BAG-domain cochaperone family act as nucleotide exchange factors that trigger ADP release and ATP rebinding, leading to the release of the client protein (Figure 2.1B) (Shomura et al. 2005; Sondermann et al. 2001). Verifying an involvement in the regulation of the ATPase cycle of

Hsc/Hsp70 therefore represents an important step during the characterization of potential cochaperones (Section 2.6).

2.4 Cochaperones determine the function of Hsc/Hsp70

Cochaperones do not only modulate the ATPase cycle of Hsc/Hsp70, but often also provide a physical link to other proteostasis components. The Hsc/Hsp70 and Hsp90 organizing cochaperone Hop, for example, can simultaneously interact with Hsc/Hsp70 and Hsp90 and thereby facilitates a cooperation of the two chaperones during protein folding (Odunuga et al. 2004). In the context of protein degradation, the cochaperone BAG-1 fulfills an important coupling function because it can associate with the proteasome through an amino-terminal ubiquitin-like domain, while the carboxy-terminal BAG-domain interacts with Hsc/Hsp70 (Lüders et al. 2000). This enables the cochaperone to facilitate the sorting of chaperone clients to the proteasome (Demand et al. 2001). BAG-1 closely cooperates with the carboxy terminus of Hsc/Hsp70-interacting protein CHIP during proteasomal sorting. CHIP indeed provides the initial trigger for this degradation pathway by mediating the ubiquitylation of the chaperone-bound client protein (Demand et al. 2001) (Figure 2.1C). It acts as a chaperone-associated ubiquitin ligase and switches the activity of the chaperone from protein folding to protein degradation (Kettern et al. 2010). Obviously, the activity of Hsc70 and Hsp70 during proteostasis in mammalian cells is regulated by an intricate network of competing and cooperating cochaperones, the full repertoire of which remains to be unraveled.

2.5 In vitro reconstitution and functional analysis of the Hsc/Hsp70 chaperone system

As pointed out in Section 2.2, a core Hsp70 chaperone system that operates in mammalian cells comprises constitutively expressed Hsc70 or stress-inducible Hsp70 together with an ATPase-stimulating J-domain protein such as Hsp40. To reconstitute this system, we recombinantly expressed rat Hsc70 in insect cells and human Hsp40 in bacteria and purified both proteins to homogeneity following the chromatography steps outlined in Section 2.5.1.

To this core chaperone system other potential cochaperones can be added to analyze their impact on the Hsc/Hsp70 ATPase cycle, on chaperone-mediated protein folding and on protein ubiquitylation. In this way, the function of novel cochaperones during chaperone-assisted folding and degradation can be determined.

2.5.1 Purification of Hsc70

For recombinant expression of rat Hsc70, Sf21 insect cells (corresponding to 30 cell culture flasks of 150 cm^2, grown to 80% confluence) were infected with a recombinant baculovirus carrying the Hsc70 cDNA under control of a baculovirus promotor (Höhfeld and Jentsch 1997). After growth for 60 h, insect cells were collected and lysed in 10 mL of lysis buffer (20 mM potassium phosphate buffer [KH_2PO_4/K_2HPO_4], pH 7.0, 1 × Complete protease inhibitor [EDTA-free, Roche], 1 mM 2-mercaptoethanol) per 1 mL wet pellet. Cell lysis was performed in a 25 mL Teflon/glass homogenizer applying

25 strokes at 800 rpm. The lysate was centrifuged at 100,000 × g for 30 min at 4°C, and the resultant supernatant was passed over a Bio-Gel HT hydoxyapatite column (Bio-Rad), equilibrated in lysis buffer. The column was washed with 3–5 column volumes (CVs) of lysis buffer until UV$_{280}$ absorption reached base level. Bound proteins were eluted from the column with a gradient comprising 3 CVs from 0 to 100% elution buffer (200 mM potassium phosphate buffer [KH$_2$PO$_4$/K$_2$HPO$_4$], pH 7.0, 1 × Complete protease inhibitor, 1 mM 2-mercaptoethanol). Obtained fractions were analyzed by SDS-PAGE, and Hsc70 containing fractions were pooled. The pooled eluate was diluted 1:8 with dilution buffer (20 mM MOPS, pH 7.2, 1 mM EDTA, 0.2 × Complete protease inhibitor, 1 mM 2-mercaptoethanol) and passed over a DEAE Sepharose column, equilibrated in 20 mM MOPS, pH 7.2, 20 mM KCl, 1 mM EDTA, 0.2 × Complete protease inhibitor, 1 mM 2-mercaptoethanol. The column was washed with the same buffer until UV$_{280}$ absorption reached base level. Bound proteins were eluted with a salt gradient from 20 to 200 mM KCl in the same buffer corresponding to 3 CVs. Following SDS-PAGE analysis, Hsc70-containing fractions were pooled, and MgCl$_2$ was added to a final concentration of 5 mM. The protein solution was passed over an ATP Sepharose column (C8-linked, Sigma) equilibrated in equilibration buffer (20 mM MOPS, pH 7.2, 50 mM KCl, 5 mM MgCl$_2$, 1 mM 2-mercaptoethanol, 0.004% phenylmethylsulphonyl fluoride [PMSF]). The column was washed with 2 CVs of the same buffer; followed by washing with 2 CVs of 20 mM MOPS, pH 7.2, 1 M KCl, 5 mM MgCl$_2$, 1 mM 2-mercaptoethanol, 0.004% PMSF; and again by washing with 2 CVs of equilibration buffer.

Bound Hsc70 was then eluted with freshly prepared ATP buffer (20 mM MOPS, pH 7.2, 50 mM KCl, 5 mM MgCl$_2$, 1 mM 2-mercaptoethanol, 0.004% PMSF, 15 mM ATP). Hsc70-containing fractions were pooled, and protein concentration was determined. For subsequent ATPase assays, buffer was exchanged by dialysis or using desalting columns to 20 mM MOPS, pH 7.2, 50 mM KCl, 1 mM EDTA, 1 mM 2-mercaptoethanol. Purified Hsc70 was stored at −80°C.

2.5.2 Purification of Hsp40

For recombinant expression of human Hsp40 in bacteria, *E. coli* BL21 (DE3) was transformed with plasmid pET-Hsp40 (Minami et al. 1996). Bacteria were grown to OD$_{600}$ 0.3–0.6 at 37°C, and protein expression was induced with 0.5 mM IPTG at 30°C. After induction for 3 h, cells were harvested by centrifugation at 4,000 × g. The cell pellet was lysed in 10 mL lysis buffer (20 mM Tris-HCl, pH 8.0, 20 mM NaCl, 1 mM EDTA, 1 mM 2-mercaptoethanol, 1 × Complete protease inhibitor) per gram wet weight. The lysate was incubated on ice for 15 min, followed by sonication to further disrupt cells and to shear DNA. The lysate was centrifuged at 100,000 × g for 30 min at 4°C, and the resultant supernatant was passed over a DEAE Sepharose column (Qiagen), equilibrated in lysis buffer. The column was washed with 3–5 CVs of washing buffer (20 mM Tris-HCl, pH 8.0, 20 mM NaCl, 1 mM EDTA, 1 mM 2-mercaptoethanol, 0.2 × Complete protease inhibitor) until UV$_{280}$ absorption reached base level. Bound proteins were eluted from the column with a gradient comprising 3 CVs from 0 to 100% elution buffer (20 mM Tris-HCl, pH 8.0, 500 mM NaCl, 1 mM EDTA, 1 mM 2-mercaptoethanol, 0.2 × Complete protease inhibitor). Obtained fractions were analyzed by SDS-PAGE, and Hsp40-containing fractions were pooled. The pooled eluate was passed over a Bio-Gel HT hydoxyapatite column (Bio-Rad), equilibrated in washing buffer (20 mM potassium phosphate buffer [KH$_2$PO$_4$/K$_2$HPO$_4$], pH 7.0, 0.2 × Complete protease inhibitor, 1 mM 2-mercaptoethanol). The column was washed with 3–5 CVs of washing buffer until UV$_{280}$ absorption reached base level. Bound proteins were eluted from the column with a gradient comprising 3 CVs from 0 to 100% elution buffer (200 mM potassium phosphate buffer [KH$_2$PO$_4$/K$_2$HPO$_4$], pH 7.0, 1 × Complete protease inhibitor, 1 mM 2-mercaptoethanol). Obtained fractions were analyzed by SDS-PAGE, and Hsp40 containing fractions were pooled. The pooled eluate was passed over a 30Q matrix (GE Lifescience), equilibrated in washing buffer (20 mM Tris-HCl, pH 8.0, 10 mM NaCl, 1 mM EDTA, 1 mM 2-mercaptoethanol). The column was washed with 3–5 CVs of washing buffer until UV$_{280}$ absorption reached base level. Bound proteins were eluted from the column with a gradient comprising 3 CVs from 0 to 100% elution buffer (20 mM Tris-HCl, pH 8.0, 500 mM NaCl, 1 mM EDTA, 1 mM 2-mercaptoethanol, 0.2 × Complete protease inhibitor). Following SDS-PAGE, Hsp40-containing peak fractions were pooled, and protein concentration was determined. Purified Hsp40 was stored at −80°C.

2.6 Measuring the ATPase activity of Hsc/Hsp70

The purification of Hsc70 and Hsp40, as described in Section 2.5, allows in vitro reconstitution of the chaperone machinery so that functional assays can be performed. Measuring the ATPase activity of Hsc70 provides a means to evaluate a role of putative cochaperones in the regulation of the ATP-driven chaperone cycle.

ATPase activity was measured in a final sample volume of 15 μL. There were 3 μM Hsc70, 0.3 μM Hsp40 and 3 μM of the potential new cochaperone added in different combinations to reaction buffer (20 mM MOPS, pH 7.2, 50 mM KCl, 2 mM MgCl$_2$, 1 mM dithiotreitol [DTT]) to give a final volume of 12 μL. The reaction was started by adding 3 μL of ATP solution (1 mM ATP, 12.5 μCi α[^{32}P]-ATP, 20 mM MOPS, pH 7.2, 50 mM KCl, 2 mM MgCl$_2$, 1 mM DTT) and subsequently incubated at 30°C. After 10 min and subsequently every 20 min (for up to 110 min), 2 μL aliquots were removed and added to 18 μL reaction buffer containing 20 mM EDTA. All samples were kept on ice until final analysis by thin layer chromatography. Then 1 μL of each sample was spotted onto a PEI Cellulose F plate (Merck). AMP/ADP was separated from ATP by chromatography in solvent (0.5 M formic acid, 0.5 M LiCl$_2$), and cellulose was subsequently air-dried. Hydrolyzed ATP was quantified with a phosphoimager system (Packard Bell).

Based on this ATPase assay, the Hsc70-interacting cochaperone Hip was shown to stabilize the ADP-bound state of the chaperone (Höhfeld et al. 1995), whereas BAG-domain cochaperones and HspBP1 were identified as nucleotide-exchange factors that facilitate ADP release and ATP rebinding (Figure 2.1B) (Alberti et al. 2004; Sondermann et al. 2001).

2.7 Determining chaperone activity

A simple assay to elucidate the chaperone activity of Hsc70 and Hsp70 involves the heat denaturation of firefly luciferase in the presence of the chaperones and diverse cochaperones. Soluble and aggregated species of luciferase are subsequently separated by centrifugation and are quantified.

First, 3 µM Hsc70, 0.3 µM Hsp40 and 3 µM cochaperones were preincubated in different combinations on ice for 10 min in 20 µL aggregation buffer (20 mM MOPS, pH7.2, 50 mM KCl, 2 mM ATP, 2 mM MgCl$_2$, 1 mM DTT). Then 0.06 µM luciferase was added to each sample, and heat denaturation was performed at 42°C for 15 min. A control reaction containing the same amount of luciferase in aggregation buffer was kept on ice. Reaction was stopped by adding EDTA to a final concentration of 5 mM, and samples were placed on ice. Protein aggregates were pelleted by centrifugation at 16,000 × g and 4°C. Pellet and supernatant were separated. The pellet was diluted in 30 µL SDS sample buffer. The supernatant was precipitated with trichloroacetic acid (TCA) added to a final concentration of 10% by incubation on ice for 30 min. Samples were centrifuged at 16,000 × g for 10 min. The supernatant was completely removed, and the TCA precipitate was dissolved in 30 µL SDS sample buffer. Equal amounts of the soluble and insoluble fraction were analyzed by SDS-PAGE and Western blotting with an anti-luciferase antibody (Promega). The ratio of soluble versus insoluble luciferase was determined.

2.8 In vitro reconstitution of chaperone-assisted ubiquitylation

Molecular chaperones are often described as cellular protein-folding factors. However, in recent years it has become increasingly clear that at least some chaperones are also very actively involved in protein degradation (Arndt et al. 2007).

Constitutively expressed Hsc70 and stress-inducible Hsp70 are particularly active in this regard and can initiate different degradation pathways for different client proteins. This includes chaperone-assisted proteasomal degradation (CAP), chaperone-assisted selective autophagy (CASA) and chaperone-mediated autophagy (CMA) (Kettern et al. 2010). Among these pathways, CAP and CASA are dependent on an initial ubiquitylation of the client, which can be mediated by CHIP or other chaperone-associated ubiquitin ligases (Figure 2.1C). The in vitro reconstitution of chaperone-assisted ubiquitylation provides a means to identify proteins that regulate this important step in the triage of chaperone clients and to elucidate underlying molecular details.

2.8.1 Purification of the E1 ubiquitin-activating enzyme

For recombinant expression of wheat E1, Sf21 insect cells (corresponding to 30 cell culture flasks of 150 cm^2, grown to 80% confluence) were infected with a recombinant baculovirus carrying the E1 cDNA under control of a baculovirus promotor. After growth for 60 h, insect cells were collected and lysed in 10 mL of lysis buffer (20 mM MOPS, pH 7.2, 20 mM KCl, 1 mM EDTA, 1 × Complete protease inhibitor, 1 mM 2-mercaptoethanol) per 1 mL wet pellet. Cell lysis was performed in a 25 mL Teflon/glass homogenizer applying 25 strokes at 800 rpm. The lysate was centrifuged at 100,000 × g for 30 min at 4°C, and the resultant supernatant was passed over DEAE Sepharose column (Qiagen), equilibrated in lysis buffer. The column was washed with 3–5 CVs of washing buffer (20 mM MOPS, pH 7.2, 20 mM KCl, 1 mM EDTA, 0.2 × Complete protease inhibitor, 1 mM 2-mercaptoethanol). Bound proteins were eluted from the column with a gradient comprising 3 CVs from 0 to 100% elution buffer (20 mM MOPS, pH 7.2, 200 mM KCl, 1 mM EDTA, 0.2 × Complete protease inhibitor, 1 mM 2-mercaptoethanol). Obtained fractions were analyzed by SDS-PAGE. E1-containing peak fractions were pooled, and protein concentration was determined. Purified E1 was stored at −80°C.

2.8.2 Purification of the E2 ubiquitin-conjugating enzyme UbcH5b

For recombinant expression of human UbcH5b in bacteria, *E. coli* BL21 (DE3) was transformed with plasmid pET-UbcH5b (Demand et al. 2001). Bacteria were grown to OD_{600} 0.3–0.6 at 37°C, and protein expression was induced with 0.5 mM IPTG at 30°C. After induction for 3 h, cells were harvested by centrifugation. The cell pellet was lysed in 10 mL lysis buffer per 1 mL cell material (20 mM MOPS, pH 7.2, 20 mM KCl, 0.5 mM EDTA, 1 × Complete protease inhibitor, 1 mM 2-mercaptoethanol). The lysate was incubated on ice for 15 min, followed by sonication to further disrupt cells and shear DNA. The lysate was centrifuged at 100,000 × g for 30 min at 4°C, and the resultant supernatant was passed over a DEAE Sepharose column (Qiagen), equilibrated in lysis buffer. The column was washed with 3–5 CVs of washing buffer (20 mM MOPS, pH 7.2, 20 mM KCl, 0.5 mM EDTA, 0.2 × Complete protease inhibitor, 1 mM 2-mercaptoethanol) until UV_{280} absorption reached base level. Bound proteins were eluted from the column with a gradient comprising 3 CVs from 0 to 100% elution buffer (20 mM MOPS, pH 7.2, 200 mM KCl, 1 mM EDTA, 0.2 × Complete protease inhibitor, 1 mM 2-mercaptoethanol). Obtained fractions were analyzed by SDS-PAGE, and UbcH5b containing fractions were pooled. The pooled eluate was passed over a Bio-Gel HT hydoxyapatite column (Bio-Rad), equilibrated in washing buffer (20 mM potassium phosphate buffer [KH_2PO_4/K_2HPO_4], pH 7.0, 0.2 × Complete protease inhibitor, 1 mM 2-mercaptoethanol). The column was washed with 3–5 CVs of washing buffer until UV_{280} absorption reached base level. Bound proteins were eluted from the column with a gradient comprising 3 CVs from 0 to 100% elution buffer (200 mM potassium phosphate buffer [KH_2PO_4/K_2HPO_4], pH 7.0, 1 × Complete protease inhibitor, 1 mM 2-mercaptoethanol). Obtained fractions were analyzed by SDS-PAGE, and UbcH5b containing fractions were pooled and diluted 1:8 with 20 mM MOPS, pH7.2. The diluted eluate was passed over a 30Q matrix (GE Lifescience), equilibrated in washing buffer (20 mM MOPS, pH 7.2, 20 mM KCl, 0.5 mM EDTA, 0.2 × Complete protease inhibitor, 1 mM 2-mercaptoethanol). The column was washed with 3–5 CVs of washing buffer until UV_{280} absorption reached base level. Bound proteins were eluted from the column with a gradient comprising 3 CVs from 0 to 100% elution buffer (20 mM MOPS, pH 7.2, 200 mM KCl, 0.5 mM EDTA, 0.2 × Complete protease inhibitor, 1 mM 2-mercaptoethanol). Following SDS-PAGE analysis, UbcH5b containing peak fractions were pooled, and protein concentration was determined. Purified UbcH5b was stored at −80°C.

2.8.3 Purification of CHIP

Human CHIP was purified after recombinant expression in *E. coli* BL21 (DE3) cells using the plasmid pET-CHIP (Demand et al. 2001). Purification was performed under the same conditions as described in Section 2.8.2 for UbcH5b.

2.8.4 Recombinant expression of Raf-1

For recombinant expression of human Raf-1 in bacteria, *E. coli* BL21 (DE3) was transformed with plasmid pET-raf1 (Demand et al. 2001). Bacteria were grown to OD_{600} 0.3–0.6 at 37°C, and protein expression was induced with 0.5 mM IPTG at 30°C. After induction for 3 h, cells were harvested by centrifugation. The cell pellet was lysed in 10 mL lysis buffer per 1 mL cell material (20 mM MOPS, pH 7.2, 20 mM KCl,

0.5 mM EDTA, 1 × Complete protease inhibitor, 1 mM 2-mercaptoethanol). The lysate was incubated on ice for 15 min, followed by sonication to further disrupt cells and shear DNA. The lysate was centrifuged at 30,000 × g for 30 min at 4°C. and the resultant supernatant was used as a Raf-1 containing extract. The supernatant was aliquoted and stored at −80°C.

2.8.5 Analyzing chaperone-assisted ubiquitylation

For in vitro reconstitution of chaperone-assisted ubiquitylation, the following samples were prepared. There were 2 µL of Raf-1 containing bacterial extract and 2 µL of ubiquitin (25 mg/mL stock solution, Sigma) added to a 1.5 mL reaction tube. Also, 0.1 µM E1 enzyme, 2 µM UbcH5b, 1 µM CHIP, 3 µM Hsc70 and 0.3 µM Hsp40 were added in varying combinations to obtain different control reactions and a complete sample. Reaction buffer (25 mM MOPS, pH 7.2, 100 mM KCl, 0.004% PMSF) was added to obtain a total volume of 18 µL. The ubiquitylation reaction was started by adding 2 µL of ATP buffer (100 mM ATP, pH7.2, 100 mM DTT, 100 mM $MgCl_2$). The samples were incubated for 2 h at 30°C. The reaction was stopped by addition of 10 µL of 3 × SDS-PAGE sample buffer and denaturation at 90°C. Samples were separated by SDS-PAGE and blotted onto a nitrocellulose membrane. Ubiquitylation of Raf-1 was detected with an anti-Raf-1 antibody.

The protocol provides a basic ubiquitylation reaction that can be modified in multiple ways to verify the involvement of newly emerging factors in chaperone-assisted ubiquitylation. This includes exchanging UbcH5b against other E2 ubiquitin-conjugating enzymes or CHIP against other E3 ubiquitin ligases with a potential ability to cooperate with Hsc70. The assay was also used to verify the impact of other cochaperones, for example, HspBP1 and BAG-2, on CHIP-mediated ubiquitylation (Alberti et al. 2004; Arndt et al. 2005). Furthermore, the assay can also be adopted for chaperone clients other than Raf-1 (Arndt et al. 2010). The obtained ubiquitylated chaperone clients can be further analyzed to determine the ubiquitin chain linkage type and to identify chain attachment sites by mass spectrometry.

2.9 Concluding remarks

Constitutively expressed Hsc70 and stress-inducible Hsp70 do not operate on their own in our cells. Only in conjunction with regulatory cochaperones are they able to fulfill their broad range of diverse functions during protein biogenesis.

Although many cochaperones have been identified, others still await detection. We hope that the methods described here provide a helpful framework for an initial characterization of potential new cochaperones. Establishing the whole cast of characters involved in Hsc/Hsp70 regulation will represent a major step toward understanding cellular protein homeostasis and toward a pharmacological modulation of this chaperone machinery in the context of protein aggregation diseases, cancer and cystic fibrosis.

References

Alberti, S., Böhse, K., Arndt, V., Schmitz, A., and Höhfeld, J. (2004). The cochaperone HspBP1 inhibits the CHIP ubiquitin ligase and stimulates the maturation of the cystic fibrosis transmembrane conductance regulator. Mol Biol Cell *15*, 4003–4010.

Arndt, V., Daniel, C., Nastainczyk, W., Alberti, S., and Höhfeld, J. (2005). BAG-2 acts as an inhibitor of the chaperone-associated ubiquitin ligase CHIP. Mol Biol Cell 16, 5891–5900.

Arndt, V., Dick, N., Tawo, R., Dreiseidler, M., Wenzel, D., Hesse, M., et al. (2010). Chaperone-assisted selective autophagy is essential for muscle maintenance. Curr Biol 20, 143–148.

Arndt, V., Rogon, C., and Höhfeld, J. (2007). To be, or not to be – molecular chaperones in protein degradation. Cell Mol Life Sci 64, 2525–2541.

Demand, J., Alberti, S., Patterson, C., and Höhfeld, J. (2001). Cooperation of a ubiquitin domain protein and an E3 ubiquitin ligase during chaperone/proteasome coupling. Curr Biol 11, 1569–1577.

Demand, J., Lüders, J., and Höhfeld, J. (1998). The carboxy-terminal domain of Hsc70 provides binding sites for a distinct set of chaperone cofactors. Mol Cell Biol 18, 2023–2028.

Hartl, F.U., and Hayer-Hartl, M. (2009). Converging concepts of protein folding in vitro and in vivo. Nat Struct Mol Biol 16, 574–581.

Höhfeld, J., and Jentsch, S. (1997). GrpE-like regulation of the hsc70 chaperone by the anti-apoptotic protein BAG-1. EMBO J 16, 6209–6216.

Höhfeld, J., Minami, Y., and Hartl, F.U. (1995). Hip, a novel cochaperone involved in the eukaryotic Hsc70/Hsp40 reaction cycle. Cell 83, 589–598.

Kettern, N., Dreiseidler, M., Tawo, R., and Höhfeld, J. (2010). Chaperone-assisted degradation: multiple paths to destruction. Biol Chem 391, 481–489.

Lüders, J., Demand, J., and Höhfeld, J. (2000). The ubiquitin-related BAG-1 provides a link between the molecular chaperones Hsc70/Hsp70 and the proteasome. J Biol Chem 275, 4613–4617.

Mayer, M.P., and Bukau, B. (2005). Hsp70 chaperones: cellular functions and molecular mechanism. Cell Mol Life Sci 62, 670–684.

Minami, Y., Höhfeld, J., Ohtsuka, K., and Hartl, F.U. (1996). Regulation of the heat-shock protein 70 reaction cycle by the mammalian DnaJ homolog, Hsp40. J Biol Chem 271, 19617–19624.

Odunuga, O.O., Longshaw, V.M., and Blatch, G.L. (2004). Hop: more than an Hsp70/Hsp90 adaptor protein. Bioessays 26, 1058–1068.

Shomura, Y., Dragovic, Z., Chang, H.C., Tzvetkov, N., Young, J.C., Brodsky, J.L., et al. (2005). Regulation of Hsp70 function by HspBP1: structural analysis reveals an alternate mechanism for Hsp70 nucleotide exchange. Mol Cell 17, 367–379.

Sondermann, H., Scheufler, C., Schneider, C., Höhfeld, J., Hartl, F.U., and Moarefi, I. (2001). Structure of a Bag/Hsc70 complex: convergent functional evolution of Hsp70 nucleotide exchange factors. Science 291, 1553–1557.

3 Membrane protein folding in detergents

David V. Tulumello and Charles M. Deber

3.1 Introduction

Membrane proteins assume critical roles in many diverse cellular processes, such as ion and solute transport, signal transduction and membrane structural integrity. Alpha-helical transmembrane (TM) proteins, which are the most abundant and ubiquitous class, are estimated to comprise 20% to 30% of open reading frames in sequenced genomes (Wallin and von Heijne 1998). Despite the importance of TM proteins, it has been soluble, globular proteins that have dominated our understanding of both protein structure and function. This discrepancy in knowledge is readily apparent in the Protein Data Bank, where currently only ~2% of the more than 70,000 deposited structures are TM proteins. This lag in our understanding is largely due to the hydrophobic nature of these molecules, which makes their production and purification technically challenging, as well as dictating that they must be studied in a lipid or detergent environment. The inclusion of such "membrane mimetic" environments introduces further hurdles for many biophysical characterization techniques or, in some cases, prevents such experiments from being feasible (particularly for systems containing lipid vesicles). Because of this situation, the choice of a suitable membrane mimetic environment is critical for the successful outcome of such experiments. To a large degree, detergents have been used to provide a hydrophobic environment similar to a natural membrane, while imparting solubility in mixed water/detergent media to membrane proteins to allow for their characterization by a broad range of techniques.

Detergents represent a very broad class of molecules, with a range of biophysical characteristics, and in many cases, the choice of the "correct" detergent for studies of membrane proteins is largely empirical. Although it would be extraordinarily beneficial if there were specified detergents that were always useful for particular applications, our understanding of these phenomena has not yet advanced to this level. In some cases, the choice of detergent is heavily dependent upon the technique. In other circumstances, several different detergents may give similar results. And in yet other cases, many detergents must be investigated before a suitable one is found. It is thus helpful to understand the characteristics and limitations of detergents as to how they are applied toward studies of membrane protein structure and folding. This chapter will focus on the challenges and strategies of studying α-helical membrane proteins, and membrane protein fragments, in detergent systems. We will discuss the nature of membrane proteins and their interactions with detergents, provide a brief overview of the most commonly used detergents and describe some common structural characterization techniques that can be performed on detergent-solubilized membrane proteins. The chapter will conclude with some examples from our lab of successful characterizations of α-helical membrane protein constructs in detergents. Beta-barrel membrane proteins, which are found in bacterial outer membranes as well as mitochondria and have differing folding behaviors and interactions with detergents (Tamm et al. 2004; Wimley 2003), will not be specifically addressed here.

3.2 Interactions of membrane proteins with detergents

3.2.1 TM protein architecture

Originally proposed in the 1990s by Popot and Engleman, the two-stage model of α-helical membrane protein folding provides a useful conceptual framework to consider the structure and assembly of these proteins (Engelman et al. 2003; Popot and Engleman 1990). In this model, TM helices first insert into the interior of the lipid bilayer as independently stable, folded domains. Following insertion, individual TM segments may then associate laterally with one another to produce the final tertiary structure. Ultimately, these steps are governed by the identities of the amino acid side chains comprising the protein sequence. For insertion into the membrane as a helix, a protein sequence must contain a stretch of ca. 20 residues that are primarily hydrophobic. In most cases, insertion occurs via a well-defined threshold based primarily upon hydrophobicity (Hessa et al. 2005; Liu et al. 1996; White and von Heijne 2008). In the second step of this model, specific helix-helix interactions between the inserted TM segments form to generate the final membrane protein fold. Such helix-helix interactions are often achieved through specific motifs including GG4, heptad, polar and multiple Ser/Thr motifs (Mackenzie 2006; Rath et al. 2009b). These motifs stabilize helix-helix interactions by maximizing van der Waals interactions and/or allow for the formation of side chain–side chain hydrogen bonds between helices.

As more detailed structural information has become available for integral membrane proteins, it is becoming increasingly apparent that there are more complex folding events occurring in TM domains other than just the lateral association of TM segments, such as binding of prosthetic groups, partial helices that do not completely span the membrane and reentrant loops. Large water-filled cavities within the membrane region formed by TM helices are also present in many transporters (Abramson et al. 2003; Stroud et al. 2003). Another complication is the common occurrence of large autonomously folded soluble domains as segments of larger membrane proteins.

3.2.2 Detergent properties

Typically, a detergent is an amphiphilic molecule that contains both polar substituents (often termed head groups) and nonpolar regions (most often an alkyl chain). The presence of two regions with differing polarity leads to detergent self-aggregation in water in an effort to minimize aqueous exposure of the nonpolar regions. Detergents differ from common biological lipids in that they typically form smaller globular assemblies termed micelles, instead of the extended bilayer structures found in membranes. In a micelle structure, a central core is composed primarily of the hydrophobic component of detergent monomers that are surrounded by an outer layer of the primarily hydrophilic chemical groups. Micelle structures can be described, on average, as roughly spherical particles, but on an individual basis, they typically have highly irregular surfaces. Notably, the micellar particle differs from cells (such as erythrocytes) and phospholipid vesicles in that the micelle does not contain an aqueous interior compartment. It is this micellar form of detergent molecules that is most relevant as a membrane mimetic for membrane proteins.

Detergent micelles are most often described by a specific aggregation number (or range) as well as a concentration above which this self-association into micelles occurs spontaneously (termed the critical micelle concentration [CMC]). These two properties are dictated by both the length of the detergent alkyl chain as well as the chemical makeup (both charge and size) of the polar head groups. These features are often used to qualitatively describe how these detergents interact with membrane proteins, where small head group size, the presence of a head group charge, and/or short alkyl tail length all increase the "harshness" of a detergent with respect to membrane protein denaturation (Prive 2007). These features are also ones that in general make detergents effective solubilizers of both membranes and membrane proteins.

3.2.3 Solubilization versus stabilization: the role of detergents in membrane protein characterization

Detergents serve a variety of different roles in the study of membrane proteins. Perhaps the most common use of detergents is to extract these proteins from their native membrane or from insoluble aggregates produced by heterologous expression (commonly referred to as inclusion bodies) (Cunningham and Deber 2007; Tate 2001). In many cases, detergents alone may accomplish this, although mechanical disruption (such as sonication) used in combination can often improve this process. Following this step, the solubilizing detergent may be exchanged for another one, or – from this detergent-solubilized state – membrane proteins may be exchanged into an artificial bilayer system. Initial solubilization protocols from membranes as well as detergent exchange have been reviewed extensively elsewhere (le Maire et al. 2000; Lin and Guidotti 2009).

The fundamental role of detergents for structural studies is to confer aqueous solubility to otherwise insoluble membrane proteins (▶Figure 3.1). In principle, the purpose of a detergent is to coat hydrophobic regions of proteins with amphipathic detergent molecules while maintaining the native tertiary and quaternary structure of the membrane protein. However, this very ability of detergents to bind to hydrophobic regions of proteins often renders them denaturing. Thus, in soluble proteins, which are stabilized by packing of hydrophobic regions into a central core, detergents compete successfully through hydrophobic protein–detergent interactions (Imamura 2006). The membrane-spanning domains of membrane proteins are likewise stabilized in large part by interactions of hydrophobic residues, and as such, detergents are also to some degree denaturing to membrane proteins. In fact, detergents have gained widespread use in membrane-protein-unfolding studies (Booth and Curnow 2006; Renthal 2006). It is this balance between solubilization versus native structure stabilization, as well as any specific characteristics required for the experimental technique used, that is at the heart of the choice of detergent for characterization.

3.2.4 Overview of commonly used detergents for membrane protein characterization

The chemical structures of some commonly used detergents are shown in ▶Figure 3.2, and a summary of their relevant characteristics is given in ▶Table 3.1. We have chosen to describe in detail a set of acyl chain detergents that covers three broad categories: ionic, zwitterionic and neutral. The set is composed of the two most widely used detergents for solution nuclear magnetic resonance (NMR) studies (SDS and *n*-dodecyl phosphocholine [DPC]); and the two most often used in X-ray crystallography

3 Membrane protein folding in detergents

A soluble globular protein **B** membrane protein **C** detergent solubilized membrane protein

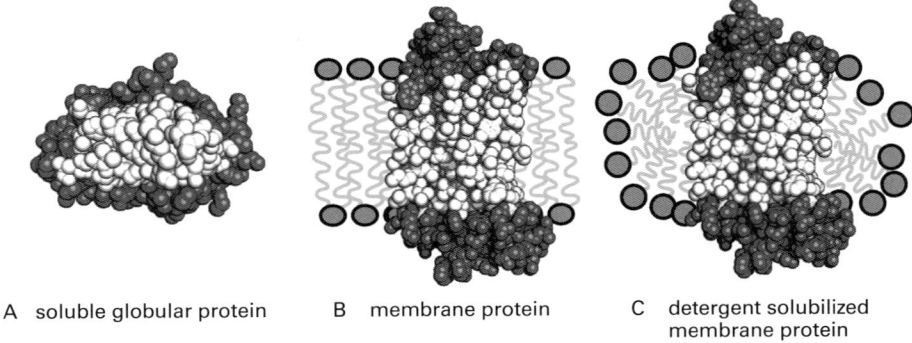

Figure 3.1: Principle of detergent solubilization of membrane proteins. (A) Typical soluble proteins fold such that hydrophobic residues form a buried, central core and are surrounded by primarily polar residues that allow for aqueous solubility. Myoglobin (PDBID: 1MBN) is used here as an example. The central core is shown in white and overlaid on the remainder of the structure, shown in gray. (B) A membrane protein typically has hydrophobic residues exposed to its environment, which is the hydrophobic lipid bilayer. The β-2-adrenergic receptor (PDBID: 2RH1) is used here as an example. Membrane-embedded residues are shown in white, while residues not found within the bilayer core are shown in gray. (C)The functional use of detergents is to confer aqueous solubility to otherwise insoluble membrane proteins. A detergent-solubilized membrane protein will have hydrophobic regions coated with amphiphilic detergent molecules, producing a polar outer surface allowing for solubility in the heterogeneous detergent-water medium (shown in right panel).

Table 3.1: Properties of commonly used detergents

Detergent	Monomer mass (Da)	Aggregation #	CMC (mM)
sodium dodecyl sulfate (SDS)	288	62–101	1.2–7.1
n-dodecyl phosphocholine (DPC)	352	50–60	1.1
n-dodecyl-β-D-maltoside (DDM)	511	110–140	0.18
n-octyl-β-D-glucoside (OG)	292	~90	19–25

Data taken from compilation in le Maire et al. (2000).

(n-dodecyl-β-D-maltoside [DDM] and n-octyl-β-D-gluocoside [OG]), according to the Membrane Protein Data Bank (available online at: http://www.mpdb.tcd.ie)(Raman et al. 2006).

3.2.4.1 Anionic – SDS

Among the most well-characterized detergents, SDS has a 12-carbon alkyl tail, along with a negatively charged sulfate head group. Like other charged detergents, its self-association and interaction with proteins is highly dependent upon the ionic strength

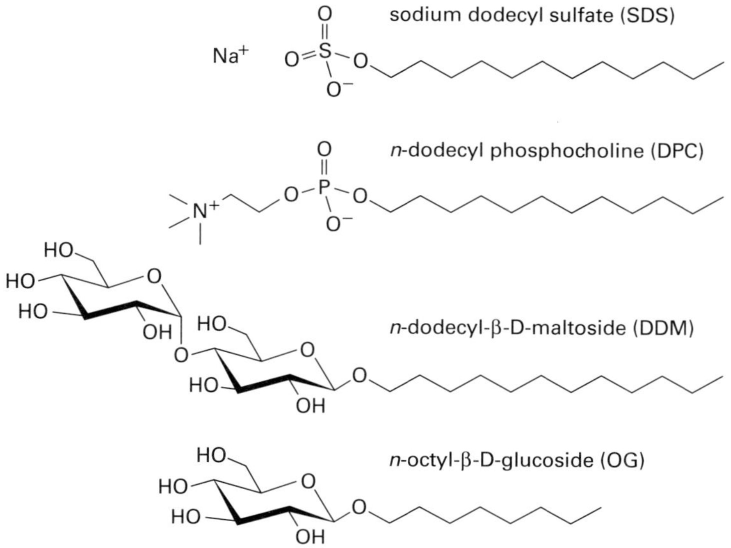

Figure 3.2: Common detergents used in membrane protein characterization. Structures were produced using ChemSketch (ACD Labs).

of the solution (Imamura 2006). As it is often considered among the "harshest" of detergents commonly used, it may seem surprising that SDS has been used in identifying helix-helix association interfaces (Arkin et al. 1994; Lemmon et al. 1992; Ng and Deber 2010; Rath et al. 2006). SDS has gained such widespread use because it is a highly affective solubilizing agent and, upon closer inspection, does not appear to denature strong helix-helix interactions. SDS has been shown in some cases to be able to recapitulate many aspects of native tertiary/quaternary folded structure (Almeida and Opella 1997; Chill et al. 2006; Howell et al. 2005).The anionic nature of SDS also allows for its use in PAGE analysis, a common molecular biology technique.

Due to its well-characterized ability to denature soluble proteins (Imamura 2006), SDS will likely denature any large extramembranous regions of membrane proteins and should not be considered a practical mimetic to preserve the overall structure of any proteins containing such domains. This renders SDS primarily useful for small membrane proteins, or fragments thereof, for which it is highly effective at producing well-solubilized, stable samples.

3.2.4.2 Zwitterionic – DPC

Zwitterionic detergents are generally considered less perturbing to membrane protein structure than ionic ones, while remaining effective membrane solubilizers (le Maire et al. 2000; Prive 2007). DPC in particular has gained increased use in structural studies due to its phosphocholine head group, which itself is one of the most prevalent among phospholipids in native biological membranes. There is also evidence that the behavior of this head group within detergent micelles is very similar to the water/membrane interface in native lipids (Beswick et al. 1999). As such, DPC is commonly used for NMR where its monodispersity, low aggregate number and small particle size make it an ideal detergent (Page et al. 2006).

3.2.4.3 Neutral – DDM

DDM is a neutral detergent with a sugar head group and a 12-carbon acyl tail. It is generally considered among the mildest detergents, making it an excellent initial choice for a wide range of applications. It is generally less efficient than other detergents at solubilizing biological membranes; however, membrane proteins are often exchanged into this detergent for characterization studies. DDM has minimal interactions with soluble domains, leading to its common use in the characterization of transport proteins containing such features (Newstead et al. 2008). In particular, DDM is often used in X-ray crystallography, albeit its relatively large alkyl chain length may encompass much of the protein, consequently reducing the protein-protein crystal contacts necessary for this technique (Newstead et al. 2008). Despite its minimally denaturing properties, the large aggregate size and poor monodispersity of DDM generally make it unsuitable for NMR studies (Sanders and Sonnichsen 2006). n-decyl-β-D-maltoside (DM), a detergent with an identical head group as DDM but containing a 10-carbon acyl tail, may also be used in similar applications.

3.2.4.4 Short tail neutral – OG

Another detergent with a neutral head group, OG, is again considered relatively "mild." It does, however, have a relatively shorter acyl chain length of 8 carbons, which can make it somewhat more denaturing than DDM (Prive 2007). Its smaller size may leave more surface area exposed for protein-protein contacts important for crystallization, potentially improving resolution of crystal structures (Sonoda et al. 2010). For purposes of NMR studies, OG has been shown to have poor sample stability and is not in widespread use for this technique (Vinogradova et al. 1998). n-nonyl-β-D-glucoside (NG), a detergent with an identical head group but a 9-carbon acyl tail is also a commonly used detergent, with very similar properties to octyl glucoside.

3.2.4.5 Other detergents

Another commonly used class of detergents is the lysophospholipids, which are derivatives of native phospholipids, but with only one fatty acid linked to the glycerol backbone rather than two. Lysophospholipids accordingly form micellar structures rather than bilayers. N,N-dimethyldodecylamine-N-oxide (lauryl dimethylamine oxide; LDAO) is a zwitterionic detergent that has been used in a number of crystallization studies of α-helical TM proteins (Andrade et al. 2005; Ye et al. 2010). Polyoxyethylene detergents, a neutral class of detergents, are also commonly used. These detergents are typically composed of an acyl chain tail, and a head group made up of polyethylene glycol, both of which may vary in length. These detergents are named based upon their composition according to the formula CxEy, where x is the number of carbons present in the acyl tail and y is the number of ethoxy groups present in the head group. A commonly used detergent in this class is C8E4. Steroid-based detergents such as 3-([3-cholamidopropyl]-dimethylammonio)-1-propanesulfonate (CHAPS), have also been used as effective solubilizers of membranes and membrane proteins. As a membrane mimetic environment for structural studies, these latter systems are most effective only in the presence of native lipids or other detergents (Sanders and Sonnichsen 2006).

Fluorinated surfactants, such as sodium perfluoroctanoate (SPFO), are similar to other detergents described here except that their hydrophobic tails contain (usually exclusively) fluorine atoms in place of hydrogen atoms. For example, the structure of SPFO is $CF_3(CF_2)_6$-COOH. Detergents in this category have lower affinity for hydrophobic amino acids and are therefore thought to be less disruptive toward protein-protein interactions, thereby preserving native structure (Melnyk et al. 2002).

3.3 Techniques to characterize TM proteins in detergents

3.3.1 General considerations

Characterizing membrane proteins in detergents presents a host of additional considerations not encountered for soluble proteins. One factor for performance of most spectroscopic techniques is light scattering. As the concentration of detergent increases, so does the associated light scattering, and although this can be background subtracted in most cases, it may increase the noise in data acquisition, thereby decreasing sensitivity. For this reason, it is often beneficial to have only a minimal level of detergent present. Yet if too little detergent is present, the membrane protein may not be fully saturated by detergent, potentially leading to insolubility or denaturation. One must therefore ensure that the protein is fully saturated with detergent, albeit the saturating concentration of detergent is not always clear. Membrane proteins have been shown in many cases to bind much more detergent than their soluble counterparts, and as such, it is better to assume a large degree of binding (Rath et al. 2009a). It is also necessary to have a detergent concentration above the CMC, to ensure the presence of micelles; in many cases, detergent monomers behave much differently than micelles themselves, a factor that can lead to membrane protein aggregation (Otzen et al. 2008; Wahlstrom et al. 2008).

In practice, for studies of membrane protein fragments consisting of 1–2 TM segments at a concentration in the μM regime, one would use an approximate 1,000-fold increase in detergent – corresponding to a mM detergent concentration. For micelle aggregates with an association number on the order of 100, this represents 10 times as many detergent micelles as TM segments. In situations where a high concentration of peptide is necessary, such as in solution NMR where peptide concentrations are in the mM range, maintaining at least a 1:1 ratio of TM segments to detergent micelles is recommended (Sanders and Sonnichsen 2006).

The final peptide-to-detergent ratio is particularly important for assessing oligomerization of TM segment peptides. As detergent concentrations are decreased, so is the total volume of the micellar phase, essentially increasing the apparent concentration of the TM segments in this phase. As such, it may be more reasonable to assess free energy of association of TM segments in terms of the peptide-to-detergent ratio (Fleming 2002). Too low a concentration of detergent may also lead to some degree of colocalization of individual TM segments in the same micelle without a specific interaction between them, thereby leading to an incorrect assessment of oligomeric state.

Another important consideration is the equilibration time required following preparation of samples. Detergent solvation generally occurs on a relatively rapid time scale (usually within minutes) from a soluble state; however, from an insoluble state this process may take much longer. The time scale between association and disassociation of TM segments is also relevant. For some TM segments, these events may occur over very long periods (Rath et al. 2009b). For techniques that require equilibration of separate fractions of TM segments, it is recommended that these fractions be mixed prior to inclusion of the final detergent environment where association occurs.

Sample stability is also an issue. For example, both NMR and, to a lesser degree, analytical ultracentrifugation (AUC) (le Maire et al. 2008) require time periods of days or longer for data collection. While lysophospholipid detergents often give high-resolution spectra, no high-resolution structures have been reported in these detergents – likely a consequence of the relatively short-term stability of the relevant samples (Page et al. 2006).

3.3.2 Detergent micelle insertion and secondary structure

Initial characterization of detergent-solubilized states typically focuses on the adoption of the correct secondary structure, as well as insertion of TM regions into detergent micelles. Helical structure should be observed for TM domains in a native state, and loss of the correct secondary structure may be used as an indication that denaturation is occurring over time.

3.3.2.1 Circular dichroism spectroscopy (CD)

CD is a common technique used to assess the secondary structure of membrane proteins. Helical TM segments display characteristic CD spectra in the far-UV region, with minima in mean residue ellipticity [θ] at 208 and 222 nm. For individual TM segments, adoption of helical structure in the presence of detergents has been used extensively to monitor insertion into a micellar environment (Liu et al. 1996). Measurements of secondary structure are also useful for monitoring the global fold of a membrane protein. For an example, see Section 4.2.2.

As CD is a spectroscopic technique where minimization of light scattering is essential, the use of lower detergent concentrations is beneficial. Another strategy to improve the sensitivity of the spectral determination is to decrease the path length of the cuvette while increasing the protein concentration, thereby reducing the scattering of the detergent, while maintaining signal from the protein. We have found that a 1 mm path length cuvette is adequate for most detergents in the mM concentration regime. In some cases, where a high CMC dictates that increased detergent concentrations be used, an 0.1 mM or smaller path length is recommended to minimize light scattering.

3.3.2.2 Fourier transform infrared spectroscopy (FTIR)

FTIR specifically probes the frequencies of molecular vibrations. In practice, this technique is in many ways complimentary to circular dichroism in that it is often used to identify the secondary structure of membrane proteins, based upon diagnostic vibration peaks of the carbonyl groups in peptide bonds. Classically, transmission FTIR of protein samples is complicated by the strong absorbance of water, which saturates the absorption spectra. Using D_2O reduces this effect; however, a very small transmission path length (>1mm) as well as relatively high protein concentrations are needed (10 to 50 times greater than that for CD) (Baenziger and daCosta 2008). The major advantage of FTIR versus CD is that FTIR is readily adaptable to studying membrane proteins in lipid bilayers. By using attenuated total reflectance FTIR on a deposited bilayer containing the membrane protein of interest, the absorbance of buffer is eliminated.

3.3.2.3 Tryptophan fluorescence

The amino acid tryptophan has a fluorescence emission spectra that is highly dependent upon the relative polarity of the local environment (Lakowicz 2006). The emission maximum of tryptophan is typically around 350 nm when its side chain is completely exposed to an aqueous environment, but it may be shifted to as low as nearly 300 nm in purely hydrophobic solvents. A central tryptophan residue in a TM segment can thus be used to monitor insertion of this segment into a hydrophobic environment. In our experience, an emission spectra maximum of around 330 nm is indicative of a tryptophan residue in the center of a detergent micelle. We have recorded emission spectra of TM segments in detergents with maxima of as low as 315 nm, which may be due to their self-association. We have also observed that while a TM fragment may be inserted into a detergent micelle, a large change in tryptophan fluorescence may not be observed if the segment adopts an "interfacial-like" conformation (Tulumello and Deber 2009).

3.3.3 TM protein tertiary and quaternary structure analysis

Although there have been recent breakthroughs in equilibrium folding and unfolding studies of membrane proteins, the field still remains very much in its infancy. There are outstanding issues that remain to be addressed, including an understanding of the nature of unfolded states in these studies. For a discussion of these types of studies, the reader is referred to Booth and Curnow (2006) and Renthal (2006). We will focus here on more classical techniques to investigate membrane protein assembly. Specifically, much work has been done in addressing the sequence determinants of TM helix association (a table of known helix-helix association motifs is found in Rath et al. [2009b]). To this end, many techniques have been developed to directly monitor the association of TM proteins and their segments, and/or to determine assembly size from which oligomeric state(s) may be inferred. These techniques have also been applied in some cases to determine structural characteristics such as compactness and structural homogeneity in the protein-detergent complex of monomeric TM proteins.

3.3.3.1 Size exclusion chromatography (SEC)

Chromatographic techniques have long been a staple of soluble protein purification and characterization. SEC has been shown to be an efficient method for determination of the presence of oligomeric states of proteins. In order to be adapted for membrane proteins, detergent must be present in the elution buffer so as to maintain a constant protein-to-detergent ratio throughout the course of the experiment. For this reason, the chromatography column must remain stable in the presence of detergent, a requirement that has been made feasible by advances in resin chemistry. Calibration of the column is commonly performed using globular, soluble protein standards, which are not always practical as detergents may denature these soluble proteins. Size exclusion should generally be treated in a qualitative manner with respect to size or oligomeric state, unless suitable standards are used. This method has also been used to screen membrane protein preparations for samples that give sharp peaks. This may indicate good "monodispersity" of the protein/detergent complex, which would thus be more likely to produce crystals (Sonoda et al. 2010). SEC has also been used to determine the amount of detergent bound to membrane proteins, where bound detergent is quantified following coelution with the protein of interest (le Maire et al. 2008; Rath et al. 2009a).

3.3.3.2 Förster resonance energy transfer (FRET)

FRET involves the nonradiative transfer of energy from an excited fluorophore (the donor) to an acceptor moiety (Lakowicz 2006). This phenomenon only occurs when the two moieties are in proximity to one another, usually less than a few nm. The degree of FRET may be monitored either by the increase of intensity of emission of the acceptor flourophore or a decrease in intensity of the emission of the donor fluorophore. An intrinsic tryptophan residue may be used as one member of a FRET pair, as its fluorescence emission of around 350 nm can be used to excite various commercial fluorophores (such as dansyl). By titrating in increasing amounts of unlabeled segment, the shape of the curve will inform as to the stoichiometry of the interaction (Adair and Engelman 1994). One main advantage of FRET is that it may be performed in the native lipid environment and be used to give estimates of the free energy of association (You et al. 2005).

3.3.3.3 Determination of helix-helix association of TM segments in detergents by FRET: experimental procedures

3.3.3.3.1 TM peptide synthesis and labeling

Peptides corresponding to TM protein segments are chemically synthesized using standard Fmoc (N-[9-fluorenyl] methoxycarbonyl) chemistry on a PAL-PEG-PS (4'-aminomethyl-3', 5'-dimethoxyphenoxyvaleric acid-poly[ethylene glycol]) polystyrene resin. In order to confer aqueous solubility to the peptides, Lys residues flanking the central TM segment are incorporated during synthesis (Melnyk et al. 2003). Fractions of the peptide are then labeled at the N terminus by incubating the resin-bound peptide with excess amine-reactive fluorophore (containing either a sulfonyl chloride or succinimide ester functional group) under basic conditions overnight. One fraction is labeled with a fluorescence donor moiety, while another fraction is labeled with a fluorescence acceptor moiety that has an absorption spectra that overlaps with the emission spectra of the donor. One suitable FRET pair is dansyl (5-dimethylamino-1-naphthalene-sufonyl chloride), a fluorescence donor, and dabsyl (4-dimethylaminoazobenzene-4'-sulfonyl chloride), a fluorescence acceptor. Following cleavage from the resin and subsequent purification, peptides are lyophilized and resuspended in ddH$_2$O.

3.3.3.3.2 Assay setup and interpretation

In order to confirm a specific oligomeric interaction, three separate samples are typically produced. Donor-only samples (D) contain a final concentration of 1 µM donor-labeled peptide and 1 µM unlabeled peptide. Donor-plus-acceptor mixtures (DA) contain 1 µM donor-labeled and 1 µM acceptor-labeled peptide. Donor-plus-acceptor and unlabeled mixtures (DAU) contain 1 µM donor-labeled, 1 µM acceptor-labeled and 4 µM unlabeled peptide. Separately labeled peptides are first mixed in a total volume of 200 µL of 20 mM Tris-HCl buffer (pH 8.0). Following mixing, 200 µL of detergent solution (with twice the desired final concentration) is added. For SDS, a final detergent concentration of 1% (w/v) is appropriate. Other detergents may be substituted and should be used at concentrations above their CMC, which ensures there is an excess of detergent micelles compared to TM segment monomers. If donor-labeled and acceptor-labeled peptides are mixed from solutions already containing detergent, energy transfer between the two populations may not observed even if an oligomeric state exists. This occurs if the dissociation of the individual peptide-detergent complexes and/or dissociation and reassociation of TM-TM complexes in detergent have not reached equilibrium (Rath et al. 2009b).

Samples are incubated overnight and dansyl emission spectra are recorded in a 1 cm excitation, 0.2 cm emission path length cuvette. Samples are excited at 341 nm with a 2 nm slit width, and emission is recorded between 450 and 650 nm with a 4 nm slit width. All spectra are background subtracted; total fluorescence is integrated and reported as a value normalized to the mean emission of samples containing only donor-labeled peptide. A decrease in fluorescence intensity upon inclusion of acceptor-labeled peptide is observed (comparing D to DA) if there is colocalization of oppositely labeled peptides within 30 Å (for the dansyl, dabsyl pair) and is indicative of oligomerization. A subsequent increase of fluorescence upon inclusion of unlabeled peptide to the donor and acceptor mix (comparing DA to DAU) indicates a decrease in FRET – implying specific oligomerization, where the unlabelled peptide is able to displace the interaction of the oppositely labelled peptides.

3.3.3.4 Thiol-disulfide interchange (TDI)

TDI, the reversible association of TM segments, is measured by the efficiency of disulfide cross-link formation between two Cys-containing segments. In principle, disulfide bonds will form more readily if the segments are in an oligomeric form. The cross-linking is then quenched, and the ratio of cross-linked species can be determined by SDS-PAGE or high-performance liquid chromatography (HPLC) (Cristian et al. 2003a). The advantage of this technique is the ability to measure a reversible association reaction at varying peptide-to-detergent ratios allowing for thermodynamic parameters to be obtained. Another advantage of this method is that it may also be used in lipid bilayers (Cristian et al. 2003b). As this method reports only on the absence or presence of dimeric cross-links, the specific oligomeric state(s) is not readily assessed, and other techniques may be required to determine thermodynamic parameters.

3.3.3.5 Sedimentation equilibrium analytical ultracentrifugation (SE-AUC)

SE-AUC provides an additional technique to determine the oligomeric state(s) of membrane proteins. In this assay, the molecular mass of the membrane protein can be directly assessed by monitoring its sedimentation in a centrifugal field. Through a "density matching" technique, the mass contributed by detergent to the membrane protein–detergent complex is ignored (Reynolds and Tanford 1976), thereby providing an estimate of oligomeric state. If the association of TM segments is reversible, this method may also be used to determine thermodynamic parameters such as the equilibrium association constant (Fleming 2008). This type of thermodynamic analysis remains subject to the long time periods required to attain equilibration (~ one day), where TM protein samples may not remain stable in certain detergents (le Maire et al. 2008).

In contrast to FRET and TDI, SE-AUC is strictly applicable to only detergents, as in lipid vesicles the mass determined will be due to all molecules within a shared vesicle. One advantage of this technique is that it does not require any additional labeling of the membrane protein or segments under investigation. Alternatively, SE-AUC may also be used to provide information about the amount of detergent bound (le Maire et al. 2008).

3.3.3.6 PAGE analysis

As mentioned in Section 3.2.4.1, SDS-PAGE is an extremely common technique in any modern protein laboratory for determination of protein molecular weights and assessment of components in a protein mixture (Reynolds and Tanford 1970). It has been shown in certain cases that SDS-PAGE may maintain membrane protein native oligomeric states (Heginbotham et al. 1997; Poulsen et al. 2009; Wegener and Jones 1984). SDS-PAGE is also sensitive to protein conformation (Dunker and Kenyon 1976; Griffith 1972), and we have previously demonstrated that gel migration rates may be in part be traceable to conformational changes in membrane protein structure (Therien et al. 2001; Wehbi et al. 2008). PAGE is perhaps the simplest and quickest way to determine membrane protein oligomeric state; however, it has the requirement of using an anionic detergent, typically SDS, which may denature all but the strongest TM-TM interactions. There have been some successes in substituting SPFO in place of SDS, which may allow weaker helix-helix interactions to persist (Melnyk et al. 2002). An additional drawback is that there is recent evidence that sequence-specific interactions of membrane proteins with SDS can skew the apparent molecular weight of the peptide-detergent complexes dramatically, making accurate assessment of precise oligomeric state(s) more complicated (Rath et al. 2009a, 2010; Tulumello and Deber 2011; Walkenhorst et al. 2009). Nevertheless, since SDS-PAGE does not require any protein labeling or sophisticated instrumentation and requires limited optimization, it should be considered the initial technique of choice to assess oligomeric state.

3.3.4 High-resolution structure determination

Importantly, the vast majority of high-resolution structures of integral membrane proteins have been obtained from a detergent solubilized state, whether obtained via X-ray crystallography or solution NMR spectroscopy.

3.3.4.1 X-ray crystallography

Although there are now examples of high-resolution structures of membrane proteins reconstituted into lipid bilayer phases (Caffrey 2009), most are obtained from detergent micelles. In many cases, major screening programs with many different detergents must be used to identify the appropriate detergent and additive combinations that will produce crystals with the necessary diffraction properties. There have been, however, some efforts to identify which detergents are expected to be most successful. In one report (Sonoda et al. 2010), the authors found that there is likely a trade-off between minimally denaturing detergents and detergents that minimally coat the protein allowing for maximal crystal contacts between the proteins themselves. The most commonly used detergents for X-ray crystallography include DDM, OG, C8E4, and LDAO, as well as related variants. Efforts are also being made to develop new classes of detergents specifically suitable to

X-ray crystallography, such as detergents without acyl tails to reduce flexibility (Hovers et al. 2011) or detergents comprised of peptide-lipid combinations (Prive 2009).

3.3.4.2 Solution NMR

The large size of vesicles prohibits their use for high-resolution solution NMR, and although there are other media such as organic solvents, amphipathic polymers and bicelles, detergents remain the classical choice of membrane mimetic (Page et al. 2006; Raschle et al. 2010; Sanders and Sonnichsen 2006). In general, for solution NMR studies, particle size is an important limiting factor for many conventional experiments, as larger particles will tumble slower and produce broader NMR lines. For this reason, detergents considered harsh by other methods (such as SDS or DPC) must be used as they form smaller micelles; nevertheless, care must be taken to ensure that they are not denaturing (Sanders and Sonnichsen 2006).

3.4 Applications of TM protein-detergent complexes

3.4.1 Comparing detergent-solubilized states with membrane proteins in lipid bilayers

In natural membranes, individual TM segments are most often inserted via the translocon. This process is largely irreversible, particularly if a TM segment is between hydrophilic domains that would be incapable of being retranslocated across the membrane (London and Shahidullah 2009). The environment of a lipid bilayer is in many ways different from a detergent micelle, as a bilayer is inherently a much more anisotropic environment and has biophysical characteristics such as lateral pressure and gradients of polarity and acyl chain order not present in micelle systems. The environment of a detergent micelle is also considerably more malleable and potentially has smaller kinetic and or energetic barriers to interconversion between fully inserted, partially inserted or noninserted (aqueous-interactive) states.

An idealized schematic of a TM segment within a detergent micelle is shown in ▶Figure 3.3A. Here the central region of the TM segment is buried within the micelle hydrophobic core where the local environment is similar to that in the bilayer with respect to a decreased dielectric constant versus water. Indeed, in many cases helix surfaces involved in protein-protein interactions in native bilayer systems may be maintained in micelles, leading to a native oligomeric state (shown in ▶Figure 3.3B). There are, however, examples of TM segments within a detergent micelle that have been observed to have some degree of exposure to the detergent head group region or to the aqueous environment (▶Figure 3.3C) (Lawrie et al. 2010; Tulumello and Deber 2009; Vincent et al. 2007).

Detergent solvation in this manner – in conjunction with the protein-protein and protein-lipid interactions that comprise the peptide-micelle complexes – may in fact constitute a highly relevant biological model. Thus, the exposure of a micelle-embedded TM segment to water may serve as a suitable surrogate for TM segments that line water-filled cavities common to membrane protein channels and pores in vivo. Accordingly, individual "faces" of TM segments can be qualitatively classified into three categories based upon their local environment: (i) TM faces exposed to an adjacent protein surface (viz., helix-helix interaction; a protein-exposed interface), (ii) TM faces interacting with membrane lipids (the protein exterior surface; a lipid-exposed face) and (iii) TM faces

3 Membrane protein folding in detergents

interacting with an aqueous environment (such as a channel or pore, or possibly a membrane surface; an aqueous-exposed face). This classification is highlighted here in the structure of the bacterial transporter lactose permease (LacY) (Abramson et al. 2003). The final tertiary/quaternary structure is dictated by the adoption of these native surfaces upon association of the individual TM helices (▶Figure 3.3D, E).

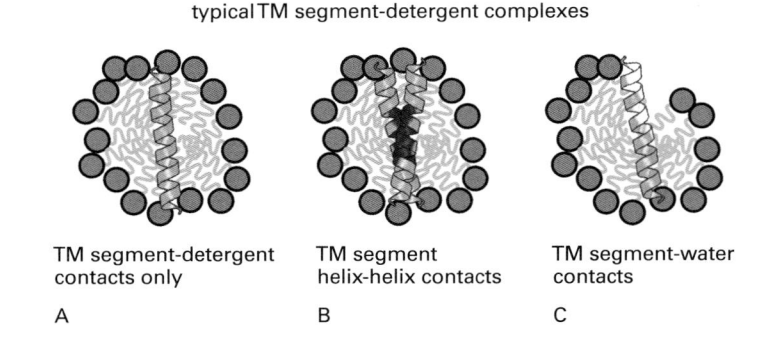

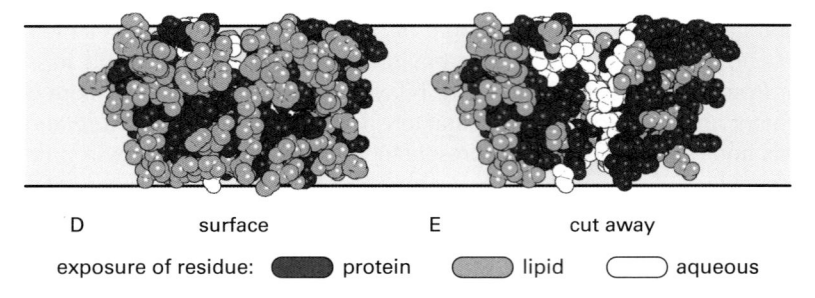

Figure 3.3: Peptide interactions within detergent micelles and comparison to native TM protein structure. Individual TM segments may exist in a variety of states in detergent micelles, as depicted schematically in (A–C). (A) A hydrophobic segment may have a strong preference for detergent solvation and exist as a monomer fully solvated by acyl tails of the detergent (lipid solvated regions shown in light gray). (B) An oligomeric segment may have a surface with a preference to form tertiary interactions with another surface, effectively producing a protein-protein interface between the interacting monomers (shown in dark gray). (C) A less hydrophobic surface arising within a TM-like segment may have a preference for aqueous solvation and partition to a state where the segment is partially solvated by detergent head groups and/or the surrounding aqueous environment (shown in white). Schematic views of the TM region of lactose permease are shown in (D–E). Individual residues are categorized and rendered in shading according to local environment, where dark gray represents protein-exposed residues involved in helix-helix contacts, light gray represents lipid-exposed residues, and white represents aqueous-exposed residues. (D) A view of TM regions only of lactose permease showing the external surface of the protein within the membrane. (E) The same view with some TM segments cutaway, showing the central water-filled cavity of the protein containing a significant number of residues that are aqueous-exposed in addition to lipid-exposed and protein-exposed residues.

It is evident that the three classifications of TM surfaces depicted in ▶Figure 3.3D,E are well-recapitulated in micelles (▶Figure 3.3A–C); in most cases, the TM surfaces present in detergent are likely to be the same as those found in the native membrane structure, albeit one cannot exclude the reality that in some cases, the physical location of TM-like segments vis-à-vis the micelle remains an artifact of detergent solubilization specifically because of protein accessibility to the micelle-water interfacial region.

We have also shown that the structure of TM segments in detergent micelles is sensitive to the position of polar residue substitutions along the length of the segment (Tulumello and Deber 2011). Such substitutions are likewise predicted to produce position-dependent effects in native bilayers, such as altered membrane insertion and transverse positioning, indicating that many of the same forces governing membrane protein folding in bilayers are indeed mirrored by detergent solvation.

3.4.1.1 Glycophorin A

Among the most well-characterized integral membrane proteins is glycophorin A (GpA). It is the major sialoglycoprotein from human erythrocytes, containing a single TM segment that mediates its association into dimers. The dimerization of the GpA TM domain has been studied in a variety of detergents as well as in bacterial membranes and in model lipid bilayers. Dimerization of GpA in all media occurs through a motif containing Gly residues spaced three residues apart (termed the GxxxG or GG4 motif) (MacKenzie et al. 1997). A large number of GpA mutant variants (~50) of this dimerization interface have been produced that have allowed for comparison of sequence-dependent folding in various environments. Differences in apparent free energies of GpA mutants as measured by an α-helix association assay in the bacterial membrane (Russ and Engelman 1999) were found to correlate well to apparent free energies obtained by SE-AUC (Fleming and Engelman 2001), as well as to relative dimerization on SDS-PAGE (Duong et al. 2007).

One observation resulting from these studies is that a decrease of hydrophobicity lowers relative dimerization in SDS, but not in membranes. Indeed, the correlation between in vivo association and SDS-PAGE has been found to be improved if loss of hydrophobicity is taken into account for the SDS-PAGE results (Duong et al. 2007). This has been interpreted as partial unfolding of the helix due to disruption of insertion into the micelle (Lawrie et al. 2010). Furthermore, while the core GxxxG motif appears to be maintained in all environments, there are some specific differences in contributions to free energy of association of additional residues that contribute to self-association of this TM dimer (Duong, et al. 2007; Hong et al. 2010).

3.4.1.2 Designed TM segments (AI series)

Our lab and others have explored the capability of SDS micelles to preserve a variety of helix-helix interactions through a series of designed TM segments; sequences are shown in ▶Figure 3.4A (Tulumello and Deber 2009). For example, we compared an in vivo *E. coli* inner membrane dimerization assay (Russ and Engelman 1999) to the association of these peptides in SDS micelles by both SDS-PAGE migration and FRET experiments and found good qualitative agreement (cf. ▶Figure 3.4B and 3.4C). The base peptide (AI10) was found to self-associate both in a native bilayer as well as in SDS micelles. A second peptide (AI5) with reduced hydrophobicity (achieved by five Ile-to-Ala substitutions), displayed a moderate decrease in apparent association in both environments. We also introduced a strong polar residue (Asn) into these two peptides and found that when placed in the more hydrophobic peptide (AI10 I17N), there was an increase in self-association, as evidenced by high association levels in the in vivo assay, and the appearance of higher order oligomers on SDS-PAGE. The polar residue, however, had the opposite effect in the less polar TM segment (AI5 I17N), reducing apparent association in both assays. In this segment, the introduction of the polar residue apparently prevents insertion of the TM segment into the membrane in the in vivo assay, while similarly altering the nature of the detergent solvation in SDS, mediating against self-association. This alteration of detergent solvation is also supported by Trp fluorescence studies (Tulumello and Deber 2009).

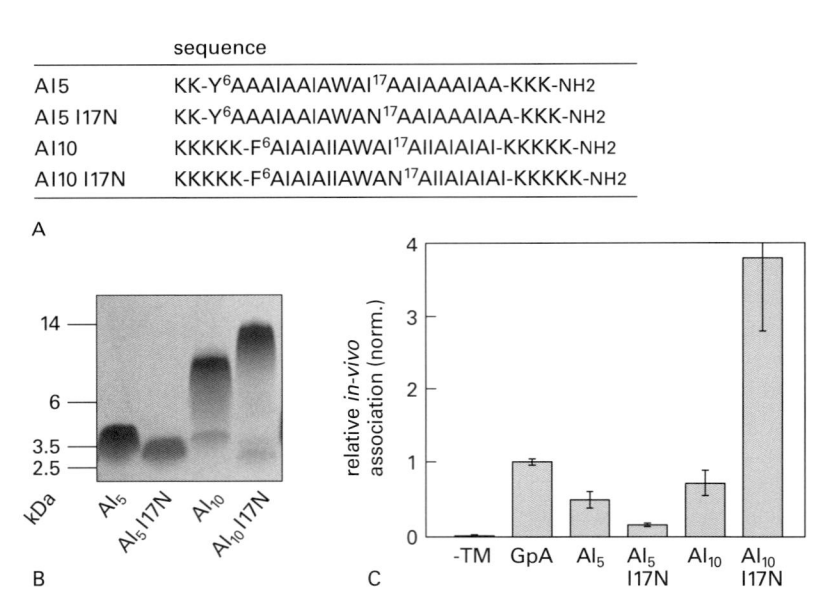

	sequence
AI5	KK-Y^{6}AAAIAAIAWAI^{17}AAIAAAIAA-KKK-NH2
AI5 I17N	KK-Y^{6}AAAIAAIAWAN^{17}AAIAAAIAA-KKK-NH2
AI10	KKKKK-F^{6}AIAIAIIAWAI^{17}AIIAIAIAI-KKKKK-NH2
AI10 I17N	KKKKK-F^{6}AIAIAIIAWAN^{17}AIIAIAIAI-KKKKK-NH2

Figure 3.4: Qualitative agreement of oliogmeric state of TM segments in detergents versus bilayers. (A) Peptide sequences of model TM segments. (B) SDS-PAGE of a series of model TM segments. Shown are the base peptide (AI10), migration rate consistent with a dimer; the base peptide with a single polar residue (AI10 I17N), migration rate consistent with a higher (than dimer) order oligomer; an analogue with decreased hydrophobicity (AI5), migration rate consistent with the presence of a monomer-dimer equilibrium; and an analogue with both decreased hydrophobicity and a polar residue (AI5 I17N), migration rate consistent with a monomer. (C) In vivo association of the same segments in a bacterial (*E. coli*) membrane. In this assay, relative association is normalized to a prototypic TM dimer, GpA. While peptides without a strongly polar residue report values consistent with dimer formation (AI5 and AI10), the inclusion of the polar residue increases relative association in the more hydrophobic sequence, while decreasing it in the less hydrophobic sequence. A negative control sequence with no TM region is shown as – TM. Figure adapted from (Tulumello and Deber 2009).

3.4.2 From fragments to full-length proteins: a "divide and conquer" strategy

While it would clearly be desirable to study membrane proteins as full-length systems, their large size and instability, and often the nonavailability of milligram quantities of pure proteins, require alternate strategies. The two-stage model predicts that, as an individual domain, each TM segment should retain its helical structure when excised from its native context. This model accordingly implies that the folding of fragments of TM proteins should provide information relevant to the full-length protein. Solid phase peptide synthesis has allowed for efficient production of single TM segments, while truncated membrane domains can be produced in adequate levels of heterologous protein expression. By using these techniques, and specifically tailoring the constructs for subsequent purification and characterization, many of the hurdles involved in production of larger membrane proteins are avoided (Cunningham and Deber 2007). Our lab has had success applying these strategies to various membrane proteins of disease relevance. In this section, we will highlight a few of these examples.

3.4.2.1 *Halobacterium salinarum* small multidrug resistance (SMR) transporter (Hsmr)

Multidrug transporters such as the SMR family of bacterial integral membrane proteins are capable of conferring clinically significant resistance to a variety of common small-molecule antibiotics. In order to efflux drug molecules, SMRs must self-assemble into homo-oligomeric structures. To investigate this, we initially synthesized peptides that correspond to each of the four wild-type TM helices of the *Halobacterium salinarum* SMR transporter protein (Hsmr). We also produced a corresponding peptide library of mutant variants of these TM segments to determine the interactive surfaces that likely contribute to protein oligomerization. Hsmr peptides were examined for strong (SDS-resistant) and weaker (SPFO-resistant) helix-helix interactions by PAGE analysis (▶Figure 3.5A) in conjunction with FRET measurements, and molecular modeling. The results identified positions that constitute a "face" of TM helix four (TM4) where mutations prevent self-assembly in SDS (Rath et al. 2006).

These results were expanded by introducing 12 single point mutants into the full-length protein that scan the central portion of the TM4 helix. These mutant variants were tested for their ability to confer resistance to the cytotoxic compound ethidium bromide (▶Figure 3.5B). The results, in conjunction with SDS-PAGE used to assess the dimerization propensity of these mutants (▶Figure 3.5C), define a minimum functional motif of $G^{90}LXLIXXGV^{98}$ within TM4, suggesting that this small residue heptad repeat sequence mediates TM4-based SMR dimerization along a single helix surface (Poulsen et al. 2009). The "strong" helix interaction surface, defined by the peptides, overlaps significantly with the functional interface defined by the full-length protein (▶Figure 3.5D), highlighting the feasibility of the "divide and conquer" approach.

3.4.2.2 Cystic fibrosis transmembrane regulator (CFTR)

Cystic fibrosis (CF), a genetic disorder common among the Caucasian population, exemplifies the membrane-protein-misfolding problem as it relates to human health. In this disease, a single amino acid deletion or point mutation in a single α-helical TM chloride channel – the 1,480-residue CFTR – will cause CF disease by altering the

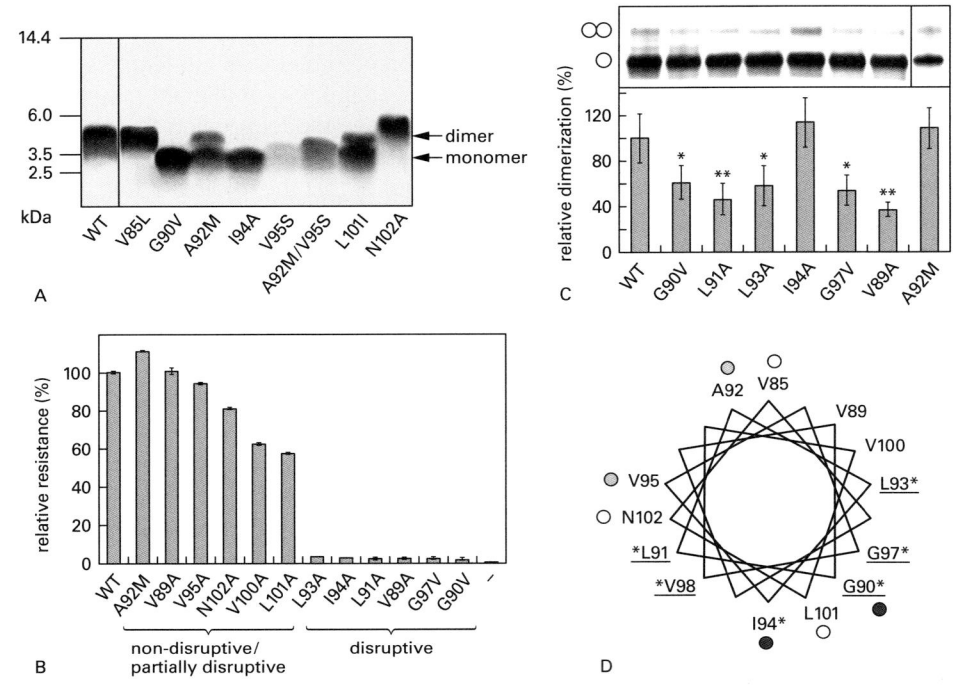

Figure 3.5: Comparison of an oligomerization motif of Hsmr (defined using a single TM segment) versus a functional activity motif (defined using the full-length protein). (A) SDS-PAGE of WT and mutant TM-4 Hsmr peptides. Oligomeric states are indicated to the right. (B) Functional assay showing the growth of *E. coli* expressing WT full-length Hsmr or the indicated mutant in the presence of 500 µM ethidium bromide. All values are normalized to WT growth values. The growth of untransformed BL21(DE3) cells (--) was used as a negative control. Error bars represent the standard deviations of at least three experiments. (C) Representative silver-stained SDS-PAGE and quantitation of dimerization efficiency of WT and mutant Hsmr proteins. Positions of monomer and dimer are indicated to the left on the gel by single and double dots, respectively. The bar graph shows the mean percentage of dimerization on SDS-PAGE normalized to WT for each mutant. Error bars represent the standard deviation of three individual gels. The significance levels of unpaired *t*-tests comparing the dimerization efficiency of each mutant with WT are as follows: *, $P \leq 0.05$; **, $P \leq 0.01$. (D) Helical wheel projection of the Hsmr TM4 sequence. Only positions that were mutated in the full-length construct are shown. Underlined residues indicate those where mutation significantly reduced dimerization as assessed by SDS-PAGE in the full-length construct (from panel C). Positions where mutations disrupted function are indicated by asterisks (from panel B). Note that the surface of the helix where dimerization was disrupted in the peptide (from panel A – indicated with black circles) was the same surface where both dimerization (underlined residues) and function (asterisks) were disrupted upon mutation. All residues that did not completely abolish dimerization when mutated (from panel A – shown in gray and white circles) also had no effect on dimerization or function of the full-length protein. Figure adapted from (Poulsen et al. 2009; Rath et al. 2006).

biosynthesis, maturation, folding and/or function of this protein (Cheung and Deber 2008). As no high-resolution structure currently exists for full-length CFTR, our lab has approached this "misfolding" problem by expression and structural analysis of the CFTR TM domains using constructs consisting of two TM segments joined by their intervening loop (i.e. helix-loop-helix constructs), which we term "helical hairpins"; these represent the minimal model of tertiary contacts between two helices in a membrane. These membrane protein fragments can be prepared in milligram amounts and may be readily manipulated by established biophysical techniques.

We have used such constructs to characterize molecular defects such as nonconservative changes in membrane-based residues that link CF disease to CF-phenotypic mutations. In a hairpin consisting of TM3-extacellular loop 2-TM4, we have addressed the structural consequence of a disease-phenotypic V232D mutation using a series of techniques such as circular dichroism, SDS-PAGE migration rate analysis, cysteine crosslinking, pyrene excimer fluorescence and SEC (Choi et al. 2004; Rath, et al. 2009a; Therien and Deber 2002; Therien et al. 2001). Through these experiments, we have demonstrated that introduction of a polar residue (a common disease-related mutation in TM domains [Partridge et al. 2004]) significantly alters the folding of the TM domain, possibly through the formation of interhelical hydrogen bonds and/or by specifically altering the native protein interaction with the lipid environment. The identity of detergent micelle used (Therien and Deber 2002), or the presence of a suitable H-bond partner on an adjacent helix (Choi et al. 2004), may dictate which of these outcomes occurs.

We have also examined the potential structural impact of CF-phenotypic mutations in the extracellular loop of this hairpin construct (including E217G and Q220R). In these constructs, we found that SDS-PAGE gel migration rates differed over a range of nearly 40% +/− the wild-type position, and that decreased migration rates correlate with increasing hairpin α-helical content as measured by CD spectra in sodium dodecyl sulfate micelles (Wehbi et al. 2008). These observations were interpreted in terms of the consequences of insertion of helix-promoting residues in a loop region evolved to reverse the chain direction. Although any and all effects observed in the various series of hairpin constructs are likely to be tempered in the context of the full protein TM domain, the potential drastic structural consequences of point mutations are more readily apparent in these detergent-based experiments than they might be in the full-length protein.

3.5 Conclusions

Even in light of the many advances described here, our understanding of the sequence-dependent structure and folding of TM proteins remains limited. Given the correct combination of protein and the choice of detergent, it is becoming increasingly possible to approximate and elucidate biologically relevant membrane protein structures in detergents.

References

Abramson, J., Smirnova, I., Kasho, V., Verner, G., Iwata, S., and Kaback, H.R. (2003). The lactose permease of *Escherichia coli*: overall structure, the sugar-binding site and the alternating access model for transport. FEBS Lett *555*, 96–101.

Adair, B.D., and Engelman, D.M. (1994). Glycophorin A helical transmembrane domains dimerize in phospholipid bilayers: a resonance energy transfer study. Biochemistry 33, 5539–5544.

Almeida, F.C., and Opella, S.J. (1997). fd coat protein structure in membrane environments: structural dynamics of the loop between the hydrophobic trans-membrane helix and the amphipathic in-plane helix. J Mol Biol 270, 481–495.

Andrade, S.L., Dickmanns, A., Ficner, R., and Einsle, O. (2005). Crystal structure of the archaeal ammonium transporter Amt-1 from Archaeoglobus fulgidus. Proc Natl Acad Sci U S A 102, 14994–14999.

Arkin, I.T., Adams, P.D., MacKenzie, K.R., Lemmon, M.A., Brunger, A.T., and Engelman, D.M. (1994). Structural organization of the pentameric transmembrane alpha-helices of phospholamban, a cardiac ion channel. EMBO J 13, 4757–4764.

Baenziger, J.E., and daCosta, C.J. (2008). Membrane protein structure and conformational change probed using Fourier transform infrared spectroscopy. In: Biophysical analysis of membrane proteins: investigating structure and function, E. Pebay-Peyroula, ed. (Weinheim, Germany: Wiley-VCH), pp. 243–288.

Beswick, V., Guerois, R., Cordier-Ochsenbein, F., Coic, Y.M., Tam, H.D., Tostain, J., et al. (1999). Dodecylphosphocholine micelles as a membrane-like environment: new results from NMR relaxation and paramagnetic relaxation enhancement analysis. Eur Biophys J 28, 48–58.

Booth, P.J., and Curnow, P. (2006). Membrane proteins shape up: understanding in vitro folding. Curr Opin Struct Biol 16, 480–488.

Caffrey, M. (2009). Crystallizing membrane proteins for structure determination: use of lipidic mesophases. Annu Rev Biophys 38, 29–51.

Cheung, J.C., and Deber, C.M. (2008). Misfolding of the cystic fibrosis transmembrane conductance regulator and disease. Biochemistry 47, 1465–1473.

Chill, J.H., Louis, J.M., Miller, C., and Bax, A. (2006). NMR study of the tetrameric KcsA potassium channel in detergent micelles. Protein Sci 15, 684–698.

Choi, M.Y., Cardarelli, L., Therien, A.G., and Deber, C.M. (2004). Non-native interhelical hydrogen bonds in the cystic fibrosis transmembrane conductance regulator domain modulated by polar mutations. Biochemistry 43, 8077–8083.

Cristian, L., Lear, J.D., and DeGrado, W.F. (2003a). Determination of membrane protein stability via thermodynamic coupling of folding to thiol-disulfide interchange. Protein Sci 12, 1732–1740.

Cristian, L., Lear, J.D., and DeGrado, W.F. (2003b). Use of thiol-disulfide equilibria to measure the energetics of assembly of transmembrane helices in phospholipid bilayers. Proc Natl Acad Sci U S A 100, 14772–14777.

Cunningham, F., and Deber, C.M. (2007). Optimizing synthesis and expression of transmembrane peptides and proteins. Methods 41, 370–380.

Dunker, A.K., and Kenyon, A.J. (1976). Mobility of sodium dodecyl sulphate – protein complexes. Biochem J 153, 191–197.

Duong, M.T., Jaszewski, T.M., Fleming, K.G., and MacKenzie, K.R. (2007). Changes in apparent free energy of helix-helix dimerization in a biological membrane due to point mutations. J Mol Biol 371, 422–434.

Engelman, D.M., Chen, Y., Chin, C.N., Curran, A.R., Dixon, A.M., Dupuy, A.D., et al. (2003). Membrane protein folding: beyond the two stage model. FEBS Lett 555, 122–125.

Fleming, K.G. (2002). Standardizing the free energy change of transmembrane helix-helix interactions. J Mol Biol 323, 563–571.

Fleming, K.G. (2008). Determination of membrane protein molecular weight using sedimentation equilibrium analytical ultracentrifugation. Curr Protoc Protein Sci, 53:7.12.1–17.12.13.

Fleming, K.G., and Engelman, D.M. (2001). Specificity in transmembrane helix-helix interactions can define a hierarchy of stability for sequence variants. Proc Natl Acad Sci U S A 98, 14340–14344.

Griffith, I.P. (1972). The effect of cross-links on the mobility of proteins in dodecyl sulphate-polyacrylamide gels. Biochem J 126, 553–560.

Heginbotham, L., Odessey, E., and Miller, C. (1997). Tetrameric stoichiometry of a prokaryotic K$^+$ channel. Biochemistry 36, 10335–10342.

Hessa, T., Kim, H., Bihlmaier, K., Lundin, C., Boekel, J., Andersson, H., et al. (2005). Recognition of transmembrane helices by the endoplasmic reticulum translocon. Nature 433, 377–381.

Hong, H., Blois, T.M., Cao, Z., and Bowie, J.U. (2010). Method to measure strong protein-protein interactions in lipid bilayers using a steric trap. Proc Natl Acad Sci U S A 107, 19802–19807.

Hovers, J., Potschies, M., Polidori, A., Pucci, B., Raynal, S., Bonnete, F., et al. (2011). A class of mild surfactants that keep integral membrane proteins water-soluble for functional studies and crystallization. Mol Membr Biol 28, 171–181.

Howell, S.C., Mesleh, M.F., and Opella, S.J. (2005). NMR structure determination of a membrane protein with two transmembrane helices in micelles: MerF of the bacterial mercury detoxification system. Biochemistry 44, 5196–5206.

Imamura, T. (Ed.) (2006). Encyclopedia of surface and colloid science (New York: Taylor & Francis).

Lakowicz, J.R. (2006). Principles of fluorescence spectroscopy. 3rd edn (New York: Springer).

Lawrie, C.M., Sulistijo, E.S., and MacKenzie, K.R. (2010). Intermonomer hydrogen bonds enhance GxxxG-driven dimerization of the BNIP3 transmembrane domain: roles for sequence context in helix-helix association in membranes. J Mol Biol 396, 924–936.

le Maire, M., Arnou, B., Olesen, C., Georgin, D., Ebel, C., and Moller, J.V. (2008). Gel chromatography and analytical ultracentrifugation to determine the extent of detergent binding and aggregation, and Stokes radius of membrane proteins using sarcoplasmic reticulum Ca2+-ATPase as an example. Nat Protoc 3, 1782–1795.

le Maire, M., Champeil, P., and Moller, J.V. (2000). Interaction of membrane proteins and lipids with solubilizing detergents. Biochim Biophys Acta 1508, 86–111.

Lemmon, M.A., Flanagan, J.M., Hunt, J.F., Adair, B.D., Bormann, B.J., Dempsey, C.E., et al. (1992). Glycophorin A dimerization is driven by specific interactions between transmembrane alpha-helices. J Biol Chem 267, 7683–7689.

Lin, S.H., and Guidotti, G. (2009). Purification of membrane proteins. Methods Enzymol 463, 619–629.

Liu, L.P., Li, S.C., Goto, N.K., and Deber, C.M. (1996). Threshold hydrophobicity dictates helical conformations of peptides in membrane environments. Biopolymers 39, 465–470.

London, E., and Shahidullah, K. (2009). Transmembrane vs. non-transmembrane hydrophobic helix topography in model and natural membranes. Curr Opin Struct Biol 19, 464–472.

Mackenzie, K.R. (2006). Folding and stability of alpha-helical integral membrane proteins. Chem Rev 106, 1931–1977.

MacKenzie, K.R., Prestegard, J.H., and Engelman, D.M. (1997). A transmembrane helix dimer: structure and implications. Science 276, 131–133.

Melnyk, R.A., Partridge, A.W., and Deber, C.M. (2002). Transmembrane domain mediated self-assembly of major coat protein subunits from Ff bacteriophage. J Mol Biol 315, 63–72.

Melnyk, R.A., Partridge, A.W., Yip, J., Wu, Y., Goto, N.K., and Deber, C.M. (2003). Polar residue tagging of transmembrane peptides. Biopolymers 71, 675–685.

Newstead, S., Ferrandon, S., and Iwata, S. (2008). Rationalizing alpha-helical membrane protein crystallization. Protein Sci 17, 466–472.

Ng, D.P., and Deber, C.M. (2010). Modulation of the oligomerization of myelin proteolipid protein by transmembrane helix interaction motifs. Biochemistry 49, 6896–6902.

Otzen, D.E., Nesgaard, L.W., Andersen, K.K., Hansen, J.H., Christiansen, G., Doe, H., et al. (2008). Aggregation of S6 in a quasi-native state by sub-micellar SDS. Biochim Biophys Acta *1784*, 400–414.

Page, R.C., Moore, J.D., Nguyen, H.B., Sharma, M., Chase, R., Gao, F.P., et al. (2006). Comprehensive evaluation of solution nuclear magnetic resonance spectroscopy sample preparation for helical integral membrane proteins. J Struct Funct Genomics *7*, 51–64.

Partridge, A.W., Therien, A.G., and Deber, C.M. (2004). Missense mutations in transmembrane domains of proteins: phenotypic propensity of polar residues for human disease. Proteins *54*, 648–656.

Popot, J.L., and Engelman, D.M. (1990). Membrane protein folding and oligomerization: the two-stage model. Biochemistry *29*, 4031–4037.

Poulsen, B.E., Rath, A., and Deber, C.M. (2009). The assembly motif of a bacterial small multidrug resistance protein. J Biol Chem *284*, 9870–9875.

Prive, G.G. (2007). Detergents for the stabilization and crystallization of membrane proteins. Methods *41*, 388–397.

Prive, G.G. (2009). Lipopeptide detergents for membrane protein studies. Curr Opin Struct Biol *19*, 379–385.

Raman, P., Cherezov, V., and Caffrey, M. (2006). The Membrane Protein Data Bank. Cell Mol Life Sci *63*, 36–51.

Raschle, T., Hiller, S., Etzkorn, M., and Wagner, G. (2010). Nonmicellar systems for solution NMR spectroscopy of membrane proteins. Curr Opin Struct Biol *20*, 471–479.

Rath, A., Glibowicka, M., Nadeau, V.G., Chen, G., and Deber, C.M. (2009a). Detergent binding explains anomalous SDS-PAGE migration of membrane proteins. Proc Natl Acad Sci U S A *106*, 1760–1765.

Rath, A., Melnyk, R.A., and Deber, C.M. (2006). Evidence for assembly of small multidrug resistance proteins by a "two-faced" transmembrane helix. J Biol Chem *281*, 15546–15553.

Rath, A., Tulumello, D.V., and Deber, C.M. (2009b). Peptide models of membrane protein folding. Biochemistry *48*, 3036–3045.

Rath, A., Nadeau, V.G., Poulsen, B.E., Ng, D.P., and Deber, C.M. (2010). Novel hydrophobic standards for membrane protein molecular weight determinations via sodium dodecyl sulfate-polyacrylamide gel electrophoresis. Biochemistry *49*, 10589–10591.

Renthal, R. (2006). An unfolding story of helical transmembrane proteins. Biochemistry *45*, 14559–14566.

Reynolds, J.A., and Tanford, C. (1970). Binding of dodecyl sulfate to proteins at high binding ratios. Possible implications for the state of proteins in biological membranes. Proc Natl Acad Sci U S A *66*, 1002–1007.

Reynolds, J.A., and Tanford, C. (1976). Determination of molecular weight of the protein moiety in protein-detergent complexes without direct knowledge of detergent binding. Proc Natl Acad Sci U S A *73*, 4467–4470.

Russ, W.P., and Engelman, D.M. (1999). TOXCAT: a measure of transmembrane helix association in a biological membrane. Proc Natl Acad Sci U S A *96*, 863–868.

Sanders, C.R., and Sonnichsen, F. (2006). Solution NMR of membrane proteins: practice and challenges. Magn Reson Chem *44*, S24–40.

Sonoda, Y., Cameron, A., Newstead, S., Omote, H., Moriyama, Y., Kasahara, M., et al. (2010). Tricks of the trade used to accelerate high-resolution structure determination of membrane proteins. FEBS Lett *584*, 2539–2547.

Stroud, R.M., Miercke, L.J., O'Connell, J., Khademi, S., Lee, J.K., Remis, J., et al. (2003). Glycerol facilitator GlpF and the associated aquaporin family of channels. Curr Opin Struct Biol *13*, 424–431.

Tamm, L.K., Hong, H., and Liang, B. (2004). Folding and assembly of beta-barrel membrane proteins. Biochim Biophys Acta *1666*, 250–263.

Tate, C.G. (2001). Overexpression of mammalian integral membrane proteins for structural studies. FEBS Lett *504*, 94–98.

Therien, A.G., and Deber, C.M. (2002). Interhelical packing in detergent micelles. Folding of a cystic fibrosis transmembrane conductance regulator construct. J Biol Chem *277*, 6067–6072.

Therien, A.G., Grant, F.E., and Deber, C.M. (2001). Interhelical hydrogen bonds in the CFTR membrane domain. Nat Struct Biol *8*, 597–601.

Tulumello, D.V., and Deber, C.M. (2009). SDS micelles as a membrane-mimetic environment for transmembrane segments. Biochemistry *48*, 12096–12103.

Tulumello, D.V., and Deber, C.M. (2011). Positions of polar amino acids alter interactions between transmembrane segments and detergents. Biochemistry, 50, 3928–3935.

Vincent, M., Gallay, J., Jamin, N., Garrigos, M., and de Foresta, B. (2007). The predicted transmembrane fragment 17 of the human multidrug resistance protein 1 (MRP1) behaves as an interfacial helix in membrane mimics. Biochim Biophys Acta *1768*, 538–552.

Vinogradova, O., Sonnichsen, F., and Sanders, C.R., II. (1998). On choosing a detergent for solution NMR studies of membrane proteins. J Biomol NMR *11*, 381–386.

Wahlstrom, A., Hugonin, L., Peralvarez-Marin, A., Jarvet, J., and Graslund, A. (2008). Secondary structure conversions of Alzheimer's Abeta(1–40) peptide induced by membrane-mimicking detergents. FEBS J *275*, 5117–5128.

Walkenhorst, W.F., Merzlyakov, M., Hristova, K., and Wimley, W.C. (2009). Polar residues in transmembrane helices can decrease electrophoretic mobility in polyacrylamide gels without causing helix dimerization. Biochim Biophys Acta *1788*, 1321–1331.

Wallin, E., and von Heijne, G. (1998). Genome-wide analysis of integral membrane proteins from eubacterial, archaean, and eukaryotic organisms. Protein Sci *7*, 1029–1038.

Wegener, A.D., and Jones, L.R. (1984). Phosphorylation-induced mobility shift in phospholamban in sodium dodecyl sulfate-polyacrylamide gels. Evidence for a protein structure consisting of multiple identical phosphorylatable subunits. J Biol Chem *259*, 1834–1841.

Wehbi, H., Gasmi-Seabrook, G., Choi, M.Y., and Deber, C.M. (2008). Positional dependence of non-native polar mutations on folding of CFTR helical hairpins. Biochim Biophys Acta *1778*, 79–87.

White, S.H., and von Heijne, G. (2008). How translocons select transmembrane helices. Annu Rev Biophys *37*, 23–42.

Wimley, W.C. (2003). The versatile beta-barrel membrane protein. Curr Opin Struct Biol *13*, 404–411.

Ye, S., Li, Y., and Jiang, Y. (2010). Novel insights into K+ selectivity from high-resolution structures of an open K+ channel pore. Nat Struct Mol Biol *17*, 1019–1023.

You, M., Li, E., Wimley, W.C., and Hristova, K. (2005). Forster resonance energy transfer in liposomes: measurements of transmembrane helix dimerization in the native bilayer environment. Anal Biochem *340*, 154–164.

4 Glycoprotein-folding quality control in the endoplasmic reticulum

Julio J. Caramelo and Armando J. Parodi

4.1 Introduction

Nearly one-third of proteins synthesized by eukaryotic cells belong to the secretory pathway, gaining access to the endoplasmic reticulum (ER) either co- or posttranslationally. In the ER, disulfide bonds are formed, proteins acquire their native tertiary fold, and, if needed, they assemble into oligomeric structures. Numerous chaperones and folding-assisting enzymes are in place to ensure the fidelity and efficiency of these processes. In addition, nearly 70% of the secretory pathway proteins are *N*-glycosylated by the oligosaccharyltransferase complex in the consensus sequence Asn-X-Ser/Thr, in which X cannot be Pro (Apweiler et al. 1999). The consensus sequences are generally modified as they emerge into the ER lumen during protein translocation, although in some cases *N*-glycosylation may occur posttranslationally (Ruiz-Canada et al. 2009). *N*-glycosylation is the most abundant and one of the more drastic protein modifications. Commonly, *N*-glycans are central players in molecular recognition events, a function particularly suitable for them given their vast compositional and structural diversity. In addition, *N*-glycans may modulate the biophysical behavior of glycoproteins. For instance, *N*-glycans may inhibit protein aggregation, may increment the resistance toward proteases and can promote the acquisition of some elements of secondary structure such as turns (Chen et al. 2010). Of particular importance is the role of *N*-glycans during glycoprotein folding in the ER (D'Alessio et al. 2010). Here, the *N*-glycans are used as an epigenetic information platform that reflects the folding status of glycoproteins. This code is generated by a family of glycosyltransferases and glycosidases that translate the structural features of glycoproteins into particular *N*-glycan structures. A set of ER-resident lectins "read" this information and react accordingly by retaining the immature species in the ER and/or promoting their degradation.

4.2 Glycoprotein-folding quality control (QC)

In most eukaryotic cells, *N*-glycosylation starts with the en bloc transfer of the glycan Glc$_3$Man$_9$GlcNAc$_2$ from a dolichol (Dol)-P-P derivative (▶Figure 4.1). There are some exceptions to this rule, since some protists transfer a shorter version of the glycan (Parodi 1993). Soon after transfer, the glycan is trimmed by the sequential action of glucosidase I (GI), which cuts the outermost Glc residue, followed by glucosidase II (GII), which cleaves the next Glc residue (▶Figure 4.1). The monoglucosylated glycan intermediates thus generated (Glc$_1$Man$_9$GlcNAc$_2$) are recognized by two ER-resident lectins, calnexin (CNX) and calreticulin (CRT) (Michalak et al. 1999) (▶Figure 4.2). These lectins retain monoglucosylated species in the ER and operate as nonclassical chaperones, preventing undesirable interactions that may lead to protein aggregation. In addition, CNX and

CRT facilitate the proper formation of disulfide bridges through the activity of ERp57, a protein disulfide isomerase (PDI) associated with them (Coe and Michalak 2010). Eventually, the remaining Glc residue is cleaved by GII, thus liberating glycoproteins from the lectin anchors. At this point, properly folded proteins can resume their travel along the secretory pathway (▶Figure 4.2). On the contrary, those species displaying a nonnative fold are recognized by the enzyme UDP-Glc:glycoprotein glucosyltransferase (UGGT). This enzyme adds back the last Glc residue removed by GII, thus regenerating the monoglucosylated glycans and initiating a new round of interactions with CNX/CRT. This mechanism, known as glycoprotein-folding QC, operates for most glycoproteins synthesized in the ER (Hebert et al. 1995). The result of cycles of deglucosylation by GII and reglucosylation by UGGT is the retention of immature species in the ER. A protein unable to acquire a native fold is eventually retrotranslocated to the cytosol, where it is degraded in the proteasome, in a process known as endoplasmic reticulum associated degradation (ERAD) (Vembar and Brodsky 2008). One of the most challenging issues faced by this system is to discriminate proteins unable to fold properly from those species that are en route to achieve their native fold. Failures in this process could lead to degradation of functional proteins or, on the contrary, could allow the escape from the ER of nonnative species. Both scenarios can have dramatic consequences on the viability of an organism.

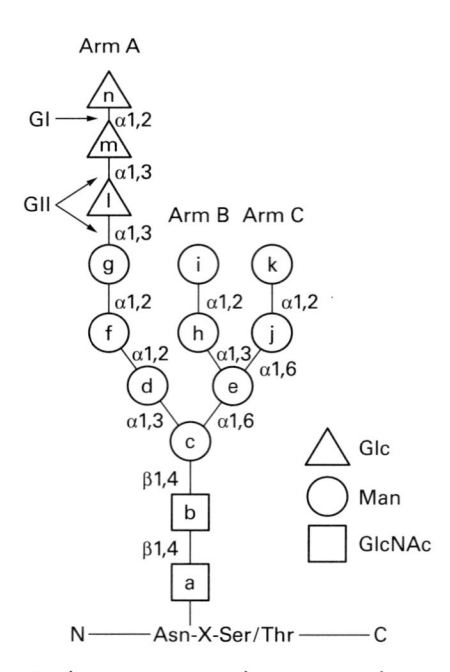

Figure 4.1: N-glycan structure. The structure depicted is that of the full length glycan transferred to Asn residues in N-glycosylation. Lettering (a, b, c…) follows the order of addition of monosaccharides in the synthesis of the dolichol-P-P derivatives.

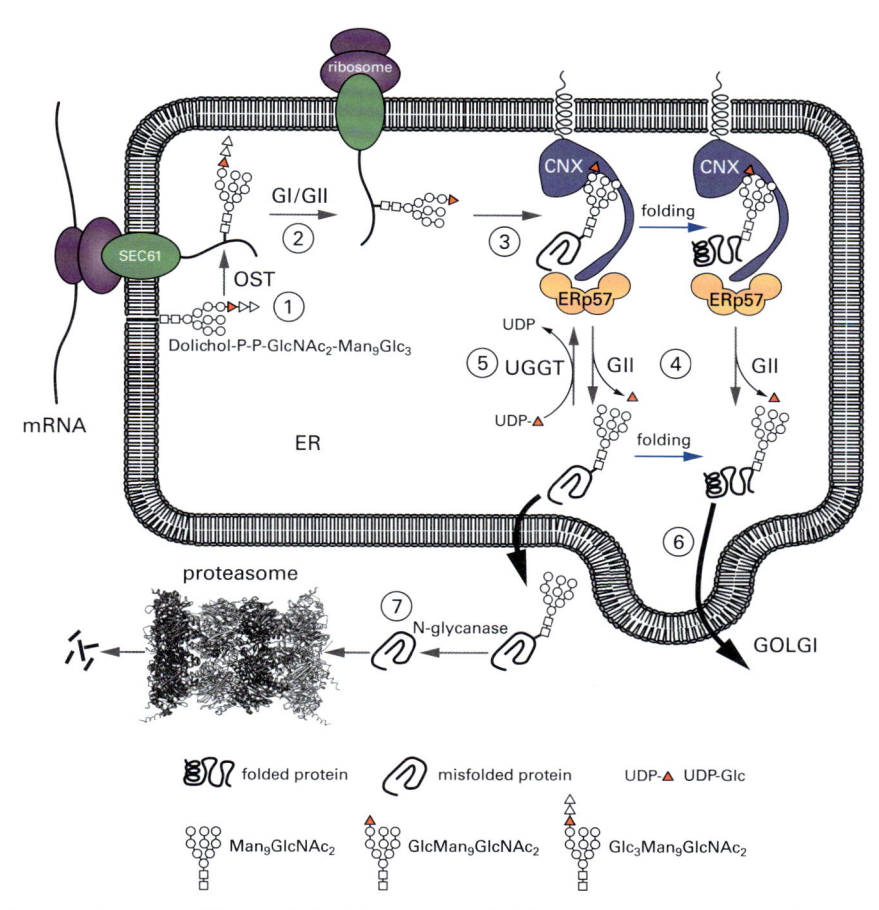

Figure 4.2: Model proposed for the QC of glycoprotein folding. (1) Proteins entering the ER are N-glycosylated by the oligosaccharyltransferase (OST) as they emerge from the translocon (Sec61). (2) Two glucoses are removed by the sequential action of GI and GII to generate monoglucosylated species (3) that are recognized by CNX and/or CRT (only CNX is shown), which are associated with ERp57. The complex formed by the lectins and folding intermediates/misfolded glycoproteins (4) dissociates upon removal of the last Glc by GII and (5) is reformed by UGGT activity. Once glycoproteins have acquired their native conformations, either free or complexed with the lectins, (5) GII hydrolyzes the remaining Glc residue and releases the glycoproteins from the lectins. (6) These species are not recognized by UGGT and are transported to the Golgi. (7) Glycoproteins unable to properly fold are retrotranslocated to the cytosol where they are deglycosylated and degraded by the proteasome.

4.3 The UGGT

The key component of the QC system is UGGT, since it is the main enzyme responsible for glycoprotein-folding status recognition (Caramelo and Parodi 2007). The enzyme is present in most eukaryotic cells that transfer high mannose glycans, with the exception of *Saccharomyces cerevisiae* (Fernandez et al. 1994). By contrast, UGGT is absent from organisms that transfer extremely short versions of the glycan such as *Giardia lamblia* and *Plasmodium falciparum* (Banerjee et al. 2007). UGGT deletion is usually tolerated in unicellular organisms such as *Schizosaccharomyces pombe* and *Trypanosoma cruzi*, although their growth is impaired under conditions that trigger ER stress (Conte et al. 2003; Fanchiotti et al. 1998). In some cases, it has been observed that UGGT deletion triggers the unfolded protein response, thus compensating for the lack of the folding sensor. UGGT minus *S. pombe* mutants grow normally, but cells are approximately 30% shorter than wild-type ones (Fernandez et al. 1996). UGGT knockout in T. cruzi results in a reduced infectivity and in an increased binding protein (BiP) level (Conte et al. 2003). UGGT knock out in *Arabidopsis thaliana* does not lead to an obvious phenotype, but several ER chaperones and folding assisting enzymes such as BiP, PDI, CNX and CRT are upregulated (Jin et al. 2007). On the other hand, UGGT knockout is embryonically lethal in mouse, although mouse embryonic fibroblasts (MEF) cells derived from those embryos are viable.

UGGT transfers a single Glc residue from UDP-Glc to the terminal Man of branch A of high mannose glycans (▶Figure 4.1). UGGT activity requires calcium concentrations in the millimolar range, a condition normally found in the ER lumen (Trombetta et al. 1991). UGGT displays an ER retention/retrieval signal at its C terminus and is mainly found in the ER and the ER-Golgi intermediate compartment (ERGIC) (Zuber et al. 2001). It is one of the few soluble glycosyltransferases of the secretory pathway (Trombetta et al. 1991). Interestingly, *T. cruzi* UGGT lacks an ER retention/retrieval sequence, and it is unknown how its ER localization is achieved (Conte et al. 2003). The protein has an approximate size of 170 kDa and is composed by at least two domains. The N-terminal domain spans 80% of the protein and has been assumed to be responsible for recognition of the misfolded polypeptide. The C-terminal domain comprises the remaining 20% of the protein, binds [β-32P]5N$_3$UDP-Glc, and displays significant similarity to members of glycosyltransferase family 8 (Tessier et al. 2000). This domain displays the conserved motif DQDXXN, where the triad DQD may serve to coordinate a divalent cation necessary for UDP-Glc binding. UGGT C-terminal domains from different species share a significant similarity (60%–70%), but much lower values occur between the N-terminal ones. For example, *S. pombe* and *Drosophila melanogaster* N-terminal domains show a low degree of similarity (16.3%), but chimeras combining the C- and N-terminal domains from both species were active in vivo, suggesting that their N-terminal domains display similar structures (Guerin and Parodi 2003). Humans have two homologous genes for UGGT sharing a 55% sequence identity, hUGGT1 and hUGGT2. While the former binds UDP-Glc and glucosylates misfolded glycoproteins such as bovine thyroglobulin, the latter is inactive when using the standard assay (see Section 4.7.1) (Arnold et al. 2000). ER stress triggered by tunicamycin or the ionophore A23187 only induces hUGGT1 expression. Interestingly, a fusion protein displaying the N-terminal domain of hUGGT1 with the C-terminal domain of hUGGT2 is enzymatically active, suggesting that hUGGT2 role may be focused on a particular set of substrates or has evolved to fulfill an alternative function (Arnold and Kaufman 2003).

4.3.1 Substrate recognition by UGGT

UGGT is a unique example of a protein combining the activity of a glycosyltransferase with the specificity of a classic chaperone. A major problem when studying the specificity of UGGT is that glycoprotein-folding intermediates (UGGT natural substrates) have an ephemeral existence, thus complicating the preparation of sufficient amounts of substrates for in vitro studies. In addition, since those species are partially misfolded, their high tendency to aggregate hinders a correct interpretation of kinetic data. By using well-characterized neoglycoproteins derived from truncated versions of chymotrypsin inhibitor 2 (CI2), it was shown that UGGT recognition mirrors the anilinonaphtalene sulfonic acid (ANS)–binding capacity of the substrates (Caramelo et al. 2003, 2004). ANS is a hydrophobic probe that binds to collapsed folding intermediates provided that they expose hydrophobic patches, while native proteins or highly disordered conformations do not bind the drug. Similar results have been reported with slightly destabilized mutants of RNAseB and β-glucanase (Ritter et al. 2005; Taylor et al. 2004). These observations show that UGGT recognizes solvent-exposed hydrophobic residues, preferentially displayed on advanced folding intermediates, focusing its activity during the last stages of protein folding. In addition, UGGT can recognize high mannose glycopeptides, provided that they are hydrophobic and long enough (Taylor et al. 2003). Besides sensing anomalies on the tertiary structure, UGGT can participate in the QC of quaternary structures. For instance, UGGT-mediated reglucosylation of the subunits of the T-cell receptor persists until its assembly is accomplished (Gardner and Kearse 1999). In vitro studies have found that UGGT recognizes incompletely assembled complexes formed by properly folded subunits of soybean agglutinin (SBA, normally a homotetramer) (Keith et al. 2005). Here, the sensing mechanism is similar to that observed with the CI2-derived family, since monomeric SBA or partially formed complexes expose hydrophobic surfaces that bind ANS and become occluded upon oligomer final assembly. The mechanism underlying the exquisite molecular reco-gnition by UGGT is yet to be unraveled.

Even though UGGT recognizes several endogenous proteins, so far only a few have been identified. The first one was cruzipain (CZ), an abundant *T. cruzi* lysosomal protease (Labriola et al. 1999). In live parasites, CZ associates with CRT at advanced folding stages, when most of its disulfide bridges are already formed (Labriola et al. 1999, 2011). Since the glycan transferred from Dol-P-P in this organism lacks Glc residues, the only pathway by which a glycoprotein may associate with CRT is that involving UGGT activity. Evidence was presented indicating that CZ-CRT association occurred on CZ molecules in which all or nearly all disulfide bridges had been formed, thus indicating, therefore, that in vivo UGGT operates mainly during the last folding stages. In addition, totally or partially reduced CZ forms are recognized mainly by BiP, the ER HSP70 chaperone (Labriola et al. 2011). These observations illustrate how the diverse folding systems present in the ER can cooperate to assist the complete folding process, starting from highly unstructured polypeptides after they enter the ER to more advanced molten globule-like intermediates before completing the folding process. Recently a protease from animal cells, named prosaposin, has been identified as an UGGT substrate (Pearse et al. 2010). Similarly to CZ, prosaposin is a lysosomal protease rich in disulfide bridges. In both cases, deletion of UGGT leads to an inefficient traffic of the protease to the lysosomes and to their retention in the ER in the form of aggregates. In plants, UGGT activity has been involved in the conformational maturation of some membrane proteins. For instance, UGGT recognition of a brassinosteroid receptor mutant (bri1–9) triggers its retention in the ER of *A. thaliana* cells. Brassinosteroid sensitivity of plants expressing mutant receptors is restored upon disruption of the UGGT encoding gene, thus showing that the structural perturbation that triggered UGGT recognition did not involve the receptor activity (Jin et al. 2007). These observations suggest that diseases caused by ER retention of defective glycoproteins might be ameliorated by inhibitors of UGGT activity. In addition, the fine sensitivity of UGGT allows it to modulate other physiological mechanisms. One beautiful example is its role during antigen presentation by the major histocompatibility complex (MHC) I complex (Zhang et al. 2011). Here, MHC I complexes loaded with suboptimal peptides are reglucosylated by UGGT, thus triggering their retention in the ER. On the contrary, complexes formed with high-affinity peptides are poor UGGT substrates and can be presented on the cell surface.

4.4 GII

GII is an ER-soluble heterodimeric protein composed of catalytic (GIIα) and regulatory (GIIβ) subunits (Trombetta et al. 1996). The last one displays two domains, one at its N terminus and the other at its C terminus, that have a relative high degree of similarity between enzymes of different species. The first one (G2B) was shown to be involved in the interaction between both subunits (Arendt and Ostergaard 2000; Stigliano et al. 2011). The C-terminal domain called MRH, for Man 6-P receptor homologous, is highly similar to the lectin domain of the mammalian mannose 6-phosphate (Man 6-P) receptor involved in the transport of glycoprotein enzymes from the Golgi to lysosomes in mammalian cells (Munro 2001). It has been shown that point mutations in amino acids involved in Man recognition in the Man 6-P receptor and conserved in GIIβ MRH domain sharply reduce in vivo deglucosylation rates, thus strongly suggesting that the MRH domain recognizes Man units in the N-glycan, thus accelerating enzymatic reactions (Stigliano et al. 2009). This result agrees with the fact that both in vivo and in cell-free assays, a decrease in the N-glycan Man content results both in a decrease in deglucosylation rates and in the affinity of GIIβ MRH domain for glycans (Grinna and Robbins 1980; Hagihara et al. 2006; Stigliano et al. 2011). It is worth mentioning that a lectin (Yos9p in yeast, OS-9 in humans) apparently involved in driving misfolded glycoproteins to proteasomal degradation also displays an MRH-like domain, but differently from what happens in that present in GIIβ, the Yos9p/OS-9 MRH-like domain shows a higher affinity for glycans with lower mannose content (Hosokawa et al. 2009; Quan et al. 2008). This finding is consistent with the model that assumes that long ER permanence of misfolded glycoproteins results in the generation of partially demannosylated molecules, thus prepared to be driven to the cytosol.

At least two mechanisms may be envisaged for the MRH domain-mediated enhancement of N-glycan deglucosylation rate. In the first one, upon binding mannose units in the B and/or C arms of the glycan (▶Figure 4.1), the MRH domain presents bonds to be cleaved to the catalytic site in GIIα. This possibility is apparently at odds with the known three-dimensional structure of $Glc_3Man_9GlcNAc_2$. As determined by NMR, the bond to be cleaved first (Glcα1,3Glc) is exposed to the external face of the A arm (residues d, f and g, ▶Figure 4.1), whereas the second bond (Glcα1,3Man) faces the internal side (that is, it faces the B and C arms, residues e and h–k in ▶Figure 4.1) (Petrescu et al. 1997). The need to reorient the substrate would make the mechanism suggested previously highly improbable as both bonds to be cleaved lie far apart in space and cannot conceivably be reached by the single GIIα catalytic site without such reorientation. However, this mechanism cannot be ruled out altogether as the reported flexibility of bonds joining mannoses c, d, f and g (▶Figure 4.1) may allow the successive presentation of both bonds in the MRH domain-bound glycan to the catalytic site (Woods et al. 1998). It is probably the required reorientation that prevents the consecutive trimming of both residues, thus allowing the glycoprotein to enter the CNX/CRT cycle when in the monoglucosylated form.

The second mechanism was proposed by A. Helenius and coworkers to explain the apparent mammalian GII requirement of two glycans in the same glycoprotein to efficiently perform the first cleavage (Deprez et al. 2005). According to this mechanism, binding of the GIIβ MRH domain to a glycan would not result in the presentation of the glycan to the catalytic site, but in a conformational change in the GIIβ subunit that in turn would modify GIIα structure, thus activating the catalytic site. Due to the structural constraints of both bonds to be cleaved mentioned previously, it was proposed that binding of the GIIβ MRH domain to a glycan would activate the first cleavage in a neighboring glycan. The second cleavage in this last glycan would be activated by the binding of its own mannoses in arms B and C to the MRH domain as the arms and the Glcα1,3Man bond would be on the same face. The necessary arm rotation in the first mechanism is replaced in the second one by the successive binding of two different glycans to the MRH domain. Nevertheless, there are exceptions, and cases in which glycoproteins bearing a single *N*-glycan interact with CNX/CRT are known. According to the proposed second mechanism, these cases may be due to transactivation of the first cleavage by an *N*-glycan in different glycoprotein molecules in the crowded ER environment or, for glycoproteins that have a rather long folding process, to a GII basal, GIIβ-independent activity. It is interesting to note that no experimental evidence for all assumptions in the proposed second mechanism were provided at that time, particularly the occurrence of two different GII activities, one of them independent (basal) and the other dependent on the presence of GIIβ and the activation of *N*-glycan deglucosylation by the interaction of the MRH domain in GIIβ and mannose units in the glycan B and/or C arms. It has been shown that GII has indeed a basal activity that is revealed when p-nitrophenyl α-D-glucopyranoside (pNPG) is used as substrate in the absence of the GIIβ subunit (Stigliano et al. 2009).

No influence of the *N*-glycan mannose content on the in vivo UGGT enzymatic activity was observed (Stigliano et al. 2011). As a result of this fact and of the neat influence of the mannose content on GII activity mentioned in this section, as demannosylation of misfolded/slow folding glycoproteins proceeds in the ER, there are increasing chances of exposure of monoglucosylated glycans, thus preventing their surreptitious exit to the Golgi and enhancing proper folding possibilities of the latter ones through interaction with CNX/CRT. The presence of the MRH domain in GIIβ confers to the enzyme, therefore, a major role in the ER glycoprotein-folding QC (Stigliano et al. 2011). It has been reported that up to three or four mannoses may be removed from misfolded glycoproteins in the mammalian cell ER (Frenkel et al. 2003). Exit of those species from futile CNX/CRT most plausibly occurs upon removal of mannose unit g in the Glc-containing Arm A (▶Figure 4.1), the last or one of the last demannosylation events, that yields glycans unable to be reglucosylated by UGGT. Alternatively, interaction of Yos9p/OS-9 MRH-like domain with partially demannosylated *N*-glycans may impede further UGGT-mediated reglucosylation.

The relevance of the possibility of displaying a slow deglucosylation mechanism that increases the window frame in which certain glycans are monoglucosylated is suggested by the fact that *Leishmania mexicana*, a protist transferring Man$_6$GlcNAc$_2$ and lacking ER mannosidase activity, has retained an ancestor-derived GIIβ with the MRH domain showing a higher specificity for high-Man-content glycans, instead of evolving it to a domain with higher affinity for low-Man *N*-glycans (Stigliano et al. 2011). It may be speculated that a MRH domain with a high specificity for low-Man glycans as Man$_6$GlcNAc$_2$ would result in high deglucosylation rates.

4.5 CNX and CRT

Glycoprotein displaying monoglucosylated oligosaccharides are retained in the ER by two lectins, CRT, a 46 kDa (400 residues) soluble ER-resident protein, and its paralogue CNX, a 65 kDa (572 residues) type I membrane protein. At variance with other chaperones that rely solely on polypeptide-based interactions for their activity, the biological function of CRT and CNX is centered on glycan-mediated interactions, although in some instances protein-protein interactions may be also involved. CRT has been localized to many subcellular compartments and has been implicated in diverse functions, such as protein folding, calcium homeostasis, gene transcription regulation, complement activation and angiogenesis. CRT was first isolated from bovine liver and characterized as a calcium-binding protein (Waisman et al. 1985). It is a typical ER-resident protein, endowed with an N-terminal signal sequence and a KDEL-like C-terminal ER retention/retrieval signal. CNX was isolated and cloned from dog pancreas and first characterized as a calcium-binding protein displaying high homology to CRT (Wada et al. 1991). Incubation of CNX with a mixture of Glc$_{0-3}$Man$_9$GlcNAc$_2$ glycans showed a marked preference to bind monoglucosylated glycans (Ware et al. 1995). A similar behavior was shortly observed with CRT (Spiro et al. 1996). The expression of these lectins is induced by several stressors (Conway et al. 1995; Heal and McGivan 1998; Llewellyn et al. 1996; Nguyen et al. 1996; Szewczenko-Pawlikowski et al. 1997), thus indicating their role in protein folding. In agreement with these results, the 5' promoter region of human CRT gene displays regulatory regions also present in the promoter region of other ER chaperones such as BiP, GRp94 and PDI (McCauliffe et al. 1992).

4.5.1 CRT and CNX structures

CNX is a type I integral membrane protein with a transmembrane segment near its C-terminal end and a RKPRRE ER-localization signal placed in a short cytosolic C-terminal tail. The crystal structure of CNX's ectodomain reveals a 140 Å long protruding arm inserted into a globular domain, which shows a fold similar to that of the legume lectin family (Schrag et al. 2001). The globular domain covers two regions of the primary sequence (residues 1–262 and 415–458) and is organized in two antiparallel β sheets, one concave (six strands) and the other convex (seven strands). The Glc unit binds to a shallow depression located on the surface of the concave β sheet, with its ring resting on Met189 and its hydroxyl groups hydrogen bonded to Tyr165, Lys186 and Glu217. Adjacent to this site, there are several hydrogen bond donors and acceptors located on the surface of the concave β sheet, which may contact the Man residues of the glycan. Also in the globular domain, but opposite to the sugar-binding face, a calcium ion is present, which accounts for the high-affinity, low-capacity calcium-binding activity of the protein (Kd ~ 2 μM) (Baksh and Michalak 1991). This calcium ion fulfills a structural role by stabilizing the lectin and, in principle, does not play a direct role in sugar binding. The arm domain, also known as P domain, is a unique feature of these proteins. It is formed by a Pro-rich tandem sequence repeats in which four copies of motif 1, IxDP(D/E)(A/D)xKP(E/D)DWD(D/E), and four copies of motif 2, GxWxxPxIxN-PxY, are arranged following a 11112222 pattern. This domain can be divided into four modules, each formed by a copy of motif 1 that interacts head-to-tail with a copy of motif 2. A similar arrangement is observed in CRT, where three tandem repeats of motifs 1 and 2 are arranged following a 111222 pattern. The structure of the CRT P domain solved by NMR shows a highly flexible hairpin structure similar (but shorter, 110 Å) to that of CNX (Ellgaard et al. 2001). It has been proposed that the P domain stabilizes the lectin-glycoprotein complex by providing additional points of contact, thus increasing the efficiency of the chaperone activity (Xu et al. 2004), and that is necessary, therefore, for optimal lectin activity (Leach et al. 2002; Vassilakos et al. 1998). Interestingly, the intermolecular contacts in the CNX crystal are mediated by the P domain, which embraces the globular domain of a symmetry-related molecule, suggesting a similar disposition for the interaction with folding glycoproteins.

The globular domain of CRT is similar to that of CNX, with some minor differences mainly in the loops connecting the strands of the β sheets (Kozlov et al. 2010). At variance with CNX, CRT has a highly acidic C-terminal domain (residues 285–400) that binds approximately 20 calcium ions with low affinity (Kd ~ 2 mM). This domain works as an ER Ca^{2+} buffer (Baksh and Michalak 1991) and is not necessary for sugar binding. Circular dichroism spectra of the isolated C-terminal domain revealed an intrinsically disordered conformation, which upon Ca^{2+} addition becomes more compact and rigid (Villamil Giraldo et al. 2010). This domain is probably loosely connected to the rest of the protein, as observed by electron microscopy (Tan et al. 2006). The C-terminal domain has been implicated in the retention of CRT in the ER in a novel mechanism that would operate in conjunction with the KDEL-based retention/retrieval system (Sonnichsen et al. 1994). In addition, this domain is necessary for the retrotranslocation of CRT to the cytosol, which seems to be regulated by the Ca^{2+} concentration in the ER lumen (Labriola et al. 2010).

4.5.2 Ligand binding to CRT and CNX

CRT and CNX specifically bind monoglucosylated high-mannose glycans (Ware et al. 1995). The terminal glucose is essential for binding, although CRT affinity for this monosaccharide is very low. In addition, the following three mannoses present in the α1,3 branch (branch A, ▶Figure 4.1) make significant contributions to complex formation (Kapoor et al. 2003). Accordingly, site-directed mutations in positions that bind directly to the Glc residue (Y109 and D135 in rat CRT) are less tolerated than changing residues that contact the following Man residues (Kapoor et al. 2004; Thomson and Williams 2005). The moderate affinity of CRT for $Glc_1Man_9GlcNAc_2$ (Kb = 4.10^5 M^{-1}) probably allows the action of sugar-modifying enzymes such as GII during the dissociation phase of the lectin-glycan equilibrium (Patil et al. 2000). The complex of CRT with the tretrasaccharide Glcα1,3Manα1,2Manα1,2Man shows the ligand bound to a long and shallow cleft formed by the concave β sheet (Kozlov et al. 2010). This complex is stabilized by a network of hydrogen bonds and hydrophobic interactions. Remarkably, the O2 atom of the Glc residue fits in a tight pocket, interacting with side chain of

Lys111 and the carbonyl of Gly124. This feature explains why Man cannot bind to this site, since the O2 in this sugar points to the opposite direction and would not fit in the pocket. This structure also explains the requirement of the Cys105-Cys137 dilsufide bridge for sugar binding, since it contacts the third and fourth sugar moieties of the tetrasaccharide.

During their transit through the ER, most glycoproteins studied so far interact with CNX and/or CRT. The reasons for an *N*-glycan to interact with a particular lectin can be partially explained from their topological differences. CNX, being attached to the membrane, interacts preferentially with glycans placed near the ER membrane. In contrast, CRT preferentially binds glycoproteins soluble in the ER lumen or *N*-glycans placed far from the membrane (Andersson et al. 1996; Daniels et al. 2003; Hebert et al. 1997; Wang et al. 2005). Furthermore, a soluble version of CNX exhibits a binding specificity similar to CRT, while the specificity of a CRT mutant endowed with a membrane anchor shifts to that observed for CNX (Danilczyk et al. 2000). Nevertheless, this interchangeability is partial since a soluble version of CNX is unable to replace CRT in assisting the peptide loading of the MHC class I complex (Gao et al. 2002). Besides, many proteins normally associated with CNX in wild-type cells associate instead with BiP and not with CRT in CNX$^{-/-}$ cells (Pieren et al. 2005). In addition, knockout of CRT- or CNX-encoding genes is lethal in mice, showing that the remaining lectin cannot fully replace the deleted one (Denzel et al. 2002; Mesaeli et al. 1999), although this conclusion may be related to the additional functions of CNX/CRT, such as CRT being a Ca^{2+} buffer, and not to its lectin role. On the other hand, certain microorganisms have only one of the lectins, as for instance *S. cerevisiae* or *S. pombe*, which only have CNX or trypanosomatid protozoa, in which only CRT is found (Labriola et al. 1999).

Glycoprotein binding to these lectins is mediated by their glycan moieties, although in some cases polypeptide-based interactions also occur. Two alternative binding models have been proposed. In the "lectin-only" model, glycoprotein binding is mediated solely by their glycan moieties, while in the "alternative" model there is a contribution of both polypeptide and glycan moieties or even of only the peptide one. Substantial experimental evidence supports both models, and probably the relative contribution of polypeptide-based contacts varies with the characteristics of a particular protein (Williams 2006). The "lectin-only" model is supported mainly by experiments using glycosylation or GI/GII inhibitors or by expressing proteins bearing site-specific mutations of consensus glycosylation sites, as protein association with the lectins is severely affected under those conditions (Branza-Nichita et al. 2001; Di Jeso et al. 2003; Hammond et al. 1994; Kang and Cresswell 2002; Keller et al. 1998; Nakhasi et al. 2001; Tatu and Helenius 1999; Zhang et al. 2003). Furthermore, protein binding to CNX is almost totally impaired in cells lacking GI or GII, as these enzymes are necessarily required for creating monoglucosylated glycans (Ora and Helenius 1995). Dissociation of the already formed lectin-glycoprotein complexes is prevented upon addition of GI/GII inhibitors, thus showing the importance of GII during the dissociating phase of the cycle (Hebert et al. 1995). Nevertheless, it has been shown that some nonglycosylated proteins remain associated with the lectins if care is taken during cell lysis in order to preserve weak complexes (Danilczyk and Williams 2001; Mizrachi and Segaloff 2004; Wanamaker and Green 2005). CNX and CRT can inhibit in vitro the aggregation of both glycosylated and nonglycosylated proteins, and they are able to form stable complexes with unfolded, nonglycosylated substrates but not with their native forms (Ihara et al. 1999; Saito et al. 1999). Interestingly, the addition of monoglucosylated glycans to these assays impaired the ability of CRT to inhibit the aggregation of glycosylated and nonglycosylated proteins (Saito et al. 1999), although the presence of a monoglucosylated glycan moiety on the substrates improved the chaperone capacity of CNX but not that of a classic chaperone such as BiP (Stronge et al. 2001). In addition, CNX and CRT mutants devoid of lectin activity inhibited the thermal aggregation of nonglycosylated substrates but not that of the monoglucosylated substrate jack bean α-mannosidase (Leach and Williams 2004; Thomson and Williams 2005). Furthermore, the CNX lectin negative mutant was able to bind MHC I heavy chain in vivo (Leach and Williams 2004). Interestingly, CNX interacts with integral membrane proteins through their transmembrane domains in a glycan-independent mode (Fontanini et al. 2005; Swanton et al. 2003). For instance, CNX associates in vivo with ER degradation enhancer mannosidase like (EDEM), a lectin that targets high-mannose glycoproteins for degradation, through their transmembrane domains, physically linking the retention and degradation of irreparably misfolded species (Oda et al. 2003).

Although CRT and CNX are proteins dedicated to deal with partially folded species, their structure is marginally stable (Tm ~ 42°C–45°C) (Bouvier and Stafford 2000; Conte et al. 2007). This property precludes a neat explanation for their precise role in many thermal-induced aggregation assays carried out near the lectin melting temperatures. Nevertheless, CRT inhibits protein aggregation of some substrates in experiments developed at temperatures far from any detectable conformational change (MHC class I and deglycosylated IgY at 37°C and 31°C, respectively) (Saito et al. 1999). Moreover, the structure of the globular domain of CRT shows a conserved patch bound to a peptide of a neighbor protein in the crystal, suggesting that this site could be responsible for its ability to interact with the protein moiety of the ligands (Chouquet et al. 2011). In addition, upon a heat shock or Ca^{2+} depletion, both CRT and CNX undergo conformational changes that induce their oligomerization and increase their ability to bind nonglycosylated substrates (Rizvi et al. 2004; Thammavongsa et al. 2005). This result suggests that changes in the ER environment may modulate their chaperone activity. The scarcity of suitable model substrates has precluded a definitive answer regarding the relative importance of polypeptide-based interactions. The current consensus holds that for most glycoproteins glycan-based interactions are determinant for the initial establishment of the complex (Ware et al. 1995; Zhang et al. 1995), while the relevance of subsequent protein-protein interactions depends on the particular gly-coprotein involved.

4.6 ERp57

CRT and CNX associate with ERp57, a member of the PDI family (Oliver et al. 1999). This association is independent of ligand binding to the lectins, since it is observed also in cells lacking GI, thus unable to generate monoglucosylated glycans. ERp57 is composed of four thioredoxin-like domains (a, b, b' and a'), in which the a and a' domains display the active – CXXC – motifs (Silvennoinen et al. 2004). The C-terminal Cys residue of these motifs form intermolecular disulfide bridges with glycoproteins bound to CRT or CNX, and usually their association with ERp57 is also abrogated when glycoprotein binding to the lectins is inhibited (Di Jeso et al. 2005; Kang and Cresswell 2002; Van der Wal et al. 1998). In addition, ERp57 activity improves when its substrates are bound to the lectins. For instance, the ability of ERp57 to catalyze disulfide bridge acquisition of denatured monoglucosylated RNase B in vitro is enhanced in the presence of CRT and CNX, while the opposite effect is observed during refolding catalyzed by PDI (Zapun et al. 1998). Conversely, BiP enhances PDI activity but exerts no effect on ERp57, suggesting that both enzymes have evolved to operate in conjunction with only one of the main ER chaperone systems (Mayer et al. 2004). Nevertheless, in some cases, ERp57 has been detected in complexes with ER proteins that are not associated with the lectins (McCormick et al. 2005; Peaper et al. 2005). The ERp57-CRT association is mediated by residues 225–251 located on the tip of CRT's P domain and the b' domain of ERp57 (Frickel et al. 2002; Kozlov et al. 2006; Russell et al. 2004), and a similar interaction has been described with respect to CNX (Pollock et al. 2004). Given the high flexibility of the P domain, ERp57 would be able to scan for disulfide bridges located at distant positions of the glycoprotein bound to the ternary complex with the lectins. The moderate stability of the ERp57–CRT complex (Kd ~ 9 µM) and its fast off-rate (koff > 1,000 s−1) probably allows ERp57 to rapidly sense for preexisting lectin-glycoprotein complexes (Kozlov et al. 2006). The domain architecture of PDI and ERp57 is similar, although both enzymes have developed different strategies to bind their substrates, since the PDI b and b' domains are able to directly contact their substrates whereas the homologous ERp57 domains have evolved to interact with the P domains of CNX and CRT, which in turn present most known substrates to ERp57. Therefore, CRT and CNX constitute a system specially fitted to present glycosylated substrates to ERp57.

4.7 Methods to study glycoprotein folding QC

4.7.1 Assay for UGGT

The assay for UGGT is based on the incorporation of radioactive Glc from the sugar donor (UDP-[^{14}C]Glc or UDP-[^{3}H]Glc) into polymannose glycans on unlabeled denatured acceptor glycoproteins. The radiolabeled reaction product is separated from the radiolabeled substrate (UDP-Glc) by trichloroacetic acid (TCA) precipitation and quantified by liquid scintillation counting. Misfolded acceptor glycoproteins (bovine thyroglobulin; SBA; bovine pancreatic RNase B) can be prepared by chemical denaturation (Trombetta and Parodi 1992; Trombetta et al. 1989). Glycoproteins are dissolved at high concentrations (20–50 mg/mL) in 10 mM Hepes, pH 7.4, 8 M urea, and the mixture is incubated at 60°C for 4 h. The samples are then exhaustively dialyzed against 10 mM Hepes, pH 7.4. Reactions are conducted in 50–100 µL final volume, containing 10 mM Hepes, pH 7.4, 10 mM CaCl$_2$, 5 mM 2-mercaptoethanol and 2.5 µM UDP-Glc (about 0.01–0.05 µCi of UDP-linked [^{14}C]Glc or [^{3}H]Glc should be added). Reactions are initiated by addition of the test sample and incubated for 5–30 min at 37°C and stopped by addition of 1 mL of 10% TCA. Next the samples are boiled for 5 min, and the precipitated proteins are recovered by low-speed centrifugation (5 min at 2,000 × g) and washed three times with 1 mL of 10% TCA. The washed pellets are resuspended in 100 µL of 1 M KOH in methanol or other commercial solubilizers, diluted with 3 mL of scintillation cocktail and quantified by liquid scintillation counting. The incorporation of radioactive

Glc into acceptor glycoproteins can also be detected by autoradiography (Caramelo et al. 2004; Ritter and Helenius 2000; Taylor et al. 2004; Trombetta et al. 1989).

4.7.2 Assay for GII

GII activity can be measured using radioactive glycans or artificial substrate analogues. Radioactive [^{14}C]Glc- or [^{3}H]Glc-labeled Glc$_1$Man$_9$GlcNAc can be prepared in vitro by incubating rat liver microsomes (20 mg/mL) with 10–40 µM UDP-Glc (0.8–3.2 µCi of UDP-linked [^{14}C]Glc or [^{3}H]Glc should be added), 1 mM castanospermine or 1-deoxynojirimycin, 10 mg/mL denatured bovine thyroglobulin, 10 mM CaCl$_2$, 1% Triton X-100, 5 mM 2-mercaptoethanol, 20 mM Hepes, pH 7.4, in a final volume of 200 µL for 60 min at 37°C. Under these conditions, the radiolabeled Glc is incorporated into N-glycans in glycoproteins directly via reglucosylation (mostly on denatured thyroglobulin) or, additionally, via the Dol-P-P pathway. At the end of the incubation, samples are extracted with chloroform/methanol/water (10:10:3), and the protein pellet successively treated with a protease and with endo-β-N-acetylglucosaminidase H (Endo H). Glc$_1$Man$_9$GlcNAc has to be chromatographically purified from Glc$_3$Man$_9$GlcNAc, Glc$_2$Man$_9$GlcNAc, Glc$_1$Man$_8$GlcNAc and Glc$_1$Man$_7$GlcNAc that might have been formed via the Dol-P-P pathway or by direct glucosylation of Man$_8$GlcNAc$_2$ and Man$_7$GlcNAc$_2$ (see Section 4.7.5.3). To measure GII activity, radiolabeled glycans are incubated for 5–60 min at 37°C with test samples in a final volume of 50–100 µL containing 10 mM Hepes, pH 7.4 (1% Triton X-100 has to be added when microsomal vesicles are used as enzyme source). At the end of the incubation, the released radioactive Glc can be detected in a number of ways. One possibility is to separate the released Glc from the intact oligosaccharide by paper chromatography (Trombetta et al. 1989; Ugalde et al. 1979).

GII activity can also be measured with the artificial substrate pNPG. GII is the main ER enzyme capable of cleaving this substrate at pH 8.0, since most lysosomal glycosidases are unstable and/or inactive at such pH. In this case, Glc removal is followed by absorbance at 405 nm to detect the free p-nitrophenol released. Assays can be conducted in a final volume of 100 µL containing 1 mM pNPG in 20 mM Tris-HCl, pH 8.0, 1 mM EDTA in a thermostatized cuvette, monitored continuously at 405 nm.

4.7.3 Purification of GII and UGGT from rat liver

Both GII and UGGT are soluble proteins of the ER lumen and are therefore soluble in the absence of detergents. Low concentration of detergents may nevertheless be used to release the soluble content of the microsomes at the start of the purification, but detergents are not further employed. UGGT and GII are minor components of the ER, and typically less than 1 mg UGGT or GII is obtained from 100 to 200 g of liver, with yields below 10%. Both enzymes are highly susceptible to proteases, and therefore it is critical to include protease inhibitors in the homogenization buffers, to maintain the pH above 7.0 and to process rapidly the microsomal extracts. Microsomes can be kept frozen for a few weeks, but it is important to go from the microsomes to the final step with minimal delays. Both enzymes are relatively stable after purification. UGGT is composed of a single polypeptide that runs at approximately 160–170 kDa in SDS-PAGE (Trombetta and Parodi 1992). GIIα runs as a 110 kDa protein, while GIIβ runs as a slightly diffuse band at approximately 80 kDa (Trombetta et al. 1996).

4.7.3.1 Buffers and chromatography media

Buffer A: 0.25 M sucrose, 2 mM EDTA, 20 mM Tris-HCl, pH 8.0 5 mM of 2-mercaptoethanol. Buffer B: 150 mM NaCl, 20 mM Tris-HCl, pH 8.0 5 mM 2-mercaptoethanol. Buffer C: 20 mM Tris-HCl, pH 8.0 and 5 mM of 2-mercaptoethanol. Buffer D: 1 M NaCl, 20 mM Tris-HCl, pH 8.0 and 5 mM 2-mercaptoethanol. Buffer E: 1 M ammonium sulfate and 5 mM of 2-mercaptoethanol. Buffer F: 0.5 M sucrose, 10 mM imidazol, pH 7.0 and 5 mM 2-mercaptoethanol. Chromatography media are from Pharmacia or Sigma.

4.7.3.2 Preparation of rat liver microsomes

Rats (male or female, 4- to 12-week-old) are starved overnight. They are euthanized, and their livers removed and rinsed on ice cold buffer A. From this point, all procedures are carried on ice or in a cold room, except for the elution step from the ConA-Sepharose column. Livers are weighted and minced in a blender with two to four volumes of buffer A containing protease inhibitors (1 mM EDTA, 10 µM leupeptin, 10 µM pepstatin, 10 µM E-64, 10 µM tosyl-lysyl-chloromethyl-ketone [TLCK], 10 µM tosyl-phenylalanyl-chloromethyl-ketone [TPCK] and 100 µM PMSF). The homogenate thus obtained is centrifuged at 10,000 × g for 10 min. The supernatant is further centrifuged at 100,000 × g for 60 min. The pellet containing the microsomal fraction is resuspended in buffer A and stored at −80 °C.

4.7.3.3 Extraction of UGGT and GII from microsomes

Since both UGGT and GII are soluble proteins, the microsomal lumenal content can be liberated with detergents at low concentrations. For this, microsomes are resuspended at 10 mg/mL in buffer C and extracted with 0.1% Triton X-100 for 30 min on ice. The homogenate is then centrifuged at 100,000 × g for 60 min, and the supernatant of the microsomal extraction containing most of UGGT and GII activity is saved.

4.7.3.4 DEAE-cellulose

The solubilized fractions are loaded onto a diethylaminoethyl (DEAE)-cellulose column equilibrated in buffer C and washed in the same buffer until the Abs280 reaches background. The column is then eluted with a gradient from 100% buffer C to 50% buffer D over 20 column volumes. The enzymatic activities for UGGT and GII are measured in the eluate of the DEAE-cellulose column (UGGT typically elutes between 300 and 400 mM NaCl, and GII elutes between 400 and 500 mM NaCl). From this point, the fractions containing UGGT activity are separated from GII activity and are pursued separately.

4.7.3.5 ConA-Sepharose

The fractions containing UGGT or GII activity eluted from the DEAE-cellulose are applied separately to Concanavalin A (ConA)-Sepharose columns (5 mL) equilibrated in buffer B supplemented with 1 mM $CaCl_2$, 1 mM $MgCl_2$ and 1 mM $MnCl_2$. After washing with buffer B until the Abs280 reaches background, the column is filled with one volume (approximately 5 mL) of buffer B supplemented with 0.5 M α-methyl-mannopyranoside prewarmed at 37°C. The column is then stopped, and kept at 37°C for 15 min. The elution continues with more prewarmed buffer B supplemented with 0.5 M α-methylmannopyranoside.

4.7.3.6 MonoQ

The fractions containing UGGT or GII activity eluted from their respective ConA-Sepharose columns are diluted 5-fold and loaded (separately) onto a MonoQ 5/5 column equilibrated in buffer C. The column is then eluted with a gradient from 100% buffer C to 50% buffer D over 20 column volumes (20 mL), and 1 or 0.5 mL fractions are collected.

4.7.3.7 Gel filtration

The fractions eluted from the MonoQ step containing UGGT or GII activity are further purified by gel filtration chromatography on a Superdex S-200 column (or equivalent) equilibrated and run in buffer B. At this stage, GII is usually homogeneous. If necessary, the MonoQ or gel filtration steps can be repeated to achieve a homogeneous preparation. GII is stored in buffer B at −80 °C.

4.7.3.8 Phenyl-Superose

After the gel filtration step UGGT usually requires further purification using hydrophobic interaction chromatography. The fractions eluted form the gel filtration column containing UGGT activity are diluted 10-fold with buffer E, filtered, and loaded onto a phenyl-Superose column (1 mL) equilibrated in buffer E. The column is eluted at 0.5 mL/min with a gradient from 100% buffer E to 100% buffer F over 20 mL, and then further eluted with another 15 mL of buffer F. UGGT is strongly retained by the column and typically elutes during the beginning of the wash with buffer F. The enzyme is usually homogeneous at this step and can be stored in buffer F at −80 °C.

4.7.4 Indirect analysis of *N*-glycans present in ^{35}S-labeled glycoproteins

The analysis of the *N*-glycans present on specific glycoproteins is a challenging task, particularly when one needs to focus on specific glycosylation sites. Such analysis requires a rather large amount of protein that has to be fragmented to isolate individual peptides containing the desired *N*-glycosylation site, from which glycans are further released and characterized. It is even more challenging to perform this kind of analysis during the folding of a glycoprotein, since only minute amounts of sample can be obtained. A first simplification is to radiolabel the *N*-glycans and subsequently isolate the protein of interest, from which glycans can be released and analyzed, providing direct structural information of the average glycan structures present in the glycoprotein studied. An even simpler approach is to evaluate *N*-glycan structures on immunoprecipitated ^{35}S-labeled proteins using glycosidases and SDS-PAGE analysis. Alternatively, the protein of interest can be detected by Western blot. These methods do not provide detailed carbohydrate structures but have several advantages: labeling is quite efficient, multiple samples can be compared simultaneously, the detection limit is very low and relatively few cells are needed. When the protein under study is synthesized in large quantities, very short pulses are sufficient to produce an intense signal, and very early events (even cotranslational) can be detected. The glycan structures are not analyzed directly, but can still be evaluated by their susceptibility to endo- and exoglycosidases, evidenced as small increment in mobility on SDS-PAGE. Each glycan removed by Endo H results in an increase in mobility of about 2 kDa, but exoglycosidases produce smaller shifts. For large proteins, the shifts are usually very small unless the glycosidases cleave N-glycans on several glycosylation sites. The most common analysis is the evaluation of whether glycoproteins contain Endo H sensitive glycans. This enzyme does not remove complex-type glycans generated in the Golgi but removes all N-glycoforms present in the ER. Therefore, when glycoproteins are sensitive to cleavage by Endo H they are considered to have N-glycan structures similar to those acquired in the ER. Some exceptions to this rule are glycoproteins that traverse the Golgi but carry N-glycans that are not modified and thus remain partially or completely sensitive to Endo H. Also, certain yeasts elongate N-glycans in the Golgi but generate structures that remain sensitive to Endo H. In these cases, Golgi modifications can be identified by mobility shifts (the Golgi modified forms run much slower and often diffuse) or with antibodies to N-glycans structures produced in the Golgi. To verify that the lack of Endo H cleavage is due to acquisition of complex-type N-glycans, parallel samples should be digested with peptide:N-glycosidase (N-glycanase). This enzyme cleaves N-glycans irrespective of their structure, and is thus insensitive to N-glycan modifications acquired in the Golgi.

For those glycoproteins that remain Endo H sensitive due to their residence in the ER, the presence of glucosylated N-glycans can be evaluated by partial resistance to digestion with α-mannosidases that only display exoglycosidase activity. Eight of the nine Man residues in ER glycoforms are linked in the α-configuration and can be removed by α-mannosidases. N-glycans that carry a terminal Glc residue contain three Man residues protected from α-mannosidases (▶Figure 4.1) (Radcliffe et al. 2002; Rodan et al. 1996; Wearsch et al. 2004). Therefore, glucosylated N-glycans are partially resistant to α-mannosidases, and a mobility difference between completely demannosylated and partially demannosylated proteins can often be detected by high resolution SDS-PAGE. It is difficult to distinguish between mono-, di- or triglucosylated N-glycans by partial resistance to α-mannosidase. Pretreatment with GII renders di- and monoglucosylated N-glycans fully sensitive to α-mannosidases. Except for GII, the other glycosidases are commercially available. When α-mannosidase is obtained as an ammonium sulfate precipitate, is should be dialyzed against 50 mM sodium acetate buffer, pH 5.0, 0.1 mM zinc acetate prior to use. The washed immunoprecipitates are resuspended in 10–50 μL of 10 mM Hepes buffer, pH 7.4, 0.5% SDS, 2 mM dithiothreitol and are heated at 95 °C for 10 min. After a short centrifugation, 1–5 μL of 10% Nonidet-P40 is added to quench the SDS, and GII, Endo H, N-glycanase or α-mannosidase is added. For α-mannosidase digestion, about 2–10 μL of 500 mM sodium acetate buffer, pH 5.0, 0.1 mM zinc acetate should be added. When appropriate, GII digestion can be performed before α-mannosidase to evaluate further the presence of glucosylated N-glycans. The samples are resuspended and incubated for 60–120 min at 37°C. Reactions are stopped by addition of SDS-PAGE sample buffer and analyzed by electrophoresis and autoradiography.

4.7.5 Direct analysis of N-glycans

4.7.5.1 Labeling of N-glycans

The most common way of radiolabeling N-glycans is by incubating cells in growth media containing radioactive Glc, Man or Gal (Varki 1991). To achieve the most efficient labeling, cells need to be depleted from unlabeled intermediates as nucleotide sugars and Dol-P derivatives. This can be done by incubating them in media devoid of those sugars for 15–30 min. If glucosidase and/or mannosidase activities need to be inhibited, suitable compounds (castanosperine and 1-deoxynojirimycin to inhibit glucosidases, or kifunensine and 1-deoxymannojirimycin to inhibit mannosidases) may be added at the onset of the starving period at 1 mM concentrations, as those inhibitors penetrate rather slowly into cells. Labeling is best performed in media containing 2–5 mM Glc, as lower concentrations may result in synthesis of truncated Dol-P-P derivatives and the formation of nonphysiological protein-linked glycans. When there are sufficient radiolabeled precursors, pulses can be kept short (5 min or less), and consequently, N-glycan processing can be followed during early stages of glycoprotein biosynthesis. When [U-^{14}C]Glc is used, radioactivity is not only incorporated onto Glc residues. [U-^{14}C]Glc can be converted in the cells into Man and GlcNAc (among other sugars), leading to Glc-, Man- and GlcNAc-labeled N-glycans. Cells also convert Glc rapidly into labeled lipids and amino acids. As a consequence, glycans have to be purified extensively to avoid detection of radioactivity in other molecules. Glc residues on N-glycans can also be labeled with radioactive Gal, since entrance of this monosaccharide to the general metabolism requires its conversion to Gal 1-P and then to UDP-Gal and UDP-Glc, which results in N-glycan efficient labeling. As a consequence, when cells are radiolabeled with [^{14}C]Gal or [^{3}H]Gal in the presence of 1–5 mM Glc, the label goes almost exclusively to Glc and/or Gal residues (Kornfeld et al. 1978; Li at al. 1978; Suh et al. 1989; Varki 1991). When N-glycans are labeled with [2-^{3}H]Man, radioactivity is confined to Man and fucose residues only, since interconversion of Man into other sugars requires oxidation of the –OH group at position 2, and therefore radiolabeled glycans retain the label exclusively in the Man or fucose. Both Glc and Man internalized by cells can be incorporated into glycoproteins following incorporation into UDP-Glc or GDP-Man and further transfer to Dol-P-Glc or Dol-P-Man (Alton et al. 1998; Panneerselvam and Freeze 1996; Panneerselvam et al. 1997).

4.7.5.2 Extraction of protein-bound N-glycans

After the pulse/chase, cells are extracted by suspension in chloroform/methanol/water (3:2:1) (this is best done in a glass tube, 12 × 75 mm). After vigorous mixing, the sample is centrifuged at 3,000 × g for 5 min. A proteinaceous interphase that forms between two liquid phases is carefully recovered discarding both the upper and lower liquid phases. The pellet is further washed twice with chloroform/methanol/water (3:2:1), followed by further washes with chloroform/methanol/water (10:10:3) to extract Dol-P-P-glycans (which can be saved for further analysis). The washed pellet is then digested exhaustively with Pronase (2 mg/mL) for 24 h in 1 mM CaCl$_2$, 200 mM Tris-HCl, pH 8.0. In some cases, it may be beneficial to predigest the Pronase solution for 30 min at 37°C before adding it to the test sample, thus allowing for the destruction of potential glycosidase activities in the pronase preparation. The digestion converts most of the insoluble pellets into amino acids, short peptides and glycopeptides. The digest is cleared by centrifugation (10,000 × g for 5 min), and the supernatant is desalted on a Sephadex G-10 column (2 × 60 cm) equilibrated and run in 7% 2-propanol, in which the glycopeptides are excluded and separated from most amino acids and monosaccharide in the hydrolysate. The isolated glycopeptides are dried and resuspended in 50 mM sodium acetate buffer, pH 5.5, and digested with Endo H for 18 h. Tubes are boiled for 5 min and then passed over an Amberlite MB3-acetate column (0.5 × 5 cm) to retain amino acids and other charged contaminants. The neutral glycans (released by Endo H digestion) are recovered in the flow through of the ion exchange column and dried. They can be analyzed by a number of chromatographic techniques.

4.7.5.3 Chromatographic analysis of *N*-glycans

The most simple analysis method is descending paper chromatography in 1-propanol/nitromethane/water (5:2:4) (Staneloni and Leloir 1978). Another simple alternative is thin layer chromatography on silica plates (Moore and Spiro 1990). *N*-glycans can also be analyzed by frontal affinity chromatography (FACE) or by HPLC, which requires dedicated equipment but provides good resolving power for free glycans, either without modification (Mellis and Baenziger 1981) or after derivatization with 2-aminopyridine (Hase 1994), 2-aminobenzamide (Bigge et al. 1995; Guile et al. 1996) or perbenzoylation (Daniel 1987).

References

Alton, G., Hasilik, M., Niehues, R., Panneerselvam, K., Etchison, J.R., Fana, F., et al. (1998). Direct utilization of mannose for mammalian glycoprotein biosynthesis. Glycobiology *8*, 285–295.

Andersson, H., Nilsson, I., and von Heijne, G. (1996). Calnexin can interact with *N*-linked glycans located close to the endoplasmic reticulum membrane. FEBS Lett *397*, 321–324.

Apweiler, R., Hermjakob, H., and Sharon, N. (1999). On the frequency of protein glycosylation, as deduced from analysis of the SWISS-PROT database. Biochim Biophys Acta *1473*, 4–8.

Arendt, C.W., and Ostergaard, H.L. (2000). Two distinct domains of the beta-subunit of glucosidase II interact with the catalytic alpha-subunit. Glycobiology *10*, 487–492.

Arnold, S.M., Fessler, L.I., Fessler, J.H., and Kaufman, R.J. (2000). Two homologues encoding human UDP-glucose:glycoprotein glucosyltransferase differ in mRNA expression and enzymatic activity. Biochemistry *39*, 2149–2163.

Arnold, S.M., and Kaufman, R.J. (2003). The noncatalytic portion of human UDP-glucose:glycoprotein glucosyltransferase I confers UDP-glucose binding and transferase function to the catalytic domain. J Biol Chem *278*, 43320–43328.

Baksh, S., and Michalak, M. (1991). Expression of calreticulin in *Escherichia coli* and identification of its Ca2+ binding domains. J Biol Chem *266*, 21458–21465.

Banerjee, S., Vishwanath, P., Cui, J., Kelleher, D.J., Gilmore, R., Robbins, P.W., et al. (2007). The evolution of *N*-glycan-dependent endoplasmic reticulum quality control factors for glycoprotein folding and degradation. Proc Natl Acad Sci U S A *104*, 11676–11681.

Bigge, J.C., Patel, T.P., Bruce, J.A., Goulding, P.N., Charles, S.M., and Parekh, R.B. (1995). Nonselective and efficient fluorescent labeling of glycans using 2-amino benzamide and anthranilic acid. Anal Biochem *230*, 229–238.

Bouvier, M., and Stafford, W.F. (2000). Probing the three-dimensional structure of human calreticulin. Biochemistry *39*, 14950–14959.

Branza-Nichita, N., Durantel, D., Carrouee-Durantel, S., Dwek, R.A., and Zitzmann, N. (2001). Antiviral effect of N-butyldeoxynojirimycin against bovine viral diarrhea virus correlates with misfolding of E2 envelope proteins and impairment of their association into E1-E2 heterodimers. J Virol *75*, 3527–3536.

Caramelo, J.J., Castro, O.A., Alonso, L.G., De Prat-Gay, G., and Parodi, A.J. (2003). UDP-Glc:glycoprotein glucosyltransferase recognizes structured and solvent accessible hydrophobic patches in molten globule-like folding intermediates. Proc Natl Acad Sci USA *100*, 86–91.

References

Caramelo, J.J., Castro, O.A., de Prat-Gay, G., and Parodi, A.J. (2004). The endoplasmic reticulum glucosyltransferase recognizes nearly native glycoprotein folding intermediates. J Biol Chem *279*, 46280–46285.

Caramelo, J.J., and Parodi, A. J. (2007). How sugars convey information on protein conformation in the endoplasmic reticulum. Semin Cell Dev Biol *18*, 732–742.

Chen, M.M., Bartlett, A.I., Nerenberg, P.S., Friel, C.T., Hackenberger, C.P., Stultz, C.M., et al. (2010). Perturbing the folding energy landscape of the bacterial immunity protein Im7 by site-specific *N*-linked glycosylation. Proc Natl Acad Sci U S A *107*, 22528–22533.

Chouquet, A., Paidassi, H., Ling, W.L., Frachet, P., Houen, G., Arlaud, G.J., et al. (2011). X-ray structure of the human calreticulin globular domain reveals a peptide-binding area and suggests a multi-molecular mechanism. PLoS One *6*, e17886.

Coe, H., and Michalak, M. (2010). ERp57, a multifunctional endoplasmic reticulum resident oxidoreductase. Int J Biochem Cell Biol *42*, 796–799.

Conte, I., Labriola, C., Cazzulo, J.J., Docampo, R., and Parodi, A.J. (2003). The interplay between folding-facilitating mechanisms in *Trypanosoma cruzi* endoplasmic reticulum. Mol Biol Cell *14*, 3529–3540.

Conte, I.L., Keith, N., Gutierrez-Gonzalez, C., Parodi, A.J., and Caramelo, J.J. (2007). The interplay between calcium and the *in vitro* lectin and chaperone activities of calreticulin. Biochemistry *46*, 4671–4680.

Conway, E.M., Liu, L., Nowakowski, B., Steiner-Mosonyi, M., Ribeiro, S.P., and Michalak, M. (1995). Heat shock-sensitive expression of calreticulin. *In vitro* and *in vivo* up-regulation. J Biol Chem *270*, 17011–17016.

D'Alessio, C., Caramelo, J.J., and Parodi, A.J. (2010). UDP-Glc:glycoprotein glucosyltransferase-glucosidase II, the ying-yang of the ER quality control. Semin Cell Dev Biol. *21*, 491–499.

Daniel, P.F. (1987). Separation of benzoylated oligosaccharides by reversed-phase high-pressure liquid chromatography: application to high-mannose type oligosaccharides. Methods Enzymol *138*, 94–116.

Daniels, R., Kurowski, B., Johnson, A.E., and Hebert, D.N. (2003). *N*-linked glycans direct the cotranslational folding pathway of influenza hemagglutinin. Mol Cell *11*, 79–90.

Danilczyk, U.G., Cohen-Doyle, M.F., and Williams, D.B. (2000). Functional relationship between calreticulin, calnexin, and the endoplasmic reticulum luminal domain of calnexin. J Biol Chem *275*, 13089–13097.

Danilczyk, U.G., and Williams, D.B. (2001). The lectin chaperone calnexin utilizes polypeptide-based interactions to associate with many of its substrates *in vivo*. J Biol Chem *276*, 25532–25540.

Denzel, A., Molinari, M., Trigueros, C., Martin, J.E., Velmurgan, S., Brown, S., et al. (2002). Early postnatal death and motor disorders in mice congenitally deficient in calnexin expression. Mol Cell Biol *22*, 7398–7404.

Deprez, P., Gautschi, M., and Helenius, A. (2005). More than one glycan is needed for ER glucosidase II to allow entry of glycoproteins into the calnexin/calreticulin cycle. Mol Cell *19*, 183–195.

Di Jeso, B., Park, Y.N., Ulianich, L., Treglia, A.S., Urbanas, M.L., High, S., et al. (2005). Mixed-disulfide folding intermediates between thyroglobulin and endoplasmic reticulum resident oxidoreductases ERp57 and protein disulfide isomerase. Mol Cell Biol *25*, 9793–9805.

Di Jeso, B., Ulianich, L., Pacifico, F., Leonardi, A., Vito, P., Consiglio, E., et al. (2003). Folding of thyroglobulin in the calnexin/calreticulin pathway and its alteration by loss of Ca2+ from the endoplasmic reticulum. Biochem J *370*, 449–458.

Ellgaard, L., Riek, R., Herrmann, T., Guntert, P., Braun, D., Helenius, A., et al. (2001). NMR structure of the calreticulin P-domain. Proc Natl Acad Sci U S A *98*, 3133–3138.

Fanchiotti, S., Fernandez, F., D'Alessio, C., and Parodi, A.J. (1998). The UDP-Glc:glycoprotein glucosyltransferase is essential for *Schizosaccharomyces pombe* viability under conditions of extreme endoplasmic reticulum stress. J Cell Biol *143*, 625–635.

Fernandez, F., Jannatipou, M., Hellman, U., Rokeach, L.A., and Parodi, A.J. (1996). A new stress protein: synthesis of *Schizosaccharomyces pombe* UDP-Glc:glycoprotein glucosyltransferase mRNA is induced by stress conditions but the enzyme is not essential for cell viability. EMBO J *15*, 705–713.

Fernandez, F.S., Trombetta, S.E., Hellman, U., and Parodi, A.J. (1994). Purification to homogeneity of UDP-glucose:glycoprotein glucosyltransferase from *Schizosaccharomyces pombe* and apparent absence of the enzyme fro *Saccharomyces cerevisiae*. J Biol Chem *269*, 30701–30706.

Fontanini, A., Chies, R., Snapp, E.L., Ferrarini, M., Fabrizi, G.M., and Brancolini, C. (2005). Glycan-independent role of calnexin in the intracellular retention of Charcot-Marie-tooth 1A Gas3/PMP22 mutants. J Biol Chem *280*, 2378–2387.

Frenkel, Z., Gregory, W., Kornfeld, S., and Lederkremer, G.Z. (2003). Endoplasmic reticulum-associated degradation of mammalian glycoproteins involves sugar chain trimming to Man$_{6-5}$GlcNAc$_2$. J Biol Chem *278*, 34119–34124.

Frickel, E.M., Riek, R., Jelesarov, I., Helenius, A., Wuthrich, K., and Ellgaard, L. (2002). TROSY-NMR reveals interaction between ERp57 and the tip of the calreticulin P-domain. Proc Natl Acad Sci U S A *99*, 1954–1959.

Gao, B., Adhikari, R., Howarth, M., Nakamura, K., Gold, M.C., Hill, A.B., et al. (2002). Assembly and antigen-presenting function of MHC class I molecules in cells lacking the ER chaperone calreticulin. Immunity *16*, 99–109.

Gardner, T.G., and Kearse, K.P. (1999). Modification of the T cell antigen receptor (TCR) complex by UDP-glucose:glycoprotein glucosyltransferase. TCR folding is finalized convergent with formation of alpha beta delta epsilon gamma epsilon complexes. J Biol Chem *274*, 14094–14099.

Grinna, L.S., and Robbins, P.W. (1980). Substrate specificities of rat liver microsomal glucosidases which process glycoproteins. J Biol Chem *255*, 2255–2258.

Guerin, M., and Parodi, A.J. (2003). The UDP-glucose:glycoprotein glucosyltransferase is organized in at least two tightly bound domains from yeast to mammals. J Biol Chem *278*, 20540–20546.

Guile, G.R., Rudd, P.M., Wing, D.R., Prime, S.B., and Dwek, R.A. (1996). A rapid high-resolution high-performance liquid chromatographic method for separating glycan mixtures and analyzing oligosaccharide profiles. Anal Biochem *240*, 210–226.

Hagihara, S., Totani, K., and Ito, Y. (2006). Exploration of oligosaccharide-protein interactions in glycoprotein quality control by synthetic approaches. Chem Rec *6*, 290–302.

Hammond, C., Braakman, I., and Helenius, A. (1994). Role of N-linked oligosaccharide recognition, glucose trimming, and calnexin in glycoprotein folding and quality control. Proc Natl Acad Sci USA *91*, 913–917.

Hase, S. (1994). High-performance liquid chromatography of pyridylaminated saccharides. Methods Enzymol *230*, 225–237.

Heal, R., and McGivan, J. (1998). Induction of calreticulin expression in response to amino acid deprivation in Chinese hamster ovary cells. Biochem J *329*, 389–394.

Hebert, D.N., Foellmer, B., and Helenius, A. (1995). Glucose trimming and reglucosylation determine glycoprotein association with calnexin in the endoplasmic reticulum. Cell *81*, 425–433.

Hebert, D.N., Zhang, J.X., Chen, W., Foellmer, B., and Helenius, A. (1997). The number and location of glycans on influenza hemagglutinin determine folding and association with calnexin and calreticulin. J Cell Biol *139*, 613–623.

Hosokawa, N., Kamiya, Y., Kamiya, D., Kato, K., and Nagata, K. (2009). Human OS-9, a lectin required for glycoprotein endoplasmic reticulum-associated degradation, recognizes mannose-trimmed N-glycans. J Biol Chem 284, 17061–17068.

Ihara, Y., Cohen-Doyle, M. F., Saito, Y., and Williams, D.B. (1999). Calnexin discriminates between protein conformational states and functions as a molecular chaperone *in vitro*. Mol Cell 4, 331–341.

Jin, H., Yan, Z., Nam, K.H., and Li, J. (2007). Allele-specific suppression of a defective brassinosteroid receptor reveals a physiological role of UGGT in ER quality control. Mol Cell 26, 821–830.

Kang, S.J., and Cresswell, P. (2002). Calnexin, calreticulin, and ERp57 cooperate in disulfide bond formation in human CD1d heavy chain. J Biol Chem 277, 44838–44844.

Kapoor, M., Ellgaard, L., Gopalakrishnapai, J., Schirra, C., Gemma, E., Oscarson, S., et al. (2004). Mutational analysis provides molecular insight into the carbohydrate-binding region of calreticulin: pivotal roles of tyrosine-109 and aspartate-135 in carbohydrate recognition. Biochemistry 43, 97–106.

Kapoor, M., Srinivas, H., Kandiah, E., Gemma, E., Ellgaard, L., Oscarson, S., et al. (2003). Interactions of substrate with calreticulin, an endoplasmic reticulum chaperone. J Biol Chem 278, 6194–6200.

Keith, N., Parodi, A.J., and Caramelo, J.J. (2005). Glycoprotein tertiary and quaternary structures are monitored by the same quality control mechanism. J Biol Chem 280, 18138–18141.

Keller, S.H., Lindstrom, J., and Taylor, P. (1998). Inhibition of glucose trimming with castanospermine reduces calnexin association and promotes proteasome degradation of the alpha-subunit of the nicotinic acetylcholine receptor. J Biol Chem 273, 17064–17072.

Kornfeld, S., Li, E., and Tabas, I. (1978). The synthesis of complex-type oligosaccharides. II. Characterization of the processing intermediates in the synthesis of the complex oligosaccharide units of the vesicular stomatitis virus G protein. J Biol Chem 253, 7771–7778.

Kozlov, G., Maattanen, P., Schrag, J.D., Pollock, S., Cygler, M., Nagar, B., et al. (2006). Crystal structure of the bb' domains of the protein disulfide isomerase ERp57. Structure 14, 1331–1339.

Kozlov, G., Pocanschi, C.L., Rosenauer, A., Bastos-Aristizabal, S., Gorelik, A., Williams, D.B., et al. (2010). Structural basis of carbohydrate recognition by calreticulin. J Biol Chem 285, 38612–38620.

Labriola, C., Cazzulo, J.J., and Parodi, A.J. (1999). *Trypanosoma cruzi* calreticulin is a lectin that binds monoglucosylated oligosaccharides but not protein moieties of glycoproteins. Mol Biol Cell 10, 1381–1394.

Labriola, C.A., Conte, I.L., Lopez Medus, M., Parodi, A.J., and Caramelo, J.J. (2010). Endoplasmic reticulum calcium regulates the retrotranslocation of *Trypanosoma cruzi* calreticulin to the cytosol. PLoS One 5, e13141.

Labriola, C.A., Giraldo, A.M., Parodi, A.J., and Caramelo, J.J. (2011). Functional cooperation between BiP and calreticulin in the folding maturation of a glycoprotein in *Trypanosoma cruzi*. Mol Biochem Parasitol 175, 112–117.

Leach, M.R., Cohen-Doyle, M.F., Thomas, D.Y., and Williams, D.B. (2002). Localization of the lectin, ERp57 binding, and polypeptide binding sites of calnexin and calreticulin. J Biol Chem 277, 29686–29697.

Leach, M.R., and Williams, D.B. (2004). Lectin-deficient calnexin is capable of binding class I histocompatibility molecules *in vivo* and preventing their degradation. J Biol Chem 279, 9072–9079.

Li, E., Tabas, I., and Kornfeld, S. (1978). The synthesis of complex-type oligosaccharides. I. Structure of the lipid-linked oligosaccharide precursor of the complex-type oligosaccharides of the vesicular stomatitis virus G protein. J Biol Chem 253, 7762–7770.

Llewellyn, D.H., Kendall, J.M., Sheikh, F.N., and Campbell, A.K. (1996). Induction of calreticulin expression in HeLa cells by depletion of the endoplasmic reticulum Ca^{2+} store and inhibition of N-linked glycosylation. Biochem J *318*, 555–560.

Mayer, M., Frey, S., Koivunen, P., Myllyharju, J., and Buchner, J. (2004). Influence of the oxidoreductase ER57 on the folding of an antibody fab fragment. J Mol Biol *341*, 1077–1084.

McCauliffe, D.P., Yang, Y.S., Wilson, J., Sontheimer, R.D., and Capra, J.D. (1992). The 5'-flanking region of the human calreticulin gene shares homology with the human GRP78, GRP94, and protein disulfide isomerase promoters. J Biol Chem *267*, 2557–2562.

McCormick, L.M., Urade, R., Arakaki, Y., Schwartz, A.L., and Bu, G. (2005). Independent and cooperative roles of N-glycans and molecular chaperones in the folding and disulfide bond formation of the low-density lipoprotein (LDL) receptor-related protein. Biochemistry *44*, 5794–5803.

Mellis, S.J., and Baenziger, J.U. (1981). Separation of neutral oligosaccharides by high-performance liquid chromatography. Anal Biochem *114*, 276–280.

Mesaeli, N., Nakamura, K., Zvaritch, E., Dickie, P., Dziak, E., Krause, K.H., et al. (1999). Calreticulin is essential for cardiac development. J Cell Biol *144*, 857–868.

Michalak, M., Corbett, E.F., Mesaeli, N., Nakamura, K., and Opas, M. (1999). Calreticulin: one protein, one gene, many functions. Biochem J *344*, 281–292.

Mizrachi, D., and Segaloff, D.L. (2004). Intracellularly located misfolded glycoprotein hormone receptors associate with different chaperone proteins than their cognate wild-type receptors. Mol Endocrinol *18*, 1768–1777.

Moore, S.E., and Spiro, R.G. (1990). Demonstration that Golgi endo-alpha-D-mannosidase provides a glucosidase-independent pathway for the formation of complex N-linked oligosaccharides of glycoproteins. J Biol Chem *265*, 13104–13112.

Munro, S. (2001). The MRH domain suggests a shared ancestry for the mannose 6-phosphate receptors and other N-glycan-recognising proteins. Curr Biol *11*, R499–501.

Nakhasi, H.L., Ramanujam, M., Atreya, C.D., Hobman, T.C., Lee, N., Esmaili, A., et al. (2001). Rubella virus glycoprotein interaction with the endoplasmic reticulum calreticulin and calnexin. Arch Virol *146*, 1–14.

Nguyen, T.O., Capra, J.D., and Sontheimer, R.D. (1996). Calreticulin is transcriptionally up-regulated by heat shock, calcium and heavy metals. Mol Immunol *33*, 379–386.

Oda, Y., Hosokawa, N., Wada, I., and Nagata, K. (2003). EDEM as an acceptor of terminally misfolded glycoproteins released from calnexin. Science *299*, 1394–1397.

Oliver, J.D., Roderick, H.L., Llewellyn, D.H., and High, S. (1999). ERp57 functions as a subunit of specific complexes formed with the ER lectins calreticulin and calnexin. Mol Biol Cell *10*, 2573–2582.

Ora, A., and Helenius, A. (1995). Calnexin fails to associate with substrate proteins in glucosidase-deficient cell lines. J Biol Chem *270*, 26060–26062.

Panneerselvam, K., Etchison, J.R., and Freeze, H.H. (1997). Human fibroblasts prefer mannose over glucose as a source of mannose for N-glycosylation. Evidence for the functional importance of transported mannose. J Biol Chem *272*, 23123–23129.

Panneerselvam, K., and Freeze, H.H. (1996). Mannose enters mammalian cells using a specific transporter that is insensitive to glucose. J Biol Chem *271*, 9417–9421.

Parodi, A.J. (1993). N-glycosylation in trypanosomatid protozoa. Glycobiology *3*, 193–9.

Patil, A.R., Thomas, C.J., and Surolia, A. (2000). Kinetics and the mechanism of interaction of the endoplasmic reticulum chaperone, calreticulin, with monoglucosylated (Glc$_1$Man$_9$GlcNAc$_2$) substrate. J Biol Chem *275*, 24348–24356.

Peaper, D.R., Wearsch, P.A., and Cresswell, P. (2005). Tapasin and ERp57 form a stable disulfide-linked dimer within the MHC class I peptide-loading complex. EMBO J *24*, 3613–3623.

Pearse, B.R., Tamura, T., Sunryd, J.C., Grabowski, G.A., Kaufman, R.J., and Hebert, D.N. (2010). The role of UDP-Glc:glycoprotein glucosyltransferase 1 in the maturation of an obligate substrate prosaposin. J Cell Biol 189, 829–841.

Petrescu, A.J., Butters, T.D., Reinkensmeier, G., Petrescu, S., Platt, F.M., Dwek, R.A., et al. (1997). The solution NMR structure of glucosylated N-glycans involved in the early stages of glycoprotein biosynthesis and folding. EMBO J 16, 4302–4310.

Pieren, M., Galli, C., Denzel, A., and Molinari, M. (2005). The use of calnexin and calreticulin by cellular and viral glycoproteins. J Biol Chem 280, 28265–28271.

Pollock, S., Kozlov, G., Pelletier, M.F., Trempe, J.F., Jansen, G., Sitnikov, D., et al. (2004). Specific interaction of ERp57 and calnexin determined by NMR spectroscopy and an ER two-hybrid system. EMBO J 23, 1020–1029.

Quan, E.M., Kamiya, Y., Kamiya, D., Denic, V., Weibezahn, J., Kato, K., et al. (2008). Defining the glycan destruction signal for endoplasmic reticulum-associated degradation. Mol Cell 32, 870–877.

Radcliffe, C.M., Diedrich, G., Harvey, D.J., Dwek, R.A., Cresswell, P., and Rudd, P.M. (2002). Identification of specific glycoforms of major histocompatibility complex class I heavy chains suggests that class I peptide loading is an adaptation of the quality control pathway involving calreticulin and ERp57. J Biol Chem 277, 46415–46423.

Ritter, C., and Helenius, A. (2000). Recognition of local glycoprotein misfolding by the ER folding sensor UDP-glucose:glycoprotein glucosyltransferase. Nat Struct Biol 7, 278–280.

Ritter, C., Quirin, K., Kowarik, M., and Helenius, A. (2005). Minor folding defects trigger local modification of glycoproteins by the ER folding sensor GT. EMBO J 24, 1730–1738.

Rizvi, S.M., Mancino, L., Thammavongsa, V., Cantley, R.L., and Raghavan, M. (2004). A polypeptide binding conformation of calreticulin is induced by heat shock, calcium depletion, or by deletion of the C-terminal acidic region. Mol Cell 15, 913–923.

Rodan, A.R., Simons, J.F., Trombetta, E.S., and Helenius, A. (1996). N-linked oligosaccharides are necessary and sufficient for association of glycosylated forms of bovine RNase with calnexin and calreticulin. EMBO J 15, 6921–6930.

Ruiz-Canada, C., Kelleher, D.J., and Gilmore, R. (2009). Cotranslational and posttranslational N-glycosylation of polypeptides by distinct mammalian OST isoforms. Cell 136, 272–283.

Russell, S.J., Ruddock, L.W., Salo, K.E., Oliver, J.D., Roebuck, Q.P., Llewellyn, D.H., et al. (2004). The primary substrate binding site in the b' domain of ERp57 is adapted for endoplasmic reticulum lectin association. J Biol Chem 279, 18861–18869.

Saito, Y., Ihara, Y., Leach, M.R., Cohen-Doyle, M.F., and Williams, D.B. (1999). Calreticulin functions *in vitro* as a molecular chaperone for both glycosylated and non-glycosylated proteins. EMBO J 18, 6718–6729.

Schrag, J.D., Bergeron, J.J., Li, Y., Borisova, S., Hahn, M., Thomas, D.Y., and Cygler, M. (2001). The structure of calnexin, an ER chaperone involved in quality control of protein folding. Mol Cell 8, 633–644.

Silvennoinen, L., Myllyharju, J., Ruoppolo, M., Orru, S., Caterino, M., Kivirikko, K.I., et al. (2004). Identification and characterization of structural domains of human ERp57: association with calreticulin requires several domains. J Biol Chem 279, 13607–13615.

Sonnichsen, B., Fullekrug, J., Nguyen Van, P., Diekmann, W., Robinson, D.G., and Mieskes, G. (1994). Retention and retrieval: both mechanisms cooperate to maintain calreticulin in the endoplasmic reticulum. J Cell Sci 107, 2705–2717.

Spiro, R.G., Zhu, Q., Bhoyroo, V., and Soling, H.D. (1996). Definition of the lectin-like properties of the molecular chaperone, calreticulin, and demonstration of its copurification with endomannosidase from rat liver Golgi. J Biol Chem 271, 11588–11594.

Staneloni, R.J., and Leloir, L.F. (1978). Oligosaccharides containing glucose and mannose in glycoproteins of the thyroid gland. Proc Natl Acad Sci U S A 75, 1162–1166.

Stigliano, I.D., Alculumbre, S.G., Labriola, C.A., Parodi, A.J., and D'Alessio, C. (2011). Glucosidase II and N-glycan mannose content regulate the half-lives of monoglucosylated species *in vivo*. Mol Biol Cell [Epub ahead of print].

Stigliano, I.D., Caramelo, J.J., Labriola, C.A., Parodi, A.J., and D'Alessio, C. (2009). Glucosidase II beta subunit modulates N-glycan trimming in fission yeasts and mammals. Mol Biol Cell *20*, 3974–3984.

Stronge, V.S., Saito, Y., Ihara, Y., and Williams, D.B. (2001). Relationship between calnexin and BiP in suppressing aggregation and promoting refolding of protein and glycoprotein substrates. J Biol Chem *276*, 39779–39787.

Suh, K., Bergmann, J.E., and Gabel, C.A. (1989). Selective retention of monoglucosylated high mannose oligosaccharides by a class of mutant vesicular stomatitis virus G proteins. J Cell Biol *108*, 811–819.

Swanton, E., High, S., and Woodman, P. (2003). Role of calnexin in the glycan-independent quality control of proteolipid protein. EMBO J *22*, 2948–2958.

Szewczenko-Pawlikowski, M., Dziak, E., McLaren, M.J., Michalak, M., and Opas, M. (1997). Heat shock-regulated expression of calreticulin in retinal pigment epithelium. Mol Cell Biochem *177*, 145–152.

Tan, Y., Chen, M., Li, Z., Mabuchi, K., and Bouvier, M. (2006). The calcium- and zinc-responsive regions of calreticulin reside strictly in the N-/C-domain. Biochim Biophys Acta *1760*, 745–753.

Tatu, U., and Helenius, A. (1999). Interaction of newly synthesized apolipoprotein B with calnexin and calreticulin requires glucose trimming in the endoplasmic reticulum. Biosci Rep *19*, 189–196.

Taylor, S.C., Ferguson, A.D., Bergeron, J.J., and Thomas, D.Y. (2004). The ER protein folding sensor UDP-glucose glycoprotein-glucosyltransferase modifies substrates distant to local changes in glycoprotein conformation. Nat Struct Mol Biol *11*, 128–134.

Taylor, S.C., Thibault, P., Tessier, D.C., Bergeron, J.J., and Thomas, D.Y. (2003). Glycopeptide specificity of the secretory protein folding sensor UDP-glucose glycoprotein:glucosyltransferase. EMBO Rep *4*, 405–411.

Tessier, D.C., Dignard, D., Zapun, A., Radominska-Pandya, A., Parodi, A.J., Bergeron, J.J., et al. (2000). Cloning and characterization of mammalian UDP-glucose glycoprotein: glucosyltransferase and the development of a specific substrate for this enzyme. Glycobiology *10*, 403–412.

Thammavongsa, V., Mancino, L., and Raghavan, M. (2005). Polypeptide substrate recognition by calnexin requires specific conformations of the calnexin protein. J Biol Chem *280*, 33497–33505.

Thomson, S.P., and Williams, D.B. (2005). Delineation of the lectin site of the molecular chaperone calreticulin. Cell Stress Chaperones *10*, 242–251.

Trombetta, E.S., Simons, J.F., and Helenius, A. (1996). Endoplasmic reticulum glucosidase II is composed of a catalytic subunit, conserved from yeast to mammals, and a tightly bound noncatalytic HDEL-containing subunit. J Biol Chem *271*, 27509–27516.

Trombetta, S.E., Bosch, M., and Parodi, A.J. (1989). Glucosylation of glycoproteins by mammalian, plant, fungal, and trypanosomatid protozoa microsomal membranes. Biochemistry *28*, 8108–8116.

Trombetta, S.E., Ganan, S.A., and Parodi, A.J. (1991). The UDP-Glc:glycoprotein glucosyltransferase is a soluble protein of the endoplasmic reticulum. Glycobiology *1*, 155–161.

Trombetta, S.E., and Parodi, A.J. (1992). Purification to apparent homogeneity and partial characterization of rat liver UDP-glucose:glycoprotein glucosyltransferase. J Biol Chem *267*, 9236–9240.

Ugalde, R.A., Staneloni, R.J., and Leloir, L.F. (1979). Microsomal glucosidases acting on the saccharide moiety of the glucose-containing dolichyl diphosphate oligosaccharide. Biochem Biophys Res Commun *91*, 1174–1181.

Van der Wal, F.J., Oliver, J.D., and High, S. (1998). The transient association of ERp57 with N-glycosylated proteins is regulated by glucose trimming. Eur J Biochem 256, 51–59.

Varki, A. (1991). Radioactive tracer techniques in the sequencing of glycoprotein oligosaccharides. FASEB J 5, 226–235.

Vassilakos, A., Michalak, M., Lehrman, M.A., and Williams, D.B. (1998). Oligosaccharide binding characteristics of the molecular chaperones calnexin and calreticulin. Biochemistry 37, 3480–3490.

Vembar, S.S., and Brodsky, J.L. (2008). One step at a time: endoplasmic reticulum-associated degradation. Nat Rev Mol Cell Biol 9, 944–957.

Villamil Giraldo, A.M., Lopez Medus, M., Gonzalez Lebrero, M., Pagano, R.S., Labriola, C.A., Landolfo, L., et al. (2010). The structure of calreticulin C-terminal domain is modulated by physiological variations of calcium concentration. J Biol Chem 285, 4544–4553.

Wada, I., Rindress, D., Cameron, P.H., Ou, W.J., Doherty, J.J., II, Louvard, D., et al. (1991). SSR alpha and associated calnexin are major calcium binding proteins of the endoplasmic reticulum membrane. J Biol Chem 266, 19599–19610.

Waisman, D.M., Salimath, B.P., and Anderson, M.J. (1985). Isolation and characterization of CAB-63, a novel calcium-binding protein. J Biol Chem 260, 1652–1660.

Wanamaker, C.P., and Green, W.N. (2005). N-linked glycosylation is required for nicotinic receptor assembly but not for subunit associations with calnexin. J Biol Chem 280, 33800–33810.

Wang, N., Daniels, R., and Hebert, D.N. (2005). The cotranslational maturation of the type I membrane glycoprotein tyrosinase: the heat shock protein 70 system hands off to the lectin-based chaperone system. Mol Biol Cell 16, 3740–3752.

Ware, F.E., Vassilakos, A., Peterson, P.A., Jackson, M.R., Lehrman, M.A., and Williams, D.B. (1995). The molecular chaperone calnexin binds Glc1Man9GlcNAc2 oligosaccharide as an initial step in recognizing unfolded glycoproteins. J Biol Chem 270, 4697–4704.

Wearsch, P.A., Jakob, C.A., Vallin, A., Dwek, R.A., Rudd, P.M., and Cresswell, P. (2004). Major histocompatibility complex class I molecules expressed with monoglucosylated N-linked glycans bind calreticulin independently of their assembly status. J Biol Chem 279, 25112–25121.

Williams, D.B. (2006). Beyond lectins: the calnexin/calreticulin chaperone system of the endoplasmic reticulum. J Cell Sci 119, 615–623.

Woods, R.J., Pathiaseril, A., Wormald, M.R., Edge, C.J., and Dwek, R.A. (1998). The high degree of internal flexibility observed for an oligomannose oligosaccharide does not alter the overall topology of the molecule. Eur J Biochem 258, 372–386.

Xu, X., Azakami, H., and Kato, A. (2004). P-domain and lectin site are involved in the chaperone function of *Saccharomyces cerevisiae* calnexin homologue. FEBS Lett 570, 155–160.

Zapun, A., Darby, N.J., Tessier, D.C., Michalak, M., Bergeron, J.J., and Thomas, D.Y. (1998). Enhanced catalysis of ribonuclease B folding by the interaction of calnexin or calreticulin with ERp57. J Biol Chem 273, 6009–6012.

Zhang, L., Wu, G., Tate, C.G., Lookene, A., and Olivecrona, G. (2003). Calreticulin promotes folding/dimerization of human lipoprotein lipase expressed in insect cells (sf21). J Biol Chem 278, 29344–29351.

Zhang, Q., Tector, M., and Salter, R.D. (1995). Calnexin recognizes carbohydrate and protein determinants of class I major histocompatibility complex molecules. J Biol Chem 270, 3944–3948.

Zhang, W., Wearsch, P.A., Zhu, Y., Leonhardt, R.M., and Cresswell, P. (2011). A role for UDP-glucose glycoprotein glucosyltransferase in expression and quality control of MHC class I molecules. Proc Natl Acad Sci U S A 108, 4956–4961.

Zuber, C., Fan, J.Y., Guhl, B., Parodi, A., Fessler, J.H., Parker, C., et al. (2001). Immunolocalization of UDP-glucose:glycoprotein glucosyltransferase indicates involvement of pre-Golgi intermediates in protein quality control. Proc Natl Acad Sci U S A 98, 10710–10715.

5 Conformational dynamics in peptides and proteins studied by triplet-triplet energy transfer

Andreas Reiner, Beat Fierz, and Thomas Kiefhaber

5.1 Introduction

Conformational dynamics play an essential role in folding and function of proteins and peptides. They represent the underlying motions in the transitions between different protein conformations, and they determine the rate at which a folding polypeptide chain can explore conformational space. Conformational search within a polypeptide chain is limited by intrachain diffusion processes (i.e. by the rate at which two points along the chain can form an interaction). The knowledge of the rates of intrachain contact formation in polypeptide chains and their dependence on amino acid sequence, chain length and local structure formation is therefore essential for an understanding of the dynamics of protein folding and for the characterization of the free energy barriers of the folding reactions. In addition, intrachain diffusion provides an upper limit for the speed at which a protein can reach its native state, just like free diffusion provides an upper limit for the rate constant of bimolecular reactions. Conformational dynamics in native proteins are essential for protein function such as catalysis, binding and signaling. A wide variety of methods exist to study protein motions. Local motions on the picosecond and on the microsecond to millisecond timescales have been extensively characterized by NMR spectroscopy (Arrington and Robertson 1997; Palmer 2004). Single-molecule spectroscopy can measure structural transitions slower than milliseconds, although fluorescence correlation spectroscopy has recently been extended to study submicrosecond reactions (Felekyan et al. 2005). Functionally important protein motions on the nanosecond to microsecond timescale are only poorly understood, due to the limitations in experimental methods. Recently, several experimental techniques using long-lived triplet states have been introduced, which directly monitor contact formation in model peptides and in protein fragments. In the following, we will present the method of triplet-triplet energy transfer (TTET), which can be applied to measure protein dynamics on a timescale of picoseconds to hundreds of microseconds. We will briefly introduce the photochemistry of the processes involved, and we will discuss the experimental requirements for direct, model-free measurements of chain dynamics. In the second part of this chapter, examples for the application of TTET to study protein and peptide dynamics will be discussed.

5.2 Concept of TTET experiments to study intrachain loop formation in polypeptide chains

▶ Figure 5.1 illustrates the concept of TTET experiments applied to measurements of protein dynamics. A triplet donor and a triplet acceptor group are introduced at specific positions in a polypeptide chain. The donor is selectively excited by a laser flash and undergoes fast intersystem crossing into the triplet state. Upon intrachain diffusion, a loop forms, which allows the triplet donor and triplet acceptor groups to meet and to transfer the triplet state to the acceptor. Since TTET is a two-electron exchange process (Dexter mechanism), it usually requires van der Waals contact to allow for electron transfer (Klessinger and Michl 1995). This should be contrasted to FRET, which occurs through dipole-dipole interactions resulting in a $1/r^6$ distance dependence and thus allows energy transfer over larger distances. A prerequisite for efficient and rapid triplet transfer is that the acceptor group has a significant lower triplet energy than the donor group.

A well-suited pair for TTET measurements is xanthone derivatives as a triplet donor combined with naphthalene as the triplet acceptor group (▶Figure 5.2A). The triplet states of these molecules have specific absorbance bands, which can be used to monitor the decay of the donor triplet states at 590 nm and the concomitant increase of acceptor triplet states at 420 nm (▶Figure 5.2B). The time constant for formation of xanthone triplet states is around 2 ps (Satzger et al. 2004). TTET between xanthone and naphthalene occurs with a time constant of about 2 ps and has a bimolecular transfer rate constant of $4 \cdot 10^9$ M^{-1} s^{-1} (Krieger et al. 2003; Satzger et al. 2004), which is the value expected for a diffusion controlled reaction between small molecules in water. Due to this fast photochemistry and the diffusion-controlled transfer process, TTET between xanthone and naphthalene allows measurements of absolute rate constants for diffusion processes slower than 10–20 ps. The upper limit of the experimental time window accessible by TTET is set by the intrinsic lifetime of the donor, which is around 30 μs for the xanthone triplet state in water. Xanthone has a high quantum yield for intersystem crossing ($\varphi_{ISC} = 0.99$, $\varepsilon \approx 4{,}000$ m^{-1} cm^{-1}), and the triplet state has a strong absorbance band with a maximum around 590 nm in water ($\varepsilon^T_{590} \approx 10{,}000$ M^{-1} cm^{-1}). This allows single pulse measurements at relatively low peptide concentrations (10–50 μM). The low concentrations applied in the experiments also rule out contributions from intermolecular transfer reactions, which would have half-times higher than 50 μs in this concentration range (Bieri et al. 1999; Krieger et al. 2003). Contributions from through-bond transfer

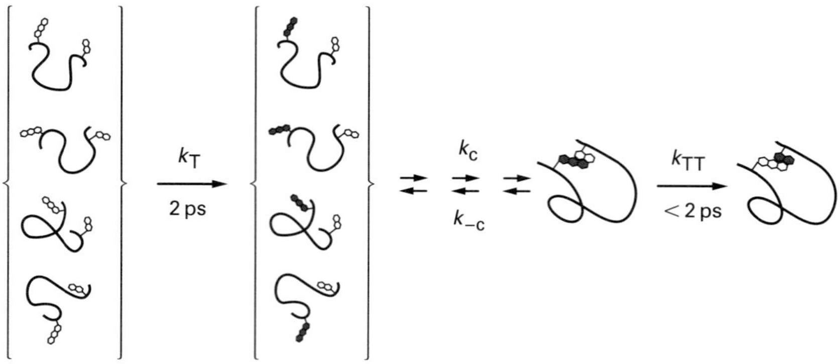

Figure 5.1: Schematic representation of the TTET experiments. In unstructured peptides, the experiments allow determination of the absolute rate constant for intrachain contact formation, k_c, if formation of the triplet state, k_T, and the transfer process, k_{TT}, are much faster than the chain dynamics ($k_{TT}, k_T \gg k_c, k_{-c}$). For the rate constants of k_T and k_{TT}, see Satzger et al. (2004).

5.2.1 Kinetics of TTET coupled to polypeptide chain dynamics

The coupling of electron transfer to chain dynamics displayed in ▶Figure 5.1 can be kinetically described by a three-state reaction if the excitation of the donor is fast compared to chain dynamics (k_c, k_{-c}) and electron transfer (k_{TT}).

$$O \underset{k_{-c}}{\overset{k_c}{\rightleftarrows}} C \overset{k_{TT}}{\longrightarrow} C^* \tag{1}$$

O and C (Equation 1) represent chain conformations with the loop opened (no contact between the labels) or closed (contact between the labels), respectively. k_c and k_{-c} are the rate constants for loop formation and breakage, respectively. If the excited state quenching or transfer (k_{TT}) reaction is on the same timescale or slower than breaking of the contact (k_{-c}), the measured rate constants (λ_i) are functions of all microscopic rate constants.

$$\lambda_{1,2} = \frac{B \pm \sqrt{B^2 - 4C}}{2}$$
$$B = k_c + k_{-c} + k_{TT}$$
$$C = k_c \cdot k_{TT} \tag{2}$$

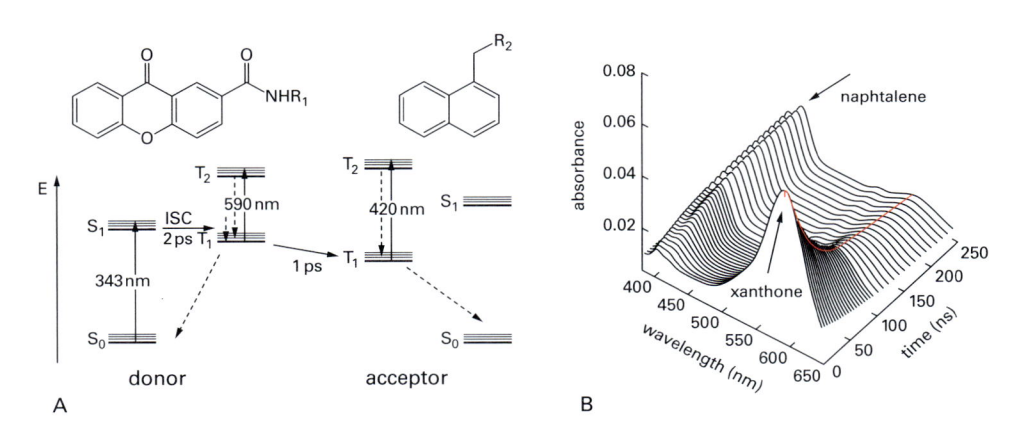

Figure 5.2: (A) Jabłoński diagram for TTET from xanthone (Xan; donor) to naphthalene (Nal; acceptor). Xan is excited to its singlet state S_1 from which it undergoes fast and efficient intersystem crossing to its triplet state T_1 (actually two states in Xan, $^3n\pi^*$ and $^3\pi\pi^*$) (Satzger et al. 2004). Van der Waals contact with Nal leads to a fast and irreversible transfer of the triplet state. Dashed arrows indicate slow internal conversion. (B) The TTET process can be monitored by a decrease in the triplet absorbance bands of Xan at 590 nm and a corresponding increase in the naphthalene triplet absorbance band at 420 nm arising from T_1–T_2 transitions. The TTET kinetics were recorded for a Xan-(Gly-Ser)$_{14}$-Nal-Ser-Gly peptide after excitation with a 4 ns laser flash at 355 nm.

If the population of contact conformations is low ($k_{-c} \gg k_c$), the kinetics will be single exponential and the observed rate constant corresponds to the smaller eigenvalue (λ_1) in Equation 2. Since this is in agreement with experimental results, we will only consider this scenario in the following. We can further simplify Equation 2 if the timescales of electron transfer and chain dynamics are well separated. In the regime of fast electron transfer compared to chain dynamics ($k_{TT} \gg k_c, k_{-c}$), Equation 2 can be approximated by

$$\lambda_1 = k_c \tag{3}$$

This is the desired case, since the observed kinetics directly reflect the dynamics of contact formation. In the regime of fast formation and breakage of the contact compared to electron transfer ($k_{TT} \ll k_c, k_{-c}$), Equation 2 can be approximated by

$$\lambda_1 = \frac{k_c}{k_c + k_{-c}} \cdot k_{TT} \cong \frac{k_c}{k_{-c}} \cdot k_{TT} = K_c \cdot k_{TT} \tag{4}$$

where K_c reflects the ratio of loop conformations (C) over open-chain conformations (O). In this limit, the chain dynamics cannot be measured, but the fraction of closed conformations can be determined. It should be kept in mind that these simplifications only hold if the formation of excited states is fast compared to the subsequent reactions. If the excitation is on the nanosecond timescale or slower, the solutions of a linear four-state model have to be used to analyze the kinetics (Szabo 1969; Kiefhaber et al. 1992).

5.2.2 Test for suitability of a donor-acceptor pair to directly measure chain dynamics

5.2.2.1 Characterization of the photochemistry

The considerations discussed in section 5.2.1 show that it is crucial to characterize the rate constants for the photochemical processes and to test for diffusion control of the system to be able to interpret the kinetic data. The photochemistry of the donor excitation and of the electron transfer process are usually characterized by investigating the isolated labels. Since the photochemical processes involved in TTET are often fast, the kinetics should be studied on the femto- to the nanosecond timescale to gain complete information on all photochemical processes in the system (Satzger et al. 2004) (cf. ▶Figure 5.2).

5.2.2.2 Test for diffusion-controlled reactions

To test whether each donor-acceptor contact leads to transfer, it should be examined whether the reaction is diffusion controlled. This can be done by studying the bimolecular transfer process from the donor to acceptor groups and determining its rate constants, its temperature dependence and its viscosity dependence.

5.2.2.3 Determination of the bimolecular transfer rate constant

The bimolecular rate constant for energy transfer (k_q) can be measured using the free labels in solution under pseudo first-order conditions. In these experiments, the concentration of the quencher or acceptor [Q] should be at least ten times higher than the concentration of the donor to be in the regime of pseudo first-order kinetics. Since [Q] is approximately constant during the experiment, the apparent first-order rate constant (k) under pseudo first-order conditions is as follows (Moore and Pearson 1981):

$$k = k_q \cdot [Q] \tag{5}$$

Thus, the bimolecular transfer constant (k_q) can be obtained by varying the quencher concentration [Q] and analyzing the data according to the Stern-Volmer equation (Stern and Volmer 1919):

$$k = k_0 + k_q \cdot [Q] \tag{6}$$

Here, k_0 denotes the rate constant for triplet decay of the donor in absence of quencher (acceptor), and k is the rate constant in presence of the acceptor or quencher. For diffusion-controlled bimolecular reactions of small molecules in water, k_q is around $5 \cdot 10^9$ M^{-1} s^{-1} according to the Smoluchowski equation,

$$k_q = \frac{N_L}{1000} \cdot 4\pi r_C (D_A + D_B) \tag{7}$$

where r_C represents the reactive boundary in cm, and D_A and D_B represent the diffusion coefficients of the interacting partners, A and B, in cm^2/s, which are given by the Stokes-Einstein equation:

$$D = \frac{k_B T}{6\pi r \eta} \tag{8}$$

5.2.2.3.1 Viscosity dependence of the transfer reaction

For a diffusion-controlled reaction, the viscosity dependence of k_q is inversely proportional to solvent viscosity ($k_q \sim 1/\eta$) (Equation 8). The viscosity is usually varied by adding cosolvents like ethylene glycol or glycerol. To determine the final macroscopic viscosity, instruments like the falling-ball viscometer or the Ubbelohde viscometer are used. The molecular size of the cosolvent used in a viscosity study is an important factor. The macroscopic viscosity of a glycerol/water mixture and a solution of polyethylene glycol (PEG) might be the same, although the microscopic properties of these solutions are very different (Kleinert et al. 1998). In the latter case, the dissolved reactant molecules do only feel a fraction of the macroscopic viscosity. Thus, viscous cosolvents of low molecular weight should be used. Glycerol/water mixtures have proved to be most useful, because the viscosity can be varied from 1 cP (pure water) over three orders of magnitude by addition of glycerol. It is imperative to control the temperature when working with high concentrations of glycerol because the viscosity is strongly temperature dependent. Note that viscous cosolvents can interfere with the reaction that is studied. In protein-folding studies, the polyols used to vary the viscosity tend to stabilize the native state of proteins (Timasheff 2002) making an analysis of the kinetic data difficult. However, for measurements of chain dynamics in unfolded polypeptide chains, this effect should not interfere with the measurements, as long as the cosolvents do not induce a structure. Specific effects of the polyols on chain conformations/dynamics can be tested by comparing the results from different cosolvents.

5.2.2.3.2 Determination of activation energy

Diffusion-controlled reactions typically have activation energies close to zero. The lack of an activation barrier in a diffusion-controlled quenching reaction can be verified by determining the temperature dependence of k_q. The diffusion coefficients of the reactants are, however, temperature dependent, mainly through the effect of temperature on solvent viscosity according to the Stokes-Einstein equation (Equation 8). After correcting for the change in viscosity of the solvent with temperature, the slope in an Arrhenius plot should not be higher than $k_B T$. As any photophysical reaction takes some time, diffusion control is only possible to a certain maximal concentration of quencher molecules. After that, the photochemistry of the quenching or transfer process becomes rate limiting. This provides an upper limit of the rate constants that can be measured, even if the system is diffusion controlled at lower quencher concentrations. For the xanthone-naphthalene system, TTET has been shown to occur faster than 2 ps, and thus this system is suitable to obtain absolute time constants for all processes slower than 10–20 ps (Satzger et al. 2004).

5.2.2.4 Interference from other amino acids

To test for possible interference from amino acid side chains with the TTET reaction, the effect of various amino acids on the donor triplet lifetimes must be tested. Bimolecular quenching and TTET rate constants for the interaction of various amino acids with xanthone measured under pseudo first-order conditions (Krieger et al. 2004). The thioether group of methionine ($k_q = [2.0 \pm 0.1] \cdot 10^9$ M^{-1} s^{-1}) and the deprotonated imidazole ring of histidine ($k_q = [1.8 \pm 0.1] \cdot 10^9$ M^{-1} s^{-1}) quench xanthone triplets very efficiently with rate constants close to the diffusion limit. The other amino acid side chains quench xanthone triplets either very inefficiently (Cys, His$^+$, N terminus) or not at all (Ala, Arg, Asn, Asp, Gly, Lys, Ser, Phe). The aromatic amino acids tryptophan, tyrosine and phenylalanine are possible acceptors for xanthone triplets in TTET reactions. TTET between xanthone and tryptophan ($k_{TTET} = [3.0 \pm 0.1] \cdot 10^9$ M^{-1} s^{-1}) and tyrosine ($k_{TTET} = [2.5 \pm 0.1] \cdot 10^9$ M^{-1} s^{-1}) occurs in diffusion-controlled reactions with virtually the same bimolecular rate constants as observed for TTET from xanthone to naphthylalanine ($k_{TTET} = [2.8 \pm 0.1] \cdot 10^9$ M^{-1} s^{-1}). However, TTET from xanthone to tryptophan and tyrosine involves complex reactions with at least two observable rate constants. For both amino acids, TTET is accompanied by radical formation (Bent and Hayon 1975b; Bent and Hayano 1975), which explains the complex kinetics and makes them not suitable for use as triplet acceptors in polypeptide chains. These results show that mainly methionine, tryptophan and tyrosine interfere with TTET from xanthone to naphthalene in intrachain diffusion experiments. Histidine-containing sequences can be measured with this donor/acceptor pair if the pH of the solution is below 5.5.

5.2.3 Practical aspects of TTET measurements

5.2.3.1 Instrumentation and data analysis

The instrumentation required to perform experiments on contact formation kinetics consists of a high-energy light source to produce excited states, and of a detection mechanism (▶ Figure 5.3). A pulsed laser is used to produce triplet donor states in TTET or triplet quenching experiments. The excitation pulse has to provide enough energy to excite a major portion of the molecules in the sample in order to generate a large signal. A pulsed Nd:YAG laser with harmonic generators and a pulse width of <5 ns and a pulse energy of ~ 100 mJ is well-suited for most purposes, although faster reactions occur during formation of short loops that require excitation by femtosecond or picosecond laser pulses (Fierz et al. 2007). Transient UV absorption is used to detect the triplet states. A pulsed flash lamp generates enough light for the absorption measurements in nanosecond laser flash setups. The lamp intensity typically remains constant for several hundred microseconds, which is sufficient to record the triplet decays. Transient spectra (▶Figure 5.2B) can be reconstructed from measurements at different wavelengths or by using a multichannel plate.

For data analysis, transient spectra may be analyzed and fitted globally to different mechanisms (see 5.4.1). This allows us to test and account for side reactions such as radical formation, which occurs in some triplet systems and interferes with the actual transfer process (Lapidus et al. 2000). If a system has been characterized by photochemistry, the kinetics are typically measured at the wavelengths of maximum signal change of donor and acceptor groups. In the case of the xanthone-naphthalene pair, kinetics are measured at 590 nm and at 420 nm (cf. ▶Figure 5.2B). Typical peptide concentrations required for Xan measurements are 30–50 μM in a cuvette with 1 cm path length. To test for possible contributions from bimolecular processes, TTET is measured in a mixture of peptides containing either donor or acceptor.

5.3 Diffusion-controlled loop formation in unstructured polypeptide chains

5.3.1 Kinetics of intrachain diffusion

Monitoring loop formation in unfolded polypeptide chains by TTET revealed single exponential kinetics on the nanosecond timescale or slower (▶Figure 5.4A) (Bieri et al. 1999; Krieger et al. 2003). The only exception is short proline containing peptides, where the kinetics of *cis* and *trans* Xaa-Pro peptide bonds could be resolved (Krieger et al. 2003). From the observation of single exponential kinetics, several conclusions

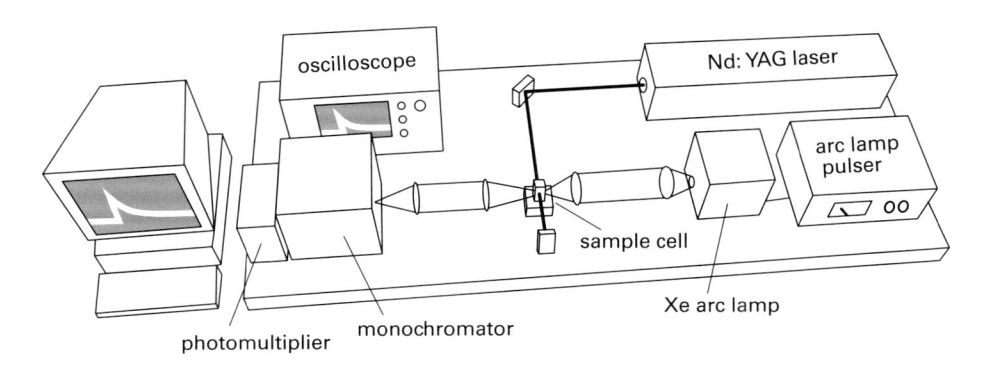

Figure 5.3: Schematic representation of the laser flash setup. The Nd:YAG laser is used for excitation of the triplet state, the pulsed Arc lamp, monochromator, and photomultiplier for measuring the transient triplet absorption bands.

can be drawn. As observed by Szabo et al. (1980) and pointed out by Zwanzig (1997), single exponential kinetics indicate fast interconversion between the different conformations in the ensemble of unfolded states, thus allowing the chain to maintain the equilibrium distribution of the ensemble of conformations that has not made contact. Fast interconversion between individual chain conformations is in agreement with results on conformational relaxation processes in strained peptides, which were shown to occur on the picosecond timescale (Bredenbeck et al. 2003; Spörlein et al. 2002). Secondly, the contact radius is small compared to the chain length, which leads to a small probability of contact formation in agreement with the observation that triplet-triplet transfer can only occur when the two labels are in van der Waals contact (Szabo et al. 1980; Zwanzig 1997).

Since the photophysics of triplet formation and TTET are fast (see 5.2), subnanosecond reactions of intrachain loop formation could be tested by a combination of femto- and nanosecond laser flash experiments (▶Figure 5.5). These experiments revealed additional faster processes on a timescale of 200 ps for formation of short loops (Fierz et al. 2007). The subnanosecond processes are most likely due to conformations that require rotation around only a single or a few bonds to enable contact between the TTET labels.

5.3.2 Effect of loop size on the dynamics in flexible polypeptide chains

The scaling of the end-to-end diffusion with loop size was measured by varying the number of amino acids between xanthone (Xan) and naphthylalanine (Nal) in flexible polypeptide chains with the repetitive sequence Xan-(Gly-Ser)$_x$-Nal-Ser-Gly, with X ranging from 1 to 28. These peptides allowed measurements of contact formation kinetics between two points on the chain separated between 3 and 57 peptide bonds, which covers the range of side chain contacts in small proteins or in protein domains. ▶Figure 5.4A displays three representative TTET kinetics for peptides from this series. As discussed in 5.3.1, single exponential kinetics for contact formation were observed for all peptides. ▶Figure 5.4B shows the effect of

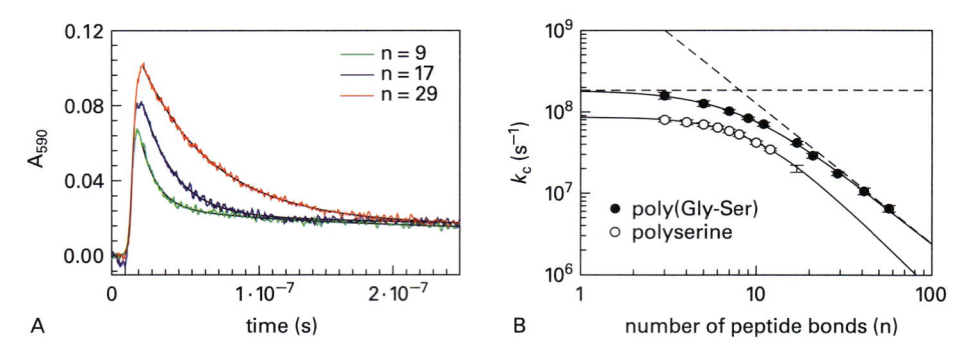

Figure 5.4: (A) Time course of formation and decay of xanthone triplets in peptides of the form Xan-(Gly-Ser)$_n$-Nal-Ser-Gly after a 4 ns laser flash at $t = 0$ measured by the change in absorbance of the xanthone triplets at 590 nm. The lines represent the results from single exponential fits. (B) Comparison of the rate constants (k_c) for end-to-end diffusion in poly(glycine-serine) (filled circles) and polyserine (open circles) peptides. The length dependence was fitted with Equation 9. For details, see Krieger et al. (2003).

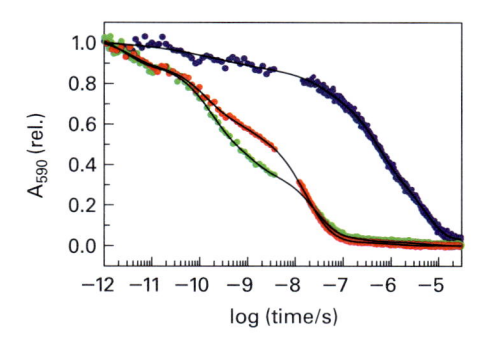

5.3 Diffusion-controlled loop formation in unstructured polypeptide chains

Figure 5.5: Kinetics of loop formation on the timescale from 1 ps to 30 μs obtained from a combination of femto- and nanosecond laser flash TTET experiments. The kinetics of Xan-Ser$_2$-Nal (red), Xan-Ser$_6$-Nal (green) and Xan-Pro$_5$-Nal (blue) are shown. For details, see Fierz et al. (2007).

increasing loop size on the rate constant for contact formation. For long loops ($n > 20$) the rate of contact formation decreases with $n^{-1.7 \pm 0.1}$, where n is the number of peptide bonds between donor and acceptor. This indicates a stronger effect of loop size on the rate of contact formation than expected for purely entropy-controlled intrachain diffusion in ideal freely jointed Gaussian chains, which should scale with $k \sim n^{-1.5}$ (Jacobsen and Stockmayer 1950; Szabo et al. 1980). However, Flory already pointed out that excluded volume effects should significantly influence the chain dimensions (Flory 1953). Accounting for excluded volume effects in the end-to-end diffusion model of Szabo et al. (1980) gives $k \sim n^{-1.8}$, which is nearly identical to the value found for the long poly(glycine-serine) chains (Krieger et al. 2003). This indicates that the dimensions and the dynamics of unfolded polypeptide chains in water are significantly influenced by excluded volume effects, which is in agreement with results on conformational properties of polypeptides derived from simplified Ramachandran maps (Pappu et al. 2000).

The observed simple scaling law breaks down for $n < 20$, and contact formation becomes virtually independent of loop size for very short loops with a limiting value of $k_a = 1.8 \cdot 10^8$ s^{-1}. As discussed in 5.2.1, the limiting rate constant for contact formation in short loops is not due to limits of TTET (k_{TT} in ▶Figure 5.1), since this process is in the time range of 2 ps. Obviously, the intrinsic dynamics of polypeptide chains are limited by different processes for motions over short and over long segments. This is in agreement with the concept of a persistence length in polymer chains and the resulting strong influence of intrinsic chain stiffness on the properties of short chains and on the formation of short loops. This leads to a breakdown of theoretically derived scaling laws for ideal chains (Flory 1969). The complete effect of loop size on intrachain contact formation can be described by a model where length-dependent diffusion processes, which scale with $k = k_b \cdot n^{-m}$, limit the kinetics of formation of long loops. Contact formation reaches a limiting rate constant (k_a) when these diffusion motions become slower than length-independent short range motions, which are probably governed by chain stiffness and steric effects. Accordingly, the effect of loop size on contact formation can be described by

$$k_c = \frac{1}{1/k_a + 1/(k_b \cdot n^{-m})} \qquad (9)$$

The fit of the experimental results to Equation 9 is shown in ▶Figure 5.4B and gives a value of $k_a = (1.8 \pm 0.2) \cdot 10^8$ s^{-1}, $k_b = (6.7 \pm 1.6) \cdot 10^9$ s^{-1} and $m = 1.72 \pm 0.08$.

5.3.3 Effect of amino acid sequence on chain dynamics

Both theoretical considerations (Brant and Flory 1965; Miller et al. 1967; Schimmel and Flory 1967) and experimental results from NMR measurements (Schwalbe et al. 1997) show that polypeptide chains are especially flexible around glycyl residues and especially rigid around prolyl residues. All other amino acids have similar chain stiffness, which is increased compared to glycine, but significantly lower than around prolyl residues. The effect of chain stiffness on peptide dynamics was tested with TTET experiments in polyserine chains (Krieger et al. 2003). ▶Figure 5.4B compares the effect of loop size on intrachain diffusion in a series of Xan-(Ser)$_x$-Nal-Ser-Gly peptides with X = 2–11 ($n = 3$–12) to the behavior of poly(glycine-serine) chains. Single exponential kinetics for contact formation were observed in all polyserine peptides. For short loops ($n < 5$), contact formation is virtually independent of loop size with a limiting value of $k_a = 8.7 \cdot 10^7$ s^{-1} (Equation 9). This indicates that the local dynamics in polyserine are about 2- to 3-fold slower than in the poly(glycine-serine) peptides, which seems to be a small effect compared to the largely different properties expected for the stiffer polyserine chains. The decreased flexibility and the longer donor-acceptor distances in the polyserine chains are probably compensated by the decreased conformational space available for polyserine compared to poly(glycine-serine) peptides (Krieger et al. 2003). For longer polyserine chains, contact formation slows down with increasing loop size. The effect of increasing loop size on the rates of contact formation seems to be slightly larger in polyserine compared to the poly(glycine-serine) series ($m = 2.1 \pm 0.3$; $k_b = (1.0 \pm 0.8) \cdot 10^{10}$ s^{-1}; Equation 9). However, due to limitations in peptide synthesis, it was not possible to obtain longer peptides, which would be required to get a more accurate scaling law for the polyserine peptides.

The kinetics of formation of short loops differ only by a factor of two for polyserine compared to poly(glycine-serine), arguing for only little effect of amino acid sequence on local chain dynamics (▶Figure 5.4B). The effect of other amino acids on local chain dynamics was measured by performing TTET experiments on short host-guest peptides of the canonical sequence Xan-Ser-Xaa-Ser-Nal-Ser-Gly using the guest amino acids Xaa = Gly, Ser, Ala, Ile, His, Glu, Arg and Pro (Krieger et al. 2003). ▶Figure 5.6 shows that the amino acid side chain indeed has only little effect on the rates of contact formation. All amino acids except proline and glycine show very similar dynamics. Interestingly, there is a small but significant difference in rate between short side chains (Ala, Ser) and amino acids with longer side chains (Ile, Glu, Arg, His). Obviously, chains that extend beyond the C$_\beta$ atom slightly decrease the rates of local chain dynamics, whereas charges do not influence the dynamics. Glycyl and prolyl residues show significantly different dynamics compared to all other amino acids, as expected from their largely different conformational properties (Schimmel and Flory 1967). As shown in Figure 5.4B, glycine accelerates contact formation about 2- to 3-fold compared to serine. Proline shows slower and more complex kinetics of contact formation with two relaxation times of $1/k_1 = 2.5 \cdot 10^8$ s^{-1} and $1/k_2 = 2 \cdot 10^7$ s^{-1} (▶Figure 5.6) and respective amplitudes of $A_1 = 20 \pm 5\%$ and $A_2 = 80 \pm 5\%$. This reflects the *cis-trans* ratio at the Ser-Pro peptide bond in the host-guest peptide, which has a *cis* content of $16 \pm 2\%$ as determined by 1D ^{1}H NMR spectroscopy using the method described by Reimer et al. (1998). Since the rate of *cis-trans* isomerization is slow ($\tau \sim 20$ s at 22°C), there is no equilibration between the two isomers on the timescale of the TTET experiments. This allows the measurement of the dynamics of both the *trans* and the *cis* form. The results show that the two isomers significantly differ in their dynamic properties of i, $i + 4$ contact formation and that the *cis*-prolyl isomer actually shows the fastest rate of local contact formation of all peptides (▶Figure 5.6).

5.3.4 Chain diffusion in natural protein sequences

End-to-end loop formation was measured in a protein fragment from carp parvalbumin to test for differences in the dynamics of loop formation in natural protein sequences compared to the homopolymer model peptides (Krieger et al. 2004). The 18-amino-acid EF loop from carp parvalbumin connects two α-helices and brings two phenylalanine residues into contact in the native protein, which were replaced by xanthone and naphthylalanine to measure the dynamics of loop formation by TTET (Krieger et al. 2004). Loop formation in this fragment can be described by single exponential kinetics with a time constant of $\tau = 50$ ns, which is comparable to the dynamics for loop formation in polyserine chains of the same length (▶Figure 5.4B). The EF loop contains several large amino acids like Ile, Val and Leu, but also four glycyl residues. Obviously, the effects of slower chain dynamics of bulkier side chains and faster dynamics around glycyl residues compensate, which leads to dynamics comparable to polyserine

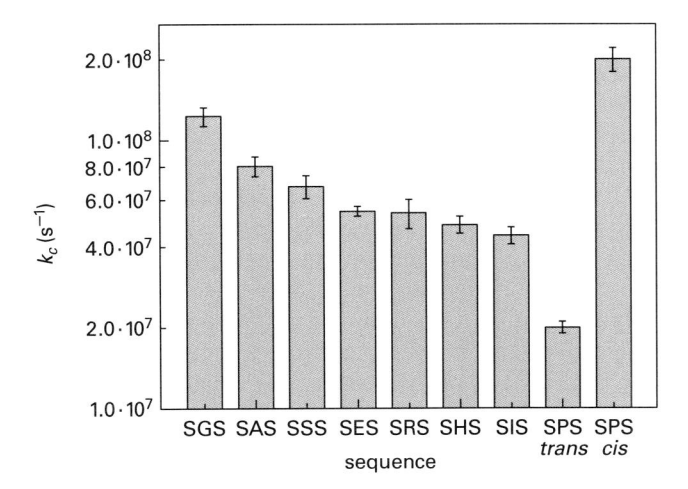

Figure 5.6: Effect of amino acid sequence on local chain dynamics measured in host-guest peptides with the canonical sequence Xan-Ser-Xaa-Ser-Nal-Ser-Gly. The guest amino acid Xaa was varied, and the rate constants for the different guest amino acids are displayed. Data taken from Krieger et al. (2003).

chains. These results indicate that polyserine is a good model to estimate the dynamics of glycine-containing loop sequences in proteins.

5.3.5 End-to-end diffusion versus intrachain diffusion

Most side chain interactions that are formed during the folding process involve residues that are located in the interior of a polypeptide chain. Thus, end-to-end loop formation represents a rather specific and rare situation during folding. Three categories of contact formation events in polymers can be distinguished based on the location of the loop ends. Type I loops correspond to end-to-end contact formation. Type II loops correspond to end-to-interior contacts, and type III loops are formed between two positions remote from the chain ends. TTET experiments in type II and III systems have shown that k_c is decreased compared to type I loops (Fierz and Kiefhaber 2007). The magnitude of the effect depends on the chain dimensions of the tails and on the nature of the loop sequence. For long and flexible loops, the dependence on tail length is weaker than for short or stiff loops. Such behavior has been predicted by theory (Perico and Beggiato 1990). A saturation of the effect is observed at about 2.5-fold decrease in k_c for type II loops and at about 4-fold decrease for type III loops. When this limit is reached, a further extension of the tails has no additional effect on the dynamics of loop formation.

5.3.6 The speed limit for protein folding

The results on the kinetics of intrachain loop formation in various unfolded polypeptide chains show that the amino acid sequence has only little effect on local dynamics of polypeptide chains. All amino acids show very similar rates of end-to-end diffusion with time constants between 12 and 20 ns for the formation of $i, i + 4$ contacts. Polypeptide chains are significantly more flexible around glycine ($\tau = 8$ ns) and stiffer around prolyl residues ($\tau = 50$ ns for the trans isomer). Due to the shorter chain dimensions resulting from a cis peptide bond, the rates of $i, i + 4$ contact formation are fastest around the cis Ser-Pro bond ($\tau = 4$ ns). These results set an upper limit for the rate constants of formation of the first productive local contacts during protein folding. They show that typical protein loops of the size of 6 to 10 amino acids (Leszcynski and Rose 1986) form with time constants of about 15 ns for glycine-rich loops and 30–40 ns for stiffer loops. Short β-hairpin loops, which are the most local structures in proteins, are often rich in glycine and proline (Wilmot and Thornton 1988). Our results indicate that the time constants for the first steps in the formation of the tightest turns with $i, i + 3$ contacts are around 5 ns for Gly and *cis* Xaa-Pro. In glycine- and proline-free turns, these rate constants are decreased to about 10 to 20 ns, depending on the amino acid sequence. Formation of α-helices is most likely initiated by formation of a helical turn, which involves formation of an $i, i + 4$ interaction (Lifson and Roig 1961; Zimm and Bragg 1959). Since helices are usually free of glycyl and prolyl residues, the initiation does not occur faster than about 12 to 20 ns. It should be noted that these values do not represent the time constants for nucleation of helices and hairpins, which most likely require formation of more than one specific interaction and thus encounter additional entropic and enthalpic barriers. The results on the dynamics of loop formation rather represent an upper limit for the dynamics of the earliest steps in secondary structure formation (i.e. for formation of the first contact during helix nucleation). As the combination of femtosecond and nanosecond laser flash experiments have shown, some chain conformations are able to form short loops on the 200 ps timescale (Fierz et al. 2007). This effect increases the complexity of the dynamics of conformational search and may lead to a fraction of unfolded molecules that can form local structures like β-hairpins and α-helices faster than the majority of unfolded conformations.

5.4 Detection of fast conformational fluctuations in folded peptides and proteins by TTET

TTET can also provide unique information about the conformational dynamics in structured peptides and in folded proteins. In general, an irreversible probing reaction coupled to an equilibrium between two states is a powerful kinetic approach to study stability and dynamics of chemical equilibria. If the equilibrium is faster than the probing reaction, the equilibrium constant (K_{eq}) can be determined if the rate constant of the irreversible reaction is known. If the equilibrium and the irreversible process are on similar timescales, both the rate constants and the equilibrium constant of the equilibrium can be obtained. A well-known example for this concept is the use of hydrogen/deuterium (H/D) exchange to gain information on dynamics and stability of individual hydrogen bonds in proteins and DNA (Arrington and Robertson 1997; Hvidt and Nielsen 1966; Linderstrøm-Lang 1955). H/D exchange occurs on the millisecond to hour timescale, depending on pH, and is thus able to measure dynamics on this timescale. As discussed in 5.3, TTET through loop formation is an irreversible process that occurs on the nanosecond to microsecond timescale, depending on loop length, amino acid sequence and the position of the loop within the chain (Bieri et al. 1999; Fierz and Kiefhaber 2007; Krieger et al. 2003, 2004). Thus, TTET coupled to a conformational equilibrium gives information on conformational dynamics that are four to five orders of magnitude faster than those accessible to H/D exchange. Here we describe the kinetic mechanism for using TTET as an irreversible probing reaction to study conformational fluctuations in folded structures, and we discuss two recent applications, in which TTET was used to study local dynamics in α-helical peptides and in the small protein domain HP35.

5.4.1 Kinetic model of TTET coupled to a conformational equilibrium

TTET coupled to a local or global conformational transition allows the determination of the equilibrium constant (K_u) and of the microscopic rate constants for the transition under equilibrium conditions (i.e. without perturbing the system). In addition, the rate constant for loop formation in the unfolded state, k_c, can be obtained (Fierz et al. 2009). Prerequisite for the application of this method is that TTET through loop formation can occur in an unfolded or partially unfolded state (U) but not in the folded state (F; Scheme 5.1).

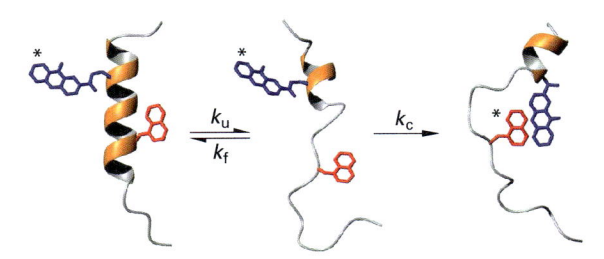

Scheme 5.1: Kinetic mechanism of TTET coupled to a fast conformational equilibrium.

This mechanism is essentially identical to H/D exchange kinetics that allow determination of the stability and dynamics of individual hydrogen bonds in proteins (Hvidt and Nielsen 1966; Linderstrøm-Lang 1955). Different kinetic regimes can be discriminated for TTET coupled to a conformational transition, depending on the relative values of k_u, k_f and k_c. If the equilibrium strongly favors the folded state (steady-state conditions; $k_f \gg k_u$), a single rate constant (λ) for TTET will be observed with

$$\lambda = \frac{k_u \cdot k_c}{k_f + k_c} \tag{10}$$

If the conformational transition is much faster than loop formation (k_f, $k_u \gg k_c$), Equation 10 simplifies to

$$\lambda = \frac{k_u}{k_f} \cdot k_c = K_u \cdot k_c \tag{11}$$

which corresponds to the EX2 limit in hydrogen exchange. If unfolding is much slower than loop formation, Equation 10 becomes

$$\lambda = k_u \tag{12}$$

which is the EX1 limit in hydrogen exchange. A shift between the EX1 and the EX2 exchange can occur if a change in experimental conditions has different effects on the different microscopic rate constants. In this case, all three microscopic rate constants can be determined.

If both F and U are significantly populated, two apparent rate constants (λ_1 and λ_2) are observed, which are described by the general solution of the three-state model shown in Scheme 5.1

$$\lambda_{1,2} = \frac{1}{2}\left(k_u + k_f + k_c \pm \sqrt{(k_u + k_f + k_c)^2 - 4k_u \cdot k_c}\right) \tag{13}$$

with the respective amplitudes

$$A_1 = \frac{1}{\lambda_1(\lambda_1 - \lambda_2)} \cdot \left([U]_0 \cdot k_c \cdot (k_u - \lambda_1) + [F]_0 \cdot k_u \cdot k_c\right) \tag{14}$$

$$A_2 = \frac{1}{\lambda_2(\lambda_1 - \lambda_2)} \cdot \left([U]_0 \cdot k_c \cdot (\lambda_2 - k_u) - [F]_0 \cdot k_u \cdot k_c\right)$$

and the relative equilibrium populations

$$[F]_{eq} = \frac{k_f}{\lambda_1 \cdot \lambda_2} \qquad [U]_{eq} = \frac{k_u}{\lambda_1 \cdot \lambda_2} \tag{15}$$

If the interconversion between the native and unfolded state is much slower than loop formation (k_f, $k_u \ll k_c$), Equation 13 simplifies to

$$\lambda_1 = k_c \qquad \lambda_2 = k_u \tag{16}$$

These considerations show that analysis of the observed rate constants and amplitudes of TTET kinetics allows the determination of all microscopic rate constants for a conformational equilibrium with dynamics on a similar timescale or slower than TTET. Even in cases when the conformational transition is much slower than the triplet lifetime of the donor, this method allows the determination of rate constant for loop formation in the unfolded state (k_c) when U is populated to as little as 3% (i.e. under native solvent conditions).

5.4.2 Experimental considerations

The kinetic approach described in 5.4.1 requires that the TTET labels are placed separated from each other in the folded state, and that the folded state is in equilibrium with a contact competent unfolded or partially unfolded state. The labels should affect neither the structure of the folded state nor the thermodynamics and kinetics of the equilibrium under study. It further has to be ensured that the triplet state is not quenched by other groups or undergoes photochemical reactions. For all donor-acceptor pairs, the corresponding donor-only peptides should be prepared to check the intrinsic donor triplet lifetime at the different positions under all experimental conditions. Fitting the obtained TTET kinetics to a three-state model (Scheme 5.1) is straightforward but requires reliable amplitude information, and the rate constants, k_c, k_u and k_f have to be of similar magnitude. However, the limiting regimes (EX1/EX2) also contain valuable information (Equations 11 and 12). The accuracy can be increased by performing measurements under different conditions, typically in the presence of various concentrations of denaturant, and fitting the data globally.

5.4.3 Dynamics in α-helical peptides

Isolated α helices are marginally stable and highly dynamic structures. The processes governing the dynamics of these structural elements remain largely elusive, mainly due to a lack of experimental methods probing structural transitions on very fast timescales. The strategy outlined in 5.4.1 was recently applied to measure local dynamics in a series of α-helical peptides prepared by solid-phase peptide synthesis. The triplet donor (Xan) and acceptor (Nal) groups were introduced as side chains with an i and $i + 6$ spacing, which places the labels on opposite sides of the helix and prevents TTET in the folded state (Scheme 5.1). Local or global unfolding (k_u) allows TTET to occur via loop formation with the rate constant k_c. To monitor the folding/unfolding at different positions in these 21 residue peptides mainly consisting of alanine, the donor-acceptor pair was moved along the helical structure (▶Figure 5.7A). The time course of TTET monitored by the decay of the Xan triplet absorption band at 590 nm revealed double exponential kinetics for all peptides (▶Figure 5.7B). This indicates that the system follows a kinetic three-state mechanism as depicted in Scheme 5.1. In

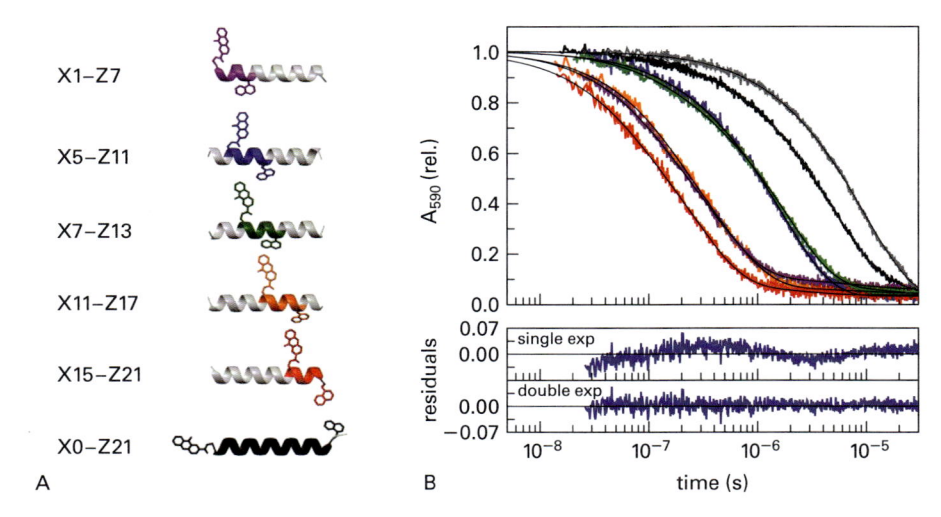

Figure 5.7: (A) Probing local dynamics and stability at different positions in a series of alanine-based helical peptides. (B) TTET kinetics for the peptides shown in (A) in absence of denaturant. All triplet decay curves can be fitted with a double exponential function, as indicated by the residuals for the X5-Z11 peptide, which is shown as an example. Adapted from Fierz et al. (2009).

order to confirm this model and to obtain more accurate data, the measurements were repeated at different denaturant concentrations to shift the conformational equilibrium from folded conformations, F, to unfolded conformations, U. The data were globally fitted with the apparent rate constants, $\lambda_{1,2}$ (▶Figure 5.8A), and their amplitudes, $A_{1,2}$ (▶Figure 5.8B), to obtain k_c, k_u and k_f, as well as their denaturant dependencies (▶Figure 5.8A) (Fierz et al. 2009).

The experiments showed that contact formation ($1/k_c$) occurs on the timescale of ~ 100 ns (at 5°C), similar to contact formation in unstructured peptides (Fierz et al. 2009). The results for the local folding and unfolding constants are summarized in ▶Figure 5.8C/D. Local helix formation ($1/k_f$) does not show any systematic position dependence and occurs with a time constant around 400 ns. In contrast, helix unfolding ($1/k_u$) is fast at the termini (~ 250 ns) and slow in the helix center (~ 1.4 μs). Determination of k_f and k_u further gives the equilibrium constants for the local helix stability ($K_{eq} = k_f/k_u$). The position dependence of the helix unfolding rate constant explains the smaller amount of helical structure toward the ends, a phenomenon known as helix fraying. In summary, TTET gave unprecedented insight into the dynamics of α helices. The comparison of the experimental results with results from dynamic simulations using a linear Ising model showed that the dynamics in these peptides are dominated by diffusion of the helix boundary (i.e. by addition and removal of helical segments at the helix ends) (Fierz et al. 2009). The comparison further revealed that elongation of the helix by one residue occurs with a time constant of about 50 ns at 5°C.

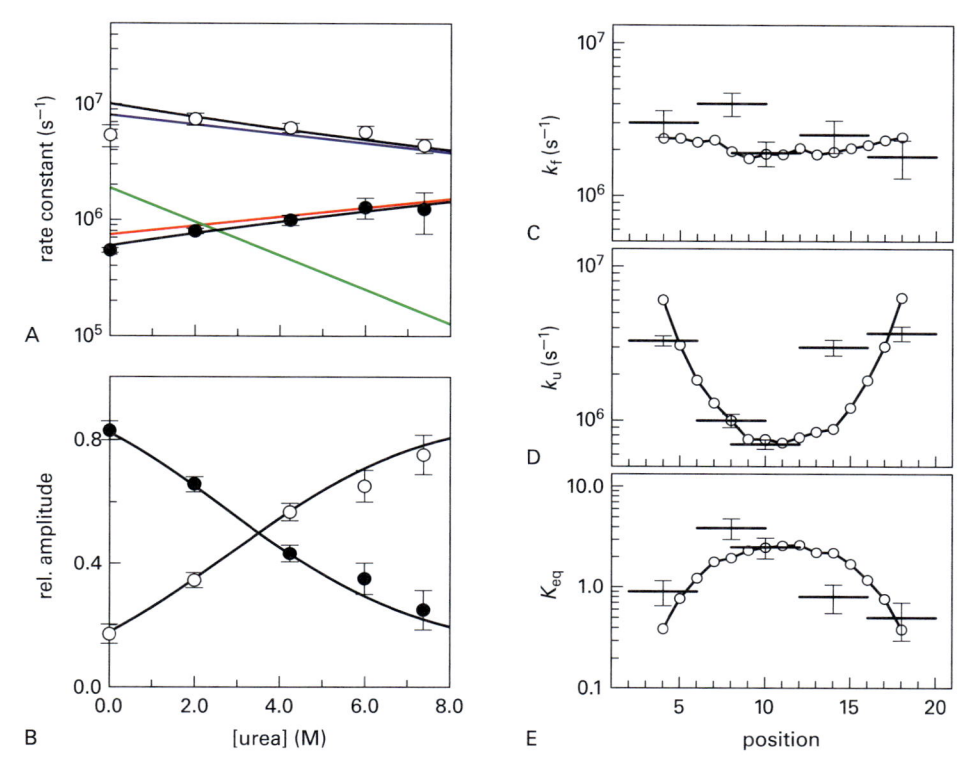

Figure 5.8: Analysis of kinetics in α-helical peptides. (A,B) TTET data from the helix carrying the labels in the center (X7-Z13). The Xan triplet decay was double exponential in all concentrations of urea. The apparent rate constants are given in (A), and the corresponding amplitudes in (B). Using the analytic solution of the three-state mechanism (Equations 13 and 14), one can globally fit the apparent rate constants and amplitudes to obtain the microscopic rate constants k_c (blue), k_u (red) and k_f (green). (C–E) Summarized results for the different positions shown in Figure 5.7. Helix folding, k_f (C), and unfolding, k_u (D), rate constants obtained from TTET measurements. (E) Local helix stability given as equilibrium constant $K_{eq} = k_f/k_u$. The bars represent the experimental data from the peptides shown in Figure 5.7, the connected points result from the simulations with a kinetic Ising model. Adapted from Fierz et al. (2009).

5.4.4 Fluctuations in the native and unfolded state of the villin headpiece subdomain (HP35)

TTET between amino acid side chains can also be used to study conformational fluctuations in both the folded and the unfolded state of proteins. In the simple but common case when global protein unfolding is too slow to occur within the triplet lifetime (for Xan, 30–70 μs, depending on the peptide/protein), the dynamics in the native state, N, and the unfolded state, U, of a protein can be probed separately. As discussed in 5.4.1, this approach allows measurement of the chain dynamics in a small fraction (≥3%) of unfolded molecules (U) under conditions that strongly favor the native state and thus gives information on residual structure in the unfolded state that affects conformational dynamics. TTET in the native fraction of molecules (N) can be used to detect heterogeneity and equilibrium dynamics in the native-state ensemble. The power of TTET to study conformational dynamics in different states of a protein under identical solvent conditions is illustrated by work on the villin headpiece subdomain (HP35) (Reiner et al. 2010). Due to its small size and fast folding, this small helical domain of 35 amino acids is a widely used model system in experimental and computational folding studies. Four different triplet donor-acceptor positions were chosen to probe dynamics in different regions of HP35 (▶Figure 5.9).

In the two donor-acceptor variants Xan0-Nal23 and Xan7-Nal23, TTET requires global unfolding of the core of HP35 (▶Figure 5.9). The experiments showed that TTET is not observed in the native state of these two variants, indicating that unfolding processes involving major parts of the protein core do not occur during the triplet lifetime of xanthone, which was determined to be 30 μs in the respective donor-only variants. This result is in accordance with previous experiments on the global unfolding reactions of HP35, which has a time constant around 300 μs under the given experimental conditions (Kubelka et al. 2003). In the variant Xan0-Nal35, which brings the TTET labels in direct contact in the native state (▶Figure 5.9), the experiments detected fast TTET in the native state. The TTET kinetics revealed heterogeneity within the native state. Around 70% of the native molecules have the labels in close proximity, leading to TTET within the 10 ns dead time of the experiment, as expected from the X-ray structure. In the remaining 30% of the native molecules, contact between the TTET labels is formed with a rate constant of about 30 ns, indicating that the termini are not in direct contact in the native state in this fraction of native molecules. Further evidence for two distinct native-state conformations in HP35 came from a variant that probes the dynamics along the C-terminal helix (Nal23-Xan35). In this variant, TTET should be prevented in the native

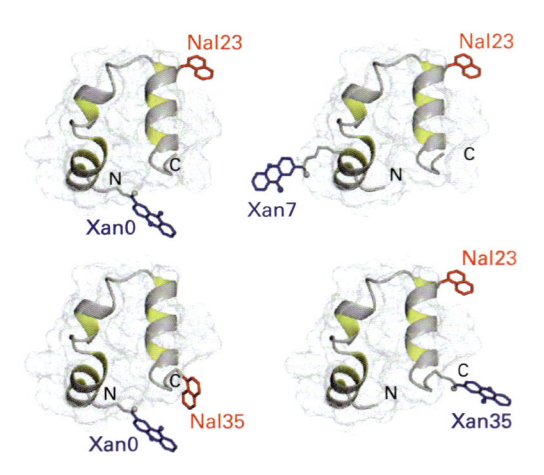

Figure 5.9: Probing dynamics in the native and unfolded state of a small protein, HP35. Triplet donor and acceptor pairs were introduced at four different positions. For details, see Reiner et al. (2010).

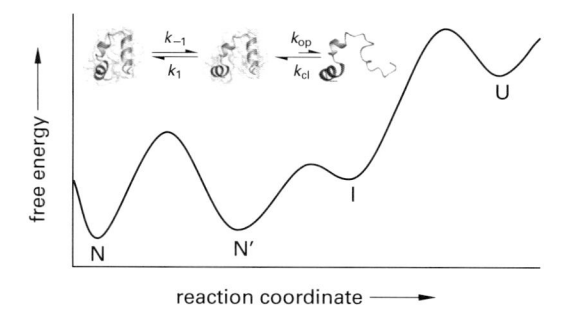

Figure 5.10: TTET revealed heterogeneity and dynamics in the native state of HP35, which is in equilibrium with an unlocked conformation, N', and a partially unfolded intermediate, I. From Reiner et al. (2010).

state, but fast, biphasic TTET kinetics are observed. Around 20% of the molecules form contact between the labels with a time constant of 170 ns. The addition of denaturant strongly accelerates this process, indicating that it involves a major exposure of solvent-accessible surface area and suggests that this step involves detachment and unfolding of helix 3 from the hydrophobic core. The remaining 80% of native molecules also form contact between Nal23 and Xan35 during the triplet lifetime of the donor, albeit in a slower process with a time constant around 1 µs. This reaction is only weakly affected by the addition of denaturant, but it is strongly accelerated by temperature. This process most likely reflects an unlocking of the native state and weakening of side-chain interactions with only minor structural changes. In the unlocked state, unfolding of helix 3 and presumably also global unfolding can occur (▶Figure 5.10). These results demonstrate the presence of alternative conformations in native HP35, which shows that native proteins can exist in different conformations that are separated by a locking/unlocking barrier. These conformations are structurally similar but differ in their dynamic properties, which may be important for protein functions like catalysis or binding (Reiner et al. 2010).

The TTET experiments on HP35 further yielded a wealth of information on the dynamics in the unfolded state (Reiner et al. 2010). In all variants and under all conditions, contact formation in the unfolded state is a single exponential process that occurs with comparable rate constants, k_c, as contact formation in unfolded model peptides. Denaturants and temperature exerted a similarly strong effect on the dynamics, as in unstructured polyserine peptides. Even at low denaturant concentrations, in which the native state is highly favored and transient structures might form, the relationships between $\ln(k_c)$ and denaturant concentration remained linear (Reiner et al. 2010).

5.5 Conclusions

Intramolecular TTET between donor and acceptor groups attached at specific positions on a polypeptide chain gives unique information on the dynamics of loop formation in unfolded polypeptide chains on a timescale of picoseconds to tens of microseconds. The results from TTET experiments contributed seminally to our current understanding of the earliest processes in protein folding and on the dynamic properties of unfolded polypeptide chains. TTET coupled to a conformational equilibrium in folded peptides and proteins is a powerful tool to detect and characterize local structural fluctuations on the nanosecond to microsecond timescale. Native-state TTET is a unique tool to distinguish between structurally similar folded states based on their different dynamic properties.

References

Arrington, C. B. and A. D. Robertson (1997). "Microsecond protein folding kinetics from native-state hydrogen exchange." *Biochemistry 36*, 8686–8691.

Bent, D. V. and E. Hayon (1975b). "Excited state chemistry of aromatic amino acids and related peptides: I. Tyrosine." *J. Am. Chem. Soc. 97*, 2599–2606.

Bent, D. V. and E. Hayon (1975). "Excited state chemistry of aromatic amino acids and related peptides: III. Tryptophan." *J. Am. Chem. Soc. 97*, 2612–2619.

Bieri, O., J. Wirz, B. Hellrung, M. Schutkowski, M. Drewello and T. Kiefhaber (1999). "The speed limit for protein folding measured by triplet-triplet energy transfer." *Proc. Natl. Acad. Sci. USA 96*, 9597–9601.

Brant, D. A. and P. J. Flory (1965). "The configuration of random polypeptide chains. II. Theory." *J. Am. Chem. Soc. 87*, 2791–2800.

Bredenbeck, J., et al. (2003). "Picosecond conformational transition and equilibration of a cyclic peptide." *Proc. Natl. Acad. Sci. U S A 100*(11), 6452–6457.

Closs, G. L., M. D. Johnson, J. R. Miller and P. Piotrowiak (1989). "A connection between intramolecular long-range electron, hole and triplet energy transfer." *J. Am. Chem. Soc. 111*, 3751–3753.

Felekyan, S., R. Kühnemuth, V. Kudryavtsev, C. Sandhagen, B. W. and C. A. M. Seidel (2005). "Full correlation from picoseconds to seconds by time-correlated single photon detection." *Rev. Sci. Instrum. 76*, 083104.

Fierz, B. and T. Kiefhaber (2007). "End-to-end vs interior loop formation kinetics in unfolded polypeptide chains." *J. Am. Chem. Soc. 129*(3), 672–679.

Fierz, B., A. Reiner and T. Kiefhaber (2009). "Local conformational fluctuations in α-helices measured by fast triplet transfer." *Proc. Natl. Acad. Sci. USA 106*, 1057–1062.

Fierz, B., H. Satzger, C. Root, P. Gilch, W. Zinth and T. Kiefhaber (2007). "Loop formation in unfolded polypeptide chains on the picoseconds to microseconds time scale." *Proc. Natl. Acad. Sci. USA 104*, 2163–2168.

Flory, P. J. (1953). *Principles of polymer chemistry*. Ithaca, Cornell University Press.

Flory, P. J. (1969). *Statistical Mechanics of Chain Molecules*. Munich, Hanser Publishers.

Hvidt, A. and S. O. Nielsen (1966). "Hydrogen exchange in proteins." *Advan. Prot. Chem.* 21, 287–386.

Jacobsen, H. and W. H. Stockmayer (1950). "Intramolecular reaction in polycondensations. I. The theory of linear systems." *J. Phys. Chem. 18*, 1600–1606.

Kiefhaber, T., H. H. Kohler and F. X. Schmid (1992). "Kinetic coupling between protein folding and prolyl isomerization. I. Theoretical models." *J. Mol. Biol. 224*(1), 217–229.

Kleinert, T., W. Doster, H. Leyser, W. Petry, V. Schwarz and M. Settles (1998). "Solvent Composition and Viscosity Effects on the Kinetics of CO Binding to Horse Myoglobin." *Biochemistry 37*, 717–733.

Klessinger, M. and J. Michl (1995). *Excited states and photochemistry of organic molecules*. Weinheim, VCH.

Krieger, F., B. Fierz, F. Axthelm, K. Joder, D. Meyer and T. Kiefhaber (2004). "Intrachain diffusion in a protein loop fragment from carp parvalbumin." *Chem. Phys. 307*, 209–215.

Krieger, F., B. Fierz, O. Bieri, M. Drewello and T. Kiefhaber (2003). "Dynamics of unfolded polypeptide chains as model for the earliest steps in protein folding." *J. Mol. Biol. 332*, 265–274.

Kubelka, J., W. A. Eaton and J. Hofrichter (2003). "Experimental tests of villin subdomain folding simulations." *J. Mol. Biol. 329*, 625–630.

Lapidus, L. J., W. A. Eaton and J. Hofrichter (2000). "Measuring the rate of intramolecular contact formation in polypeptides." *Proc. Natl. Acad. Sci. USA 97*, 7220–7225.

Leszcynski, J. F. and G. D. Rose (1986). "Loops in globular proteins: a novel category of secondary structure." *Science 234*, 849–855.

Lifson, S. and A. Roig (1961). "On the theory of helix-coil transition in polypeptides." *J. Chem. Phys. 34*, 1963–1974.

Linderstrøm-Lang, K. (1955). "Deuterium exchange between peptides and water." *Chem. Soc. Spec. Publ. 2*, 1–20.

Miller, W. G., D. A. Brant and P. J. Flory (1967). "Random coil configurations of polypeptide chains." *J. Mol. Biol. 23*, 67–80.

Moore, J. W. and R. G. Pearson (1981). *Kinetics and Mechanisms*. New York, John Wiley & Sons.

Palmer III, A. G. (2004). "NMR characterization of the dynamics of biomacromolecules." *Chem. Rev. 104*, 3623–3640.

Pappu, R. V., R. Srinivasan and G. D. Rose (2000). "The Flory isolated-pair hypothesis is not valid for polypeptide chains: implications for protein folding." *Proc. Natl. Acad. Sci. USA 97*(23), 12565–12570.

Perico, A. and M. Beggiato (1990). "Intramolecular Diffusion-Controlled Reactions in Polymers in the Optimized Rouse Zimm Approach .1. The Effects of Chain Stiffness, Reactive Site Positions, and Site Numbers." *Macromolecules 23*(3), 797–803.

Reimer, U., G. Scherer, M. Drewello, S. Kruber, M. Schutkowski and G. Fischer (1998). "Sidechain effects on peptidyl-prolyl cis/trans isomerisation." *J. Mol. Biol. 279*(2), 449–460.

Reiner, A., P. Henklein and T. Kiefhaber (2010). "An Unlocking/Relocking Barrier in Conformational Fluctuations of Villin Headpiece Subdomain." *Proc. Natl. Acad. Sci. U S A 107*, 4955–4960.

Satzger, H., B. Schmidt, C. Root, W. Zinth, B. Fierz, F. Krieger, T. Kiefhaber and P. Gilch (2004). "Ultrafast quenching of the xanthone triplet by energy transfer: new insight into the intersystem crossing kinetics." *J. Phys. Chem. A 108*, 10072–10079.

Schimmel, P. R. and P. J. Flory (1967). "Conformational energy and configurational statistics of poly-L-proline." *Proc. Natl. Acad. Sci. USA 58*, 52–59.

Schwalbe, H., K. M. Fiebig, M. Buck, J. A. Jones, S. B. Grimshaw, A. Spencer, S. J. Glaser, L. J. Smith and C. M. Dobson (1997). "Structural and Dynamical Properties of a Denatured Protein. Heteronuclear 3D NMR Experiments and Theoretical Simulations of Lysozyme in 8 M Urea." *Biochemistry 36*, 8977–8991.

Spörlein, S., H. Carstens, H. Satzger, C. Renner, R. Behrendt, L. Moroder, P. Tavan, W. Zinth and J. Wachtveitl (2002). "Ultrafast spectroscopy reveals sub-nanosecond peptide conformational dynamics and validates molecular dynamics simulation." *Proc. Natl. Acad. Sci. USA 99*, 7998–8002.

Stern, O. and M. Volmer (1919). "Über die Abklingungszeit der Fluoreszenz." *Physik. Z. 20*, 183–188.

Szabo, A., K. Schulten and Z. Schulten (1980). "First passage time approach to diffusion controlled reactions." *J. Chem. Phys. 72*, 4350–4357.

Szabo, Z. G. (1969). Kinetic characterization of complex reaction systems. *Comprehensive chemical kinetics*. C. H. Bamford and C. F. H. Tipper. Amsterdam, Elsevier Publishing Company. *2*, 1–80.

Timasheff, S. N. (2002). "Protein hydration, thermodynamic binding and prefential hydration." *Biochemistry 41*, 13473–13482.

Wagner, P. J. and P. Klán (1999). "Intramolecular triplet energy transfer in flexible molecules, electronic, dynamic, and structural aspects." *J. Am. Chem. Soc. 121*, 9626–9635.

Wilmot, C. M. and J. M. Thornton (1988). "Analysis and prediction of the different types of β-turns in proteins." *J. Mol. Biol. 203*, 221–232.

Zimm, B. H. and J. K. Bragg (1959). "Theory of phase transition between helix and random coil in polypeptide chains." *J. Chem. Phys. 31*, 526–535.

Zwanzig, R. (1997). "Two-state models for protein folding." *Proc. Natl. Acad. Sci. USA 94*, 148–150.

6 Protein import into the intermembrane space of mitochondria

Jan Riemer and Johannes M. Herrmann

6.1 Introduction

Even simple eukaryotic cells like those of baker's yeast contain several thousand different proteins. Following their synthesis, the majority these proteins are translocated across cellular membranes to reach their specific locations in the cell (Huh et al. 2003; Kumar et al. 2002). In many cases, aminoterminal targeting signals are employed to direct polypeptides during or following their synthesis into their appropriate target organelles (Wickner and Schekman 2005). Examples for these aminoterminal targeting sequences are signal sequences for targeting to the endoplasmic reticulum (ER), presequences that commit proteins to the import into the mitochondrial matrix or transit peptides that direct plant proteins to chloroplasts (Emanuelsson et al. 2007). In most cases, these aminoterminal targeting sequences are proteolytically removed by processing proteases in the target organelles. Alternatively, internal signals in the mature sequence of proteins can be used as targeting information. These internal signals are more difficult to identify, in particular since they are often scattered over the protein sequence and only come into spatial proximity when a protein is folded into its three-dimensional structure. Moreover, these internal signals often cannot be transplanted from one protein to another, which makes the analysis of their mechanistic role in protein translocation difficult. Examples for such internal signals are nuclear localization sequences (Lange et al. 2007), some peroxisomal targeting signals (Lazarow 2006; Girzalsky et al. 2010) and the signals that direct proteins to the intermembrane space (IMS) of mitochondria (Hell 2008; Herrmann and Hell 2005; Milenkovic et al. 2009; Sideris et al. 2009). For protein targeting to the IMS, cysteine residues often play a critical role (Allen et al. 2003; Hofmann et al. 2002; Koehler 2004; Koehler et al. 1999; Lutz et al. 2003). These residues are oxidized during or directly following the translocation across the mitochondrial outer membrane. The players and the mechanism of this redox-dependent import process into the mitochondrial IMS are described in Section 6.2.

6.2 The mitochondrial IMS

The IMS of mitochondria (Herrmann and Riemer 2010) is enclosed by the outer and the inner membranes of the organelle. During evolution, this compartment originated from the periplasm of the bacterial ancestors of mitochondria. It differs, however, structurally from the bacterial periplasm since the mitochondrial inner membrane forms many invaginations, called cristae. On the one hand, these cristae increase the area size of the inner membrane to accommodate a large number of respiratory chain complexes, and, on the other hand, they allow the separation of the external IMS from the intracristae space (Frey and Mannella 2000; Frezza et al. 2006; Zick et al. 2009). Both subcompartments are separated by rather tight openings at the necks of the cristae, so-called cristae junctions.

About 50 to 100 different proteins were identified in the IMS (Herrmann and Riemer 2010). These factors fulfill a variety of crucial functions in different processes, like in the transport of proteins, electrons or metal ions, in the assembly of inner membrane proteins, in cellular respiration and other metabolic processes. In addition, several apoptotic components are sequestered in the IMS until their release triggers the programmed cell death (Koehler and Tienson 2009; van Gurp et al. 2003; Wang and Youle 2009).

Little information exists about the physicochemical properties of IMS. The outer membrane of mitochondria contains porins, which are membrane-embedded β barrels that facilitate the diffusion of molecules up to a molecular mass of about 2–6 kDa between the cytosol and the IMS (Benz 1994). As a consequence, the concentration of small molecules like glutathione in the IMS is therefore expected to be similar to that of the cytosol, suggesting that the IMS is a strongly reducing environment. However, experiments with redox-sensitive fluorescent proteins suggest that the IMS is more oxidizing than the cytosol (Hu et al. 2008). Moreover, a number of stable disulfide bridges have been reported in proteins of the IMS, which are formed by a specialized oxidation machinery (Koehler and Tienson 2009; Riemer et al. 2009). This is reminiscent of the situation of the bacterial periplasm where many proteins are stabilized by disulfide bonds (Collet and Bardwell 2002; Herrmann et al. 2009; Inaba and Ito 2008; Kadokura et al. 2003).

6.3 The mitochondrial disulfide relay

The thiol groups of cysteine residues are highly reactive and are modified in various ways under physiological conditions. Its property to form disulfide bonds in the presence of oxidizing conditions can be used to sense, for example, the presence of oxygen or hydrogen peroxide (Barford 2004; Reddie and Carroll 2008). In most cellular compartments including the cytosol, the formation of disulfide bonds is prevented by the presence of millimolar amounts of reduced glutathione. Under these reducing conditions, cysteine residues are largely present in the reduced state. In some compartments, however, the formation of disulfide bonds is catalyzed by enzymes, and there, cysteine residues are predominantly oxidized. Examples for oxidizing compartments are the bacterial periplasm, the ER and the IMS of mitochondria (Riemer et al. 2009).

The catalytic formation of disulfide bonds is typically mediated by two kinds of components: sulfhydryl oxidases and oxidoreductases. Sulfhydryl oxidases can form disulfide bonds de novo, which are then transferred to the final substrates by one or several oxidoreductases. Since oxidoreductases can shuffle disulfide bonds between different proteins or different sites in a given protein, they may also serve as protein disulfide isomerases.

6.4 The sulfhydryl oxidase Erv1

The Erv1 protein (▶Figure 6.1) was initially identified in *Saccharomyces cerevisiae* as a protein essential for respiration and vegetative growth (Lisowsky 1992). It has homologs in almost all eukaryotes that contain respiration-competent mitochondria (Lill and Mühlenhoff 2005). The human homolog of Erv1 is also referred to as ALR (augmenter of liver regeneration). All homologs share a conserved flavin adenine dinucleotide

(FAD)-binding domain that contains a sequence of two cysteine residues spaced by two amino acid residues. This CxxC motif is essential for Erv1 function (Bien et al. 2010; Hofhaus et al. 1999). FAD is noncovalently bound in an unusual horseshoe-like conformation with the flavin and isoalloxazine parts buried in the interior of the domain (Daithankar et al. 2009; Gross et al. 2002; Wu et al. 2003). The FAD cofactor forms planar stacking interactions with aromatic side chains that are conserved among Erv1 homologs.

This FAD-binding domain is flanked by less structured regions that are located N terminally (animals, fungi) or C terminally (plants) to the FAD domain. These tail domains also contain pairs of cysteine residues, which play a critical role for the interaction of Erv1 with its substrate proteins (Ang and Lu 2009; Bien et al. 2010; Lionaki et al. 2010). This region presumably forms a flexible "lever arm" that is characterized by a high content in proline and glycine residues. Erv1 proteins form homodimers in head-to-tail orientation in which the FAD domains are positioned in close proximity to the opposing subunit allowing an efficient intermolecular disulfide transfer. Mutations that destabilize the dimer were shown to cause a recessive myopathy in humans (Daithankar et al. 2009; Di Fonzo et al. 2009).

In the IMS, Erv1 interacts with cytochrome c, which serves as its electron acceptor under physiological conditions (Allen et al. 2005; Bihlmaier et al. 2007; Dabir et al.

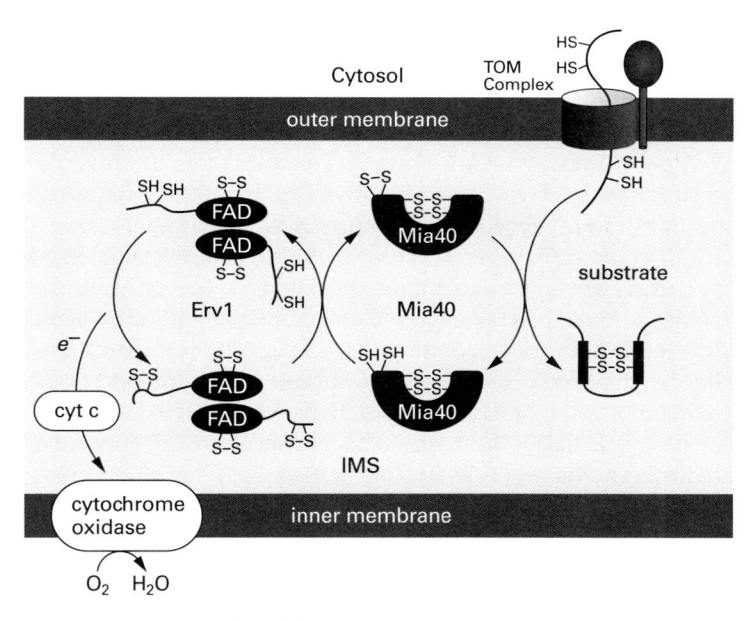

Figure 6.1: The mitochondrial disulfide relay. IMS proteins are imported through the protein-conducting channel of the translocase of the outer membrane of mitochondria (TOM) complex. During or directly following membrane translocation they interact with Mia40 to be transferred into their oxidized form. Mia40 serves as an oxidoreductase that is maintained in its oxidized, active conformation by the sulfhydryl oxidase Erv1. Erv1 is a homodimeric, FAD-bound enzyme that shuttles electrons from Mia40 to cytochrome c of the respiratory chain. See text for details.

2007; Farrell and Thorpe 2005). This interaction allows the rapid and efficient channeling of electrons from the disulfide relay system via cytochrome c and cytochrome c oxidase to molecular oxygen (▶Figure 6.1). Thus protein oxidation in mitochondria is coupled to the reduction of oxygen to water. However, the number of electrons that flow from Erv1 through the respiratory chain are presumably negligible in comparison to the large amounts that originate from NADH and succinate.

6.5 The oxidoreductase Mia40

Mia40 is an essential mitochondrial protein that is conserved between fungi, plants and animals including humans. Mia40 was initially identified in mitochondria of baker's yeast (Sickmann et al. 2003), and what we currently know about this protein is almost entirely based on studies of the yeast homolog. All Mia40 homologs share a highly conserved C-terminal domain containing six invariant cysteine residues. Fungal members of the Mia40 family, but not those of animals or plants, are synthesized with N-terminal mitochondrial targeting signals followed by hydrophobic membrane anchor domains (Naoe et al. 2004; Terziyska et al. 2005). While it was initially proposed that this region represents a bipartite presequence that targets the protein into the IMS (Chacinska et al. 2004), more recent studies suggested that the hydrophobic region remains present on the mature protein and anchors it permanently to the inner membrane. The stable association of Mia40 with the inner membrane is, however, not essential for its function; Mia40 mutants in which the conserved C-terminal domain was fused to the bipartite presequence of cytochrome b_2 were still viable (Naoe et al. 2004).

The conserved domain that is present in all Mia40 homologs comprises roughly 60 amino acid residues. It contains six cysteine residues present in the pattern CPC-x_8-Cx_9C-x_{12}-Cx_9C. The CPC motif is redox active and shuttles electrons from the lever arm of Erv1 to the substrate proteins of Mia40 (Banci et al. 2009; Kawano et al. 2009; Terziyska et al. 2009). The latter four cysteine residues resemble the twin Cx_9C motif (see Section 6.7.4) and stabilize the structure of Mia40. They have a highly negative redox potential and are presumably permanently oxidized. The structure of Mia40 was recently solved by NMR and crystallization (Banci et al. 2009; Kawano et al. 2009); the twin Cx_9C region forms a dish-like structure that presumably represents a hydrophobic substrate-binding cleft. The CPC active site is solvent accessible and adjacent to the cleft in direct proximity to the bound substrate (Banci et al. 2010; Kawano et al. 2009).

The loss of Mia40 is lethal both in yeast (Chacinska et al. 2004; Naoe et al. 2004; Terziyska et al. 2005) and human cells (Hofmann et al. 2005). Mia40 serves as an intramitochondrial receptor protein that interacts with newly synthesized substrates during or after their translocation across the mitochondrial outer membrane (Mesecke et al. 2005; Müller et al. 2008; Sideris and Tokatlidis 2007; Tienson et al. 2009) (▶Figure 6.1). In addition, it appears to play a role in the folding of its substrate proteins (Banci et al. 2010; Sideris and Tokatlidis 2007).

6.6 Substrates of the mitochondrial disulfide relay

All so far identified substrates of the mitochondrial disulfide relay system contain cysteine residues. In many cases, the cysteine residues form characteristic patterns.

The best characterized substrates of the Mia40 pathway are members of the small Tim proteins. These proteins are characterized by the presence of an invariant pattern of four cysteine residues forming two Cx_3C motifs and are therefore often called twin Cx_3C proteins. Mitochondria of fungi, animals and plants contain typically five members of this family, which in yeast were named Tim8, Tim9, Tim10, Tim12 and Tim13 (for review, see Bauer et al. 2000; Koehler 2004). They form hexameric complexes in the IMS (Webb et al. 2006) that play an essential role in the transfer of hydrophobic carrier proteins from the TOM complex to their insertion sites in the inner membrane.

The second group of Mia40-dependent substrate proteins is characterized by the presence of two pairs of cysteine residues, each spaced by short helices formed by nine variable residues. In these twin Cx_9C proteins, these helices interact in an antiparallel orientation and are stabilized by intramolecular disulfide bridges between the flanking cysteine residues (Abajian et al. 2004; Arnesano et al. 2005). Yeast contains 14 different twin Cx_9C proteins (Gabriel et al. 2007; Longen et al. 2009), and mammals about 30 (Cavallaro 2010). The best characterized representative of this family is Cox17, a copper-binding protein required for the assembly of cytochrome oxidase (Glerum et al. 1996). The function of most of the other twin Cx_9C proteins is unclear, and it remains to be shown whether the twin Cx_9C motif is simply a structural element of these proteins or whether it contributes to the biological activity of these components.

Mia40 appears to recognize specific sites in the sequences of the twin Cx_3C and the twin Cx_9C proteins, which were named MISS (Milenkovic et al. 2009) or ITS (Sideris et al. 2009) signals. These short stretches consist of hydrophobic patterns flanking one of the cysteine residues and presumably mediate the binding of these proteins into the hydrophobic cleft of Mia40.

6.7 Methods to study mitochondrial protein translocation

Most of what we know about mitochondrial protein import is deduced from studies with yeast mitochondria. Mitochondria can be relatively easily isolated from yeast and maintain their ability to import precursor proteins even after prolonged storage in a −80°C freezer. Import assays with isolated yeast mitochondria were successfully used to analyze the properties of translocation of proteins into the matrix and also the IMS. In this chapter, a number of techniques will be introduced that can be employed to follow the import of radiolabelled proteins into the IMS and the redox state of their cysteine residues at different steps of this process.

6.7.1 Growth of yeast cells

To induce the formation of mitochondria, yeast cells should be grown under respirative conditions. For the preparation of mitochondria, the growth in liquid lactate medium is recommended. Lactate medium contains per liter 3 g Bacto-yeast extract, 1 g KH_2PO_4, 1 g NH_4Cl, 0.5 g $CaCl_2·2H_2O$, 0.5 g NaCl, 0.6 g $MgSO_4·H_2O$, 3 mg $FeCl_3$ (or 0.3 mL from a 1% w/v stock solution) and 22 mL 90% lactic acid (2% v/v final concentration). The pH has to be adjusted to 5.5 with KOH before autoclaving.

6.7.2 Isolation of mitochondria from yeast cells

Mitochondrial can be easily isolated from yeast cells. The cultures should grow for 2 to 4 days and be repeatedly diluted so that they does not reach stationary phase (OD_{600} of 5 or more). The growth rate depends notably on the yeast strain. Therefore, it is recommended to measure the generation time for each strain before inoculating a larger culture (2 to 10 L) that will be used for mitochondria isolation. Cultures should be grown over night until they reach an OD_{600} of about 1. The cells are collected at 3,000 × g for 5 min, and wet weight is measured. The pellet should be washed in dH_2O and resuspended in MP1 buffer (100 mM Tris base, 10 mM dithiotreitol [DTT], pH not adjusted; 2 mL/g of wet weight). Following incubation of the suspension for 10 min at 30°C under agitation, the cells are collected by centrifugation at 2,000 × g for 5 min. The cell pellet is washed with 1.2 M sorbitol (5 mL/g of wet weight) and resuspended in MP2 buffer (1.2 M sorbitol, 20 mM phosphate buffer, pH 7.4; 6.7 mL/g of wet weight), to which 2 mg/g Zymolyase 20T is added. Upon incubation under agitation (ca. 140 rpm) for 30 to 60 min at 30°C, the cell wall is removed. The sphaeroplasts are harvested by centrifugation at 2,000 × g for 5 min and are resuspended in ice-cold buffer MP3 (0.6 M sorbitol, 10 mM Tris-HCl, pH 7.4, 1 mM EDTA, 0.2% (w/v) fatty acid free bovine serum albumin [BSA, essentially fatty acid free], 1 mM phenylmethylsulphonyl fluoride [PMSF]; 6.7 mL/g of wet weight). The suspension is homogenized in a 50 mL dounce homogenizer (tight-fitting glass pistil) with 15 strokes and diluted with MP3 buffer (again 6.7 mL/g of wet weight). The homogenate is centrifuged at 2,000 × g for 5 min to spin down cell debris. The supernatant is transferred to a fresh centrifugation bucket and centrifuged 12 min at 17,000 × g for 12 min. After discarding the supernatant, the pellet is carefully resuspended in 10 mL of SEH (0.6 M sorbitol, 1 mM EDTA, 20 mM Hepes, pH 7.4) buffer. Mitochondria are collected by centrifugation for 12 min at 17,000 × g for 12 min and resuspended in 0.5 mL of SEH buffer. The protein concentration is adjusted to a final concentration of 10 mg/ mL. Aliquots of 50 µL have to be immediately frozen in liquid nitrogen and can be stored at −80°C for months.

6.7.3 Synthesis of radiolabeled proteins

Radiolabeled preproteins can be easily synthesized in cell-free transcription systems (Pelham and Jackson 1976). It should be noted that the protein concentrations obtained by this method are very low and in most cases are below the nanomolar range. However, since the proteins are typically associated with chaperones, they are often highly import competent. The high import competence, and the simple production of these radiolabeled proteins, made this method the standard procedure for the generation of preproteins.

Proteins can be easily produced in cell-free protein expression systems that are commercially available. For example, the TNT-coupled reticulocyte lysate system from Promega can be used according to the protocol supplied with the reagents. This system allows the expression of radiolabeled proteins in a coupled transcription/translation system. The DNA sequence for the protein of the interest has to be introduced into an expression plasmid (like pGEM3), which allows expression from an SP6 promoter. Alternatively, transcription and translation can be carried out separately (Bihlmaier et al. 2008).

6.7.4 Import of twin CX$_9$C and twin CX$_3$C proteins into isolated mitochondria

The import of proteins can be analyzed using radioactively labeled proteins and isolated mitochondria (▶Figure 6.2). The import kinetics of twin CX$_9$C and twin CX$_3$C proteins is strongly influenced, for example, by the presence of metal ions and glutathione in the import buffer or by import into mitochondria containing differing amounts of Mia40 (Bien et al. 2010; Chacinska et al. 2004; Mesecke et al. 2008; Naoe et al. 2004; Terz-iyska et al. 2005).

Per import reaction, 10 μL isolated mitochondria (concentration 10 μg/μL) were mixed with 100 μL import buffer (100 mM Hepes/KOH, pH 7.2, 1.2 M Sorbitol, 160 mM KCl, 20 mM Mg acetate, 4 mM KH$_2$PO$_4$, 2 mM MnCl$_2$). Then, 100 μL of glutathione dissolved in import buffer was added to final concentrations between 0 and 30 mM. This mix was incubated for 5 min at 25°C. Subsequently, radioactively labeled proteins were added, and the import reaction was allowed to proceed for 5 to 10 min. Import was stopped, and nonimported protein proteolytically digested by the addition of 800 μL stop buffer (0.6 M Sorbitol, 20 mM Hepes, pH 7.2, proteinase K [0.1 mg/ mL final concentration]) and incubation for 20 min on ice. Addition of 18 μL PMSF solution (200 mM in ethanol) stopped the activity of proteinase K. Then, mitochondria were pelleted by centrifugation and washed in PMSF-containing stop buffer without proteinase K. Mitochondria were resuspended in SDS-loading buffer and analyzed by SDS-PAGE and autoradiography.

6.7.5 Purification of the components of the mitochondrial disulfide relay

Most of the previous studies on the disulfide relay system were performed with crude mitochondrial fractions or whole cells. However, in these systems, many other factors might influence disulfide bond formation. The reconstitution of the pathway with purified components therefore is a critical tool to assess the biochemical details and the minimal requirements for disulfide bond formation in the IMS (Ang and Lu 2009; Bien et al. 2010; Lionaki et al. 2010; Tienson et al. 2009). To this end, full-length Erv1, the two domains of Erv1 (the flexible domain Erv1$^{(N)}$ spanning the amino acids 1–83 and the FAD domain Erv1$^{(C)}$ containing the amino acids 83–189 of the *S. cerevisiae* protein), Mia40, and Cox19 as a classical twin CX$_9$C substrate were recombinantly expressed in *E. coli* and purified to homogeneity .

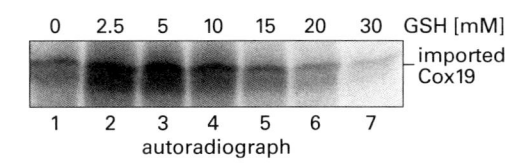

Figure 6.2: Glutathione dependence of mitochondrial import. Radioactively labeled Cox19 was imported into isolated mitochondria in the presence of different amounts of glutathione (GSH). Notably, low amounts of glutathione accelerate mitochondrial import up to an optimal concentration of about 5 to 10 mM glutathione (lanes 3 and 4), while higher glutathione concentrations interfere with import (lanes 5 to 7).

6.7.5.1 Expression of recombinant proteins from *E. coli*

Recombinant proteins with an hexahistidine tag or a glutathione S-transferase (GST) tag were expressed in *E. coli* BL21 cells from the pET-24a(+) vector (Novagen) or the pGEX-6P-1 (GE Healthcare) vector, respectively. Then 100 mL of an overnight preculture was diluted into 1 L fresh Luria-Bertani (LB) medium containing 30 μg/mL kanamycin (pET24a[+]) or 100 μg/mL ampicillin (pGEX-6P-1) and grown until an OD_{600} of 0.5–0.8 at 37°C with moderate shaking (140 rpm). Expression of the recombinant proteins was induced by addition of 0.5–1 mM isopropyl-β,D-thiogalactopyranoside (IPTG), and the culture was grown for 16–18 h at 28°C (Erv1 variants) or for 4–6 h at 37°C for Cox19 and Mia40. Subsequently, cells were harvested by centrifugation (5,000 × g, 10 min, RT). Cell pellets were stored at −20°C before purification. Expression and purification were controlled by SDS-PAGE and Coomassie staining.

6.7.5.2 Purification of recombinant Erv1-His$_6$, Erv1$^{(N)}$-His$_6$ and Erv1$^{(C)}$-His$_6$ from *E. coli*

Cell pellets containing the Erv1 proteins were resuspended in 20 mL binding buffer (50 mM sodium phosphate [NaP$_i$], pH 8.0, 50 NaCl, 10 mM imidazole, 1 mM PMSF) and incubated at 4°C for 15 min. Cells were broken by sonication, and the lysate was centrifuged for 20 min at 16,000 rpm (JA 25.50 rotor, Beckman coulter, Avanti J-25 or Avanti J-26XP). The supernatant was filtered through a 0.45 μm filter and then applied to a 1 mL volume Ni-NTA Superflow cartridge (Qiagen). This column had been preequilibrated with 10 mL of binding buffer. After washing the column with 10 mL washing buffer (50 mM NaP$_i$, pH 8.0, 50 mM NaCl, 20 mM imidazole), bound proteins were eluted in 500 μL fractions using a linear increasing gradient of the elution buffer (50 mM NaP$_i$, pH 8.0, 50 mM NaCl, 500 mM imidazole). Fractions containing Erv1 protein variants were collected and dialyzed against 20 mM NaP$_i$ buffer pH 8.0. The purified protein was stored at 4°C.

6.7.5.3 Purification of recombinant GST-Mia40ΔTM from *E. coli*

For in vitro experiments, a truncated variant of Mia40 lacking the transmembrane segment of the protein was expressed and purified (Grumbt et al. 2007). Cell pellets were resuspended in binding buffer (20 mM Tris, pH 7.0, 200 mM NaCl, 1 mM PMSF and 0.2 mg/mL lysozyme) and incubated for 10 min at 4°C. Subsequently, cells were broken by sonication, and the cell lysate was cleared by centrifugation for 20 min at 16,000 rpm (JA 25–50 rotor, Beckman coulter, Avanti J-25 or Avanti J-26XP). The supernatant containing the protein was loaded onto 0.5 mL preequilibrated GSH-Sepharose beads and incubated for 1.5 h at 4°C. Beads were washed three times with 10 mL binding buffer without lysozyme, and the protein was eluted in 5 mL binding buffer containing 0.05 mg/mL Prescission protease at 4°C overnight. In a second step, the protein was further purified on a Hi Load 16/60 Superdex 200 column preequilibrated with gelfiltration buffer (100 mM NaCl, 20 mM Tris, pH 8.0, 1 mM EDTA). Fractions containing Mia40 were collected and concentrated in Amicon Ultra-15 Centrifugal Filter Unit (Millipore).

6.7.5.4 Purification of recombinant GST-Cox19 from *E. coli*

For purification of Cox19, cell pellets from 1 L expression culture were resuspended in 20 mL binding buffer (50 mM Tris-HCl, pH 8.0, 100 mM NaCl, 1 mM PMSF, 0.2 mg/mL lysozyme, 1 mM tris(2-carboxyethyl) phosphine [TCEP]). Cells were incubated in the buffer for 10 min at 4°C before. After centrifugation of the cell lysate for 20 min at 16,000 rpm (JA 25–50 rotor, Beckman coulter, Avanti J-25 or Avanti J-26XP), the supernatant was filtered through a 0.45 µm filter and applied to 0.5 mL GSH-Sepharose beads that were preequilibrated with the binding buffer without DTT and lysozyme. Binding of GST-Cox19 was allowed to proceed for 1.5 h at 4°C. After binding of the proteins, the beads were washed three times with 10 mL binding buffer without TCEP and lysozyme followed by elution of the protein with 500 µL 0.05 mg/ mL Prescission protease in binding buffer without TCEP and lysozyme overnight. For further purification and TCEP removal, the protein was subjected to gel filtration using a Hi Load 16/60 Superdex 200 column. The reduced state of the protein was verified by AMS gel shift assays.

6.7.6 In vitro reconstitution of the mitochondrial disulfide relay with purified proteins

Disulfide bond formation works as a cascade of electron transfer steps (Riemer et al. 2009; Stojanovski et al. 2008b; Tokatlidis 2005; Tu and Weissman 2004). The beginning of the cascade is constituted by an electron donor; in most cases, this is the reduced substrate. For the stepwise characterization of the disulfide relay pathway, it is also possible to provide either reduced Mia40 or parts of Erv1 as electron donors.

6.7.6.1 Preparation of electron donors (reduction of proteins)

Cox19 was purified under reducing conditions in the presence of 1 mM TCEP. After cleavage of the GST-tag, 1 mM TCEP was added and the protein was incubated for 15 min at room temperature. TCEP was removed by gel filtration using a Hi Load 16/60 Superdex 200 column (GE Healthcare). **Mia40** was reduced by incubation with 5 mM DTT for 15 min at 25°C. DTT was removed by gel filtration using a PD MiniTrap G-10 column (GE Healthcare). Recombinant **Erv1**$^{(N)}$ was incubated with 1 mM TCEP for 30 min at RT. TCEP was removed by gel filtration using a Hi Load 16/60 Superdex 200 column (GE Healthcare). Protein content and content of free thiol groups were measured using 5,5′-dithiobis-(2-nitrobenzoic acid) (DTNB, Ellman's reagent) (Ellman 1958).

6.7.6.2 Kinetic assays

To analyze the kinetics of the electron transfer reactions in the Mia40-Erv1 disulfide system, different assays have been applied that rely on monitoring redox state changes in the different components of the system by thiol alkylation and subsequent Western blot analysis (*kinetic alkylation shift assays*). A further powerful readout is the consumption of the terminal electron acceptors cytochrome c (*cytochrome c reduction assay*) and oxygen, or the concomitant production of H_2O_2 in the case that Erv1 transfers electrons directly to oxygen.

6.7.6.2.1 Alkylation shift assays with purified proteins

Kinetic alkylation shift assays usually consist of two parts: the kinetic reaction, which is started by mixing the respective components and stopped by rapid acid quenching (trichloroacetic acid [TCA] precipitation), and a modification step in which the redox state of the respective protein is fixed and analyzed.

6.7.6.2.2 Kinetic assay setup

The reduced electron donors (Cox19, Mia40 or Erv1$^{(N)}$, respectively) were incubated with cytochrome *c* and the respective enzymes required for reoxidation (▶Table 6.1). Samples were removed at different time points, TCA-precipitated and modified with an alkylating agent (see Section 6.7.6.2.4).

6.7.6.2.3 TCA precipitation of proteins

Proteins were precipitated by addition of 72% TCA to a final concentration of 12%. The samples were incubated at –20°C for 2 h or overnight and centrifuged for 30 min at 27,000 × g. The precipitated proteins were washed once with acetone (–20°C) and centrifuged again (27,000 × g, 30 min). Protein pellets were dried shortly at room temperature and dissolved in sample buffer.

6.7.6.2.4 Thiol modification with 4-acetoamido-4'-maleimidylstilbene-2, 2'-disulphonic acid (AMS)

Modification with AMS was used for the redox state analysis of Cox19, Mia40 and Erv1$^{(N)}$. AMS is a thiol-specific agent that alkylates free cysteine residues and thereby adds 0.5 kDa per bound AMS. Modifications of the respective proteins were performed after TCA precipitation. TCA pellets were resuspended in 20 μL AMS buffer (20 mM Tris, pH 7.0, 2% SDS, 15 mM AMS) and incubated at room temperature for 1 h. Then 20 μL 2× SDS sample buffer was added, and samples were incubated at 96°C for 3 min and further analyzed by SDS-PAGE or Tris-Tricine gel electrophoresis and Western blotting.

6.7.6.2.5 Alkylation shift assays with radiochemical protein amounts

A further possibility to analyze the kinetics of the mitochondrial disulfide relay is the use of in vitro translated radiolabeled Cox19 instead of purified Cox19 (▶Figure 6.3A). This Cox19 was already in a reduced state since the translation mix contains 5 mM DTT. The radioactive protein was diluted 60-fold and incubated with 30 μM of purified Mia40 and/or Erv1 under hypoxic conditions. Samples were removed at the indicated time points, TCA precipitated and modified with AMS (see Section 6.7.6.2.4). Samples were analyzed by SDS-PAGE or Tris-Tricine electrophoresis and autoradiography.

Table 6.1: Concentrations of recombinant proteins

Electron donor	Cyt c	Erv1	Erv1$^{(C)}$	Erv1$^{(N)}$	Mia40	Cox19
Cox19	40 μM	4 μM			4 μM	50 μM
Mia40	40 μM	4 μM			100 μM	
Mia40	40 μM		4 μM	4 μM	100 μM	
Erv1$^{(N)}$	40 μM		4 μM	100 μM		

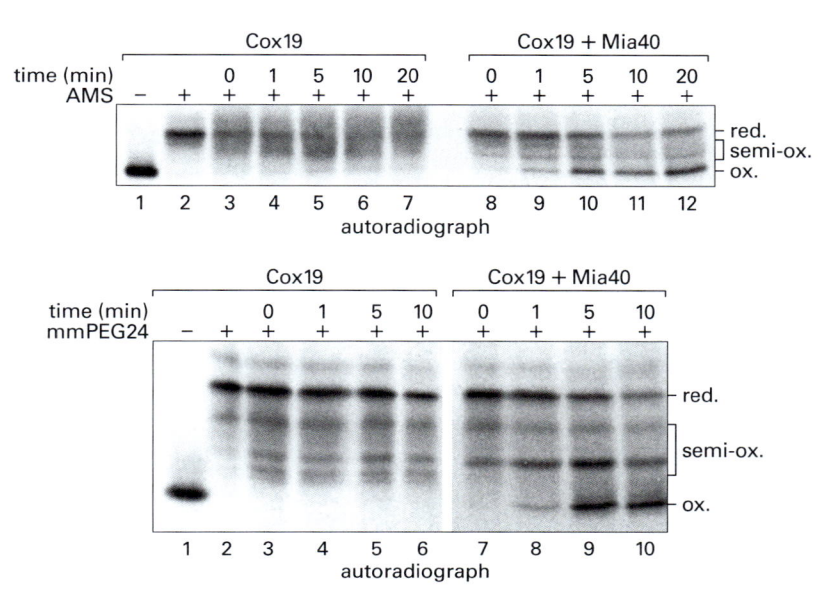

Figure 6.3: In vitro kinetic analyses of disulfide bond formation by Mia40 using AMS and mmPEG24. Radioactively labeled Cox19 was either incubated with buffer or with purified Mia40 for the indicated times. Subsequently, the reaction mix was TCA precipitated and modified with AMS (A) or mmPEG24 (B). Notably, alkylation with mmPEG24 results in a significantly increased migration shift of the reduced proteins compared to the shifts caused by treatment with AMS.

6.7.6.3 Cytochrome *c* reduction assays

In the disulfide relay system electrons are shuffled from Erv1 to cytochrome *c*. The reduction of cytochrome *c* can be monitored measuring the absorbance at 550 nm UV/visible light spectrophotometer (Jasco V650). Reduced cytochrome *c* has a maximum in absorbance at this wavelength, while the oxidized form has not. This can be used in a cytochrome *c* reduction assay, where electron transfer in the disulfide relay system is measured by monitoring the reduction of cytochrome *c*. The assay was performed in a 200 µL scale using the same concentrations as given in ▶Table 6.1, and the changes in cytochrome *c* absorption were followed online.

6.8 General comments to the analysis of thiol-disulfide redox states

6.8.1 Stop of thiol-disulfide exchange

A crucial step in working with thiol redox biology is the rapid "freezing" of the cellular thiol-disulfide redox state. The step is important to prevent postlysis air oxidation of the respective sample and to avoid the perturbation of thiol-disulfide redox equilibria in samples that contain redox-active enzymes. It is therefore recommendable to utilize a combined protein denaturation–thiol modification approach. To this end, proteins are first denatured and precipitated by addition of 10%–20% TCA or sulfosalicylic acid (SSA) (Jiang et al. 2004; Sivaraman et al. 1997). Under these low pH conditions, cysteine residues bear a thiol instead of the reactive thiolate anion group. This thiol only possesses a very low reactivity toward other thiols and is less susceptible to metal-catalyzed oxidation. Moreover, a denaturation ensures that previously buried cysteines become exposed and can subsequently be irreversibly modified by alkylation. Upon precipitation, low-molecular-weight compounds (e.g. reductants) remain soluble and thus can be efficiently separated from the precipitated proteins.

6.8.2 Visualization of redox states by alkylation and choice of alkylation agent

Thiol-alkylating agents are central to experimental redox biology and are used for a variety of applications in the characterization of thiols and disulfides. Here, we will only discuss the thiol alkylation by different maleimide-based modification reagents (for other reagents, refer to e.g. Hansen and Winther 2009). They all change the electrophoretic mobility of a protein and thus can be used to estimate the redox state of a thiol-disulfide couple in SDS-PAGE-based analyses.

In the alkylation reaction, the thiolate anion is the reactive nucleophile, and thus the reaction rate is dependent on thiol deprotonation. Consequently, it is necessary to raise the pH after TCA precipitation for alkylation to take place. Maleimides are generally thiol-specific below pH 7 and at concentrations in the range of 1–20 mM. However, reactions with protein amines have been observed at pH values above 7 if maleimides are in large excess or incubation times are prolonged to more than 2 h. Size shift-inducing maleimides include polyethylene glycol derivatives of N-ethyl maleimide (NEM) with molecular masses of 0.7 to 5 kDa (mmPEG's 12, 24, 2,000 and 5,000; see e.g. the Pierce homepage for further information on these compounds, http://www.piercenet.com) that have been used to determine the in vivo redox state of various proteins like members of the protein disulfide isomerase family (cf. ▶Figure 6.3B). In addition, the NEM derivative AMS, with a molecular mass of ca. 540 Da, has been applied extensively in a similar manner. AMS is light sensitive, so reactions should be carried out in the dark.

For the analysis of specific proteins, it is first necessary to choose the appropriate alkylating agent. Therefore, two factors have to be considered: First, an alkylating agent does not induce the same shift with all proteins. Good examples are the two oxidoreductases ERp57 and protein disulfide isomerase (PDI) that have the same size (ca. 55 kDa) and domain composition (four domains) as well as the same cysteine-containing active sites (CGHC). However, while the migration behavior of ERp57 can easily be influenced by AMS modification, PDI has to be analyzed using larger maleimides like mPEG 5000 (Appenzeller-Herzog and Ellgaard 2008). The second parameter that has to be taken into account when choosing an alkylating agent is that with increasing size of the maleimide variant, the efficiency of protein transfer upon Western blot analysis decreases. Thus, the selection of the alkylating agent always has to be a compromise between the induced size shift and the transferability during Western blot.

6.8.3 Reduction of disulfides

The specific detection of disulfide bonds requires the efficient and quantitative reduction of disulfides prior to alkylation. Disulfide reduction is typically flanked by two thiol derivatization steps – the former to modify free (reduced) cysteines and the latter to induce a detectable size shift of the protein by modifying previously oxidized cysteines. In addition to being an efficient reductant, the reducing agent must therefore be removable or unable to cross-react with the derivatization agent in the second step. The most commonly used disulfide reductants are thiols themselves. The reaction follows a thiol-disulfide exchange mechanism. The main advantage of thiol exchange reductants is the high degree of specificity, meaning that only disulfides are chemically affected. The main disadvantage of thiol-containing reductants is that their thiol groups compete directly with the protein thiols for attachment of thiol modifying reagents. In addition, the reagents require a reaction pH above 7, where oxidation by ambient molecular oxygen can occur. The reagents themselves are particularly sensitive to oxidation, and consequently all solutions should be maintained in an oxygen-free state. The most commonly used reductant is the dithiol DTT, which reduces disulfides in a reaction where DTT itself is converted into a stable intramolecular cyclic disulfide. The formation of oxidized DTT is driven by favorable steric and entropic effects, and thus DTT is a strong reducing agent (K_{eq} = of 1.3×10^4 M) (Cleland 1964). Accordingly, mixed disulfides between thiols and DTT do not accumulate, and a large excess of reagent is not required for complete disulfide reduction. The thiol pK_a values of DTT are 9.2 and 10.1 (Whitesides et al. 1977), and reduction is typically carried out at or above pH 8. A further important advantage of DTT is its membrane permeability, and it is widely used as a reductant in cellular systems, where it reversibly inhibits protein disulfide bond formation without hindering protein synthesis. Typically, 5–10 mM DTT for 5–10 min is used for the reduction of proteins in whole cells.

Nonthiol reductants include trialkylphosphines. For these compounds, the first and rate-limiting step is attack of the disulfide bond by the phosphine nucleophile, forming a thiophosphonium salt. Next, rapid hydrolysis releases the second thiol fragment and the phosphine oxide. Importantly, the hydrolysis step renders the reaction irreversible. Thus, contrary to DTT, the oxidized phosphine cannot participate in further thiol-disulfide reactions (Burns et al. 1991). Trialkylphosphines react specifically with disulfides and do not react with other functional groups commonly found in proteins. Phosphines do, however, react rapidly with maleimide compounds even under acidic conditions. Thus, as in the case of thiol reductants, reactions with phosphines are best carried out in a step separate from alkylation with NEM, and their derivatives. TCEP is a highly water-soluble reagent and the most commonly used phosphine for disulfide bond reduction. The pK_a value of the phosphorus is 7.6 (Krezel et al. 2003), and TCEP is significantly more effective in reducing low-molecular-weight disulfides and surface-exposed protein disulfides than is DTT, particularly at pH below 7. TCEP is highly acidic when dissolved in water (four protons accompany each TCEP molecule), so samples must be adequately buffered after the addition of this reagent.

6.9 Outlook

The mitochondrial disulfide relay system was identified only recently. Since then, this system was analyzed mainly by variations of the protocols described in this chapter. In addition both structural analyses (Banci et al. 2009; Gross et al. 2002; Kawano et al. 2009; Wu et al. 2003) and the use of blue native gels (Milenkovic et al. 2009; Rissler et al. 2005; Stojanovski et al. 2008a) proved to be very powerful to explore the mechanism by which Mia40 and Erv1 mediate protein import. In the future, these strategies should be complemented on the one hand by biophysical methods to unravel the molecular details of the system and, on the other hand, by studies in cell culture and knock-out animals to learn more about its physiological relevance in higher eukaryotes.

References

Abajian, C., Yatsunyk, L.A., Ramirez, B.E., and Rosenzweig, A.C. (2004). Yeast cox17 solution structure and copper(I) binding. J Biol Chem *279*, 53584–53592.

Allen, S., Balabanidou, V., Sideris, D.P., Lisowsky, T., and Tokatlidis, K. (2005). Erv1 mediates the Mia40-dependent protein import pathway and provides a functional link to the respiratory chain by shuttling electrons to cytochrome *c*. J Mol Biol *353*, 937–944.

Allen, S., Lu, H., Thornton, D., and Tokatlidis, K. (2003). Juxtaposition of the two distal CX3C motifs via intrachain disulfide bonding is essential for the folding of Tim10. J Biol Chem *278*, 38505–38513.

Ang, S.K., and Lu, H. (2009). Deciphering structural and functional roles of individual disulfide bonds of the mitochondrial sulfhydryl oxidase Erv1p. J Biol Chem *284*, 28754–28761.

Appenzeller-Herzog, C., and Ellgaard, L. (2008). In vivo reduction-oxidation state of protein disulfide isomerase: the two active sites independently occur in the reduced and oxidized forms. Antioxid Redox Signal *10*, 55–64.

Arnesano, F., Balatri, E., Banci, L., Bertini, I., and Winge, D.R. (2005). Folding studies of Cox17 reveal an important interplay of cysteine oxidation and copper binding. Structure *13*, 713–722.

Banci, L., Bertini, I., Cefaro, C., Cenacchi, L., Ciofi-Baffoni, S., Felli, I.C., et al. (2010). Molecular chaperone function of Mia40 triggers consecutive induced folding steps of the substrate in mitochondrial protein import. Proc Natl Acad Sci U S A *107*, 20190–20195.

Banci, L., Bertini, I., Cefaro, C., Ciofi-Baffoni, S., Gallo, A., Martinelli, M., et al. (2009). Mia40 is an oxidoreductase that catalyzes oxidative protein folding in mitochondria. Nat Struct Mol Biol *16*, 198–206.

Barford, D. (2004). The role of cysteine residues as redox-sensitive regulatory switches. Curr Opin Struct Biol *14*, 679–686.

Bauer, M.F., Hofmann, S., Neupert, W., and Brunner, M. (2000). Protein translocation into mitochondria: the role of TIM complexes. Trends Cell Biol *100*, 25–31.

Benz, R. (1994). Permeation of hydrophilic solutes through mitochondrial outer membranes: review on mitochondrial porins. Biochim Biophys Acta *1197*, 167–196.

Bien, M., Longen, S., Wagener, N., Chwalla, I., Herrmann, J.M., and Riemer, J. (2010). Mitochondrial disulfide bond formation is driven by intersubunit electron transfer in Erv1 and proof read by glutathione. Mol Cell *37*, 516–528.

Bihlmaier, K., Bien, M., and Herrmann, J.M. (2008). In vitro import of proteins into isolated mitochondria. Methods Mol Biol *457*, 85–94.

Bihlmaier, K., Mesecke, N., Terzyiska, N., Bien, M., Hell, K., and Herrmann, J.M. (2007). The disulfide relay system of mitochondria is connected to the respiratory chain. J Cell Biol *179*, 389–395.

Burns, J.A., Butler, J.C., Moran, J., and Whitesides, G.M. (1991). Selective reduction of disulfides by tris(2-carboxyethyl)phosphine. J Org Chem *56*, 2648–2650.

Cavallaro, G. (2010). Genome-wide analysis of eukaryotic twin CX9C proteins. Mol Biosyst *6*, 2459–2470.

Chacinska, A., Pfannschmidt, S., Wiedemann, N., Kozjak, V., Sanjuan Szklarz, L.K., Schulze-Specking, A., et al. (2004). Essential role of Mia40 in import and assembly of mitochondrial intermembrane space proteins. EMBO J *23*, 3735–3746.

References

Cleland, W.W. (1964). Dithiothreitol, a new protective reagent for Sh groups. Biochemistry *3*, 480–482.

Collet, J.F., and Bardwell, J.C. (2002). Oxidative protein folding in bacteria. Mol Microbiol *44*, 1–8.

Dabir, D.V., Leverich, E.P., Kim, S.K., Tsai, F.D., Hirasawa, M., Knaff, D.B., et al. (2007). A role for cytochrome c and cytochrome c peroxidase in electron shuttling from Erv1. EMBO J *26*, 4801–4811.

Daithankar, V.N., Farrell, S.R., and Thorpe, C. (2009). Augmenter of liver regeneration: substrate specificity of a flavin-dependent oxidoreductase from the mitochondrial intermembrane space. Biochemistry *48*, 4828–4837.

Di Fonzo, A., Ronchi, D., Lodi, T., Fassone, E., Tigano, M., Lamperti, C., et al. (2009). The mitochondrial disulfide relay system protein GFER is mutated in autosomal-recessive myopathy with cataract and combined respiratory-chain deficiency. Am J Hum Genet *84*, 594–604.

Ellman, G.L. (1958). A colorimetric method for determining low concentrations of mercaptans. Arch Biochem Biophys *74*, 443–450.

Emanuelsson, O., Brunak, S., von Heijne, G., and Nielsen, H. (2007). Locating proteins in the cell using TargetP, SignalP and related tools. Nat Protoc *2*, 953–971.

Farrell, S.R., and Thorpe, C. (2005). Augmenter of liver regeneration: a flavin-dependent sulfhydryl oxidase with cytochrome c reductase activity. Biochemistry *44*, 1532–1541.

Frey, T.G., and Mannella, C.A. (2000). The internal structure of mitochondria. Trends Biochem Sci *25*, 319–324.

Frezza, C., Cipolat, S., Martins de Brito, O., Micaroni, M., Beznoussenko, G.V., Rudka, T., et al. (2006). OPA1 controls apoptotic cristae remodeling independently from mitochondrial fusion. Cell *126*, 177–189.

Gabriel, K., Milenkovic, D., Chacinska, A., Muller, J., Guiard, B., Pfanner, N., et al. (2007). Novel mitochondrial intermembrane space proteins as substrates of the MIA import pathway. J Mol Biol *365*, 612–620.

Girzalsky, W., Saffian, D., and Erdmann, R. (2010). Peroxisomal protein translocation. Biochim Biophys Acta *1803*, 724–731.

Glerum, D.M., Shtanko, A., and Tzagoloff, A. (1996). Characterization of *COX17*, a yeast gene involved in copper metabolism and assembly of cytochrome oxidase. J Biol Chem *271*, 14504–14509.

Gross, E., Sevier, C.S., Vala, A., Kaiser, C.A., and Fass, D. (2002). A new FAD-binding fold and intersubunit disulfide shuttle in the thiol oxidase Erv2p. Nat Struct Biol *9*, 61–67.

Grumbt, B., Stroobant, V., Terziyska, N., Israel, L., and Hell, K. (2007). Functional characterization of Mia40p, the central component of the disulfide relay system of the mitochondrial intermembrane space. J Biol Chem *282*, 37461–37470.

Hansen, R.E., and Winther, J.R. (2009). An introduction to methods for analyzing thiols and disulfides: reactions, reagents, and practical considerations. Anal Biochem *394*, 147–158.

Hell, K. (2008). The Erv1-Mia40 disulfide relay system in the intermembrane space of mitochondria. Biochim Biophys Acta *1783*, 601–609.

Herrmann, J.M., and Hell, K. (2005). Chopped, trapped or tacked – protein translocation into the IMS of mitochondria. Trends Biochem Sci *30*, 205–212.

Herrmann, J.M., Kauff, F., and Neuhaus, H.E. (2009). Thiol oxidation in bacteria, mitochondria and chloroplasts: common principles but three unrelated machineries? Biochim Biophys Acta *1793*, 71–77.

Herrmann, J.M., and Riemer, J. (2010). The intermembrane space of mitochondria. Antioxid Redox Signal *13*, 1341–1358.

Hofhaus, G., Stein, G., Polimeno, L., Francavilla, A., and Lisowsky, T. (1999). Highly divergent amino termini of the homologous human ALR and yeast scERV1 gene products define species specific differences in cellular localization. Eur J Cell Biol *78*, 349–356.

Hofmann, S., Rothbauer, U., Muhlenbein, N., Baiker, K., Hell, K., and Bauer, M.F. (2005). Functional and mutational characterization of human MIA40 acting during import into the mitochondrial intermembrane space. J Mol Biol *353*, 517–528.

Hofmann, S., Rothbauer, U., Muhlenbein, N., Neupert, W., Gerbitz, K.D., Brunner, M., et al. (2002). The C66W mutation in the deafness dystonia peptide 1 (DDP1) affects the formation of functional DDP1.TIM13 complexes in the mitochondrial intermembrane space. J Biol Chem *277*, 23287–23293.

Hu, J., Dong, L., and Outten, C.E. (2008). The redox environment in the mitochondrial intermembrane space is maintained separately from the cytosol and matrix. J Biol Chem *283*, 29126–29134.

Huh, W.K., Falvo, J.V., Gerke, L.C., Carroll, A.S., Howson, R.W., Weissman, J.S., et al. (2003). Global analysis of protein localization in budding yeast. Nature *425*, 686–691.

Inaba, K., and Ito, K. (2008). Structure and mechanisms of the DsbB-DsbA disulfide bond generation machine. Biochim Biophys Acta *1783*, 520–529.

Jiang, L., He, L., and Fountoulakis, M. (2004). Comparison of protein precipitation methods for sample preparation prior to proteomic analysis. J Chromatogr A *1023*, 317–320.

Kadokura, H., Katzen, F., and Beckwith, J. (2003). Protein disulfide bond formation in prokaryotes. Annu Rev Biochem *72*, 111–135.

Kawano, S., Yamano, K., Naoe, M., Momose, T., Terao, K., Nishikawa, S., et al. (2009). Structural basis of yeast Tim40/Mia40 as an oxidative translocator in the mitochondrial intermembrane space. Proc Natl Acad Sci U S A *106*, 14403–14407.

Koehler, C.M. (2004). The small Tim proteins and the twin Cx3C motif. Trends Biochem Sci *29*, 1–4.

Koehler, C.M., Leuenberger, D., Merchant, S., Renaold, A., Junne, T., and Schatz, G. (1999). Human deafness dystonia syndrome is a mitochondrial disease. Proc Natl Acad Sci U S A *96*, 2141–2146.

Koehler, C.M., and Tienson, H.L. (2009). Redox regulation of protein folding in the mitochondrial intermembrane space. Biochim Biophys Acta *1793*, 139–145.

Krezel, A., Latajka, R., Bujacz, G.D., and Bal, W. (2003). Coordination properties of tris(2-carboxyethyl)phosphine, a newly introduced thiol reductant, and its oxide. Inorg Chem *42*, 1994–2003.

Kumar, A., Agarwal, S., Heyman, J.A., Matson, S., Heidtman, M., Piccirillo, S., et al. (2002). Subcellular localization of the yeast proteome. Genes Dev *16*, 707–719.

Lange, A., Mills, R.E., Lange, C.J., Stewart, M., Devine, S.E., and Corbett, A.H. (2007). Classical nuclear localization signals: definition, function, and interaction with importin alpha. J Biol Chem *282*, 5101–5105.

Lazarow, P.B. (2006). The import receptor Pex7p and the PTS2 targeting sequence. Biochim Biophys Acta *1763*, 1599–1604.

Lill, R., and Mühlenhoff, U. (2005). Iron-sulfur-protein biogenesis in eukaryotes. Trends Biochem Sci *30*, 133–141.

Lionaki, E., Aivaliotis, M., Pozidis, C., and Tokatlidis, K. (2010). The N-terminal shuttle domain of Erv1 determines the affinity for Mia40 and mediates electron transfer to the catalytic Erv1 core in yeast mitochondria. Antioxid Redox Signal *13*, 1327–1339.

Lisowsky, T. (1992). Dual function of a new nuclear gene for oxidative phosphorylation and vegetative growth in yeast. Mol Gen Genet *232*, 58–64.

Longen, S., Bien, M., Bihlmaier, K., Kloeppel, C., Kauff, F., Hammermeister, M., et al. (2009). Systematic analysis of the twin cx_9c protein family. J Mol Biol *393*, 356–368.

Lutz, T., Neupert, W., and Herrmann, J.M. (2003). Import of small Tim proteins into the mitochondrial intermembrane space. EMBO J *22*, 4400–4408.

Mesecke, N., Bihlmaier, K., Grumbt, B., Longen, S., Terziyska, N., Hell, K., et al. (2008). The zinc-binding protein Hot13 promotes oxidation of the mitochondrial import receptor Mia40. EMBO Rep *9*, 1107–1113.

Mesecke, N., Terziyska, N., Kozany, C., Baumann, F., Neupert, W., Hell, K., et al. (2005). A disulfide relay system in the intermembrane space of mitochondria that mediates protein import. Cell *121*, 1059–1069.

Milenkovic, D., Ramming, T., Muller, J.M., Wenz, L.S., Gebert, N., Schulze-Specking, A., et al. (2009). Identification of the signal directing Tim9 and Tim10 into the intermembrane space of mitochondria. Mol Biol Cell 20, 2530–2539.

Müller, J.M., Milenkovic, D., Guiard, B., Pfanner, N., and Chacinska, A. (2008). Precursor oxidation by mia40 and erv1 promotes vectorial transport of proteins into the mitochondrial intermembrane space. Mol Biol Cell *19*, 226–236.

Naoe, M., Ohwa, Y., Ishikawa, D., Ohshima, C., Nishikawa, S., Yamamoto, H., et al. (2004). Identification of Tim40 that mediates protein sorting to the mitochondrial intermembrane space. J Biol Chem *279*, 47815–47821.

Pelham, H.R.B., and Jackson, R.J. (1976). An efficient mRNA-dependent translation system from reticulocyte lysates. Eur J Biochem *67*, 247–256.

Reddie, K.G., and Carroll, K.S. (2008). Expanding the functional diversity of proteins through cysteine oxidation. Curr Opin Chem Biol *12*, 746–754.

Riemer, J., Bulleid, N., and Herrmann, J.M. (2009). Disulfide formation in the ER and mitochondria: two solutions to a common process. Science *324*, 1284–1287.

Rissler, M., Wiedemann, N., Pfannschmidt, S., Gabriel, K., Guiard, B., Pfanner, N., et al. (2005). The essential mitochondrial protein Erv1 cooperates with Mia40 in biogenesis of intermembrane space proteins. J Mol Biol *353*, 485–492.

Sickmann, A., Reinders, J., Wagner, Y., Joppich, C., Zahedi, R., Meyer, H.E., et al. (2003). The proteome of *Saccharomyces cerevisiae* mitochondria. Proc Natl Acad Sci U S A *100*, 13207–13212.

Sideris, D.P., Petrakis, N., Katrakili, N., Mikropoulou, D., Gallo, A., Ciofi-Baffoni, S., et al. (2009). A novel intermembrane space-targeting signal docks cysteines onto Mia40 during mitochondrial oxidative folding. J Cell Biol *187*, 1007–1022.

Sideris, D.P., and Tokatlidis, K. (2007). Oxidative folding of small Tims is mediated by site-specific docking onto Mia40 in the mitochondrial intermembrane space. Mol Microbiol *65*, 1360–1373.

Sivaraman, T., Kumar, T.K., Jayaraman, G., and Yu, C. (1997). The mechanism of 2,2,2-trichloroacetic acid-induced protein precipitation. J Protein Chem *16*, 291–297.

Stojanovski, D., Milenkovic, D., Muller, J.M., Gabriel, K., Schulze-Specking, A., Baker, M.J., et al. (2008a). Mitochondrial protein import: precursor oxidation in a ternary complex with disulfide carrier and sulfhydryl oxidase. J Cell Biol *183*, 195–202.

Stojanovski, D., Muller, J.M., Milenkovic, D., Guiard, B., Pfanner, N., and Chacinska, A. (2008b). The MIA system for protein import into the mitochondrial intermembrane space. Biochim Biophys Acta *1783*, 610–617.

Terziyska, N., Grumbt, B., Kozany, C., and Hell, K. (2009). Structural and functional roles of the conserved cysteine residues of the redox-regulated import receptor Mia40 in the intermembrane space of mitochondria. J Biol Chem *284*, 1353–1363.

Terziyska, N., Lutz, T., Kozany, C., Mokranjac, D., Mesecke, N., Neupert, W., et al. (2005). Mia40, a novel factor for protein import into the intermembrane space of mitochondria is able to bind metal ions. FEBS Lett *579*, 179–184.

Tienson, H.L., Dabir, D.V., Neal, S.E., Loo, R., Hasson, S.A., Boontheung, P., et al. (2009). Reconstitution of the mia40-erv1 oxidative folding pathway for the small tim proteins. Mol Biol Cell *20*, 3481–3490.

Tokatlidis, K. (2005). A disulfide relay system in mitochondria. Cell *121*, 965–967.

Tu, B.P., and Weissman, J.S. (2004). Oxidative protein folding in eukaryotes: mechanisms and consequences. J Cell Biol *164*, 341–346.

van Gurp, M., Festjens, N., van Loo, G., Saelens, X., and Vandenabeele, P. (2003). Mitochondrial intermembrane proteins in cell death. Biochem Biophys Res Commun *304*, 487–497.

Wang, C., and Youle, R.J. (2009). The role of mitochondria in apoptosis. Annu Rev Genet *43*, 95–118.

Webb, C.T., Gorman, M.A., Lazarou, M., Ryan, M.T., and Gulbis, J.M. (2006). Crystal structure of the mitochondrial chaperone TIM9–10 reveals a six-bladed alpha-propeller. Mol Cell *21*, 123–133.

Whitesides, G.M., Lilburn, J.E., and Szajewski, R.P. (1977). Rates of thiol-disulfide interchange reactions between mono- and dithiols and Ellmans reagent. J Org Chem *42*, 332–338.

Wickner, W., and Schekman, R. (2005). Protein translocation across biological membranes. Science *310*, 1452–1456.

Wu, C.K., Dailey, T.A., Dailey, H.A., Wang, B.C., and Rose, J.P. (2003). The crystal structure of augmenter of liver regeneration: a mammalian FAD-dependent sulfhydryl oxidase. Protein Sci *12*, 1109–1118.

Zick, M., Rabl, R., and Reichert, A.S. (2009). Cristae formation-linking ultrastructure and function of mitochondria. Biochim Biophys Acta *1793*, 5–19.

7 On-membrane identification of gel-resolved proteins by matrix-assisted laser desorption ionization mass spectrometry (MALDI-MS)

Yoko Ino, Reiko Kazamaki, and Hisashi Hirano

7.1 Introduction

In proteome research, two methods are frequently used to identify proteins: two-dimensional gel electrophoresis/mass spectrometry (2-DE/MS) and shotgun analysis. In 2-DE/MS, proteins are separated by 2-DE and digested in gels with a protease such as trypsin; the digests are then analyzed by MS or tandem mass spectrometry (MS/MS). In shotgun analysis, protein mixtures are digested with a protease such as trypsin, and the digests are separated by liquid chromatography (LC) and analyzed by MS/MS. When the same sample is analyzed by both methods, shotgun analysis usually results in the identification of more proteins than 2-DE/MS. On the other hand, proteins such as isoforms and posttranslationally modified proteins that cannot be detected by shotgun analysis can often be identified by 2-DE/MS. However, in 2-DE/MS, in-gel digestion is laborious and time consuming and requires a certain amount of skill if an automated in-gel digestion instrument is not available. To address these problems, experts are working on developing simple and efficient techniques for analyzing gel-resolved proteins.

It is well known that proteins separated by electrophoresis can be transferred electrophoretically from gels onto membranes such as polyvinylidene difluoride (PVDF) membranes. From a population of immobilized proteins, antigens can be detected using specific antibodies, and posttranslationally modified proteins such as glycoproteins and phosphoproteins can be identified using certain ligands. Also, the N-terminal amino acid sequence of the immobilized proteins can be determined by a gas-phase sequencer (Matsudaira 1987), allowing identification.

If proteins transferred from gels onto membranes are analyzed directly by MS on the same membranes, we can easily obtain the information about the protein structure. This technique avoids the laborious and time-consuming in-gel digestion that is ordinarily required for identification. In addition, proteins immobilized on the membranes can easily be desalted, resulting in a high signal-to-noise ratio in MS.

Studies of the electroblotting and MS analysis of proteins separated by gel electrophoresis have been performed since the 1990s (▶Table 7.1).

Eckerskorn et al. (1992) electroblotted SDS-PAGE gel-resolved proteins onto PVDF membrane, cut out pieces of the membrane containing proteins and analyzed protein molecular masses by MALDI MS. This is the first report on mass spectrometric analysis of blotted proteins after gel electrophoresis by MALDI MS. Subsequently, Strupat et al. (1994) separated six proteins by SDS-PAGE and electroblotted them onto five types of PVDF membranes: Transblot (140 μm thick, 0.3 μm pore), Immobilon PSQ (195 μm thick, 0.25 μm pore), Fluorotrans (140 μm thick, 0.33 μm pore), Westran (150 μm thick, 0.78 μm pore), and Immobilon P (130 μm thick, 0.69 μm pore). In their experiment, pieces of the membrane containing a protein of interest were incubated

Table 7.1: Mass spectrometric analysis of proteins electroblotted onto membranes

Reference	Electrophoresis	Membrane for Blotting	Cleavage	Target for MALDI MS
Eckerskorn et al. (1992)	SDS-PAGE	PVDF (Fluoretrans)	None	Proteins on the membrane
Sprupat et al. (1994)	SDS-PAGE	PVDF (Transblott, Immobilon PSQ, Fluorotrans, Westran, Immobillon-P)	None	Proteins on the membrane
Patterson (1995)	SDS-PAGE	PVDF (Immobilon CD)	Lys-C	Peptides eluted from the membrane
Blackledge and Alexander (1995)	SDS-PAGE	Polyethylene	None	Proteins on the membrane
Sutton et al. (1995)	2-DE	PVDF (Fluorotrans)	Trypsin	Peptides eluted from the membrane
Gharahdaghi et al. (1996)	SDS-PAGE	PVDF (Immobilon PSQ)	Lys-C	Peptides eluted from the membrane
Schreiner et al. (1996)	SDS-PAGE	PVDF (Immobilon PSQ, Trans-Blot, Immobilon CD)	None	Proteins on the membrane
Eckerskorn et al. (1997)	2-DE	PVDF (Immobilon PSQ)	None	Proteins on the membrane
Fernandez et al. (1998)	SDS-PAGE	PVDF (not described)	Lys-C	Peptides on the membrane
Schleuder et al. (1999)	SDS-PAGE	PVDF (Immobilon CD)	Trypsin	Peptides on the membrane
Sloane et al. (2002)	SDS-PAGE/2-DE	PVDF (Immobilon PSQ)	Trypsin, Glu-C	Peptides on the membrane
Bunai et al. (2003)	Tricine SDS-PAGE/2-DE	PVDF (0.45 µm pore)	Lys-C	Peptide eluted from the membrane
Methogo et al. (2005)	SDS-PAGE/2-DE	PVDF (not described)	Trypsin	Peptides eluted from the membrane
Pham et al. (2005)	SDS-PAGE	PVDF (ProBlot)	Trypsin, Lys-N, Lys-C, Glu-C	Peptides eluted from the membrane

Nakanishi et al. (2005)	SDS-PAGE/2-DE	PVDF (Immobilon PSQ)	Trypsin	Peptides on the membrane
Chang et al. (2006)	SDS-PAGE	PVDF (Hybond-P)	Chymotrypsin	Peptides eluted from the membrane
Luque-Garcia et al. (2006)	SDS-PAGE	Nitrocellulose (0.2 μm pore)	Trypsin	Peptides eluted from the membrane
Kimura et al. (2007)	SDS-PAGE/2-DE	PVDF (Immobilon PSQ)	Trypsin	Glycopeptides on the membrane
Nakanishi et al. (2007)	2-DE	PVDF (Immobilon FL)?	Trypsin	Phosphoproteins on the membrane

with the matrix, cut out and mounted on the sample support with a conductive double-sided adhesive tape; then, the masses of *proteins*, but not peptides, were determined by infrared–matrix-assisted laser desorption ionization–time-of-flight mass spectrometry (IR-MALDI-TOF MS). The authors reported that better results were obtained from membranes that have highly specific surfaces and low mean pore size. Similar studies were performed to determine the masses of gel-resolved proteins on membranes (Blackledge and Alexander 1995; Eckerskorn et al. 1997; Schreiner et al. 1996).

This technique was used to identify gel-resolved proteins by MALDI MS. Patterson (1995) digested gel-resolved proteins on PVDF membrane with Lys-C endopeptidase and *eluted* the digests from the membrane for analysis by MALDI MS in order to identify the proteins. Patterson used Immobilon CD PVDF membrane with quaternary amines linked to the membrane surface, which might be useful for efficient recovery of the electroblotted proteins (peptides) from the membrane. Bunai et al. (2003), Methogo et al. (2005), Pham et al. (2005), Chang et al. (2006) and Luque-Garcia et al. (2006) identified gel-resolved proteins by similar techniques but used different types of membranes.

Schleuder et al. (1999) *directly* analyzed peptides on the membrane by MALDI MS. After electroblotting, they used trypsin to digest proteins contained in pieces of PVDF membrane and secured them on the MALDI sample plate by conductive double-sided adhesive tape. Next, they introduced the membrane into MS in order to identify the proteins by mass fingerprinting. Later, similar experiments were performed by Sloane et al. (2002) and Nakanishi et al. (2005). They immobilized gel-resolved proteins onto PVDF membranes using a piezoelectric chemical inkjet printer, and subsequently digested them with protease(s). Sloane et al. applied this technique to the identification of antigens and glycoproteins, which can be detected by Western blotting and lectin blotting, respectively. Kimura et al. (2007) and Nakanishi et al. (2007) used the same technique to identify glycochains of proteins and phosphoproteins, respectively.

Techniques for membrane-based protein identification techniques following 2-DE protein separation are an area of interest for many researchers. However, there are a number of limitations that hinder widespread use of these techniques. First, these methods have not always yielded robust and interpretable results. One of the reasons for this poor outcome is low ionization efficiency, or low ion yield of peptides and proteins that are immobilized on PVDF membranes. Therefore, it was necessary to develop new membranes from which digests of the immobilized protein could be ionized efficiently. Without this improvement in membranes, the peptides obtained by on-membrane digestion of gel-resolved proteins would have to be eluted from the membrane before being subjected to MALDI for identification.

Previously, PVDF membranes with 0.2~0.45 µm pore have been used for electroblotting and subsequent mass spectrometric analysis of gel-resolved proteins. Recent work explored other membranes that might be used for this purpose; it was found that the smaller pore membranes, such as 0.1 µm pore PVDF membranes (Millipore, Bradford, MA, USA) are suitable for highly efficient and effective electroblotting and mass spectrometric analysis. In this chapter, we describe methods for electroblotting and mass spectrometric analysis using 0.1 µm pore PVDF membranes.

7.2 Methods for identifying proteins electroblotted onto the PVDF membrane

▶ Figure 7.1 shows the experimental procedure of mass spectrometric analysis of proteins transferred from gels to the PVDF membranes.

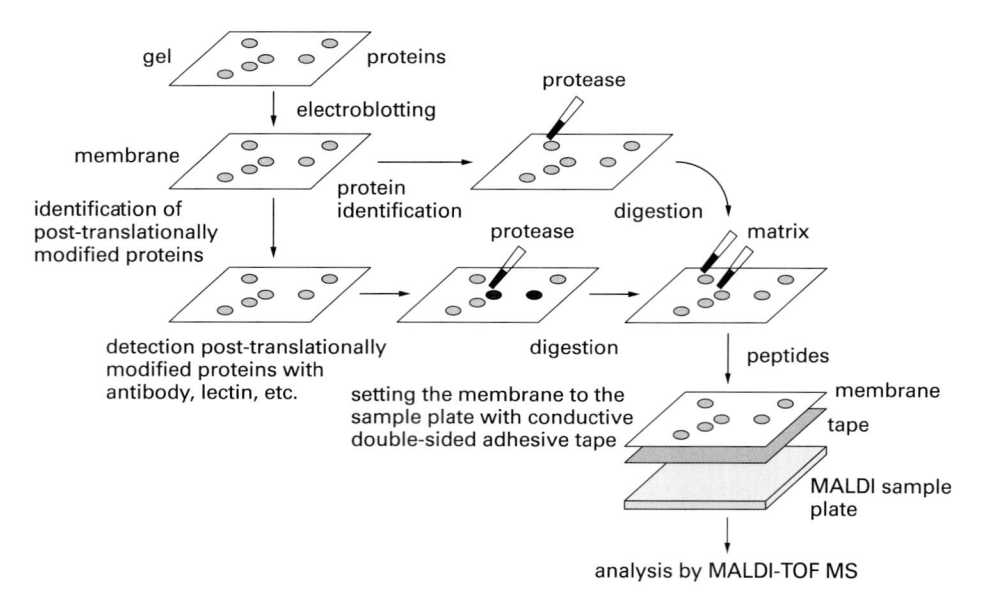

Figure 7.1: Experimental procedure of mass spectrometric analysis for proteins transferred from gels to PVDF membranes.

7.2.1 Materials

Low-molecular-weight marker proteins for use as standard proteins can be obtained from GE Healthcare (Milwaukee, WI, USA); this mixture contains rabbit muscle phosphorylase *b*, bovine serum albumin, bovine carbonic anhydrase, soybean trypsin inhibitor and α-lactalbumin. Des-Arg-bradykinin, angiotensin I, Glu-fibrinopeptide B, ACTH (1-17 clip), ACTH (18-39 clip), can be obtained from Applied Biosystems (Foster City, CA, USA). Adrenocorticotropic hormone (ACTH) 18–39 (human) and insulin β chain (bovine) can be obtained from Sigma Aldrich (Ontario, Canada).

Model proteins are extracted from yeast (*Saccharomyces cerevisiae* B-8052) grown in YPD medium at 30°C, as previously described (Iwafune et al. 2007).

7.2.2 Electrophoresis and electroblotting

SDS-PAGE is performed using 12%T and 2.6%C polyacrylamide gels as previously described (Laemmli 1970); 2-DE using immobilized pH gradient gel strips (pH 3–10, 13 cm)(Bio-Rad Laboratories, Hercules, CA, USA) is performed by the method as previously described (Gorg et al. 1988). Proteins separated on the gels are electroblotted onto the membranes using a semidry blotting apparatus (Nihon Eido, Tokyo, Japan) as previously described (Hirano 1989). Proteins on the PVDF membranes are stained with Direct Blue 71 (Sigma Aldrich, St. Louis, MO, USA) and subjected to MS analysis.

7.2.3 Quantification of proteins electroblotted onto membranes

The amount of protein electroblotted onto PVDF membranes can be determined using fluorescent dyes; proteins such as phosphorylase *b*, albumin, carbonic anhydrase, trypsin inhibitor and α-lactalbumin are labeled with Cy3 separated by SDS-PAGE or 2-DE and electroblotted onto various membranes using a semidry blotting apparatus. The Cy3-labeled proteins are detected on the membranes using a Typhoon 9410 laser scanner (GE Healthcare) with an emission wavelength of 580 nm and excitation wavelength of 532 nm. Then these proteins are quantitatively analyzed using ImageQuant software (GE Healthcare).

7.2.4 On-membrane tryptic digestion and mass spectrometric analysis

After electroblotting, the membranes are rinsed with distilled water and pasted to a MALDI target plate using electrically conductive double-sided adhesive tape (Sumitomo 3M, Tokyo, Japan). An aliquot (0.5 µL) of the protease solution containing 100 µg/mL trypsin (Promega, Madison, WI, USA) in 25 mM NH_4HCO_3 (pH 8.0)/40% (v/v) acetonitrile is spotted on each protein position. Proteins on the membranes are digested with trypsin overnight at 37°C in a humid reaction chamber. After digestion, 0.5 µL of matrix solution saturated with 10 mg/mL 2,5-dihydroxy benzoic acid in 30% (v/v) acetonitrile/0.1% trifluoroacetic acid is spotted on the protein positions. The digests are analyzed using matrix-assisted laser desorption ionization–time-of-flight mass spectrometry (MALDI-TOF MS) (AXIMA-CFR, Kyoto, Shimadzu), and proteins are identified by peptide mass fingerprinting. In MALDI-TOF MS, analysis of peptides is performed in the reflector TOF detection mode. Otherwise, after tryptic digestion of proteins, 0.5 µL of the matrix solution saturated with α-cyano-4-hydroxycinnamic acid (CHCA) in 30% (v/v) acetonitrile/0.1% trifluoroacetic acid is spotted on the protein positions. The digests are analyzed using matrix-assisted laser desorption ionization–time-of-flight/time-of-flight mass spectrometry (MALDI-TOF/TOF MS) (4800, Applied Biosystems), and proteins are identified by MS/MS analysis. The TOF/TOF MS is operated in positive ion reflectron mode. MS spectra are acquired at 800 and 4,000 (m/z) using 1,000 laser shots. The top ten peaks in each MS spectrum are selected for MS/MS analysis. MS/MS spectra are collected from 3,000 laser shots using air as the collision gas and collision energy of 1 kV.

7.2.5 Protein identification by MS

The monoisotopic masses of peptides detected by MALDI-TOF MS or MALDI-TOF/TOF MS are analyzed by the MASCOT search engine (version 2.2.04). Spectrum processing is done with Kratos Launchpad (version 2.7) software (Shimadzu) or 4000 series explore (version 3.5) software (Applied Biosystems). MS peak-list files are submitted to the MASCOT search engine using parameters as follows. For the peptide mass fingerprinting of carbonic anhydrase: database, SwissProt 57.4 (470,369 sequences; 166,709,521 residues); taxonomy, mammal (64,643 sequences); enzyme, trypsin; variable modifications, acetyl (K), acetyl (Protein N-term), oxidation (M); peptide mass tolerance, ±0.5 Da; peptide charge state, 1+; max missed cleavages, 1. For the peptide mass fingerprinting of yeast proteins: database, SwissProt 57.4 (470,369 sequences; 166,709,521 residues); taxonomy, yeast (6,858 sequences); enzyme, trypsin; variable modifications, carbamidomethylation (C), oxidation (M); peptide mass tolerance, ±0.5 Da; peptide charge state, 1+; max missed cleavages, 1. For the MS/MS analysis of yeast proteins: database, SWISSPROT; taxonomy, yeast (6,858 sequences); enzyme, trypsin; variable modifications, carbamidomethylation (C), oxidation (M); peptide mass tolerance, ±0.3 Da; MS/MS mass tolerance, ±0.3 Da; peptide charge state, 1+; max missed cleavages, 1.

7.3 General comments to the analysis of proteins on membranes

7.3.1 Advantages of gel electrophoresis

The resolution of MS, including MALDI-TOF/TOF MS, electrospray ionization-quadrupole/time-of-flight mass spectrometry (ESI-Q/TOF MS), electrospray ionization-linear ion trap/time-of-flight mass spectrometry (ESI-LIT/TOF MS), and even linear ion trap Orbitrap mass spectrometry (LTQ-Orbitrap MS), is often insufficient to directly measure the masses of high-molecular-weight proteins; therefore, proteins must be digested with proteases to obtain peptides that can be analyzed with these mass spectrometric approaches. These methods often miss information about protein isoforms, splicing variants and posttranslational modifications that might be obtained by analyzing intact proteins. In contrast, gel electrophoresis, particularly 2-DE using isoelectric focusing in the first dimension and SDS-PAGE in the second dimension (Klose 1975; O'Farrell 1975), can easily discriminate isoforms, products of splicing variants and posttranslationally modified proteins. Once separated by gel electrophoresis, these variant proteins can be identified by mass spectrometric analysis; in contrast, it is not easy to identify them using MS shotgun analysis.

In addition to O'Farrell's 2-DE, several useful electrophoresis techniques are available. For example, 2-DE using Phos-tag gel electrophoresis instead of SDS-PAGE in the second dimension is suitable for analysis of the phosphorylation status of phosphoproteins, which is impossible to perform by the shotgun analysis (Kinoshita et al. 2009).

Applying 2-DE using blue native electrophoresis in the first dimension and SDS-PAGE in the second dimension can separate proteins within a protein complex, and this allows identification of complex components by MS (Camacho-Carvajal et al. 2004).

7.3.2 Efficiency of electroblotting

The electroblotting efficiency was compared between the PVDF membranes and the other 21 different types of membranes including nitrocellulose, nylon, polyethersulfone, polytetrafluoroethylene, polycarbonate and polypropylene membranes (▶Table 7.2).

Table 7.2: Blotting efficiency and ion yield of membrane filters in MS

Membrane	Material (pore size, μm)	Manufacturer	Hydro-phobicity[1]	Blotting	MS
Fluorotrans G	PVDF (0.2)	Pall	+	+++	-
Fluorotrans W	PVDF (0.2)	Pall	+	+++	-
Immobilon CD	PVDF (0.1)	Millipore	-	++	-
Immobilon P^{SQ}	PVDF (0.2)	Millipore	+	+++	+
Problot	PVDF (0.22)	Applied Biosystems	+	+++	+
Immobilon P	PVDF (0.22)	Millipore	+	+++	+
Immobilon P	PVDF (0.45)	Millipore	+	+++	+
PVDF	PVDF (0.1)	Millipore	+	+++	+++
Durapore	PVDF (0.22)	Millipore	-	+	-
Hyperbond	PVDF (0.22)	Porton	+	+	-
Milli Wrap	PTFE (0.45)	Millipore	+	+	-
Teflon tape	PTFE (0.2)	Unknown	+	++	-
Sure Vent PTFE	PTFE (0.22)	Millipore	+	+	-
Gore Tex	PTFE (0.1)	Goretex	+	+	-
Gore Tex	PTFE (0.2)	Goretex	+	+	-
Cellulosenitrat	Nitrocellulose (0.1)	BoD	-	++	-
HF240	Nitrocellulose	Millipore	-	+	-
Immobilon NY+	Nylon (0.45)	Millipore	-	+	-
Express Plus PES	PES (0.22)	Millipore	-	+	-
MPES	PES (0.03–0.2)	Millipore	-	+	-
Isopore	Polycarbonate (0.4)	Millipore	-	+	-
AN06	Polypropylene (0.6)	Millipore	+	+	-

[1] +, hydrophobic; -, hydrophilic. PTFE, polytetrafluoroethylene; PES, polyethersulfone.

In this experiment, first CyDye-labeled phosphorylase *b*, albumin, carbonic anhydrase, trypsin inhibitor and α-lactalbumin were separated by SDS-PAGE and electroblotted from the gels onto these membranes using a semidry electroblotting apparatus. The fluorescence intensity of each band on the membranes was measured by a fluorescence image analyzer. In this analysis, the PVDF membranes were found to have a higher electroblotting efficiency than the other membranes regardless of protein species. Nitrocellulose membranes also had a high electroblotting efficiency, but it varied depending on protein species.

The electroblotting efficiency was also examined in the PVDF membranes with various pore sizes; 0.1, 0.22 and 0.45 μm; however, there are no great differences among the membranes. In the case of 0.1 and 0.22 μm pore membranes, no proteins were observed passing through during electroblotting. However, proteins partially passed through the 0.45 μm pore membrane. The longer the blotting time, the fewer proteins remained immobilized in the 0.45 μm pore PVDF membrane. It was therefore concluded that the 0.1 and 0.22 μm pore PVDF membranes most efficiently immobilize various proteins.

7.3.3 Efficiency of mass spectrometric analysis

To examine the relationship between pore size of PVDF membranes and MS efficiency, five peptides des-Arg-bradykinin, angiotensin I, neurotensin, Glu-fibrinopeptide B, ACTH (1-17 clip), ACTH (18-39 clip) were spotted onto the 0.1, 0.22 and 0.45 μm pore PVDF membranes. After the addition of the matrix to the peptides, the PVDF membranes were mounted onto MALDI target plates with conductive double-sided adhesive tape, and the peptides were analyzed by MALDI-TOF MS. In this experiment, a stronger mass signal was detected on PVDF membranes with smaller pore size (▶Fig. 7.2).

Next, carbonic anhydrase was separated by SDS-PAGE and electroblotted onto the 0.1 and 0.22 μm pore PVDF membranes. The immobilized proteins were stained with Ponceau S, and the membranes were mounted onto MALDI target plates with conductive double-sided adhesive tape. A trypsin solution was added to the protein bands and incubated overnight at 37°C with moisture. Then matrix solution was added, and the digests were identified by MALDI-TOF MS. In the analyses, a strong mass signal was detected on the 0.1 μm pore PVDF membranes. However, the mass signal of cabonic anhydrase on the 0.22 μm pore PVDF membranes was weak. Moreover, the number of peptides detected by MASCOT search was higher for the 0.1 μm pore PVDF membranes than the 0.22 μm pore PVDF membranes.

It is likely that proteins cannot penetrate very small pores during electroblotting, and that most proteins are retained on the surface of 0.1 μm pore membrane. Additionally, peptides generated by tryptic digestion may not easily spread into such small membrane pores. Proteins and peptides on the membrane surface are easily irradiated by MS lasers, resulting in increased ion yield. Furthermore, the 0.1 μm pore PVDF membranes could prevent the diffusion of the matrix solution that was added to the samples.

In another experiment, yeast proteins were separated by 2-DE and electroblotted onto the 0.1 and 0.22 μm pore membranes. The immobilized proteins were stained with Ponceau S, and the membranes were mounted onto MALDI target plates with conductive double-sided adhesive tape. A trypsin solution was added to the protein spots and incubated overnight at 37°C with moisture. Then matrix solution was added

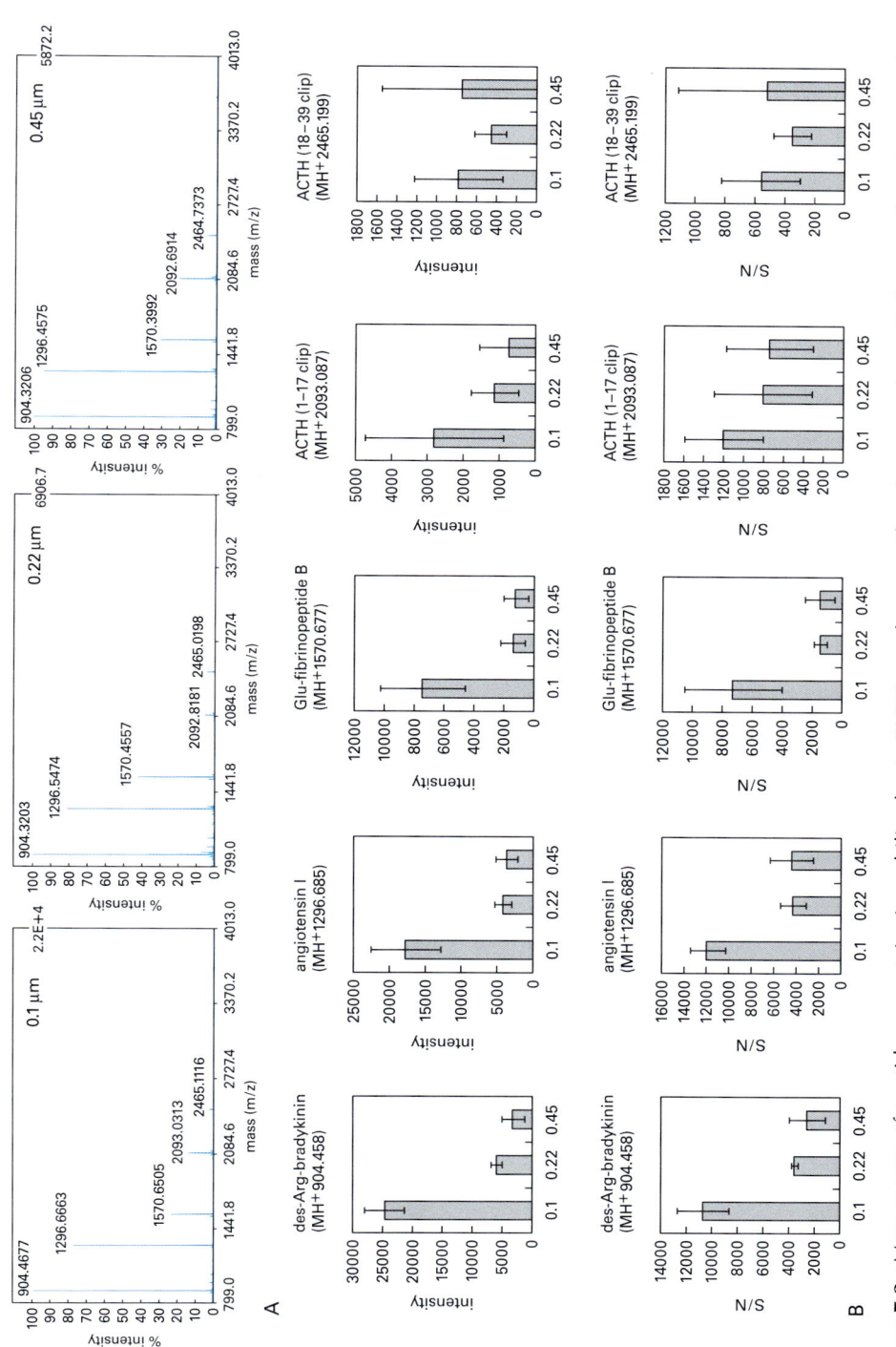

Figure 7.2 : Mass spectra of peptides or proteins immobilized onto PVDF membranes with various pore sizes. (A) Spectra of peptides immobilized onto the membranes: des-Arg-bradykinin; angiotensin I, Glu-fibrinopeptide B, ACTH 18-39 (1-17 clip), ACTH (18-39 clip). (B) Electrophoresis patterns of carbonic anhydrase electroblotted onto PVDF membranes from SDS-PAGE gels. (C) Spectra of peptides generated from tryptic digestion of carbonic anhydrase electroblotted onto PVDF membranes.

to spots containing four known proteins (enolase 2, fructose-bisphosphate aldolase, phosphoglycerate kinase and glyceraldehyde-3-phosphate dehydrogenase 3), and digests of these proteins were analyzed by MALDI-TOF MS. In the analyses of these proteins, stronger mass signals were detected in the PVDF membranes with the smallest pore size (▶Fig. 7.3).

In most cases, from these results, it was determined that 0.1 µm pore PVDF membranes ensure most efficient MS analysis.

7.3.4 Large scale of mass spectrometric analysis

The possibility of high throughput analysis of many proteins separated by 2-DE was investigated. First, yeast proteins were separated by 2-DE and electroblotted onto 0.1 µm pore PVDF membranes (▶Fig. 7.4).

A number of proteins were immobilized on the membranes. Among them, 38 proteins with different physicochemical characteristics, including high- and

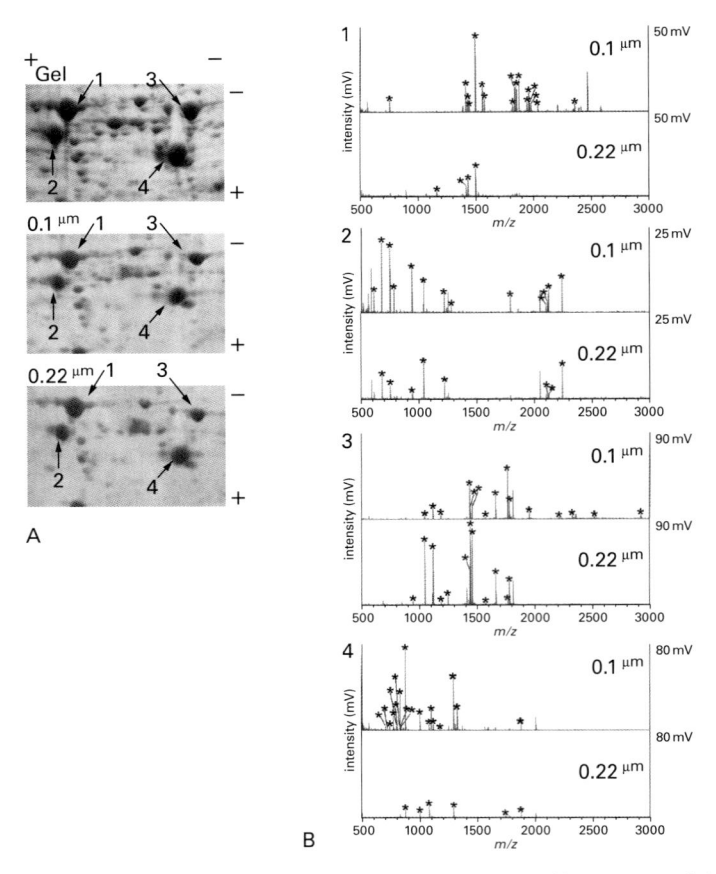

Figure 7.3 : Mass spectra of yeast proteins that were separated by 2-DE and then electroblotted onto PVDF membranes with different pore sizes. Spots 1–4 were analyzed by MALDI-MS.

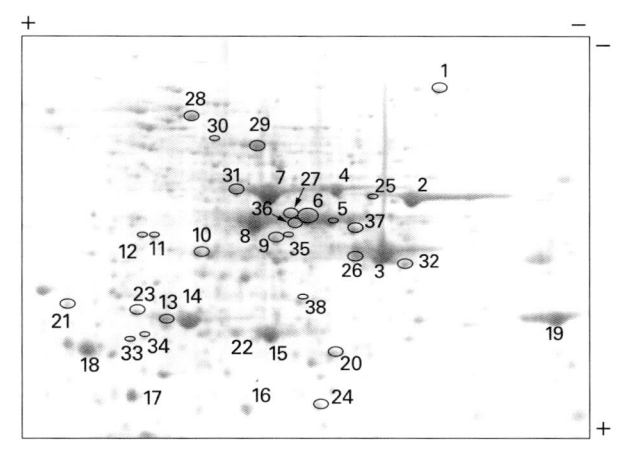

7.3 General comments to the analysis of proteins on membranes

Figure 7.4: The 2-DE patterns of yeast proteins on the gel and PVDF membrane. Yeast proteins were separated by 2-DE and electroblotted onto the PVDF membrane with 0.1 µm pores, and then 38 distinctive proteins were digested on the membrane using trypsin. The digests were analyzed by MALDI-TOF/TOF MS. Proteins identified (score and number of matched peptides): spot 1, aconitate hydratase (115, 4); 2, phosphoglycerate kinase (108, 3); 3, glyceraldehyde-3-phosphate dehydrogenase 3 (106, 6); 4, enolase 1 (149, 7); 5, alcohol dehydrogenase 2 (297, 7); 6, alcohol dehydrogenase 1 (103, 3); 7, enolase 2 (71, 2); 8, fructose-bisphosphate aldolase (47, 1); 9, enolase 2 (145, 5); 10, fructose-bisphosphate aldolase (290, 4); 11, adenosine kinase (157, 3); 12, adenosine kinase (161, 3); 13, heat-shock protein 26 (68, 3); 14, heat-shock protein 26 (127, 4); 15, triosephosphate isomerase (251, 6); 16, superoxide dismutase [Cu-Zn] (179, 6); 17, peroxiredoxin type-2 (26, 1); 18, protein GRE1 (215, 5); 19, phosphoglycerate mutase 1 (49, 2); 20, glyceraldehyde-3-phosphate dehydrogenase 2 (233, 7); 21, cochaperone protein SBA1 (28, 1); 22, triosephosphate isomerase (207, 8); 23, nascent polypeptide-associated complex subunit alpha (238, 4); 24, glyceraldehyde-3-phosphate dehydrogenase 3 (121, 5); 25, phosphoglycerate kinase (895, 5); 26, glyceraldehyde-3-phosphate dehydrogenase 3 (51, 3); 27, alcohol dehydrogenase 1 (97, 5); 28, heat-shock protein SSB1 (129, 5); 29, pyruvate decarboxylase isozyme 1 (105, 4); 30, magnesium-activated aldehyde dehydrogenase (85, 4); 31, enolase 2 (142, 4); 32, glyceraldehyde-3-phosphate dehydrogenase 3 (63, 4), 33, heat-shock protein SSA1 (419, 7); 34, heat-shock protein SSA2 (366, 8); 35, phosphoglycerate kinase (175, 5); 36, phosphoglycerate kinase (51, 1); 37, ketol-acid reductoisomerase (91, 3); 38, glyceraldehyde-3-phosphate dehydrogenase 3 (31, 2).

low-molecular-weight proteins, basic and acidic proteins, and high and low abundance proteins were selected. After tryptic digestion, they were analyzed by MS, as described in Section 7.3.3. In this analysis, all 38 proteins were successfully identified, showing the possibility of high throughput analysis of many proteins separated by 2-DE.

In parallel, these proteins were digested in-gel and analyzed by the same MS in order to compare the efficiency of MS analysis. For many proteins, the MASCOT scores from the MS/MS analysis were higher for MS of the in-gel digested samples than for on-membrane MS. However, for some spots (4, 5, 10, 11, 12, 18, 20, 33, 34 and 35), the scores were lower for in-gel digests than on-membrane MS. The proteins in these spots were not abundant, and their spot sizes were comparatively small and faint. However, this analysis shows that various kinds of proteins can be identified by these methods.

7.4 PVDF membranes or diamond-like carbon-coated (DLC) stainless steel plates?

Iwafune et al. (2007) previously developed a DLC stainless steel plate that was chemically modified with an N-hydroxysuccinimide ester. This ester reacts with primary amines on proteins and covalently attaches them to the surface of the DLC plate. Because the DLC plate is conductive, it can be used to electroblot proteins from gels after gel electrophoresis. Moreover, this plate can be used as a target plate for MALDI-TOF MS. Using the DLC plate and MALDI-TOF MS, protein digestion on the DLC plate and peptide mass fingerprinting analysis were performed in order to identify proteins. However, special devices are required to electroblot the proteins from the gels onto the DLC plate. For example, 0.3 mm thick SDS-PAGE gels must be used to increase the blotting efficiency, and the gels must be handled carefully after electrophoresis. It might be necessary to use a special instrument to automatically transfer gels from the gel plate to the DLC plate. In addition, the DLC plates are very expensive because they require complicated processing, such as surface grinding and chemical modifications to activate the surface. Therefore, it would be ideal if 0.1 μm pore membranes could be used instead of DLC plates.

7.5 Concluding remarks

Various biochemical characteristics of proteins separated by gel electrophoresis can be analyzed on the PVDF membranes. The technique, in which the electroblotted proteins are subjected to on-membrane protease digestion and the resultant peptides are analyzed directly by MALDI-MS on the membranes, has been studied and improved for more than 20 years and is now usable in practice. Further improvement in membrane filters, electroblotting conditions, mass spectrometers and MS analysis methods will be particularly important in order to achieve more efficient and effective on-membrane identification of gel-resolved proteins by MALDI-MS.

References

Blackledge, J.A., and Alexander, A.J. (1995). Polyethylene membrane as a sample support for direct matrix-assisted laser desorption/ionization mass spectrometric analysis of high mass proteins. Anal Chem 67, 843–848.

Bunai, K., Nozaki, M., Hamano, M., Ogane, S., Inoue, T., Nemoto, T., et al. (2003) Proteomic analysis of acrylamide gel separated proteins immobilized on polyvinylidene difluoride membranes following proteolytic digestion in the presence of 80% acetonitrile. Proteomics 3, 1738–1749.

Camacho-Carvajal, M.M., Wollscheid, B., Aebersold, R., Steimle, V., and Schamel, W.W. (2004). Two-dimensional blue native/SDS gel electrophoresis of multi-protein complexes from whole cellular lysates: a proteomics approach. Mol Cell Proteomics 3, 176–182.

Chang, C.Y., Liao, H.K., Juo, C.G., Chen, S.H., and Chen, Y.J. (2006). Improved analysis of membrane protein by PVDF-aided, matrix-assisted laser desorption/ionization mass spectrometry. Anal Chim Acta 556, 237–246.

Eckerskorn, C., Strupat, K., Karas, M., Hillenkamp, F., and Lottspeich, F. (1992). Mass spectrometric analysis of blotted proteins after gel electrophoretic separation by matrix-assisted laser desorption/ionization. Electrophoresis *13*, 664–665.

Eckerskorn, C., Strupat, K., Schleuder, D., Hochstrasser, D., Sanchez, J., Lottspeich, F., et al. (1997). Analysis of proteins by direct-scanning infrared- MALDI mass spectrometry after 2D-PAGE separation and electroblotting. Anal Chem *69*, 2888–2892.

Fernandez, J., Gharahdaghi, F., and Mische, S.M. (1998). Routine identification of proteins from sodium dodecyl sulfate-polyacrylamide gel electrophoresis (SDS-PAGE) gels or polyvinyl difluoride membranes using matrix assisted laser desorption/ionization-time of flight-mass spectrometry (MALDI-TOF-MS). Electrophoresis *19*, 1036–1045.

Gharahdaghi, F., Kirchner, M., Fernandez, J., and Mische, S.M. (1996). Peptide-mass profiles of polyvinylidene difluoride-bound proteins by matrix-assisted laser desorption/ionization time-of-flight mass spectrometry in the presence of nonionic detergents. Anal Biochem *233*, 94–99.

Gorg, A., Postel, W., and Gunther, S. (1988). The current state of two-dimensional electrophoresis with immobilized pH gradients. Electrophoresis *9*, 531–546.

Hirano, H. (1989). Microsequence analysis of winged bean seed proteins electroblotted from two-dimensional gel. J Protein Chem *8*, 115–130.

Iwafune, Y., Tan, J.Z., Ino, Y., Okayama, A., Ishigaki, Y., Saito, K., et al. (2007). On-chip identification and interaction analysis of gel-resolved proteins using a diamond-like carbon-coated plate. J Proteome Res *6*, 2315–2322.

Kimura, S., Kameyama, A., Nakaya, S., Ito, H., and Narimatsu, H. (2007). Direct on-membrane glycoproteomic approach using MALDI-TOF mass spectrometry coupled with microdispensing of multiple enzymes. J Proteome Res *6*, 2488–2494.

Kinoshita, E., Kinoshita-Kikuta, E., and Koike, T. (2009). Separation and detection of large phosphoproteins using Phos-tag SDS-PAGE. Nat Protoc *4*, 1513–1521.

Klose, J. (1975). Protein mapping by combined isoelectric focusing and electrophoresis of mouse tissues. A novel approach to testing for induced point mutations in mammals. Humangenetik *26*, 231–243.

Laemmli, U.K. (1970). Cleavage of structural proteins during the assembly of the head of bacteriophage T4. Nature *227*, 680–685.

Luque-Garcia, J.L., Zhou, G., Sun, T.T., and Neubert, T.A. (2006). Use of nitrocellulose membranes for protein characterization by matrix-assisted laser desorption/ionization mass spectrometry. Anal Chem *78*, 5102–5108.

Matsudaira, P. (1987). Sequence from picomole quantities of proteins electroblotted onto polyvinylidene difluoride membranes. J Biol Chem *262*, 10035–10038.

Methogo, R.M., Dufresne-Martin, G., Leclerc, P., Leduc, R., and Klarskov, K. (2005). Mass spectrometric peptide fingerprinting of proteins after Western blotting on polyvinylidene fluoride and enhanced chemiluminescence detection. J Proteome Res *4*, 2216–2224.

Nakanishi, T., Ando, E., Furuta, M., Tsunasawa, S., and Nishimura, O. (2007). Direct on-membrane peptide mass fingerprinting with MALDI-MS of tyrosine-phosphorylated proteins detected by immunostaining. J Chromatogr B Anal Technol Biomed Life Sci *847*, 24–29.

Nakanishi, T., Ohtsu, I., Furuta, M., Ando, E., and Nishimura, O. (2005). Direct MS/MS analysis of proteins blotted on membranes by a matrix-assisted laser desorption/ionization-quadrupole ion trap-time-of-flight tandem mass spectrometer. J Proteome Res *4*, 743–747.

O'Farrell, P.H. (1975). High resolution two-dimensional electrophoresis of proteins. J Biol Chem *250*, 4007–4021.

Patterson, S.D. (1995). Matrix-assisted laser-desorption/ionization mass spectrometric approaches for the identification of gel-separated proteins in the 5–50 pmol range. Electrophoresis *16*, 1104–1114.

Pham, V.C., Henzel, W.J., and Lill, J.R. (2005). Rapid on-membrane proteolytic cleavage for Edman sequencing and mass spectrometric identification of proteins. Electrophoresis 26, 4243–4251.

Schleuder, D., Hillenkamp, F., and Strupat, K. (1999). IR-MALDI-mass analysis of electroblotted proteins directly from the membrane: comparison of different membranes, application to on-membrane digestion, and protein identification by database searching. Anal Chem 71, 3238–3247.

Schreiner, M., Strupat, K., Lottspeich, F., and Eckerskorn, C. (1996). Ultraviolet matrix assisted laser desorption ionization-mass spectrometry of electroblotted proteins. Electrophoresis 17, 954–961.

Sloane, A.J., Duff, J.L., Wilson, N.L., Gandhi, P.S., Hill, C.J., Hopwood, F.G., et al. (2002). High throughput peptide mass fingerprinting and protein macroarray analysis using chemical printing strategies. Mol Cell Proteomics 1, 90–99.

Strupat, K., Karas, M., Hillenkamp, F., Eckerskorn, C., and Lottspeich, F. (1994). Matrix-assisted laser desorption ionization mass spectrometry of proteins electroblotted after polyacrylamide gel electrophoresis. Anal Chem 66, 464–470.

Sutton, C.W., Pemberton, K.S., Cottrell, J.S., Corbett, J.M., Wheeler, C.H., Dunn, M.J., et al. (1995). Identification of myocardial proteins from two-dimensional gels by peptide mass fingerprinting. Electrophoresis 16, 308–316.

8 Analysis of protein complexes using chemical cross-linking and mass spectrometry

Alexander Leitner

8.1 Introduction

The last two decades have undoubtedly seen major advances in the large-scale qualitative and quantitative analysis of proteins, spawning a research field that is now commonly described as proteomics. However, it is obvious that in order to draw functional conclusions about the role of proteins in biological systems, one cannot look at them as isolated players but must also consider their interactions. Proteins typically exert their function in protein complexes and/or in larger protein networks; therefore, tools to study these interactions are of essential interest. Recent advances in proteomics, the large-scale study of the protein complement of a cell or an organism, have provided some, but not all of these tools.

Protein interactions can be studied at different levels of detail (Alber et al. 2008). At the most fundamental level, interaction partners are identified, from which crude networks may be constructed. More detailed functions to be considered may include the determination of complex stoichiometries, the dynamics of interactions (e.g. in dependence on cellular states) and, at a structural level, the binding sites that are involved in these interactions and the resulting structure and topology of the complexes.

To probe the structure of protein complexes at different levels of resolution, a number of technologies have been employed. Some of the currently most relevant ones include X-ray crystallography, nuclear magnetic resonance spectroscopy and various electron microscopy (EM) techniques. Additionally, in recent years, mass spectrometry (MS) has contributed significantly to the protein interactions toolbox (Choudhary and Mann 2010; Gavin et al. 2011; Gingras et al. 2007; Pflieger et al. 2011; Vermeulen et al. 2008). Affinity purification–MS, for example, is now widely adopted to identify interaction partners of protein targets of interest (Collins and Choudhary 2008; Gavin et al. 2011; Pflieger et al. 2011; Vermeulen et al. 2008), but the concept can also be applied to study proteins as drug targets (Rix and Superti-Furga 2009). Entire intact protein complexes can be directly subjected to mass spectrometric analysis in what has been termed "native MS" and their properties studied in the gas phase (Heck 2008). The focus of this article, however, will be on chemical cross-linking in combination with MS, where the formation of covalent bonds by means of dedicated reagents is used to infer information about the composition and structure of protein complexes.

8.2 Reagents for chemical cross-linking

Historically, protein cross-linking has been applied for different purposes, for example, the immobilization of target proteins (Hermanson 2008), the stabilization of protein interactions during the isolation of complexes and, more specifically, as a tool for probing the structure of proteins and protein complexes by providing direct evidence of a physical contact between two functional groups (Fabris and Yu 2010; Lee 2008; Leitner et al. 2010; Sinz 2010). In the latter case, homobifunctional cross-linking reagents have been most frequently applied.

Petrotchenko and Borchers (2010b) give a more detailed overview of different reagents. Here, I will focus on the most commonly applied groups, lysine-reactive disuccinimidyl esters of dicarboxylic acids. The primary amines of lysine side chains (together with free protein N termini) are attractive targets that are commonly abundant and surface exposed and can be relatively specifically addressed by succinimide and sulfosuccinimide compounds. They are most reactive at neutral to slightly basic pH, which makes them ideally compatible for working with protein complexes under native-like conditions, although they are subject to ester hydrolysis in aqueous solutions. This competitive reaction limits the applicable time frame for the cross-linking process.

Reagent hydrolysis is also one of the reasons why not every reaction of the cross-linking reagent with the target moiety will lead to the formation of a cross-link. If after the first reaction step no other cross-linkable site is in the vicinity or the second reactive group is deactivated, a "mono-link" (or type 0 cross-link according to nomenclature proposed by Schilling et al. [2003]) is formed. That means the reaction product can be treated like a conventional covalent side-chain modification. If a second reactive site is, however, approachable, different cross-linking products may be formed, which are usually differentiated according to the fact that an enzymatic digestion step into peptides is being performed prior to mass spectrometric analysis: If both reactive sites reside on the same peptide after digestion, this is being classified as a "loop-link" (or type 1 cross-link); if the linker connects two independent peptide chains, this is usually referred to as an actual "cross-link" (type 2 cross-link). Only the latter type provides useful distance information, with the exception of two adjacent peptides being connected. Such cross-links may be further grouped into *intra*protein and *inter*protein cross-links if more than one protein is present in the sample. Conceptually, also higher-order reaction products are possible, for example cross-linked peptides carrying an additional mono-link, although these species are harder to identify and seldom reported in the literature. An overview of the different reaction products is given is ▶Figure 8.1.

8.2.1 Conventional homobifunctional cross-linking reagents

The most commonly used succinimide-type reagents are depicted in ▶Figure 8.2. Disuccinimidyl glutarate (DSG) and its sulfonated analog, bis(sulfosuccinimidyl) glutarate (BS^{2}G), possess a spacer with three -(CH$_2$)- groups and a total spacer length of approximately 7.7 Å. Disuccinimidyl suberate (DSS) and bis(sulfosuccinimidyl) suberate (BS^3) contain six -(CH$_2$)- groups in the spacer and span 11.4 Å.

cross-link
(type 2 cross-link)

loop-link
(type 1 cross-link)

mono-link
(type 0 cross-link)

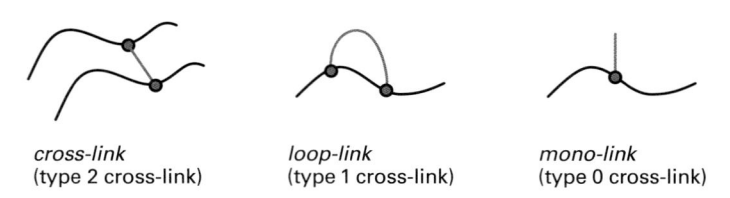

Figure 8.1: Nomenclature of possible products from a cross-linking reaction.

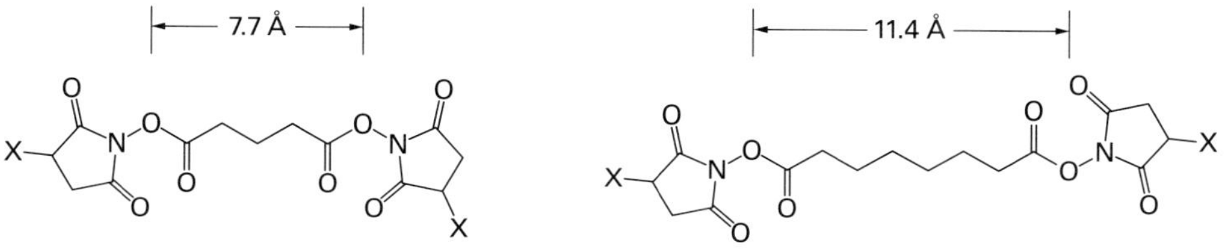

X = H, DSG; X = SO$_3$H, BS^{2}G   X = H, DSS; X = SO$_3$H, BS^3

Figure 8.2: Structure of commonly used reagents for cross-linking of amine groups.

The sulfonated cross-linkers are more hydrophilic which gives them better solubility, although these reagents are mostly membrane impermeable, which limits their applicability on live cells. Unsulfonated reagents need to be dissolved in apolar organic solvents such as dimethylformamide (DMF) or dimethyl sulfoxide (DMSO) prior to the reaction (Section 8.2.2).

Besides other experimental variables, also discussed in Section 8.2.2, the spacer length of the cross-linker is an influential factor for the yield of cross-links. The longer the spacer, the larger the distance between two functional groups it is able to bridge. However, reagents with very long spacers reduce the accuracy in distance information and may have unfavorable properties for further analysis. The reagents DSS and BS^3 are usually a good compromise for routine application.

8.2.2 Functionalized cross-linking reagents and associated workflows

The downstream analysis of cross-linked samples is quite challenging. Because of the low reaction yields, cross-linked peptides are present after a digestion step in a large excess of unmodified peptides and peptides carrying mono-links. Therefore various strategies have been developed to facilitate the detection and identification of cross-linked peptides, and many of those are reflected in the reagent design (▶Figure 8.3).

Isotope-coded cross-linking reagents are used in two forms ("light" and "heavy") where in one version of the reagent several atoms are replaced by their heavy isotope version (Müller et al. 2001; Seebacher et al. 2006). Currently, most reagents use the replacement of hydrogen by deuterium, although ^{13}C-labeled linkers are now also emerging. These avoid the partial chromatographic separation of hydrogen/deuterium-labeled pairs (Zhang et al. 2001), although their synthesis is usually more demanding and expensive. Most frequently, the light and heavy reagents are used in a 1:1 ratio; therefore, they appear as doublets in the mass spectrum with a mass shift corresponding to the number of heavy isotopes incorporated (commonly 4 to 12 Da). This facilitates the identification of modified peptides, although isotope coding by itself does not allow the differentiation among mono-, loop- and cross-links because each form will contain an identical mass shift. Many succinimide reagents are now available in their isotope-coded form from commercial sources (e.g. ThermoFisher/Pierce) or specialist suppliers such as Creative Molecules.

Cross-linkers carrying affinity tags have emerged shortly after the first reagents of this sort for general proteomic applications (Gygi et al. 1999). In this design, the linker region carries an additional functional group that serves as a handle for affinity purification

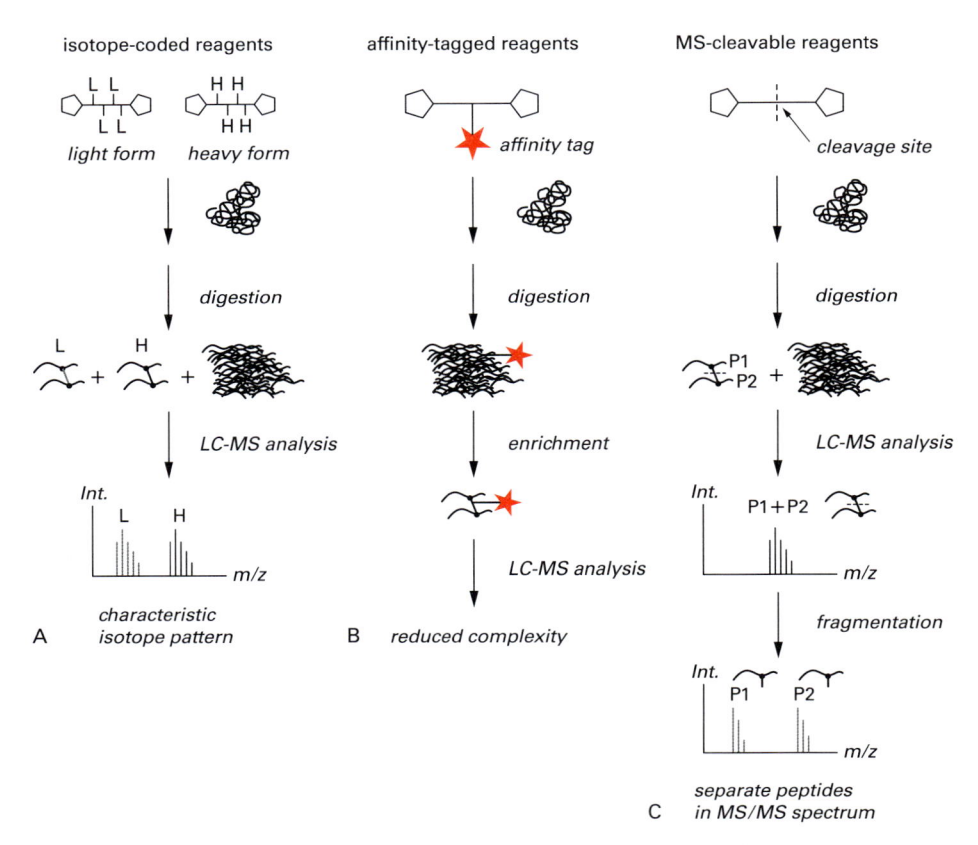

Figure 8.3: Functionalized cross-linking reagents and associated workflows. (A) isotope-coded reagents, (B) affinity-tagged reagents, (C) MS-cleavable reagents. Int., intensity; *m/z*, mass-to-charge ratio.

(Chu et al. 2006; Petrotchenko et al. 2011; Trester-Zedlitz et al. 2003) or a moiety where such a group can be attached after the cross-linking reaction (Chowdhury et al. 2009; Nessen et al. 2009; Vellucci et al. 2010). Frequently, biotin is used as the affinity tag, enabling the enrichment of modified peptides by avidin affinity chromatography. Although this seems a very attractive means to reduce the complexity of cross-link digests, there are noticeable drawbacks, including tedious synthesis and suboptimal properties for MS analysis. Therefore, although the first reagents have been reported already a decade ago, applications have been mostly restricted to proof-of-principle studies.

Cleavable cross-linker designs incorporate at least one functional group in the linker that deliberately affects the fragmentation behavior of modified peptides. Usually, a very labile bond is introduced that is the preferred fragmentation site in a tandem mass spectrometry (MS/MS) experiment. This way, in the case of cross-linked peptides, the primary fragmentation event will be the cleavage in the linker, so that modified versions of single peptides will be formed as fragments. These primary fragments are then selected in a second fragmentation stage (MS^3), enabling their identification. A reagent class based on this approach is the protein interaction reporters (PIRs) introduced by the group of Bruce (Tang and Bruce 2010). These reagents also carry a biotin affinity tag for enrichment purposes, and the identification of more than 20 protein interactions from cross-linking of living cells has been reported using this method (Zhang et al. 2009).

Independent of the use of a particular reagent design, cross-links may also be enriched by chromatographic techniques. Examples include strong cation exchange (SCX) chromatography, taking advantage of the higher charge of cross-linking products (Rinner et al. 2008; Chen et al. 2010) and size-exclusion chromatography (Bohn et al. 2010). In the mass spectrometer, highly charged precursors may also be selected preferentially for fragmentation, thereby increasing the odds for identification.

8.3 The chemical cross-linking workflow

For practical purposes, especially concerning the study of complex systems such as large protein assemblies composed of tens of subunits, cross-linking reagents with a defined, restricted specificity are used. In analogy to standard proteomic workflows, cross-linked proteins are proteolytically cleaved and analyzed by MS on the peptide level. Therefore, a general cross-linking protocol consists of the following steps (▶ Figure 8.4): Isolating the complex of interest; performing the actual cross-linking reaction; processing of the sample (enzymatic digestion, enrichment and/or fractionation steps); mass spectrometric analysis, typically following a liquid chromatographic separation; and finally bioinformatic data analysis (identification of cross-linking sites, modeling of structures).

The tedious task to isolate protein complexes and to purify them to homogeneity shall not be discussed here any further. It suffices to say that for cross-linking applications, high sample purity is desirable, although not an absolute necessity. Samples that are used for other structural characterization techniques such as X-ray crystallography

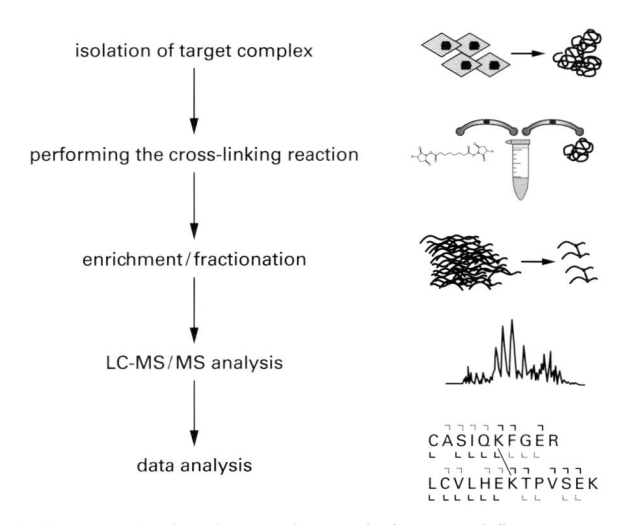

Figure 8.4: Key steps in the chemical cross-linking workflow.

or cryoEM are usually relatively stable, provided in sufficient amounts (tens of micrograms or more) and at high concentrations (mg per mL range) and are considered ideal cross-linking targets. However, such yields may not be achievable for many interesting complexes because of lower expression levels. Also, the stability of complexes may be limited so that lengthy purification steps are not advisable, resulting in the presence of contaminant proteins. This complicates both the downstream mass spectrometric and bioinformatic analysis as will be shown later. To date, almost all successful cross-linking studies have been carried out using highly purified recombinant complexes.

Key experimental parameters in carrying out the actual cross-linking reaction include: Reaction buffer and pH, incubation time and temperature, protein and cross-linker concentration and, of course, choice of the cross-linking reagent. Buffer systems need to be selected in order to reflect native conditions as much as possible while at the same time allowing for an efficient cross-linking reaction. The commonly used amine-reactive succinimide esters work well in a pH range of 7.5 to 8.3, which is compatible with many protein complexes. Buffer constituents must not contain functionalities that would also react with the cross-linker. Therefore, buffer salts such as Tris cannot be used with succinimides due to the presence of primary amines. As alternatives, 4-(2-hydroxyethyl)-1-piperazineethanesulfonic acid (HEPES) or phosphate buffers are popular. Typical incubation times range in the tens of minutes when the reaction is carried out at room temperature or at elevated temperature (37°C), but reaction times need to be extended when working at 4°C. Obviously, the reaction kinetics will depend on the concentration of both protein and cross-linking reagent. Our group has been most successful when working at protein concentrations of 0.5 to 2 mg per mL, although ideally this should be optimized for each sample and progress of the reaction monitored by SDS-PAGE. Reagents with different functional group specificities may require different protocols, but emphasis must always be given to preserving the native state of the complex. A typical protocol for amine-reactive linkers is given in the textbox. Detailed experimental conditions may be found in the recent literature, see for example the references given in Section 8.5.

A cross-linking protocol for amine-reactive cross-linking reagents

- Prepare protein complex sample (ideally, at least 25 µg) at a total concentration of 0.5–2 mg mL^{-1} in an amine-free buffer (e.g. 20 mM HEPES, pH 8.2).
  Note: If necessary, perform buffer exchange and concentrate or dilute the sample accordingly.
- Prepare a stock solution of the amine-reactive cross-linker (e.g. DSS/BS^3, DSG/BS^{2}G). For example, prepare a 25 mM DSS stock in dry DMF or DMSO.
  Note: Always prepare fresh stock solutions to prevent hydrolysis prior to cross-linking.
- Combine protein and cross-linker solutions, mix briefly and incubate at the desired temperature.
  Note: Good starting conditions for DSS are 1 mM final cross-linker concentration and 30 min incubation at 37°C, but adjusting the conditions to the particular target may be required.
- Add ammonium bicarbonate solution to a final concentration of 50 mM and incubate for further 20 min (quenching step).
  Note: This will deactivate all remaining cross-linking reagent and effectively stop the reaction.

- Process the sample according to established protocols for proteomics sample preparation.
  Note: This may include steps such as reduction of disulfide bonds, alkylation of free cysteines, enzymatic digestion with a protease (usually trypsin) and purification of the resulting peptides.
- Analyze the sample by liquid chromatography (LC)–MS and perform data analysis.

8.4 MS and data analysis

8.4.1 Mass spectrometric analysis of cross-linked samples

Historically, the analysis of cross-linked samples with MS has followed the developments in MS instrumentation for the general analysis of biomacromolecules. Nowadays, many cross-linking studies are carried out using linear ion trap Orbitrap (LTQ OT) hybrid instruments (Makarov et al. 2006; Olsen et al. 2009) that have become widespread in academic laboratories and core facilities (Section 8.4.2). Other platforms include quadrupole-time-of-flight, linear ion trap–Fourier transform ion cyclotron resonance and tandem-time-of-flight (TOF/TOF) devices. With the exception of the latter, which is usually combined with matrix-assisted laser desorption ionization (MALDI) as the ionization technique, all other instruments typically use electrospray ion sources. As a front-end separation technique, reversed-phase LC is the technique of choice, which separates peptides according to their hydrophobicity.

Compared to conventional MS-based proteomics workflows, the importance of generating high-resolution/high-mass-accuracy data is even greater in the case of cross-linking studies, which is due to the enormous search space encountered (discussed in Section 8.4.2). Therefore, low-resolution analyzers such as stand-alone ion traps are not commonly used. However, in combination with high-resolution analyzers, ion traps are ideal for sequencing cross-linked peptides that have been connected with a gas-phase cleavable linker such as the PIR approach. In this case, the first, low-energy, fragmentation attempt causes dissociation of one or several bonds in the linker region, thereby liberating the two initially connected peptides. In a subsequent fragmentation stage, the individual peptides are sequenced separately. While this approach overcomes problems with incomplete fragmentation of cross-linked peptides, the lower sequencing speed and losses in sensitivity due to multiple fragmentations may be considered drawbacks.

In absence of a cross-linker with specific fragmentation properties, MS/MS spectra of cross-linked peptides are highly complex. Bond cleavages can occur on either of the two peptides and usually follow the general b- and y-pathways that are also observed for linear peptides (Paizs and Suhai 2005; Steen and Mann 2004). However, a more pronounced formation of internal ions as a result of two fragmentation events in the same molecule has been noted, although large data sets suitable for a detailed statistical evaluation do not yet exist. Additional spectral congestion is caused by the high charge states of the precursors, commonly 3+ to 6+, which results in highly charged fragment ions, frequently also in several charge states.

It is not uncommon to observe unequal fragmentation between the two cross-linked peptides, so that most peptide bond cleavages occur on one peptide only. This can be either a general pattern of cross-linked peptides, or it might be associated with higher than expected false-positive rates during data analysis (e.g. a presumed cross-link assignment where one peptide is correct, but not the other). To overcome problems with poor fragmentation, different fragmentation techniques than the commonly used collision-induced dissociation have been proposed. Among those, electron transfer dissociation appears to be the most promising, although to date only few examples have appeared in the literature (Chowdhury et al. 2009; Trnka and Burlingame 2010).

8.4.2 Data analysis

Once MS and MS/MS information has been acquired using the workflows outlined in Section 8.4.1, the resulting spectra must be identified. Over the years, different informatics strategies have been established, often with similarities to conventional proteomics data analysis. Mayne and Patterton (2011) have recently provided an excellent overview of relevant software, and I refer the interested reader to their review for more technical details than can be provided herein.

Regardless of which approach for data analysis is being followed, the major difference between conventional peptide sequencing and the sequencing of cross-linked peptides is the difference in search space. Cross-links are always binary combinations of peptides that may be derived from the same protein or from different proteins. So unless specific preliminary information is available, a cross-link data analysis must consider all possible peptide combinations. This means that the number of possible combinations scales roughly with $n^2/2$, where n is the number of peptides in the database. For an organism with a relatively small proteome, such as yeast (*S. cerevisiae*), the number of potential peptide combinations already reaches roughly 10^{11} if searches against a complete protein database would be carried out (Leitner et al. 2010). This is usually impractical because search times are extremely long and the statistical certainty of the results is usually very low due to large numbers of random false positives. Therefore, when working with protein complexes, it is advisable to restrict the database to the most likely sample constituents, which puts emphasis on sample purity.

The most straightforward option for data analysis relies on accurate mass information on the MS^1 level alone, without acquiring or interpreting any fragmentation data. Conceptually, this approach can be considered similar to peptide mass fingerprinting (PMF) for normal peptide identification. In the case of cross-linking analysis, all potential binary combinations of peptide masses must be considered (in addition to the possible presence of unmodified peptides or mono-links). Therefore, this approach is even more restricted to small databases than PMF to allow unequivocal assignment of cross-links even when mass errors are low.

More commonly, information from MS/MS spectra are nowadays interpreted by a variety of software solutions that compare experimental and theoretical spectra and score the degree of similarity between the two spectra, similar to the conventional sequence database searches for general proteomic workflows (Matthiesen et al. 2011; Nesvizhskii 2010). In the last few years, two major strategies have been followed to assign MS/MS spectra of cross-linked peptides: One uses conventional search engines, and the other one dedicated programs for cross-linking data. Both approaches will be briefly discussed in the following paragraphs.

Identifying cross-links using conventional search engines: In order to deal with the structural complexity of cross-links, conventional search engines must be adapted at least at one point in the workflow. One possibility is to construct special sequence databases, where each entry corresponds to a theoretically calculated cross-link (within a peptide or between two peptides) rather than a whole protein sequence. This protocol, originally introduced by Rappsilber and coworkers (Maiolica et al. 2007), can be used with a number of commercial and free search engines such as Mascot, Sequest, X!Tandem and others. Specialized software to construct such databases is available (Panchaud et al. 2010). Alternatively, cross-links can be assigned in a way that only one peptide is directly identified by the search engine from its fragment ions, and the other one is treated as a modification with a variable ("open") mass shift. Such a workflow has been implemented in the search engine ProteinProspector (Singh et al. 2008).

One limitation of both methods is that the scoring schemes that were initially developed for linear peptide identification are not necessarily equally suited to judge the quality of cross-linked peptide assignments. Furthermore, the sheer size of the search space makes cross-link databases very large and restricts the approach to samples of limited complexity. However, in contrast to dedicated search engines, the demand for bioinformatics expertise is somewhat limited.

Identifying cross-links using dedicated search engines: These programs have been typically developed for particular cross-linking workflows such as those relying on isotope-coded (Petrotchenko and Borchers 2010a) cross-linkers or cleavable cross-linkers (Hoopmann et al. 2010). In some cases, however, the software is easily adaptable to other workflows. xQuest, a relatively widely used open program developed at the author's institution shall serve as an example for such a specialized tool (Rinner et al. 2008).

xQuest is especially powerful when used together with isotope-coded cross-linking reagents, although it can process data from unlabeled linkers as well. It contains a module that identifies pairs of MS/MS spectra with a given mass difference, based on the isotope shift between the two forms. Therefore, only a subset of the MS/MS spectra needs to be searched – although as a drawback this approach requires the acquisition of MS/MS spectra from both isotopic states of the candidates. Light and heavy spectra are then compared, and fragment ions are differentiated based on the absence or presence of an isotopic shift between the two spectra: Fragment ions containing the cross-linking region will be different when peptides are connected with a cross-linker either in its light or its heavy form. In contrast, fragment ions that do not contain the cross-link region will be identical in both forms. xQuest creates subspectra from these two types of fragment ions and uses them independently for searching. Similar to the approaches that use conventional search engines, it creates theoretical spectra (cross-links, mono-links) from the proteins of interest contained in a sequence database after filtering for likely candidates based on accurate precursor mass and compares these with the experimental spectra. Because the prefiltering step takes advantage of isotope coding, the program is relatively "economical" in its use of computational resources and can therefore be used for searches in large databases, as has been demonstrated by identifying cross-links from an *E. coli* lysate (Rinner et al. 2008).

Like other dedicated cross-link search engines, xQuest provided a scoring model that is optimized for MS/MS spectra of cross-linked peptides. Frequently, this class of software is trained on a data set of known cross-link spectra to derive cutoff levels. Because of the high spectral complexity and the lack of large-scale data on the fragmentation of cross-linked peptides, the scoring models may still be imperfect. Therefore, it is still common practice that spectra are manually evaluated, validated and filtered after the automated scoring.

8.5 Practical examples

In the past, chemical cross-linking has been applied to a number of smaller, mainly binary complexes, but applications to larger macromolecular assemblies have been limited. Here, I will highlight several recent examples where data derived from cross-linking studies have helped to elucidate the structure of larger complexes.

8.5.1 RNA polymerase II (Pol II)

Rappsilber and coworkers have employed cross-linking technology to study the 12-subunit Pol II complex as well as the binding of the initiation factor TFIIF to Pol II (Chen et al. 2010). In this study, BS^3 was used as the cross-linking reagent, cross-linked samples were separated by SDS-PAGE and bands of interest were subjected to in-gel digestion. Cross-linked peptides were enriched by SCX chromatography, analyzed by LC-MS/MS on an LTQ OT instrument and identified using proprietary in-house software. In the Pol II complex alone, 106 cross-links were assigned, of which 99 conformed to the known structure (up to a distance of 27.4 Å) and a further 6 were detected in flexible regions. In the 15-subunit, 670 kDa Pol II–TFIIF complex, 182 contacts between the polymerase and the initiation factor were detected, making this the largest data set of cross-linking constraints reported to date.

To make use of the cross-linking data, TFIIF domains were modeled based on structural information from human homologs. Subunits were then assembled based on intra-TFIIF contacts, and the complex was positioned on Pol II based on further spatial constraints that pointed toward interactions between the Tfg1 and Tfg2 domains of TFIIF and the Rpb2 subunit of Pol II. The results obtained provided further insight into the function of TFIIF during transcription initiation.

8.5.2 The 26S proteasome

Our group has applied chemical cross-linking to study the 26S proteasome from *S. pombe* for some time. In combination with low-resolution structural information from cryoEM, the arrangement of subunits within the AAA-ATPase module in the 19S regulatory particle of the proteasome has recently been elucidated (Bohn et al. 2010). The proteasome was cross-linked using isotope-coded DSS (d_0/d_{12}), and the cross-linked complex was digested in solution using trypsin as the protease. Following an enrichment step for cross-linked peptides using size exclusion chromatography, samples were also analyzed by LTQ OT MS. Data analysis was carried out using xQuest, taking advantage of the isotope-coding concept as described in Section 8.2.2.

Structural elucidation of the intact proteasome has been hampered by its size and enormous complexity, being composed of more than 50 individual subunits. To date, a high-resolution X-ray structure is only available for the 20S core module that consists of four heteroheptameric rings in a stacked orientation (Groll et al. 1997). For the 19S module that is essential for the recognition of ubiquitinated substrates, only partial structures have been tentatively assigned. Based on the cross-linking data obtained from this study, the proposed topology of the hexameric subcomplex of AAA-ATPases was confirmed based on 15 intersubunit contacts. Furthermore, the orientation of the ATPase module toward the 20S core module was defined based on four 19S-20S cross-links.

Although to date sufficient information to discern the topology of the whole proteasome has not yet been obtained, the results from this study already allow considerable insight into the processing of protein substrates within the proteasome.

8.5.3 Bacterial needle complex

The *Salmonella typhimurium* needle complex is a highly relevant macromolecular machine responsible for the transfer of effector proteins through the bacterial envelope. Two groups recently reported structural studies that involved chemical cross-linking as part of the experimental toolbox (Sanowar et al. 2010; Schraidt et al. 2010).

Marlovits and coworkers combined cross-linking with cryoEM, bacterial genetics and chemical labeling to study the organization of the needle complex as part of the type III secretion system in *Salmonella* (Schraidt et al. 2010). Domain interactions within the needle complex were elucidated using isotope-coded BS^{2}G (d_0/d_4) as the reagent; acquired LTQ OT MS data were analyzed by xQuest. The results confirmed several interactions between subunits near the base of the complex, on the one hand between InvG and PrgH (with a single cross-link identified) as well as between PrgH and PrgK (13 cross-links in total). In the latter case, cross-links were only identified between the N-terminal domain of PrgH and the C-terminal region of PrgK, in agreement with localization information from concomitant cryoEM experiments using nanogold labeling.

In a second study, Miller and colleagues obtained identical subunit contacts (InvG-PrgH and PrgH-PrgK) using a combination of various cross-linkers (Sanowar et al. 2010). BS^{2}G for Lys-Lys cross-linking, *N*-Ethyl-*N'*-(3-dimethylaminopropyl)carbodiimide (EDC) for zero-length cross-linking between carboxyl groups and amines and a nonspecific succinimidyldiazirine reagent were employed. In contrast to the work discussed in the previous paragraph, the authors used the CXMS cross-linking/MS pipeline (Singh et al. 2008) and employed Phenyx as the primary search engine for their LTQ OT data. The cross-linking data were combined with computational approaches and existing structural information to build a refined model of the needle complex assembly.

8.6 The use of spatial constraints for modeling

The studies presented in Section 8.5 represent excellent examples for the potential of chemical cross-linking in structural biology. However, currently the direct use of cross-linking data for computational purposes is far from routine and requires specialist expertise. Several groups have argued (Fabris and Yu 2010; Leitner et al. 2010; Mayne and Patterton 2011; Rappsilber 2011) that the field would benefit enormously from improved software tools to feed experimentally derived constraints into modeling pipelines.

Historically, already the first cross-linking studies on individual, small proteins were aimed at providing structural information for use in modeling approaches. Young et al. (2000) cross-linked bovine basic fibroblast growth factor FGF-2 using DSS and successfully used the thus obtained distance information (15 through-space cross-links) for fold prediction. Since these initial attempts, computational modeling has advanced considerably, although to date there is still no integrated software suite available that makes direct use of the output of cross-linking studies.

In order to evaluate the potential of the relatively low-resolution constraints obtained from cross-linking experiments, we recently performed simulations on how the quality and quantity of various parameters of these constraints influence the modeling of protein complexes (Leitner et al. 2010). Using actual structures that were used as benchmarks in a protein-protein docking study, it was shown that already relatively small numbers of these constraints (three to five) could help in discarding a large fraction (70% to 95%) of predicted conformations and lead to more accurate structures (in terms of root mean square distance from the actual residue positions). However, in practice not all complexes will yield sufficient numbers of cross-links, for example, because the binding interface is too small or no cross-linkable residues are present that can be connected by common reagents. On the other hand, the few studies reported to date that have produced intersubunit cross-links that number in the tens or even hundreds (see Section 8.5) show that with recent instrumental advances (and more expected to come) such large-scale data sets are more likely than ever to be obtainable.

Rappsilber (2011: 536) has therefore concluded enthusiastically that cross-linking "yields constraints more plentiful and easier that any low-resolution biophysical method." To live up to these expectations, it will be crucial in the coming years to establish chemical cross-linking as a relevant and accurate method at the interface between classical low- and high-resolution structural determination techniques. For this purpose, experimentalists need to work together with bioinformaticians in order to provide robust and efficient protocols for generating information-rich primary data. These data should be fed into transparent (and ideally publicly available) software tools, resulting in a list of validated constraints with a known error rate. As I have discussed in Section 8.4.2, such tools are now beginning to emerge.

Informaticians and structural biologists need to deal with cross-linking data not as absolute distance information, but with associated appropriate experimental uncertainty. In addition to the spacer length of the cross-linking reagent and the length of the amino acid side chain, researchers typically include an additional hardly quantifiable "safety margin" to the distance constraint. This value should compensate for any conformational flexibility in the protein backbone, so in practice typical distance constraints reported for the lysine-reactive reagents DSS and BS^3, both with a spacer length of 11.4 Å, range from 27 to 30 Å. Furthermore, modeling approaches need to be error tolerant as any MS data set will include some low level of false positives. Even if an accurate estimate of the false-positive rate can be made, such an error rate is always associated with the whole data set and not with individual assignments.

With the growing record of large-scale data sets, other aspects of the cross-linking reaction that have been discussed in the literature are more likely to be subject to further scrutiny. On the protein level, these include the coexistence of multiple conformations that cannot be discriminated on the individual cross-link basis or the "freezing" of poorly populated conformational states by forming cross-links. On the complex level, heterogeneity of the complex assembly needs to be taken into account as well, as in most cases the complex under study will be isolated directly from the biological source, where multiple states may exist.

8.7 Conclusion and outlook

Here, I have shown that chemical cross-linking is emerging as a viable tool for analyzing the topology and low-resolution structure of protein complexes. The use of MS as a well established technology for the readout of cross-linking experiments provides sufficient sample throughput, and the method has benefited enormously from instrumental advances, leading to significantly higher sensitivity. Although bioinformatic analysis of cross-linking data is still a bottleneck, a number of software solutions are now available that are relatively user friendly and accessible to the beginner. In contrast, the interface to computational and structural biology is not yet as widely developed. However, the wealth of recently published data on highly relevant complexes (some of which discussed herein) demonstrates that cross-linking is a rapidly evolving and highly promising technique that will certainly have a growing impact in the future.

References

Alber, F., Förster, F., Korkin, D., Topf, M., and Sali, A. (2008). Integrating diverse data for structure determination of macromolecular assemblies. Annu Rev Biochem 77, 443–477.

Bohn, S., Beck, F., Sakata, E., Walzthoeni, T., Beck, M., Aebersold, R., et al. (2010). Structure of the 26S proteasome from *Schizosaccharomyces pombe* at subnanometer resolution. Proc Natl Acad Sci U S A 107, 20992–20997.

Chen, Z.A., Jawhari, A., Fischer, L., Buchen, C., Tahir, S., Kamenski, T., et al. (2010). Architecture of the RNA polymerase II–TFIIF complex revealed by cross-linking and mass spectrometry. EMBO J 29, 717–726.

Choudhary, C., and Mann, M. (2010). Decoding signalling networks by mass spectrometry-based proteomics. Nat Rev Mol Cell Biol 11, 427–439.

Chowdhury, S.M., Du, X.X., Tolic, N., Wu, S., Moore, R.J., Mayer, M.U., et al. (2009). Identification of cross-linked peptides after click-based enrichment using sequential collision-induced dissociation and electron transfer dissociation tandem mass spectrometry. Anal Chem 81, 5524–5532.

Chu, F.X., Mahrus, S., Craik, C.S., and Burlingame, A.L. (2006). Isotope-coded and affinity-tagged cross-linking (ICATXL): an efficient strategy to probe protein interaction surfaces. J Am Chem Soc 128, 10362–10363.

Collins, M.O., and Choudhary, J.S. (2008). Mapping multiprotein complexes by affinity purification and mass spectrometry. Curr Opin Biotechnol 19, 324–330.

Fabris, D., and Yu, E.T. (2010). Elucidating the higher-order structure of biopolymers by structural probing and mass spectrometry: MS3D. J Mass Spectrom 45, 841–860.

Gavin, A.C., Maeda, K., and Kuhner, S. (2011). Recent advances in charting protein-protein interaction: mass spectrometry-based approaches. Curr Opin Biotechnol 22, 42–49.

Gingras, A.C., Gstaiger, M., Raught, B., and Aebersold, R. (2007). Analysis of protein complexes using mass spectrometry. Nat Rev Mol Cell Biol 8, 645–654.

Groll, M., Ditzel, L., Lowe, J., Stock, D., Bochtler, M., Bartunik, H.D., et al. (1997). Structure of 20S proteasome from yeast at 2.4 angstrom resolution. Nature 386, 463–471.

Gygi, S.P., Rist, B., Gerber, S.A., Turecek, F., Gelb, M.H., and Aebersold, R. (1999). Quantitative analysis of complex protein mixtures using isotope-coded affinity tags. Nat Biotechnol 17, 994–999.

Heck, A.J.R. (2008). Native mass spectrometry: a bridge between interactomics and structural biology. Nat Methods 5, 927–933.

Hermanson, G.T. (2008). Bioconjugate techniques. 2nd ed. (London: Academic Press/Elsevier).

Hoopmann, M.R., Weisbrod, C.R., and Bruce, J.E. (2010). Improved strategies for rapid identification of chemically cross-linked peptides using protein interaction reporter technology. J Proteome Res 9, 6323–6333.

Lee, Y.J. (2008). Mass spectrometric analysis of cross-linking sites for the structure of proteins and protein complexes. Mol Biosyst 4, 816–823.

Leitner, A., Walzthoeni, T., Kahraman, A., Herzog, F., Rinner, O., Beck, M., et al. (2010). Probing native protein structures by chemical cross-linking, mass spectrometry, and bioinformatics. Mol Cell Proteomics 9, 1634–1649.

Maiolica, A., Cittaro, D., Borsotti, D., Sennels, L., Ciferri, C., Tarricone, C., et al. (2007). Structural analysis of multiprotein complexes by cross-linking, mass spectrometry, and database searching. Mol Cell Proteomics 6, 2200–2211.

Makarov, A., Denisov, E., Kholomeev, A., Baischun, W., Lange, O., Strupat, K., et al. (2006). Performance evaluation of a hybrid linear ion trap/orbitrap mass spectrometer. Anal Chem 78, 2113–2120.

Matthiesen, R., Azevedo, L., Amorim, A., and Carvalho, A.S. (2011). Discussion on common data analysis strategies used in MS-based proteomics. Proteomics 11, 604–619.

Mayne, S.L.N., and Patterton, H.-G. (2011). Bioinformatics tools for the structural elucidation of multi-subunit protein complexes by mass spectrometric analysis of protein-protein cross-links. Briefings Bioinformatics DOI: 10.1093/bib/bbq087 (in press).

Müller, D.R., Schindler, P., Towbin, H., Wirth, U., Voshol, H., Hoving, S., et al. (2001). Isotope tagged cross linking reagents. A new tool in mass spectrometric protein interaction analysis. Anal Chem 73, 1927–1934.

Nessen, M.A., Kramer, G., Back, J., Baskin, J.M., Smeenk, L.E.J., et al. (2009). Selective enrichment of azide-containing peptides from complex mixtures. J Proteome Res 8, 3702–3711.

Nesvizhskii, A.I. (2010). A survey of computational methods and error rate estimation procedures for peptide and protein identification in shotgun proteomics. J Proteomics 73, 2092–2123.

Olsen, J.V., Schwartz, J.C., Griep-Raming, J., Nielsen, M.L., Damoc, E., Denisov, E., et al. (2009). A dual pressure linear ion trap Orbitrap instrument with very high sequencing speed. Mol Cell Proteomics 8, 2759–2769.

Paizs, B., and Suhai, S. (2005). Fragmentation pathways of protonated peptides. Mass Spectrom Rev 24, 508–548.

Panchaud, A., Singh, P., Shaffer, S.A., and Goodlett, D.R. (2010). xComb: a cross-linked peptide database approach to protein-protein interaction analysis. J Proteome Res 9, 2508–2515.

Petrotchenko, E.V., and Borchers, C.H. (2010a). ICC-CLASS: isotopically-coded cleavable crosslinking analysis software suite. BMC Bioinformatics 11, 64 (article number).

Petrotchenko, E.V., and Borchers, C.H. (2010b). Crosslinking combined with mass spectrometry for structural proteomics. Mass Spectrom Rev 29, 862–876.

Petrotchenko, E.V., Serpa, J.J., and Borchers, C.H. (2011). An isotopically coded CID-cleavable biotinylated cross-linker for structural proteomics. Mol Cell Proteomics 10, M110.001420 (article number).

Pflieger, D., Gonnet, F., de la Fuente van Bentem, S., Hirt, H., and de la Fuente, A. (2011). Linking the proteins – elucidation of proteome-scale networks using mass spectrometry. Mass Spectrom Rev 30, 268–297.

Rappsilber, J. (2011). The beginning of a beautiful friendship: cross-linking/mass spectrometry and modelling of proteins and multi-protein complexes. J Struct Biol 173, 530–540.

Rinner, O., Seebacher, J., Walzthoeni, T., Mueller, L.N., Beck, M., Schmidt, A., et al. (2008). Identification of cross-linked peptides from large sequence databases. Nat. Methods 5, 315–318.

Rix, U., and Superti-Furga, G. (2009). Target profiling of small molecules by chemical proteomics. Nat Chem Biol 5, 616–624.

Sanowar, S., Singh, P., Pfuetzner, R.A., André, I., Zheng, H., Spreter, T., et al. (2010). Interactions of the transmembrane polymeric rings of the *Salmonella enterica* serovar Typhimurium type III secretion system. mBio 1, e00158–10.

Schilling, B., Row, R.H., Gibson, B.W., Guo, X., and Young, M.M. (2003). MS2Assign, automated assignment and nomenclature of tandem mass spectra of chemically crosslinked peptides. J Am Soc Mass Spectrom 14, 834–850.

Schraidt, O., Lefebre, M.D., Brunner, M.J., Schmied, W.H., Schmidt, A., Radics, J., et al. (2010). Topology and organization of the *Salmonella typhimurium* type III secretion needle complex components. PLoS Pathogens 6, e1000824 (article number).

Seebacher, J., Mallick, P., Zhang, N., Eddes, J.S., Aebersold, R., and Gelb, M.H. (2006). Protein cross-linking analysis using mass spectrometry, isotope-coded cross-linkers, and integrated computational data processing. J Proteome Res 5, 2270–2282.

Singh, P., Shaffer, S.A., Scherl, A., Holman, C., Pfuetzner, R.A., Larson Freeman, T.J., et al. (2008). Characterization of protein cross-links via mass spectrometry and an open-modification search strategy. Anal Chem 80, 8799–8806.

Sinz, A. (2010). Investigation of protein-protein interactions in living cells by chemical crosslinking and mass spectrometry. Anal Bioanal Chem 397, 3433–3440.

Steen, H., and Mann, M. (2004). The ABC's (and XYZ's) of peptide sequencing. Nat Rev Mol Cell Biol 5, 699–711.

Tang, X., and Bruce, J.E. (2010). A new cross-linking strategy: protein interaction reporter (PIR) technology for protein-protein interaction studies. Mol Biosyst 6, 939–947.

Trester-Zedlitz, M., Kamada, K., Burley, S.K., Fenyö, D., Chait, B.T., and Muir, T.W. (2003). A Modular cross-linking approach for exploring protein interactions. J Am Chem Soc 125, 2416–2425.

Trnka, M. J., and Burlingame, A. L. (2010). Topographic studies of the GroEL-GroES chaperonin complex by chemical cross-linking using diformyl ethynylbenzene. Mol Cell Proteomics 9, 2306–2317.

Vellucci, D., Kao, A., Kaake, R.M., Rychnovski, S.D., and Huang, L. (2010). Selective enrichment and identification of azide-tagged cross-linked peptides using chemical ligation and mass spectrometry. J Am Soc Mass Spectrom 21, 1432–1445.

Vermeulen, M., Hubner, N.C., and Mann, M. (2008). High confidence determination of specific protein-protein interactions using quantitative mass spectrometry. Curr Opin Biotechnol 19, 331–337.

Young, M.M., Tang, N., Hempel, J.C., Oshiro, C.M., Taylor, E.W., Kuntz, I.D., et al. (2000). High throughput protein fold identification by using experimental constraints derived from intramolecular cross-links and mass spectrometry. Proc Natl Acad Sci U S A 97, 5802–5806.

Zhang, H., Tang, X., Munske, G.R., Tolic, N., Anderson, G.A., and Bruce, J.E. (2009). Identification of protein-protein interactions and topologies in living cells with chemical cross-linking and mass spectrometry. Mol Cell Proteomics 8, 409–420.

Zhang, R.J., Sioma, C.S., Wang, S.H., and Regnier, F.E. (2001). Fractionation of isotopically labeled peptides in quantitative proteomics. Anal Chem 73, 5142–5149.

9 Single-crystal spectroscopy correlated with X-ray crystallography provides complementary perspectives on macromolecular function

Allen M. Orville

> *Dedicated to my father, Dr. Harold (Harry) D. Orville (1932–2011), whose lifelong love of science inspired me. He was a gentleman, a scholar and a father who spread out computer graphs on the dining room table and let his young children "help" by coloring them between the lines. His patient answer to my persistent question, "Why . . .?" was often, "because no one has figured this out yet, but I think we can." Thank you for your love, generosity and wonderful, curious outlook on life.*

9.1 Introduction

Macromolecular crystallography (MX) is the preeminent method for structural biology worldwide. The intrinsic value of MX is recognized by Nobel Prize selection committees as evidenced by five Nobel Prizes in Chemistry in recent years: Walker (1997), MacKinnon (2003), Kornberg (2006), Tsien (2008) and Ramakrishnan, Steitz and Yonath (2009). These research teams relied heavily upon synchrotron radiation for their research. Thus, synchrotron-based methods have driven many new developments in the MX field, especially in the past two decades. The success of these efforts is also revealed by the fact that approximately 80% of the crystal structures deposited to the Protein Data Bank (PDB) annually are now solved with data collected at synchrotron X-ray sources (see ▶Table 9.1).

Since its inception, more than 70,000 structures have been deposited to and released by the PDB to date. This is truly a worldwide effort with active researcher groups on every continent (with the probable exception of Antarctica). To support these efforts, there are more than two-dozen synchrotron light sources located across Europe, the Americas, Asia and Australia (see ▶Figure 9.1). Each of these highly optimized facilities is ideal for MX and delivers tunable, intense X-ray photons to one or more beamlines dedicated to structural biology. At most synchrotrons, structural biologists represent the largest group of users and typically publish more than half of highest-profile papers highlighting results from the facilities.

With outstanding successes come more difficult problems. Thus, the frontier challenges for MX now include structures of enzymes catalyzing reactions or trapped in reactive intermediate states, large macromolecular machines and viruses, membrane proteins, protein-protein complexes, and protein–nucleic acid complexes. Furthermore, the first results from free electron laser X-ray sources have recently been published with data obtained from submicron-sized crystals with diffraction acquired using femtosecond pulses (Chapman et al. 2011; Seibert et al. 2011). Clearly, the future remains bright for structural biology, and MX will remain a critical tool for generations of scientists to come.

Table 9.1: Statistical summary of X-ray crystal structures curated by the "PDB[a]"

Year	Total X-ray structures	Total synchrotron structures	Sync. structures (% total)	Sync. Americas[b]	Sync. Americas (% sync)	Sync. Europe[c]	Sync. Europe (% sync)	Sync. Asia[d]	Sync. Asia (% sync)
2005	5,131	3,872	75.5	1,887	48.7	1,390	35.9	599	15.5
2006	5,664	4,464	78.8	2,373	53.2	1,477	33.1	624	14.0
2007	6,695	5,439	81.2	2,860	52.6	1,659	30.5	928	17.1
2008	6,115	5,141	84.1	2,808	54.6	1,667	32.4	666	13.0
2009	7,357	6,249	84.9	3,403	54.5	1,982	31.7	869	13.9
2010	8,002	6,889	86.1	3,535	51.3	2,306	33.5	1,052	15.3
Total[e]	65,984	48,582	73.6	25,150	51.8	16,829	34.6	6,643	13.7

[a] Source: BioSync (http://biosync.sbkb.org).
[b] Americas synchrotron facilities: ALS, APS, CAMD, CHESS, CLSI (Canada), LNLS (Brazil), NSLS, SSRL.
[c] European synchrotron facilities: BESSY, DIAMOND, ELETTRA, EMBL/DESY, ESRF, KURCHATOV SNC, LURE, MAX II, MPG/DESY, SLS, SOLEIL, SRS.
[d] Asian/Oceanic synchrotron facilities: AUSTRALIAN SYNCHROTRON, BSRF, NSRRC, PAL/PLS, PHOTON FACTORY, SPRING-8, SSRF.
[e] Summation over the period 1975 to May 2011.

9.1 Introduction | 145

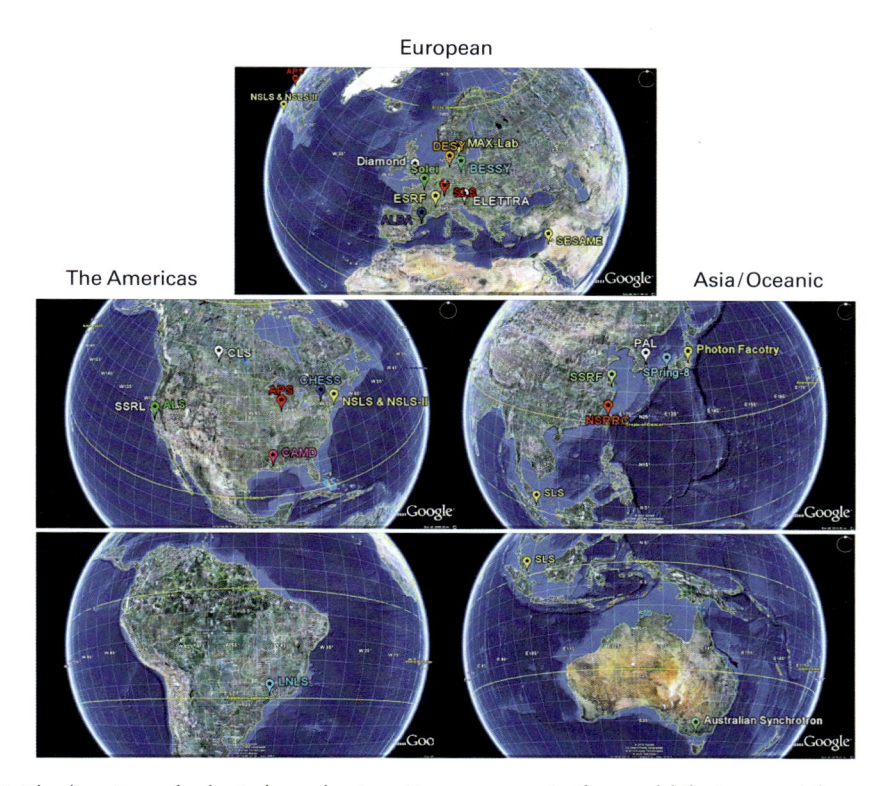

Figure 9.1: The location of selected synchrotron X-ray sources in the world that support the majority of structural biology research (Google Earth; for more details, visit BioSync [http://biosync.sbkb.org]). **Europe:** ALBA is located in Cerdanyola del Vallès (near Barcelona), Spain; BESSY, the Berliner Elektronenspeicherring-Gesellschaft für Synchrotronstrahlung, is located in Berlin, Germany; Diamond Light Source is located in South Oxfordshire, UK; ELETTRA is located in Trieste, Italy; ESRF, the European Synchrotron Radiation Facility, is located in Grenoble, France; DESY, the Deutsches Elektronen-Synchrotron (including PETRA III), is located in Hamburg, Germany; MAXLab is located in Lund University, Sweden; SESAME, the Synchrotron-light for Experimental Science and Applications in the Middle East, under the auspices of UNESCO, is located at Al-Balqa' Applied University in Jordan; SLS, the Swiss Light Source, is located in Villigen, Switzerland; SOLEIL synchrotron is located in Saint-Aubin, France. **The Americas:** ALS, the Advanced Light Source, is located at Lawrence Berkeley National Laboratory, University of California at Berkeley in Berkeley, CA; APS, the Advanced Photon Source, is located at Argonne National Laboratory in Argonne, IL; CAMD, the Center for Advanced Microstructures and Devices, is located at Louisiana State University in Baton Rouge, LA; CHESS, the Cornell High Energy Synchrotron Source, is located at Cornell University in Ithaca, NY; CLSI, the Canadian Light Source Inc., is located in Saskatchewan, Canada; LNLS, the Brazilian Synchrotron Light Laboratory, is located in Campinas, Brazil; NSLS and NSLS-II, the National Synchrotron Light Source and NSLS-II, are located at Brookhaven National Laboratory in Upton, NY; SSRL, the Stanford Synchrotron Radiation Laboratory, is located at the Stanford Linear Accelerator Center (SLAC), Stanford University in Menlo Park, CA. **Asia/Oceanic:** The Australian Synchrotron is located in Melbourne, Australia; NSRRC, the National Synchrotron Radiation Research Center, is located in Taipei, Taiwan; PAL, the Pohang Accelerator Laboratory, is located in Pohang, Korea;

(*Continued*)

Figure 9.1: (*Continued*)

the Photon Factory is located in Tsukuba, Japan; Spring-8, the Super Photon ring 8 GeV, is located in Aioi, Japan; SSRF, the Shanghai Synchrotron Radiation Facility, is located in Shanghai, China; SLS, the Singapore Synchrotron Light Source, is located at the National University of Singapore.

9.2 Ionizing radiation: essential for crystal structures; a problem and a reagent

The focus of this chapter is motivated, in part, by our increasing understanding of the consequences of the interaction of X-rays with macromolecular crystals. Simply put, MX provides three-dimensional structures – often at atomic resolution – of nucleic acids, proteins and viruses. These macromolecules range in size from a few hundred to hundreds of thousands of atoms. They are folded into discrete three-dimensional shapes that evolved to support their function. Despite their complexity, they can be "coaxed" into crystals by manipulating the chemical and physical conditions such as salt concentration, temperature and numerous other parameters. The resulting crystal may contain around 10^{15} copies of the macromolecule arranged in a regular three-dimensional lattice that also includes about 50% solvent by volume.

Electrons bound to ordered atoms within each unit cell of the crystal lattice diffract X-rays. Where these photon waves add constructively, a reflection is observed on the X-ray detector as a discrete spot. Bragg's law indicates that diffraction is a function of the incident X-ray wavelength and the spacing and angle between Miller planes that yield the reflection. A typical diffraction pattern is composed of thousands to millions of X-ray reflections, the number of which in the complete data set depends upon the resolution range and the unit cell dimensions within the crystal. The diffraction pattern, therefore, contains information about the crystal lattice and its ordered contents. The electron density for the molecule, and thus its atomic structure, is determined by Fourier transform of the diffraction spot intensities (structure factors) and positions (Miller indices) of all the X-ray reflections, with experimentally determined phase information for each diffracted X-ray.

Crystallographers have known for years that the exposure of biological crystals to X-rays eventually leads to a degradation of the quality of the diffraction data (Garman 2010; Holton 2009). This is most often observed as a fading of the high-resolution spots in the diffraction pattern and indicates a loss of crystalline order within the sample. The reason why relates to the fact that for typical MX measurements using 12.4 keV (1 Å) X-rays, approximately 2% of the photons interact with the sample. However, of these photons only 8% gives rise to the diffraction pattern (Thomson, coherent scattering), whereas 84% of them are totally absorbed by the sample (photoelectric effect) and the other 8% experience inelastic scattering (Compton effect). The latter two phenomena account for greater than 90% of the interacting photons and deposit energy into the crystal but do not contribute significantly to the intensity of the diffraction pattern. The photoelectric effect dominates and is a resonance phenomenon that ejects an inner shell electron upon photon absorption (Neutze et al. 2000). The loss of a K-shell electron from a carbon, nitrogen, oxygen, phosphorous or sulfur atom yields an unstable ion with a "hole" that is filled by an electron from a higher shell falling into the vacant orbital. For these elements, the falling electron typically gives up its energy to another electron, which is ejected (the Auger effect). Consequently, the photoelectric effect liberates two electrons from biological elements. Moreover, they have sufficient energy to induce several hundred further ionization events and can propagate for up to 4 microns (Sanishvili et al. 2011). Many of these events form radical species that originate on the macromolecule itself or by the radiolysis of the bulk solvent molecules within the crystal volume. Together, this energy loss is the X-ray "dose" measured per mass of the sample, which is given in SI units of grays (Gy; 1 Gy = 1 J kg^{-1}). Since these events can break bonds by radical initiated processes, the term *radiation damage* is often applied to describe the complex chemistry caused by X-ray photons depositing energy directly into the sample.

The solvated electrons generated in biological samples by X-ray irradiation are liberated on the timescale of electronic transitions (e.g. Xu and Chance 2007 and references therein). Rapid electron transfer over significant distances can be facilitated by aromatic side chains, protein backbone atoms and protein secondary structure (Gray and Winkler 1996). Hydroxyl and superoxide radicals, as well as other reaction products, are also generated in biological samples by X-ray exposure, but on a time scale of microseconds to milliseconds (Xu and Chance 2007). However, the migration of hydroxyl and superoxide radicals, dioxygen and protons within proteins at cryogenic temperatures is less well understood than electron transfer processes. At cryogenic temperature, these species can migrate within and between macromolecular cavities. For example, geminate recombination has been observed with low-temperature crystallographic studies of CO dissociation from myoglobin (Austin et al. 1975; Bourgeois et al. 2003; Chu et al. 2000; Srajer et al. 1996; Teng et al. 1994; Tomita et al. 2009).

Now consider that many macromolecules evolved to facilitate and/or control the transfer of electrons from one moiety to another. Therefore, important functional aspects of proteins and enzymes may be promoted during the X-ray diffraction experiment, even at cryogenic temperature. Moreover, because the X-ray photons themselves generate solvated electrons in situ, these should be considered as potential reagents for reaction cycles. This is particularly true for redox-active reactions typified by oxidoreductases that evolved to catalyze the transfer of electrons from one molecule to another. In these cases, it is possible that the data collection process may advance normal and/or off-path reaction cycles.

9.3 Cofactors in biology provide spectroscopic access to reaction cycles

Recent bioinformatics surveys demonstrate that approximately 47% of all enzymes with known three-dimensional structures contain metal(s) (Holm et al. 1996; Waldron et al. 2009). Indeed, 41% of all enzymes, from all six Enzyme Commission classes, incorporate metal centers into their active sites (e.g. oxidoreductases [44%], transferases [40%], hydrolases [39%], lyases [36%], isomerases [36%] and ligases [59%]). These enzymes evolved to exploit metal ions and/or organic cofactors to achieve catalysis because their electronic properties provide more strategies to lower transition-state barriers than macromolecules composed of only H, C, N, O, P and S atoms.

Selected examples of proteins containing colored cofactors, and their typical optical absorption characteristics, are listed in ▶Table 9.2. These types of simple queries to the PDB return more than 12,000 entries, which is about 17% of the current PDB archive. These are of particular interest because it is common for the spectroscopic signature(s) of metalloproteins to change during catalysis. This includes proteins and enzymes involved in photosynthesis and H_2O oxidation, N_2 fixation and NO_x metabolism, C1 metabolism (methane and CO_2), H_2 metabolism, O_2 activation and the sulfur

Table 9.2: Summary of selected cofactors within structures deposited in the PDB archive (May 2011)

Prototypical protein example	Cofactor or chromophore	λ_{max} (nm)	Typical ε (mM^{-1} cm^{-1})	PDB ligand ID or name	No. PDB structures
Aspartate aminotransferase	Pyridoxal-5'-phosphate	350, 412	6.6, 9.5	PLP	596
Lactate dehydrogenase	NAD(H) or NADP(H)	340	6.2	NAD, NDP, NAP	1,671
Dihydrofolate reductase	Pterin	350	4.5	Pterin	294
p-Hydroxybenzoate hydroxylase	Flavin adenine dinucleotide (FAD) or Flavin mononucleotide (FMN)	450	11	FAD, FMN, flavin	1,665
Myoglobin	Heme	409	170	Hem	2,814
Mn superoxide dismutase	Mn(III)-O/N	480	0.91	Mn	1,536
Rubredoxin	Fe (nonheme)	430–500	3–5	Fe	835
Ferredoxin	Fe$_2$/S$_2$ Cluster	~400	30.6	FeS	352
High potential iron protein	Fe$_4$/S$_4$ Cluster	~400	15	SF4	441
Methylmalonyl-coenzyme A (CoA) mutase	Co or B$_{12}$	360	27.5	Cobalt	406
Urease	Ni(II)-O/N	407	0.21	Nickel	546
Azurin or plastocyanin	Cu	600	5	Cu	839
Photosynthetic reaction center	Chlorophyll	465, 665	>100	Bacteriochlorophyll A (BCL), Bacteriopheophytin A (BPH), photosynthesis	176
Bacteriorhodopsin	Retinal	480	27	Ret	123
Green florescent protein	Posttranslation modification	350–700	6–94.5	Luminescent proteins	193

cycle (S$_0$ and SO$_x$ metabolism). Thus, the electronic properties of metal centers and organic cofactors provide spectroscopic access to chemical reactions, but this may also make them sensitive to alteration by X-ray-derived photoelectrons. Consequently, unanticipated electron density or "observations" of color changes during X-ray diffraction studies may be reported but are often difficult to explain based upon the structure alone. Therefore, a powerful strategy to promote, trap and study reaction intermediates at low

9.3 Cofactors in biology provide spectroscopic access to reaction cycles | 149

temperature within the macromolecular crystal is enabled by spectroscopic analysis of the crystals themselves before, during and after X-ray diffraction.

Whereas the three-dimensional structure of a molecule is defined by the atomic coordinates for each atom, the electronic structure relates to electron orbitals at the subatomic level (▶Figure 9.2). Furthermore, because the overlap and mixing of molecular orbitals results in chemical reactions, macromolecular function is the direct result of the

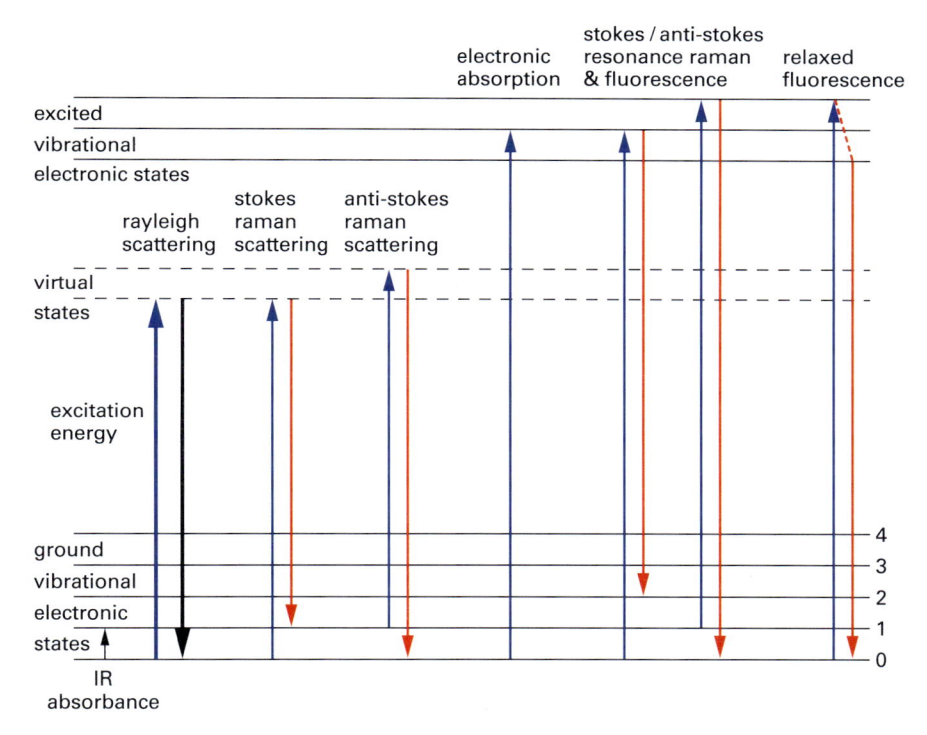

Figure 9.2: Energy level diagram illustrating various spectroscopic techniques. Electronic absorption spectroscopy typically uses UV/vis wavelength light to promote allowed transitions from ground vibration states to elicited vibration states. Raman spectroscopy often relies on laser excitation in the visible, near infrared (e.g. promotion to virtual states) or near ultraviolet range and detects inelastic scattering of these photons at either higher (anti-Stokes) or lower (Stokes) energy than the excitation photons. Those photons with wavelengths close to the laser line (elastic Rayleigh scattering) are filtered out. The Raman-scattered photons give information about the particular molecular vibrations in the system. Infrared spectroscopy yields similar, but complementary, information about the ground-state vibrational modes. In resonance Raman spectroscopy, the laser excitation energy is matched to an electronic transition of the molecule, so that vibrational modes associated with the excited electronic state are greatly enhanced. However, Raman scattering and fluorescence emission are two competing phenomena, which have similar origins and differ in their lifetimes (fluorescence ≥ 10^{-9} s; Raman ~ 10^{-12} s) and the nature of their so-called intermediate states. The Raman spectrum tends to be more informative about the molecule in question, but fluorescence is often much more intense. Consequently, the Raman spectroscopist tends to search for sample or exitation conditions that avoid fluorescence.

electronic structure. This type of information cannot be probed with crystallography, but rather with optical spectroscopy. For example, the electronic absorption spectroscopic analysis of myoglobin provides insights into the iron redox state, its spin state and its coordination environment – especially exogenous ligand coordination. Although it can be difficult to differentiate diatomic ligands bound to iron based upon moderate electron density maps, the spectroscopic signature is often dramatically different for different ligands. Thus, the often stated "structure-function" relationship derives from both the atomic and electronic structures of the molecule in question.

Inelastic scattering processes also provide a means to observe molecular vibrations that often reveal insights into function (Carey 2006). For example, in Raman scattering an absorbed photon is reemitted with lower energy (Stokes scattering) that corresponds to the energy difference between different molecular vibrational modes. Thus, the Raman effect involves interactions between atomic positions, electron distributio, and intermolecular forces. Consequently, it is ideally suited to provide insights into function that a crystal structure alone cannot. Indeed, the information available from Raman spectroscopy can be very detailed, exceeding the level of resolution found in all but the highest-resolution X-ray crystal structures. It can also reveal changes in the distribution of electrons in a bound ligand as well as details about hydrogen bonding strengths between active-site residues and substrate, inhibitor or solvent molecules.

Despite these desirable characteristics, Raman spectroscopy has not been widely exploited by scientists interested in complex macromolecular systems because of three inherent difficulties: (a) low Raman sensitivity, (b) interference from fluorescence background and (c) difficult data interpretation. The latter problem becomes readily apparent when one considers that the number of vibrational modes present in a molecule is $3n - 6$, where n is the number of atoms; this is quite large for macromolecules. Therefore, the complete Raman spectrum of a macromolecule is very complex; it probes vibration bands inherent to all the chemical bonds throughout the macromolecule. Consequently, to be most effective, Raman spectroscopy is often applied to systems where one can focus upon a small region of interest, such as an enzyme active site. For example, isotopic editing – wherein the ligand of interest is selectively or uniformly labeled with stable isotopes such as ^2H, ^{13}C, ^{15}N, ^{17}O, ^{18}O, ^{19}F, ^{34}S, and so forth – is a powerful method to identify the Raman bands of interest. Another effective strategy is to use a laser excitation energy that is tuned to the vicinity of an allowed electronic transition (▶Figure 9.2). This resonance Raman effect yields greatly enhanced Raman vibration bands that are directly related to the chromophore(s) within the macromolecule. Moreover, resonance Raman coupled with isotopic editing is a very powerful method to define otherwise very complex systems.

9.4 Single-crystal spectroscopy correlated with X-ray diffraction

Almost none of the crystal structures listed in ▶Table 9.2 were correlated with any type of single-crystal spectroscopy. Not because this technique is particularly difficult, but more likely because the opportunity or the expertise was not available. Therefore, fundamental questions remain about many important structures in the PDB, such as the redox state of the metalloprotein in its corresponding structure. In contrast, single-crystal spectroscopy correlated with X-ray diffraction provides data to address these types of issues. Without question, the best correlations of atomic and electronic structure derive

9.4 Single-crystal spectroscopy correlated with X-ray diffraction | 151

from results that are obtained from the same crystalline samples. The general concept of single-crystal spectroscopy integrated with X-ray diffraction is illustrated in ▶Figure 9.3. The essence of the concept is to focus each microscope objective used for spectroscopy so that it corresponds with the crystal eucentric rotation point and the X-ray beam. Thus, the same volume of the crystal is probed by X-ray diffraction and the photons used for spectroscopic analysis. By integrating the beamline controls for spectroscopic analysis and X-ray diffraction, one has the ability to collect spectroscopic data as a function of X-ray dose. Research of this type is currently ongoing at several synchrotron facilities as discussed in Sections 9.5.1 and 9.5.2, including at the NSLS at beamline X26-C (Héroux et al. 2009; Major et al. 2009; Orville et al. 2009, 2011; Stoner-Ma et al. 2011; Yi et al. 2010); it is also proposed for NSLS-II beamline SM3.

Modern conventional MX methods involve the following: (a) growing and mounting single crystals that are typically ~100 μm along an edge; (b) cryopreserving and placing one crystal into a monochromatic X-ray beam; (c) correlating precise crystal rotation with hundreds of individual X-ray diffraction images; (d) integrating and processing the raw data, which are typically composed of tens of thousands to several million

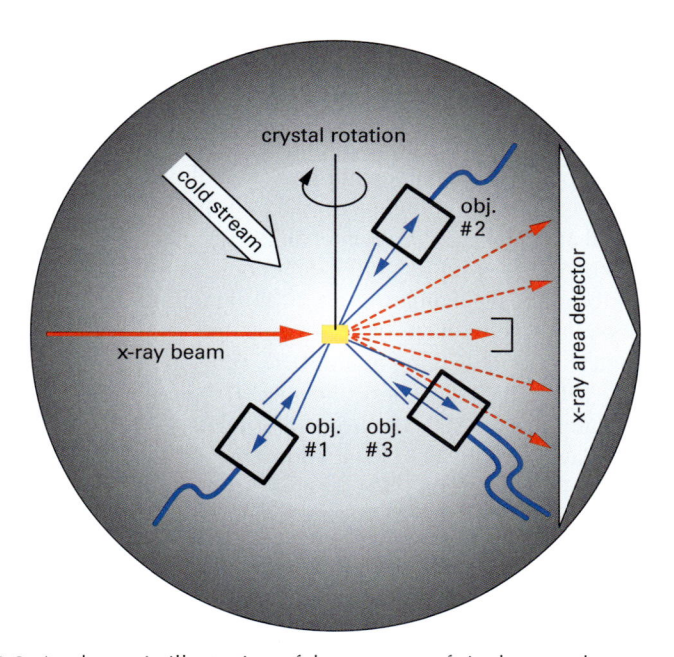

Figure 9.3: A schematic illustration of the concept of single-crystal spectroscopy correlated with X-ray diffraction. In the X-ray diffraction experiment, the crystal is held in a loop within a cryostream and rotated about an axis that brings the diffraction pattern onto the area detector under well-controlled conditions. Spectroscopy photons are delivered to and collected from the crystal by optical fibers and long-working-distance microscopic objectives. For example, the linear geometry of objectives #1 and #2 are ideal for electronic absorption spectroscopy, whereas Raman spectroscopy is supported by excitation and collection from the same objective in back-scatter geometry (objective #3). The focal point of each objective used for spectroscopy must overlap with the crystal eucentric rotation point and the X-ray beam.

reflections, into a complete data set; (e) solving the phase problem to locate the position of every atom within the macroscopic crystal and (f) refining an atomic model to fit the electron density maps of the macromolecule comprising the crystal. It is in this last step that chemical information and other biophysical data are often used to help interpret some features of the structure. This is particularly true for active-site features such as metal centers of organic cofactors.

Perhaps the best way to correlate electronic and atomic structure is to collect single-crystal spectroscopy and X-ray diffraction data from the same crystal during the X-ray diffraction experiment. To that end, some of the first single-crystal microspectrophotometers for use at synchrotron X-ray sources were developed in Europe by Hadfield and Hajdu (1993) and by Bourgeois et al. (2002). More recent examples include the Cryobench and SNBL installations at the ESRF (Carpentier et al. 2007; McGeehan et al. 2009; Røhr et al. 2010; Royant et al. 2007) and similar efforts at the SLS (Owen et al. 2009), at the APS (De la Mora-Rey and Wilmot 2007; Pearson et al. 2007) and at the SSRL (Meharenna et al. 2010). These types of instruments, at the beamline as well as off-line as in the Cryobench setup at the ESRF, have proved to be pivotal in correlating crystal structures to "radiation damage" (Garman 2010) or to trap intermediates with characteristic spectroscopic signals for reactive species (reviewed in Beitlich et al. 2007; Bourgeois and Royant 2005; De la Mora-Rey and Wilmot 2007; Hajdu et al. 2000; Meharenna et al. 2010; Pearson and Owen 2009). Other reports have documented these methods applied to organic-cofactor-containing proteins or those with posttranslational modifications: for example, FAD (Barends et al. 2009; Churbanova et al. 2010; Héroux et al. 2009; Jung et al. 2006; Major et al. 2009; Orville et al. 2009; Røhr et al. 2010), the blue-light-sensing, FMN-dependent photoreceptor present in the LOV (light, oxygen, or voltage) domain (Fedorov et al. 2003), photoactive yellow protein (Imamoto et al. 2001; Kort et al. 2004; Rajagopal and Moffat 2003; Ren et al. 2001), green fluorescent protein (Adam et al. 2008, 2009; Andresen et al. 2005; Lelimousin et al. 2009; Rosell and Boxer 2003) and bacteriorhodopsin (Edman et al. 2004; Hajdu et al. 2000; Schertler et al. 1991). Several metalloproteins have been investigated with correlated studies, including those that contain copper (Chang et al. 2010; Pearson et al. 2007; Solomon et al. 2004; Wilmot et al. 1999), nickel (Cedervall et al. 2010), nonheme, mononuclear iron (Adam et al. 2004; Katona et al. 2007) or iron-sulfur centers (Karlsson et al. 2000).

Heme-based enzymes have perhaps the oldest and richest history in protein crystallography (Kendrew et al. 1958; Perutz 1963; Perutz et al. 1968; Bolton and Perutz 1970) and exhibit a molar absorptivity (▶Table 9.2) that is among the highest for all protein cofactors. Research over the past 50 years into the structure and function of heme proteins has produced several iconic results. However, correlated crystallographic and spectroscopic studies are only just now providing critical new insights into the nature of reactive intermediates along the reaction coordinate for only a few of these proteins. For example, high-valent intermediates for several heme-containing proteins have been generated and studied by correlated methods, including cytochrome P450$_{cam}$ (Beitlich et al. 2007; Denisov et al. 2005; Makris et al. 2006; Schlichting et al. 2000), cytochrome c peroxidase (Meharenna et al. 2010), chloroperoxidase (Beitlich et al. 2007; Kühnel et al. 2007), horseradish peroxidase (Berglund et al. 2002), catalase (Gouet et al. 1996; Jouve et al. 1997) and the peroxidase activity of myoglobin (Hersleth et al. 2007). Moreover, the ability to distinguish between Fe(II)-OH$_2$, Fe(III)-OH, Fe(III)-O-O(H) and Fe(IV)=O species based on electron density maps is nearly impossible in the absence of sub-Ångstrom resolution data. In contrast, the spectroscopic features of these species are dramatically different. This was recently demonstrated with Fe(III)-NO$_2$ and Fe(II)-NO$_2$ complexes of myoglobin (Yi et al. 2010). Consequently, the use of correlated spectroscopic and crystallographic analysis is essential to assign chemical and mechanistic insight(s) to typical resolution structures.

9.5 Correlated studies at beamline X26-C of the NSLS

The NSLS at Brookhaven National Laboratory is a second-generation synchrotron. It started operations in late 1982 and has been a very productive facility for nearly three decades. However, its days are numbered as evidenced by the construction of the NSLS-II very nearby, which is on schedule to begin full operations in 2014. The NSLS-II will be a new, state-of-the-art, medium-energy electron storage ring (3 GeV) designed to deliver world-leading intensity and brightness, and it will produce X-rays more than 10,000 times brighter than the current NSLS (Ozaki et al. 2007).

As indicated in Section 9.4, only rarely are complementary methods coupled to the standard MX experiments. Therefore, we are continuing to develop a national facility at beamline X26-C at the NSLS for single-crystal spectroscopy correlated with X-ray diffraction and will implement a similar facility at beamline SM3 of the NSLS-II. We integrated several complementary techniques so that single-crystal electronic absorption spectroscopy, Raman spectroscopy and X-ray diffraction can be collected from the same sample and under nearly identical experimental conditions (Orville et al. 2011; Stoner-Ma et al. 2011). This beamline is dedicated full time to multidisciplinary studies and is available to the general user population at the NSLS.

Inside the hutch at beamline X26-C, the X-ray beam travels along x, the crystal rotation axis parallel to y, and the electronic absorption and Raman spectroscopy axis along z (▶Figure 9.4). The collimator and final X-ray slits are upstream from the crystal, whereas the X-ray detector occupies the region immediately downstream from the sample. A telescope is mounted on the diffractometer in order to visualize the crystal with a line of sight in the plane of the X-ray beam, from the upstream side, and 35° below the horizontal plane. The photons for optical absorption spectroscopy travel along the vertical axis and are focused to a point that intersects the crystal rotation point of the diffractometer and the X-ray beam. The Raman spectroscopy system, from Horiba Jobin-Yvon Inc., consists of (a) three diode lasers (473 nm, 532 nm and 785 nm); (b) three Raman microprobe heads, each containing an edge filter specific for one of the laser excitation energies to block the Raleigh scattered photons; (c) an IHR 550 spectrometer and (d) a Synapse CCD detector (charge-coupled device). The Raman data are collected in back-scatter geometry, and optical fibers are used to connect the Raman lasers to the probe heads and back to the spectrometer. Each microprobe head uses an Olympus SLMPlanN 50x/0.35 objective to focus the laser beam to a ~25 μm spot size and includes a video camera that allows one to align the laser focal spot to the same region of the crystal as probed by UV/vis spectroscopy and X-ray diffraction. The lower objective for electronic absorption and the Raman microprobe head are mounted on a motorized stage below the sample to facilitate switching between electronic absorption and Raman modes of operation.

For electronic absorption spectroscopy, the two reflective microscope objectives (telescope objectives) are based upon the Schwarzschild parabolic mirror design. They were taken from an XSPECTRA system supplied by 4DX-ray Systems AB, Uppsala, Sweden.

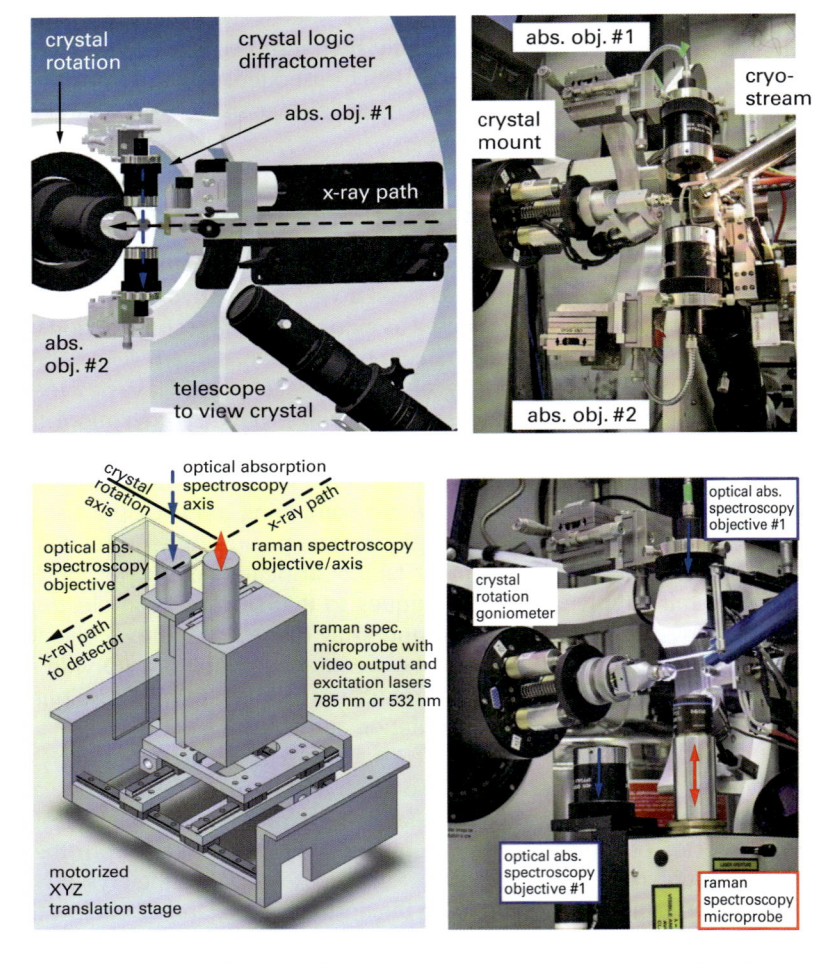

Figure 9.4: First- and second-phase single-crystal spectroscopy components installed at beamline X26-C. (Top) A schematic (left) and photograph (right) of the perpendicular alignment of the objectives for electronic absorption spectroscopy correlated with X-ray diffraction. (Bottom) A schematic (left) of the lower objectives for electronic absorption objective #2 and the Raman microprobe head with a close-up view aligned for Raman spectroscopy (right). To achieve phase 2, the lower arm of the arc was replaced with the motorized XYZ translation stage.

The light is focused to a 25 μm diameter spot using a 50 μm optical fiber and collected from a 75 μm diameter region and focused into a 400 μm optical fiber. The parabolic objectives provide a 24 mm working distance, which allows for cryocooling and crystal mounting, and so forth. The incident light (350–850 nm) is typically from a 75W Xe research arc lamp (Newport Corp.). An Ocean Optics USB 4000 spectrophotometer is used to collect electronic absorption spectra, which is under beamline instrumentation control with EPICS (Experimental Physics and Industrial Control System, a set of Open

Source software tools, libraries, and applications developed collaboratively and used worldwide to create distributed soft real-time control systems for scientific instruments such as particle accelerators, telescopes, and other large scientific experiments).

The typical experiment at beamline X26-C involves mounting a crystal and centering it to the X-ray beam within the cold stream. Next, 72 digital images of the loop/crystal are recorded, one every 5° for a full 360° rotation. At each of the 72 orientations, an optical absorption spectrum is also collected. Our beamline software program, CBASS, then uses the C3D algorithm (Lavault et al. 2006) which determines the broadest, flat face of the loop/crystal, directs the goniometer to rotate the crystal so that this face is presented to the incident objective lens for spectroscopy and records the motor positions for the lower spectroscopy components. If desired, the lower stage is shifted to bring the Raman spectroscopy microprobe into focus. Thus, electronic absorption and Raman spectroscopic data can be collected from the same region of the crystal that intersects the X-ray beam. With the lower objective in place for absorption mode, electronic absorption spectra are collected between nearly every X-ray diffraction image. For example, during the readout loop of the X-ray detector and while the X-ray shutter is closed, (a) the crystal is rotated back to the flattest orientation, (b) another optical spectrum is recorded, (c) the result is plotted as an overlay on the first spectrum, then (d) the X-ray diffraction data collection strategy continues at the appropriate rotation angle. This creates a family of optical spectra that is obtained as a function of X-ray dose, wherein each spectrum differs from the others by the cumulative X-ray exposure. Upon completion of the X-ray diffraction data set, the lower objective is then realigned for Raman spectroscopy based upon prerecorded motor positions. All the spectroscopic and X-ray diffraction data are recorded in a database tracking system. The spectroscopic data are often submitted to the PDB in addition to coordinates supporting publications resulting from work performed at beamline X26-C.

9.5.1 An example of correlated studies of an FAD-dependent system

An example of how correlated studies at beamline X26-C were used to trap a reactive intermediate was published recently in *Biochemistry* by Orville in collaboration with Gadda and Prabhakar (Orville et al. 2009). Flavoproteins were first discovered in the 1930s and derive from riboflavin, or vitamin B_2 (Massey 2000). These proteins are now known to catalyze a wide range of biochemical reactions, including those that use molecular oxygen (O_2) to help convert food into energy in animals, plants, fungi and some types of bacteria – a process known as oxygen activation. As documented in ▶ Table 9.2, more than 1,650 crystal structures of flavoproteins have been determined. However, prior to our results, no one had ever determined a structure of a reactive oxygen intermediate in any flavoprotein. This type of FAD-O_2 molecular complex often forms halfway through their particular reaction cycle. These intermediates possess high chemical potential energy, which is necessary to complete many critical, but difficult-to-catalyze reactions in biology. Therefore, they typically have a lifetime of only a few milliseconds and are consequently very hard to observe using traditional synchrotron methods.

Choline oxidase (CHO) from *Arthrobacter globiformis* is an FAD-dependent enzyme that catalyzes the two-step, four-electron oxidation of choline to glycine betaine, with betaine aldehyde as a two-electron oxidized intermediate. In the two oxidative half-reactions, two molecules of O_2 are converted into two H_2O_2 molecules. We performed several spectroscopic measurements on a number of choline oxidase crystals including before and after X-ray exposure. ▶Figure 9.5 illustrates some of the key results. Importantly, the difference spectra (after – before) clearly show a spectrum with λ_{max} at 400 nm that is nearly identical to spectra obtained from flavin C4a-OOH or C4a-OH enzyme reaction intermediates. Typically these reactive oxygen intermediates exhibit half-lives of only several milliseconds in solution, but remarkably, it is stable in the crystal at 100 K. The time-dependent results show that this species is generated with a $t_{1/2}$ of approximately 40 seconds upon X-ray exposure. Moreover, it is formed commensurate with

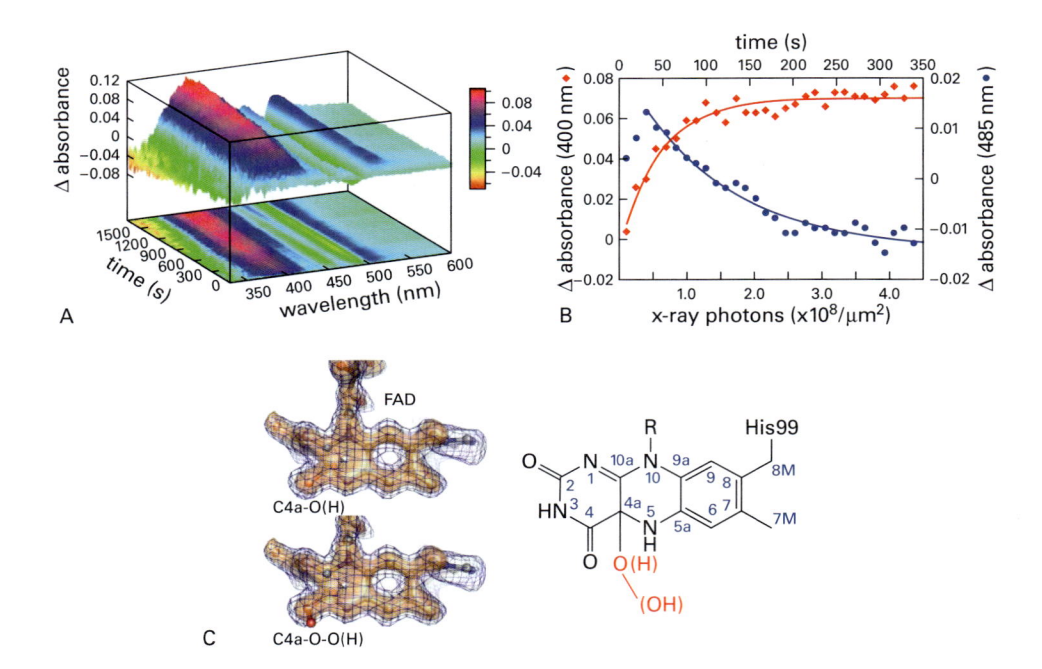

Figure 9.5: Spectroscopic changes observed in single crystals of choline oxidase upon X-ray irradiation at 100 K provides evidence for the flavin C4a-oxygen adduct in choline oxidase. (A, B) The time-dependent change of the electronic absorption spectrum of a choline oxidase crystal held stationary in the X-ray beam. Spectra were recoded every 10 seconds; six reference spectra were recoded before the X-ray shutter was opened, averaged and then subtracted from each spectrum obtained after the X-ray shutter was opened. The lower panel of the three-dimensional plot is a projection of the time-dependent difference spectra. (B) Difference features at 400 and 485 nm from (A) were fit to the single exponential equation: $y = a*e^{(bt + c)}$, where t = time in seconds and for the 400 nm data points $a = -0.074012$, $b = 0.021617$ and $c = 0.070048$, whereas the first two data points were excluded for the 485 nm and fit with $a = 0.036755$, $b = 0.0087557$ and $c = -0.0144$. (C) The atomic structure of the adduct is consistent with either one or two oxygen atoms bound to C4a. The $mF_o - DF_c$ difference map (+3 σ) is shown with orange, translucent surface contours, and the $2mF_o - DF_c$ map (1 σ) is displayed as a blue mesh. The FAD, C4a and adduct were omitted from the model. For comparison, the refined C4a-OH and C4a-OO(H) atomic models are shown superimposed with C, N and O atoms colored gray, blue and red, respectively.

the decrease of the 460 and 485 nm features attributed to oxidized FAD. The resulting electron density maps (unbiased simulated annealing at 1.8 Å resolution), and the interpretation of the atomic structure is consistent with two possible reactive oxygen species, namely, a covalent flavin C4a-OOH and/or C4a-OH adducts. Either structure correlates well with the spectroscopic observations. Moreover, the complementary nature of the two types of data reinforces the conclusions, whereas either type of data alone yields somewhat ambiguous results.

9.5.2 An example of correlated studies from a metalloenzyme

An ongoing collaboration between P. Liu and K. Allen (Boston University) and D. Stoner-Ma and A. Orville (Brookhaven National Laboratory) involves the investigating stachydrine demethylase (Stc2) from *Sinorhizobium meliloti* 1021. These symbiotic, N_2-fixing bacteria can inhabit the root nodules of legumes such as alfalfa. Legumes such as these produce a number of betaines that impact soil microorganisms, are important in the establishment of symbiosis and are necessary for normal root nodulation. For example, the plant roots secret flavonoids and betaines (e.g. stachydrine, *N,N*-dimethylproline) into the rhizosphere to attract the beneficial bacteria. *S. meliloti* responds because under free-living conditions, it can utilize stachydrine as an osmoprotectant and as a source of carbon and nitrogen. The catabolism is initiated by Stc2, a new member of the Rieske-type oxygenase family similar to those used in aromatic hydrocarbon degradation in *Pseudomonas* (Burnet et al. 2000; Ferraro et al. 2005). These types of enzymes catalyze their reactions with an active site mononuclear iron atom. However, they also require electrons to complete their reaction, which are ultimately provided from NADH via an electron transport chain and a Rieske [2Fe-2S] cluster that spans the enzyme subunit interface. Thus, Stc2 catalyzes the two-electron and O_2-dependent cleavage of a stachydrine C-N bond to yield N-methylproline, formaldehyde and water.

As illustrated in ▶Figure 9.6, we applied three complementary methods to study crystals of Stc2 as isolated and in an aerobic substrate complex. Our crystal structures reveal an α3 holoenzyme in which the Rieske cluster and the mononuclear iron interact with each other across the subunit interfaces. Oxidized enzyme crystals reveal two absorption peaks at 460 nm and 550 nm, which correlate well with solution UV/vis spectroscopy for a [2Fe-2S]+ state in Stc2. After the X-ray exposure required for each diffraction image, the intensity of the peak at 460 nm decreased and the peak at 550 nm reduced in intensity and shifted to 525 nm. This observation was further supported by extracting the change in absorption at 460 nm as a function of absorbed X-ray dose. The final absorption spectrum collected after the diffraction data set was complete is very similar to that obtained by chemical reduction of the Stc2 Rieske cluster to the [2Fe-2S]+ state in solution.

To further address the redox state of the Rieske cluster, we used a 532 nm excitation laser to collect resonance Raman spectra from the same single crystals at 100 K. Comparison of these results with published solution Raman spectra of other Rieske-type [2Fe-2S] proteins allows us to assign the majority of the low-frequency Raman features (Kuila et al. 1987, 1992; Rotsaert et al. 2003). Thus, the Raman spectra reveal various Fe-S(Cys), Fe-S(bridge) and Fe-N(His) stretching modes. Moreover, after X-ray data collection, the electronic absorption and the Raman spectroscopic data are consistent with a reduction of Rieske cluster from the [2Fe-2S]$^{2+}$ state to [2Fe-2S]$^{+}$ state. Indeed, the

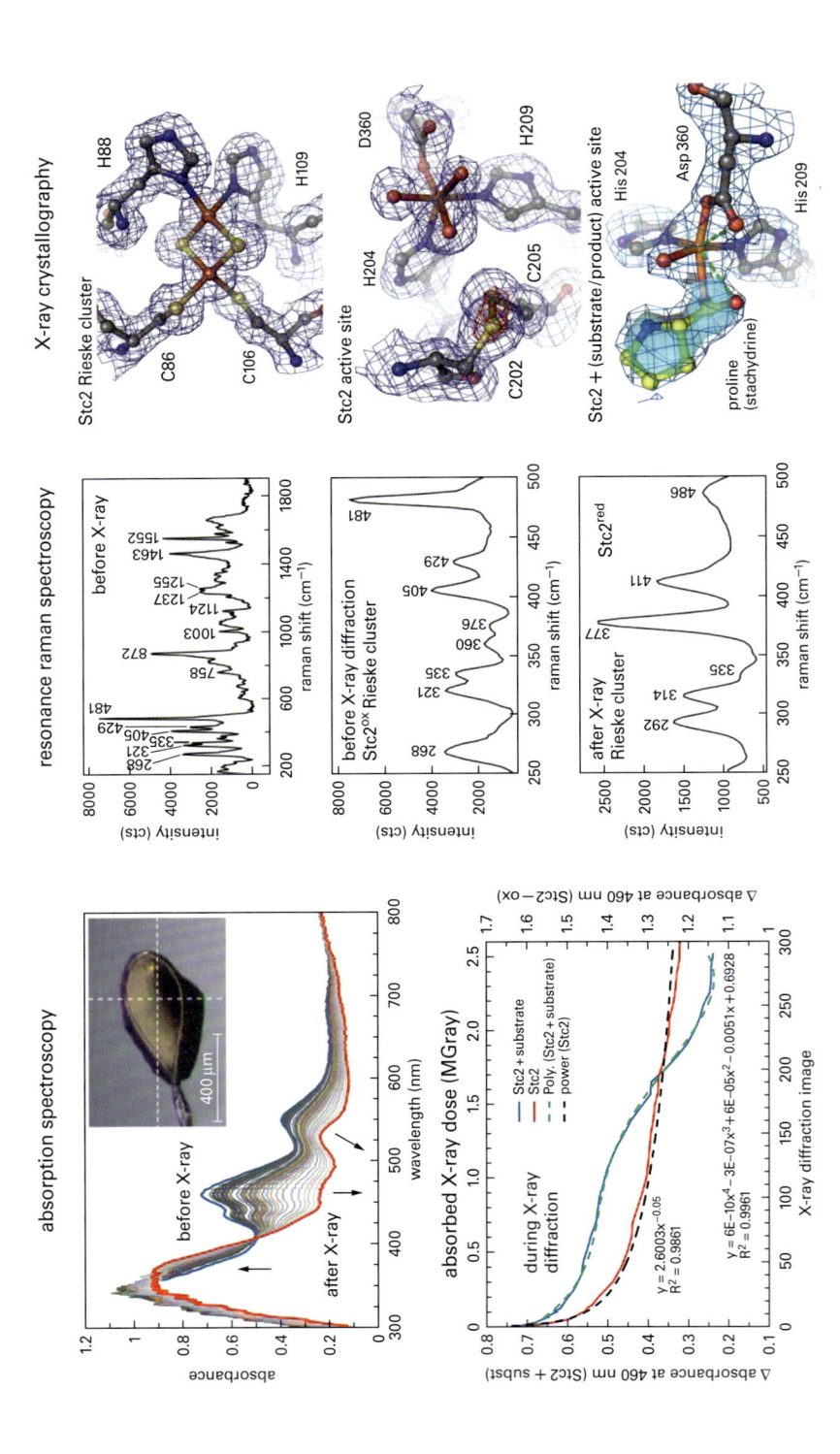

Figure 9.6: Spectroscopic and structural analysis of stachdrine demethylase from *Sinorhizobium meliloti* 1021. (Left, top) Electronic absorption spectra of the aerobic Stc2-stachydrine complex (Stc2-Subst) collected between X-ray diffraction frames indicate that the Rieske cluster is reduced by the X-ray beam. (Left, bottom) The change in absorption at 460 nm as a function of X-ray image or absorbed X-ray dose (blue solid lines) fit to a polynomial series (green dashed lines) for the aerobic Stc2-stachydrine complex. For comparison, the change in absorbance as a function of dose for oxidized Stc2 (Stc2) is shown with the solid red line, and the power-law fit with a dashed black line. Comparison of these results shows that the presence of substrate yields a different X-ray dose-dependent reduction process. (Center three panels) Resonance Raman spectra collected from single crystals of Stc2 at 100 K using a 532 nm excitation laser. The whole spectrum is shown in the top panel, whereas the low-frequency region corresponding to Rieske cluster stretching modes are shown before (middle) and after X-ray data collection (bottom). Analysis of the data suggests that the iron coordinated to the two histidine residues is reduced by solvated electrons created by the X-ray dose. (Right) The resulting $2mF_o − DF_c$ electron density maps for the Rieske cluster (top) and the active-site mononuclear iron center for resting enzyme (middle) and the Stc2-substrate/product complex. Together, the data suggest that the enzyme-dependent demethylation reaction cycle is promoted by solvated electrons at cryogenic temperature.

Raman data support a model in which the histidine-liganded iron of the Rieske cluster is reduced, whereas the cysteine-liganded iron remains oxidized. Thus, the two irons form a spin localized, Fe(II)/Fe(III) cluster. Importantly, the Fe(II) ion is "connected" to the mononuclear iron via only three hydrogen bonds.

Catalysis by Stc2 is a two-electron and O_2-dependent process. Moreover, it is likely that O_2 reactivity at the mononuclear iron site is gated by both organic substrate binding and by reduction of the Rieske cluster. This helps ensure that a reactive oxygen species is not generated unless all of the reagents needed for catalysis are present in the active site. Consequently, stachydrine binding to oxidized Stc2 should not result in catalysis until or unless electrons are provided to the system. Therefore, we crystallized oxidized Stc2 in the presence of 1 mM substrate, N,N-dimethylproline under aerobic conditions to trap the Michaelis complex.

The spectroscopy shows that the Rieske cluster within these **aerobic** enzyme-substrate crystals is reduced by the X-ray beam. However, the X-ray dose-dependent kinetics of the reduction process for the complex are dramatically different compared to oxidized Stc2. The absorption peak at 460 nm is quickly reduced, then exhibits a more stable phase between ~0.33 and 1.25 MGr. The latter phase is followed by a second rapid phase that subsequently tapers off. The kinetic trace, the final electronic absorption spectrum and the Raman spectra all suggest that full reduction of the Rieske cluster was achieved by the end of the diffraction data set. We speculate that the plateau between ~0.33 and 1.25 MGr may be due to the consumption of electrons in the demethylation process. In support of this hypothesis, the resulting electron density maps for the Stc2 substrate are consistent with product in the active site. Thus, it appears that X-rays and solvated electrons may promote the reaction cycle in situ that converts stachydrine into proline at 100 K.

9.6 Future prospects

Many research collaborations are creating pipelines for the structural and biophysical analysis of the "most important" macromolecules from nearly every major life-form on earth. An example includes our ongoing efforts with *S. meliloti* 1021 and its symbiotic plant host(s). The N_2-fixing bacterium clearly has many requirements for metal ions, as does the symbiotic legume species. Moreover, these requirements change as the microbe and plant alter their survival strategies based upon a free-living or symbiotic relationship. The sequence of *S. meliloti* genome is known (Galibert et al. 2001), and all of its open reading frames (ORFs) have been transformed into a variety of expression vectors (Becker et al. 2009). In collaboration with structural genomics initiatives, these proteins will be expressed, purified and crystallized. Many of these will be characterized by complementary methods including single-crystal spectroscopy correlated with X-ray crystallography, small angle scattering, and X-ray absorption spectroscopy. The initial targets for such a pipeline will include ORFs selected because their annotation suggest that the protein binds either iron (~144 ORFs), heme (63 ORFs) or copper (28 ORFs) or is from the oxidoreductase family (535 ORFs). Collaborative teams such as these will, therefore, probe reaction mechanisms of metals at the microbe-plant interface in unprecedented detail and with high throughput methods. Because I believe that correlated, complementary results are more significant than results obtained by individual techniques, the philosophy of the team members operating beamlines X26-C and SM3 at the NSLS and the NSLS-II is to support single-crystal spectroscopy and X-ray diffraction on a full-time basis.

References

Adam, V., Carpentier, P., Violot, S., Lelimousin, M., Darnault, C., Nienhaus, G.U., et al. (2009). Structural basis of X-ray-induced transient photobleaching in a photoactivatable green fluorescent protein. J Am Chem Soc *131*, 18063–18065.

Adam, V., Lelimousin, M., Boehme, S., Desfonds, G., Nienhaus, K., Field, M.J., et al. (2008). Structural characterization of IrisFP, an optical highlighter undergoing multiple photo-induced transformations. Proc Natl Acad Sci U S A *105*, 18343–18348.

Adam, V., Royant, A., Niviere, V., Molina-Heredia, F.P., and Bourgeois, D. (2004). Structure of superoxide reductase bound to ferrocyanide and active site expansion upon X-ray-induced photo-reduction. Structure *12*, 1729–1740.

Andresen, M., Wahl, M.C., Stiel, A.C., Grater, F., Schafer, L.V., Trowitzsch, S., et al. (2005). Structure and mechanism of the reversible photoswitch of a fluorescent protein. Proc Natl Acad Sci U S A *102*, 13070–13074.

Austin, R.H., Beeson, K.W., Eisenstein, L., Frauenfelder, H., and Gunsalus, I.C. (1975). Dynamics of ligand binding to myoglobin. Biochemistry *14*, 5355–5373.

Barends, T.R., Hartmann, E., Griese, J.J., Beitlich, T., Kirienko, N.V., Ryjenkov, D.A., et al. (2009). Structure and mechanism of a bacterial light-regulated cyclic nucleotide phosphodiesterase. Nature *459*, 1015–1018.

Becker, A., Barnett, M.J., Capela, D., Dondrup, M., Kamp, P.B., Krol, E., et al. (2009). A portal for rhizobial genomes: RhizoGATE integrates a *Sinorhizobium meliloti* genome annotation update with postgenome data. J Biotechnol *140*, 45–50.

Beitlich, T., Kuhnel, K., Schulze-Briese, C., Shoeman, R.L., and Schlichting, I. (2007). Cryoradiolytic reduction of crystalline heme proteins: analysis by UV-Vis spectroscopy and X-ray crystallography. J Synchrotron Radiat *14*, 11–23.

Berglund, G.I., Carlsson, G.H., Smith, A.T., Szoke, H., Henriksen, A., and Hajdu, J. (2002). The catalytic pathway of horseradish peroxidase at high resolution. Nature *417*, 463–468.

Bolton, W., and Perutz, M.F. (1970). Three dimensional Fourier synthesis of horse deoxyhaemoglobin at 2.8 Angstrom units resolution. Nature *228*, 551–552.

Bourgeois, D., and Royant, A. (2005). Advances in kinetic protein crystallography. Curr Opin Struct Biol *15*, 538–547.

Bourgeois, D., Vallone, B., Schotte, F., Arcovito, A., Miele, A.E., Sciara, G., et al. (2003). Complex landscape of protein structural dynamics unveiled by nanosecond Laue crystallography. Proc Natl Acad Sci U S A *100*, 8704–8709.

Bourgeois, D., Vernede, X., Adam, V., Fioravanti, E., and Ursby, T. (2002). A microspectrophotometer for UV-visible absorption and fluorescence studies of protein crystals. J Appl Crystallogr *35*, 319–326.

Burnet, M.W., Goldmann, A., Message, B., Drong, R., El Amrani, A., Loreau, O., et al. (2000). The stachydrine catabolism region in *Sinorhizobium meliloti* encodes a multienzyme complex similar to the xenobiotic degrading systems in other bacteria. Gene *244*, 151–161.

Carey, P.R. (2006). Raman crystallography and other biochemical applications of Raman microscopy. Annu Rev Phys Chem *57*, 527–554.

Carpentier, P., Royant, A., Ohana, J., and Bourgeois, D. (2007). Advances in spectroscopic methods for biological crystals. 2. Raman spectroscopy. J Appl Crystallogr *40*, 1113–1122.

Cedervall, P.E., Dey, M., Pearson, A.R., Ragsdale, S.W., and Wilmot, C.M. (2010). Structural insight into methyl-coenzyme M reductase chemistry using coenzyme B analogues. Biochemistry *49*, 7683–7693.

Chang, C.M., Klema, V.J., Johnson, B.J., Mure, M., Klinman, J.P., and Wilmot, C.M. (2010). Kinetic and structural analysis of substrate specificity in two copper amine oxidases from *Hansenula polymorpha*. Biochemistry 49, 2540–2550.

Chapman, H.N., Fromme, P., Barty, A., White, T.A., Kirian, R.A., Aquila, A., et al. (2011). Femtosecond X-ray protein nanocrystallography. Nature 470, 73–77.

Chu, K., Vojtchovsky, J., McMahon, B.H., Sweet, R.M., Berendzen, J., and Schlichting, I. (2000). Structure of a ligand-binding intermediate in wild-type carbonmonoxy myoglobin, Nature 403, 921–923.

Churbanova, I.Y., Poulos, T.L., and Sevrioukova, I.F. (2010). Production and characterization of a functional putidaredoxin reductase-putidaredoxin covalent complex. Biochemistry 49, 58–67.

De la Mora-Rey, T., and Wilmot, C.M. (2007). Synergy within structural biology of single crystal optical spectroscopy and X-ray crystallography. Curr Opin Struct Biol 17, 580–586.

Denisov, I.G., Makris, T.M., Sligar, S.G., and Schlichting, I. (2005). Structure and chemistry of cytochrome P450. Chem Rev 105, 2253–2277.

Edman, K., Royant, A., Larsson, G., Jacobson, F., Taylor, T., van der Spoel, D., et al. (2004). Deformation of helix C in the low temperature L-intermediate of bacteriorhodopsin. J Biol Chem 279, 2147–2158.

Fedorov, R., Schlichting, I., Hartmann, E., Domratcheva, T., Fuhrmann, M., and Hegemann, P. (2003). Crystal structures and molecular mechanism of a light-induced signaling switch: the Phot-LOV1 domain from *Chlamydomonas reinhardtii*. Biophys J 84, 2474–2482.

Ferraro, D.J., Gakhar, L., and Ramaswamy, S. (2005). Rieske business: structure-function of Rieske non-heme oxygenases. Biochem Biophys Res Commun 338, 175–190.

Galibert, F., Finan, T.M., Long, S.R., Puhler, A., Abola, P., Ampe, F., et al. (2001). The composite genome of the legume symbiont *Sinorhizobium meliloti*. Science 293, 668–672.

Garman, E.F. (2010). Radiation damage in macromolecular crystallography: what is it and why should we care? Acta Crystallogr Sect D Biol Crystallogr 66, 339–351.

Gouet, P., Jouve, H.M., Williams, P.A., Andersson, I., Andreoletti, P., Nussaume, L., et al. (1996). Ferryl intermediates of catalase captured by time-resolved Weissenberg crystallography and UV-VIS spectroscopy. Nat Struct Biol 3, 951–956.

Gray, H.B., and Winkler, J.R. (1996). Electron transfer in proteins. Annu Rev Biochem 65, 537–561.

Hadfield, A., and Hajdu, J. (1993). A fast and portable microspectrophotometer for protein crystallography. J Appl Crystallogr 26, 839–842.

Hajdu, J., Neutze, R., Sjogren, T., Edman, K., Szoke, A., Wilmouth, R.C., et al. (2000). Analyzing protein functions in four dimensions. Nat Struct Biol 7, 1006–1012.

Héroux, A., Bozinovski, D.M., Valley, M.P., Fitzpatrick, P.F., and Orville, A.M. (2009). Crystal structures of intermediates in the nitroalkane oxidase reaction. Biochemistry 48, 3407–3416.

Hersleth, H.P., Uchida, T., Rohr, A.K., Teschner, T., Schunemann, V., Kitagawa, T., et al. (2007). Crystallographic and spectroscopic studies of peroxide-derived myoglobin compound II and occurrence of protonated Fe^{IV}-O$^-$. J Biol Chem 282, 23372–23386.

Holm, R.H., Kennepohl, P., and Solomon, E.I. (1996). Structural and functional aspects of metal sites in biology. Chem Rev 96, 2239–2314.

Holton, J.M. (2009). A beginner's guide to radiation damage. J Synchrotron Radiat 16, 133–142.

Imamoto, Y., Mihara, K., Tokunaga, F., and Kataoka, M. (2001). Spectroscopic characterization of the photocycle intermediates of photoactive yellow protein. Biochemistry 40, 14336–14343.

Jouve, H.M., Andreoletti, P., Gouet, P., Hajdu, J., and Gagnon, J. (1997). Structural analysis of compound I in hemoproteins: study on *Proteus mirabilis* catalase. Biochimie 79, 667–671.

Jung, A., Reinstein, J., Domratcheva, T., Shoeman, R.L., and Schlichting, I. (2006). Crystal structures of the AppA BLUF domain photoreceptor provide insights into blue light-mediated signal transduction. J Mol Biol *362*, 717–732.

Karlsson, A., Parales, J.V., Parales, R.E., Gibson, D.T., Eklund, H., and Ramaswamy, S. (2000). The reduction of the Rieske iron-sulfur cluster in naphthalene dioxygenase by X-rays. J Inorg Biochem *78*, 83–87.

Katona, G., Carpentier, P., Niviere, V., Amara, P., Adam, V., Ohana, J., et al. (2007). Raman-assisted crystallography reveals end-on peroxide intermediates in a nonheme iron enzyme. Science *316*, 449–453.

Kendrew, J.C., Bodo, G., Dintzis, H.M., Parrish, R.G., Wyckoff, H., and Phillips, D.C. (1958). A three-dimensional model of the myoglobin molecule obtained by x-ray analysis. Nature *181*, 662–666.

Kort, R., Hellingwerf, K.J., and Ravelli, R.B. (2004). Initial events in the photocycle of photoactive yellow protein. J Biol Chem *279*, 26417–26424.

Kühnel, K., Derat, E., Terner, J., Shaik, S., and Schlichting, I. (2007). Structure and quantum chemical characterization of chloroperoxidase compound 0, a common reaction intermediate of diverse heme enzymes. Proc Natl Acad Sci U S A *104*, 99–104.

Kuila, D., Fee, J.A., Schoonover, J.R., Woodruff, W.H., Batie, C.J., and Ballou, D.P. (1987). Resonance Raman spectra of the [2Fe-2S] clusters of the Rieske protein from *Thermus* and phthalate dioxygenase from *Pseudomonas*. J Am Chem Soc *109*, 1559–1561.

Kuila, D., Schoonover, J.R., Dyer, R.B., Batie, C.J., Ballou, D.P., Fee, J.A., et al. (1992). Resonance Raman studies of Rieske-type proteins. Biochim Biophys Acta *1140*, 175–183.

Lavault, B., Ravelli, R.B., and Cipriani, F. (2006). C3D: a program for the automated centering of cryocooled crystals. Acta Crystallogr Sect D Biol Crystallogr *62*, 1348–1357.

Lelimousin, M., Noirclerc-Savoye, M., Lazareno-Saez, C., Paetzold, B., Le Vot, S., Chazal, R., et al. (2009). Intrinsic dynamics in ECFP and Cerulean control fluorescence quantum yield. Biochemistry *48*, 10038–10046.

Major, D.T., Héroux, A., Orville, A.M., Valley, M.P., Fitzpatrick, P.F., and Gao, J. (2009). Differential quantum tunneling contributions in nitroalkane oxidase catalyzed and the uncatalyzed proton transfer reaction. Proc Natl Acad Sci U S A *106*, 20734–20739.

Makris, T.M., von Koenig, K., Schlichting, I., and Sligar, S.G. (2006). The status of high-valent metal oxo complexes in the P450 cytochromes. J Inorg Biochem *100*, 507–518.

Massey, V. (2000). The chemical and biological versatility of riboflavin. Biochem Soc Trans *28*, 283–296.

McGeehan, J., Ravelli, R.B., Murray, J.W., Owen, R.L., Cipriani, F., McSweeney, S., et al. (2009). Colouring cryo-cooled crystals: online microspectrophotometry. J Synchrotron Radiat *16*, 163–172.

Meharenna, Y.T., Doukov, T., Li, H., Soltis, S.M., and Poulos, T.L. (2010). Crystallographic and single-crystal spectral analysis of the peroxidase ferryl intermediate. Biochemistry *49*, 2984–2986.

Neutze, R., Wouts, R., van der Spoel, D., Weckert, E., and Hajdu, J. (2000). Potential for biomolecular imaging with femtosecond X-ray pulses. Nature *406*, 752–757.

Orville, A.M., Buono, R., Cowan, M., Héroux, A., Shea-McCarthy, G., Schneider, D.K., et al. (2011). Correlated single-crystal electronic absorption spectroscopy and X-ray crystallography at NSLS beamline X26-C. J Synchrotron Radiat *18*, 358–366.

Orville, A.M., Lountos, G.T., Finnegan, S., Gadda, G., and Prabhakar, R. (2009). Crystallographic, spectroscopic, and computational analysis of a flavin C4a-oxygen adduct in choline oxidase. Biochemistry *48*, 720–728.

Owen, R.L., Pearson, A.R., Meents, A., Boehler, P., Thominet, V., and Schulze-Briese, C. (2009). A new on-axis multimode spectrometer for the macromolecular crystallography beamlines of the Swiss Light Source. J Synchrotron Radiat *16*, 173–182.

Ozaki, S., Bengtsson, J., Kramer, S.L., Krinsky, S., and Litvinenko, V.N. (2007). Philosophy for NSLS-II design with sub-nanometer horizontal emittance. Particle Accelerator Conference, 2007. PAC IEEE, pp. 77–79.

Pearson, A.R., and Owen, R.L. (2009). Combining X-ray crystallography and single-crystal spectroscopy to probe enzyme mechanisms. Biochem Soc Trans 37, 378–381.

Pearson, A.R., Pahl, R., Kovaleva, E.G., Davidson, V.L., and Wilmot, C.M. (2007). Tracking X-ray-derived redox changes in crystals of a methylamine dehydrogenase/amicyanin complex using single-crystal UV/Vis microspectrophotometry. J Synchrotron Radiat 14, 92–98.

Perutz, M.F. (1963). X-ray analysis of hemoglobin. Science 140, 863–869.

Perutz, M.F., Muirhead, H., Cox, J.M., and Goaman, L.C. (1968). Three-dimensional Fourier synthesis of horse oxyhaemoglobin at 2.8 Å resolution: the atomic model. Nature 219, 131–139.

Rajagopal, S., and Moffat, K. (2003). Crystal structure of a photoactive yellow protein from a sensor histidine kinase: conformational variability and signal transduction. Proc Natl Acad Sci U S A 100, 1649–1654.

Ren, Z., Perman, B., Srajer, V., Teng, T.Y., Pradervand, C., Bourgeois, D., et al. (2001). A molecular movie at 1.8 A resolution displays the photocycle of photoactive yellow protein, a eubacterial blue-light receptor, from nanoseconds to seconds. Biochemistry 40, 13788–13801.

Røhr, A.K., Hersleth, H.P., and Andersson, K.K. (2010). Tracking flavin conformations in protein crystal structures with Raman spectroscopy and QM/MM calculations. Angew Chem Int Ed Engl 49, 2324–2327.

Rosell, F.I., and Boxer, S.G. (2003). Polarized absorption spectra of green fluorescent protein single crystals: transition dipole moment directions. Biochemistry 42, 177–183.

Rotsaert, F.J., Pikus, J.D., Fox, B.G., Markley, J.L., and Sanders-Loehr, J. (2003). N-isotope effects on the Raman spectra of Fe(2)S(2) ferredoxin and Rieske ferredoxin: evidence for structural rigidity of metal sites. J Biol Inorg Chem 8, 318–326.

Royant, A., Carpentier, P., Ohana, J., McGeehan, J., Paetzold, B., Noirclerc-Savoye, M., et al. (2007). Advances in spectroscopic methods for biological crystals. 1. Fluorescence lifetime measurements. J Appl Crystallogr 40, 1105–1112.

Sanishvili, R., Yoder, D.W., Pothineni, S.B., Rosenbaum, G., Xu, S., Vogt, S., et al. (2011). Radiation damage in protein crystals is reduced with a micron-sized X-ray beam. Proc Natl Acad Sci U S A 108, 6127–6132.

Schertler, G.F., Lozier, R., Michel, H., and Oesterhelt, D. (1991). Chromophore motion during the bacteriorhodopsin photocycle: polarized absorption spectroscopy of bacteriorhodopsin and its M-state in bacteriorhodopsin crystals. EMBO J 10, 2353–2361.

Schlichting, I., Berendzen, J., Chu, K., Stock, A.M., Maves, S.A., Benson, D.E., et al. (2000). The catalytic pathway of cytochrome P450$_{cam}$ at atomic resolution. Science 287, 1615–1622.

Seibert, M.M., Ekeberg, T., Maia, F.R., Svenda, M., Andreasson, J., Jonsson, O., et al. (2011) Single mimivirus particles intercepted and imaged with an X-ray laser. Nature 470, 78–81.

Solomon, E.I., Szilagyi, R.K., DeBeer George, S., and Basumallick, L. (2004). Electronic structures of metal sites in proteins and models: contributions to function in blue copper proteins. Chem Rev 104, 419–458.

Srajer, V., Teng, T., Ursby, T., Pradervand, C., Ren, Z., Adachi, S., et al. (1996). Photolysis of the carbon monoxide complex of myoglobin: nanosecond time-resolved crystallography. Science 274, 1726–1729.

Stoner-Ma, D., Skinner, J.M., Schneider, D.K., Cowan, M., Sweet, R.M., and Orville, A.M. (2011). Single-crystal Raman spectroscopy and X-ray crystallography at beamline X26-C of the NSLS. J Synchrotron Radiat 18, 37–40.

Teng, T.Y., Srajer, V., and Moffat, K. (1994). Photolysis-induced structural changes in single crystals of carbonmonoxy myoglobin at 40 K. Nat Struct Biol 1, 701–705.

Tomita, A., Sato, T., Ichiyanagi, K., Nozawa, S., Ichikawa, H., Chollet, M., et al. (2009). Visualizing breathing motion of internal cavities in concert with ligand migration in myoglobin. Proc Natl Acad Sci U S A *106*, 2612–2616.

Waldron, K.J., Rutherford, J.C., Ford, D., and Robinson, N.J. (2009). Metalloproteins and metal sensing. Nature *460*, 823–830.

Wilmot, C.M., Hajdu, J., McPherson, M.J., Knowles, P.F., and Phillips, S.E. (1999). Visualization of dioxygen bound to copper during enzyme catalysis. Science *286*, 1724–1728.

Xu, G., and Chance, M.R. (2007). Hydroxyl radical-mediated modification of proteins as probes for structural proteomics. Chem Rev *107*, 3514–3543.

Yi, J., Orville, A.M., Skinner, J.M., Skinner, M.J., and Richter-Addo, G.B. (2010). Synchrotron X-ray-induced photoreduction of ferric myoglobin nitrite crystals gives the ferrous derivative with retention of the O-bonded nitrite ligand. Biochemistry *49*, 5969–5971.

10 Wide-angle X-ray solution scattering (WAXS)

Lee Makowski, Oleg A. Makarov, Shenglan Xu, and Robert F. Fischetti

10.1 Introduction

X-ray scattering studies of proteins in solution generate information about protein structure, flexibility and dynamics that complement and extend high-resolution structural information from other techniques. Whereas crystallography and NMR is capable of producing high-resolution "snapshots" of protein structure, solution scattering has the capability of providing information about conformational changes, intermolecular interactions, structural fluctuations and dynamics. It is particularly effective in characterization of large-scale motions that are difficult to track by other techniques. It can be utilized on virtually any macromolecule or macromolecular assembly that can be purified and prepared at a concentration of ~ 5 mg/mL or more.

Small-angle X-ray solution scattering (SAXS) has been used for decades (Luzzati and Tardieu 1980) to determine the radius of gyration (Rg), oligomerization state and pair-distance distribution function – literally a histogram of the interatomic vector lengths – of macromolecules (Putnam et al. 2007). More recently, it has proved possible to generate directly from SAXS data information about protein shapes (Svergun 1999; Walther et al. 2000). In studies of intrinsically disordered proteins, SAXS has proved uniquely valuable for deducing structural information (Eliezer 2009).

WAXS is analogous to SAXS but encompasses the collection and analysis of scattering data to much higher scattering angles. Scattered intensities from proteins can be collected to atomic resolution, and these intensities have proved highly sensitive to small changes in structure or structural ensemble (Makowski 2010). Although the information content of these data is inadequate for determining high-resolution molecular structures, WAXS can be used to carry out experimental tests of molecular models. The power of the method is substantially enhanced by the capability of using atomic coordinate sets to accurately predict WAXS data.

The increased availability of high-brilliance synchrotron X-ray sources has encouraged rapid growth of solution scattering studies extending to wide angles (WAXS) where scattered intensities are orders of magnitude weaker than in the small-angle regime, and background scattering from buffer and sample chamber is considerably stronger than scattering from protein. Use of a synchrotron source can overcome these obstacles to produce excellent quality data.

The benefits of extending solution scattering to wide angles are substantial and continue to be explored. WAXS is highly sensitive to changes in protein secondary, tertiary and quaternary structures (Hirai et al. 2002; Makowski et al. 2008a), and to small structural changes, including those generated by ligand binding (Fischetti et al. 2004b; Rodi et al. 2007), heme iron oxidation state (Tiede et al. 2002) and single site mutations (Makowski et al. 2011b). Detailed information about the process of protein folding and unfolding can be extracted from WAXS data (Hirai et al. 2004). Time-resolved studies have been used to track structural changes in hemoglobin from the subnano- to millisecond timescales (Cammarata et al. 2008; Cho et al. 2010). Structural fluctuations as reflected in changes in the protein ensemble of states accessible to the protein can also be studied with WAXS and have proved to be sensitive to protein concentration (Makowski et al. 2008b), mutations and ligand binding (Makowski et al. 2011a). The linkage between structural ensemble and protein function is being actively explored in systems such as hemoglobin where the relationship of structure to function was once thought to be well established (Makowski et al. 2011a).

Here, the methods required to produce, analyze and interpret WAXS data will be reviewed. Requirements for sample preparation and handling are outlined, including use of a plate-based sample-handling robot designed for rapid data collection. Protocols for efficient data collection are outlined as well as data reduction, background subtraction and error analysis. Methods of data interpretation are briefly surveyed, and a short discussion of the future outlook and applications of these techniques is provided.

10.2 Sample preparation

As in all biophysical experiments, the data produced are only as good as the sample being analyzed. Samples should be prepared using the strictest biochemical standards and should be well characterized prior to data collection.

Protein concentration: WAXS patterns from many proteins have proved to be sensitive to protein concentration, reflecting increased amplitude structural fluctuations at lower concentrations. Fluctuations may mute the features in a scattering pattern, potentially obscuring information that could be obtained in their absence. Molecular crowding appears to suppress these fluctuations at higher concentrations (Makowski 2008b). Because fluctuations tend to mute the features of a diffraction pattern (see Section 10.13), data sets are frequently of higher quality when protein concentration is increased. Although acceptable data have been occasionally obtained using protein concentrations of ~2 mg/mL, data quality is often greatly improved in samples concentrated to 5 mg/mL, and 10–20 mg/mL appears optimal for many proteins. It is worth noting that SAXS data often utilize protein concentrations of 2 mg/mL or less for several reasons: (i) scattered intensity is much greater in the SAXS regime, (ii) intermolecular interference effects in the data are strongest in the SAXS regime and can be avoided by using very dilute solutions and (iii) data in the SAXS regime appear less sensitive to changes in protein concentration.

Sample volume: Although a few microliters of sample should, in principle, be adequate to collect a WAXS data set, radiation damage may require use of a flow cell, substantially increasing the amount of sample required. WAXS experiments typically utilize a very intense X-ray beam, and some proteins exhibit radiation damage after less than 1 second of exposure (Fischetti et al. 2003). In order to avoid radiation damage, use of a flow cell is strongly recommended. At BioCAT, operating at an X-ray energy of 12 keV and a flux density of 10^{12} photons/second with a beam size of 40 × 120 µm^2, we typically use a flow rate that corresponds to each protein receiving no more than 100 milliseconds of exposure to X-rays. A full data set, as described in Section 10.4, typically takes ~100 microliters of protein solution using this flow rate. At 10 mg/mL, that corresponds to about 1 mg of protein per sample. For samples where availability is limited, it is possible to work with ~10-microliter samples by oscillating the protein in front of the beam. Using this protocol, additional computational checks to evaluate potential radiation damage should be carried out.

Sample purity and homogeneity: All efforts should be made to prepare samples that are highly purified and in a well-characterized buffer. WAXS has been shown to be very sensitive to small changes in protein structure or structural fluctuations, which are, in turn, sensitive to temperature, protein concentration, pH, buffer and potentially other environmental factors. Therefore, reproducibility demands complete control of the protein environment. Purity, as assessed by gel electrophoresis, is adequate. If a sample appears homogeneous and pure on a gel, it is probably adequately homogeneous for a scattering experiment.

10.3 Sample-handling robot

The rate-limiting step in data collection is the cycle time required for changing samples and carrying out the essential security interlock procedure to close the experimental hutch prior to exposure. To minimize this time and eliminate the need for opening the hutch between each sample, a sample-handling robot was designed and constructed to utilize samples arrayed in 96-well plates.

The sample-handling robot is part of an integrated WAXS Data Collection Platform that provides portability with the integration of beamline elements, sample reservoirs, sample holder and control systems. The elements of the platform are as follows: (i) Beamline elements including a pinhole scatter guard; helium tunnel and beam stop. The pinhole and beam stop are both mounted on independent *x-z* stages to make possible remote alignment. (ii) Sample reservoirs include a 96-well plate (for holding of protein and buffer solutions). These are maintained at a fixed temperature by mounting on a copper block that accommodates the flow of fluid from a recirculating bath through small channels drilled into the block. The block is positioned on an *x-y-z* platform designed with adequate movement that each well in the 96-well plate can be moved to the position of the stationary sample feed tube for automatic aspiration of the sample to the sample capillary. (iii) Sample holder made of a copper block, thermostatically controlled using fluid flow from the recirculating bath. It accommodates a thin-walled quartz capillary within which sample flows past the X-ray beam. A syringe pump with a multiway value controls all fluid flow. Protein and buffer are aspirated through the plastic feed tube and on the sample holder where the sample flows through the capillary during X-ray exposure. The sample continues on to a syringe and then is dispensed to a waste bucket. Cleaning solutions such as a 10% bleach solution, deionized water or ethanol for drying are aspirated from remote reservoirs and then dispensed in the opposite flow direction through the capillary. The carrier for the 96-well plate has a drain that is positioned under the pickup tube during the cleaning process. The pump can be programmed for various cleaning protocols that can be automatically implemented. (iv) A control system based on a commercial multiaxis motor drive system is installed in a dedicated computer. This system is programmed to control (a) automatic sample feeds from the 96-well plate, (b) washing sequences, (c) alignment of capillary and beam stop with X-ray beam, (d) communication with the detector and (e) sequencing of the shutter, detector and pump control. A rendering of the experimental arrangement and sample-handling robot are in ▶Figure 10.1.

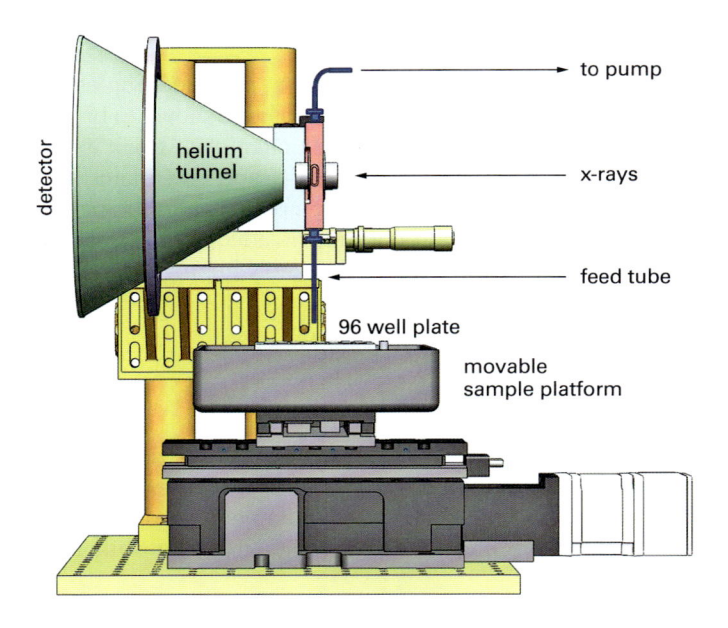

Figure 10.1: Sample-handling robot. A movable 96-well plate-based sample platform is positioned below the sample chamber to provide automated sample feed to the WAXS camera. Samples, buffers and washing solutions are automatically repositioned to allow aspiration by a stationary pipette that feeds a thin-walled quartz capillary enclosed within a temperature-controlled copper block sample chamber. X-rays enter from the right, scatter from the sample, pass through a helium tunnel and are recorded by a two-dimensional CCD detector at the left edge of this diagram.

10.4 Data collection

Although, in principle, WAXS can be carried out on a conventional lab source, a synchrotron source is required for generating the highest-quality data. All data reported here were collected at the BioCAT beamline (18-ID) at the Advanced Photon Source at Argonne National Laboratory (Fischetti et al. 2004a). The X-ray source and camera are arranged as previously described (Fischetti et al. 2004b; Makowski et al. 2008b) and are diagrammed in ▶Figure 10.2. The X-ray scattering pattern is recorded with a MAR165 2k X 2k CCD detector or appropriate alternative. Hybrid detector arrangements providing for simultaneous collection of SAXS and WAXS data have now been implemented at several beamlines. The advantages of collecting SAXS and WAXS from the same sample are likely to encourage development of efficient data collection schemes to make this possible. The data reported here were collected with a specimen-to-detector distance of approximately 170 mm, calibrated using powder diffraction rings from either silicon or silver-behenate. This arrangement allows data collection over a range of scattering angles corresponding to $0.06 < q < 2.5$ Å^{-1}. where q is the momentum transfer and $q = 2\pi/d = 4\pi \sin(\theta)/\lambda$. Different investigators use q or $1/d$ or occasionally other units for reporting of data. Data in the figures in this chapter are plotted as a function of $1/d$

(= $q/2\pi$). This measure is favored by some investigators because d is a direct measure of the spatial periodicity that gives rise to scattering at a spacing of $1/d$. For instance, a reflection at $1/d = 0.1$ Å^{-1} arises from a spatial periodicity with a wavelength of $1/0.1 = 10$ Å in the sample.

The BioCAT beamline is capable of delivering approximately 2×10^{13} photons/second per 100 mA of beam current into a 40×120 µm² beam at 12 keV. Experience on the Bio-CAT beamline has demonstrated that most proteins are damaged after exposure times of a few tenths of a second to a few seconds at these intensity levels (Fischetti et al. 2003). It is often necessary to attenuate the intensity of the X-ray beam ~ 10-fold by inserting as many as thirty-two 20 µm thick aluminum foils. In most cases, a data set consists of a series of 1- to 2-second exposures with 4 exposures from buffer-filled capillary, 7 to 15 from protein solution-filled capillary and 5 from the empty capillary. Multiple short exposures are used to minimize data loss due to sporadic bubbles in the sample moving through the flow cell and to provide a measure of statistical variation. Exposure sequences from sample and buffer are alternated to minimize the possible effects of drift in any experimental parameter. Incident beam flux is monitored using nitrogen-gas-filled ion chambers. Integrated beam flux during each exposure is used to scale scattering from protein and solutions.

10.5 Data processing

Scattering data are typically collected from protein-solution-filled capillary, buffer-filled capillary and empty capillary using a two-dimensional detector as described in Section

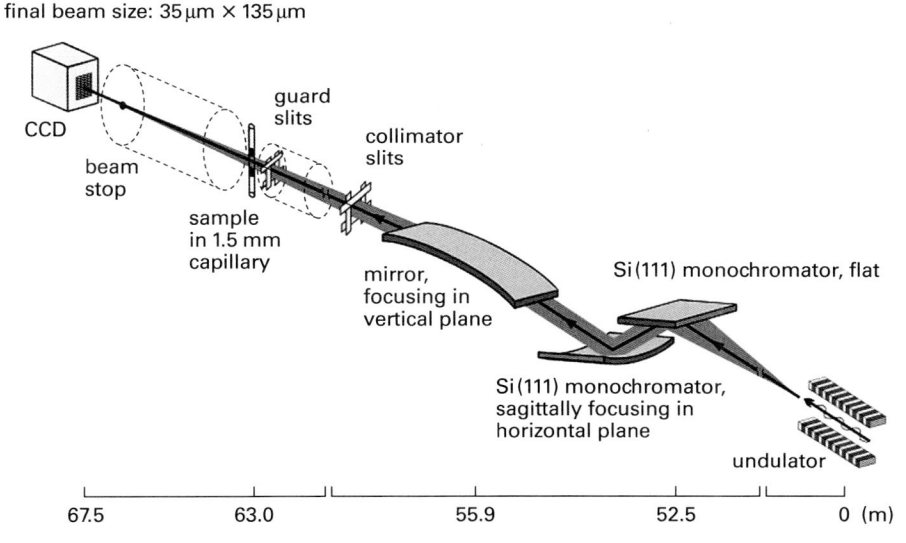

Figure 10.2: Schematic drawing of the beamline optics at BioCAT, 18ID at the Advanced Photon Source. The X-ray beam travels from the right to the left in this figure. The second monochromator with the Si(400) crystals is omitted for clarity and is located at 53.5 m from the source. The experimental arrangement shown is for a standard SAXS/WAXS camera with a flow cell for solution scattering. (Reproduced from Fischetti et al. [2004a] with permission.)

10.4. The two-dimensional scattering patterns are circularly integrated to generate one-dimensional scattering intensity profiles using the program Fit2D (Hammersley 1997, 1998; Hammersley et al. 1996) or equivalent software. The origin of the diffraction pattern is determined from the powder diffraction rings from silver-behenate and/or silicon powder. ▶Figure 10.3 contains a detector image of a WAXS pattern from hemoglobin (▶Figure 10.3A) and the circularly averaged intensity obtained from that pattern as well as intensity from patterns collected from buffer-filled and empty capillaries (▶Figure 10.3B). A WAXS pattern can be roughly divided into regimes that provide information on the overall protein size and shape (SAXS regime), the quaternary and tertiary structure and the secondary structure. The approximate (and overlapping) extents of these regimes are indicated in ▶Figure 10.3B.

Scattering from protein is estimated according to:

$$I_{prot} = I_{obs} - I_{cap} - (1 - v_{ex}) I_{solvent} \tag{1}$$

where I_{obs} is the measured scattering from the protein sample, I_{cap} is the measured scattering from the empty capillary, v_{ex} is the estimated proportion of the solution taken up by the protein (excluded volume) and $I_{solvent}$ is estimated by

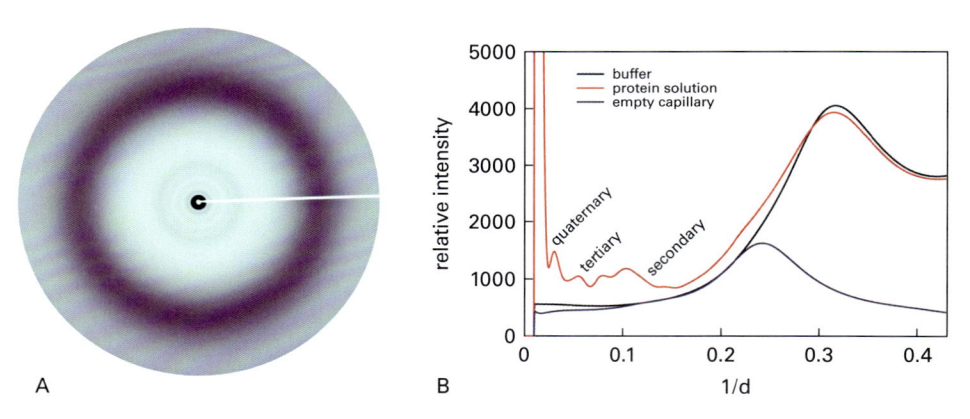

$$I_{solvent} = I_{bkgd} - I_{cap} \tag{2}$$

Figure 10.3: WAXS from a solution of carbonmonoxy human hemoglobin A (HbCO A) at 150 mg/mL. (A) Detector image of the scattering pattern. In this image, dark regions correspond to the highest intensity. The pattern is circularly symmetric except for the shadow of the horizontal arm that holds the beam stop. The beam stop at the center is required to prevent damage to the detector from the direct beam. Highest intensity is observed at small angles, in a tight circle about the beam stop; and at wide angles, about two-thirds of the way to the edge of the pattern. This diffuse circle of scattered intensity is from water, which constitutes more than 85% of the sample. (B) Circularly averaged data from protein solution, buffer solution and empty capillary. The origin of this plot corresponds to the center of the diffraction pattern. Intensity at increasingly wide angles corresponds to progressively smaller structural details. Quaternary structure is reflected most strongly at relatively small angles, followed by tertiary and secondary structures. A peak at $1/d = 0.1$ Å^{-1} ($d = 10$ Å) is often seen in scattering from α-helical proteins since α helices pack approximately 10 Å center-to-center.

where I_{bkgd} is the measured scattering from the capillary containing buffer. Scattering from empty capillary does not adequately reflect the scattering it contributes when the capillary is filled with buffer or protein solution due to absorption by buffer and/or protein. Absorption is accounted for by modeling the capillary scatter as

$$I_{cap}(\text{corrected}) = (\text{scale factor})*I_{cap}(\text{observed}) + \text{constant (absorption correction)} \quad (3)$$

The scale factor and constant are selected by empirically fitting the capillary scatter to the scatter from a buffer-filled capillary in the scattering range $0.3 < q < 1.2$ Å^{-1} (or $0.05 < 1/d < 0.20$ Å^{-1}) where scattered intensity from buffer is negligible (▶Figure 10.3).

10.6 Structural information

The scattered intensity, $I(q)$, from a protein solution can be calculated, in principle, from the position of all atoms in the protein using the Debye formula,

$$I(q) = \sum_{i=1}^{n}\sum_{j=1}^{n} f_i f_j \frac{\sin 2\pi q r_{ij}}{2\pi q r_{ij}} \quad (4)$$

where f_i is the scattering factor from the ith atom, and r_{ij} is the distance between atom i and atom j. The obstacles to accurately carrying out this calculation arise from the fact that the protein is immersed in water. Strategies for dealing with these obstacles are described in Section 10.12. The equation reflects the fact that the scattering does not depend on the relative orientations of interatomic vectors in the protein and, thus, cannot contain sufficient information to reconstruct the entire atomic structure of the protein.

The distribution of interatomic vector lengths (pair-distribution function), $P(r)$, can be obtained directly from the measured intensity using the relationship (Luzzati and Tardieu 1980; Putnam et al. 2007),

$$P(r) = \frac{r}{2\pi^2} \int_0^{\infty} I(q) q \sin(qr) \, dq. \quad (5)$$

The calculation of this function is subject to large errors due to truncation of the intensity at both high and low q and requires an indirect Fourier approach (Svergun 1992) to prevent the introduction of spurious features into the resulting distribution. An example of the use of $P(r)$ for analysis of protein unfolding is given in Section 10.11.

10.7 Size and shape

The size and shape of a macromolecule or macromolecular complex can usually be determined from data in the SAXS regime. There are a number of excellent recent reviews of SAXS (e.g. Doniach 2001; Hura et al. 2009; Petoukhov and Svergun 2007; Putnam et al. 2007). Here, we only indicate what can be learned from that data, and note that extending data to higher resolution does not necessarily provide improved accuracy or resolution for the information most often derived from SAXS data.

Data from the SAXS regime are often used to estimate the Rg, of a protein, literally, the average radius of scattering density from the center of mass (analogous to the moment of inertia about an axis of rotation). Data at higher resolution do not improve the estimate of Rg because the assumptions necessary to derive Rg become increasingly inaccurate at higher scattering angles. In fact, extending data to smaller angles is more important for improving the accuracy of the estimate of Rg. The detector and beam stop arrangements required for collection of WAXS data place limits on the minimum scattering angle to which data can be collected. For the data collection arrangement described in Section 10.4, Rg can be estimated accurately only for proteins smaller than about 100 Å in diameter. A hybrid SAXS/WAXS detector scheme can provide for collection of both simultaneously.

SAXS data can also be used to calculate a shape envelope – a low-resolution rendering of the protein (Chacon et al. 1998; Svergun 1999; Svergun et al. 2001; Walther et al. 2000). Data used for this calculation are limited to $q < \sim 0.3$ Å^{-1}($1/d < 0.05$ Å^{-1}), and extending the data to higher q is likely to result in spurious features in the reconstruction. The algorithms used to obtain shape from SAXS data implicitly assume that the scattering density within the protein is a constant. This is approximately true to about 20 Å resolution but becomes increasingly less true at higher resolutions. Features in the scattering pattern at spacings greater than $(1/d) \sim 1/20$ Å^{-1} ($q \sim 0.3$ Å^{-1}) are most often due to internal structural features, not shape, and including them in a shape-determining calculation will lead to artifactual features in the reconstruction.

WAXS data can improve the accuracy and resolution of $P(r)$. Intensity in a WAXS pattern can be thought of as a band-limited function where the band limit is the length of the longest interatomic vector in the protein. Larger proteins will give rise to scattering patterns with sharper peaks and troughs because of the higher-frequency terms introduced by the longest interatomic vectors. An estimate of the longest interatomic vector in the protein can be made from the pair-distribution function, $P(r)$, since that length corresponds to the radius, r, at which $P(r)$ falls to zero (in the absence of intermolecular interference effects).

10.8 Secondary and tertiary structure

Equations 4 and 5 indicate that the measured intensity will be a function of the distribution of interatomic vector lengths in the protein. Secondary structures have, by definition, characteristic patterns of interatomic vector length. For instance, α helices pack approximately 10 Å apart, so α-helical proteins have a large number of ~ 10 Å long interatomic vectors and in most cases give rise to strong scattering at $1/d \sim (1/10)$ Å^{-1}. Similarly, the strands of β sheets lie ~ 4.7 Å apart and give rise to characteristic peaks at $1/d \sim (1/4.7)$ Å^{-1}. As an example, ▶Figure 10.4A shows ribbon diagrams of myoglobin (an alpha-helical protein) and ubiquitin (a largely β-sheet protein). ▶Figure 10.4B contains WAXS patterns from these two proteins, which exhibit peaks characteristic of their secondary structures.

An analysis of the amount of information that can be extracted from a WAXS pattern was used to demonstrate that it is possible to estimate the secondary structure content of a protein from a WAXS pattern and to strongly limit its possible tertiary structures (Makowski et al. 2008a). In that study, WAXS patterns computed from the atomic coordinates of a set of 498 protein domains representing all of known fold space were used as the basis for constructing a multidimensional space of all corresponding WAXS patterns ("WAXS space"). Within WAXS space, each scattering pattern is represented by a single vector. A principal components analysis (PCA) was carried out to identify

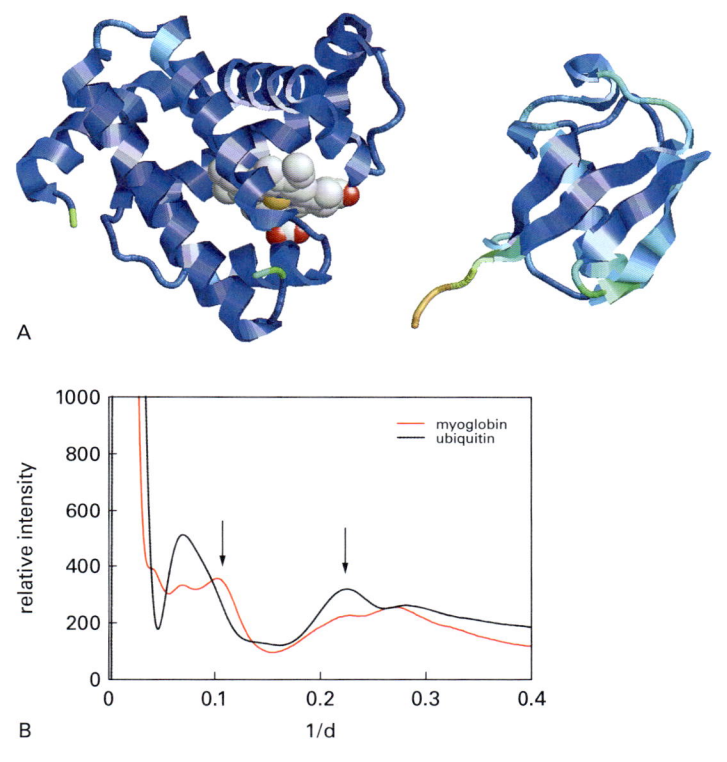

Figure 10.4: WAXS patterns from myoglobin and ubiquitin. (A) Ribbon diagrams of myoglobin (left) and ubiquitin (right). (B) Circularly averaged WAXS data from myoglobin (red) and ubiquitin (black). Arrows indicate the positions of the peak at 0.1 Å^{-1} in the myoglobin pattern due to packing of α helices, and at 0.21 Å^{-1} in ubiquitin corresponding to the separation of β strands in a β-sheet structure.

those directions in WAXS space that provide the greatest discrimination among patterns. Estimates of the relative abundances of secondary structures were made using training/test sets derived from this data set. The average error in the estimate of α-helical content was 11% and of β-sheet content was 9%. The distribution of proteins that are members of the four structure classes, α, β, α/β and α+β, are well separated in WAXS space when data extending to a spacing of 2.2 Å are used. Data limited to ~ 10 Å spacing exhibit little power to discriminate among secondary or tertiary structures.

10.9 Quaternary structure

Quaternary structure involves the relative positions and orientations of domains in a structure, information that is embedded in the longest interatomic vectors in a protein, and thereby makes a substantial contribution to scattering at relatively low angles as indicated in ▶Figure 10.3B. It also gives rise to higher-frequency oscillations in the intensity than would be observed in a smaller, monomeric protein. This effect can be readily observed in a comparison of scattering from myoglobin with that from hemoglobin. Hemoglobin is a tetramer of subunits each of which has high structural homology to myoglobin. Very roughly, hemoglobin looks like four myoglobin molecules stuck together. ▶Figure 10.5A shows that the scattered intensity from hemoglobin is very similar to that from myoglobin, but with a stronger high-frequency component. ▶Figure 10.5B is the ratio of the hemoglobin data to the myoglobin data. The position of the first subsidiary, marked by the vertical arrow, corresponds to a spacing of d ~ 29.4 Å, which is very close to the average distance between the centers of mass of subunits in the hemoglobin molecule. The quaternary structure leads to the introduction of the high-frequency components of the scattering that dominate the ratio displayed in ▶Figure 10.5B.

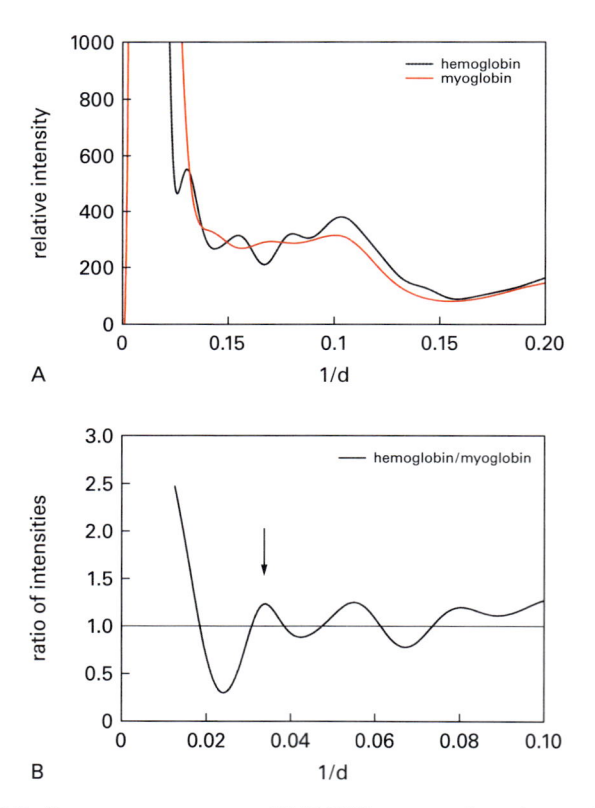

Figure 10.5: Quaternary structure. (A) WAXS patterns from hemoglobin and myoglobin and (B) ratio of the two patterns (hemoglobin divided by myoglobin). Hemoglobin is a tetramer of subunits that are structurally very similar to myoglobin, so the WAXS pattern from hemoglobin resembles that of myoglobin but is modulated by an "interference" term that corresponds to the distance between subunits in hemoglobin. The vertical arrow points to the first subsidiary peak in the intensity ratio. The position of that peak is at a spacing of 29.4 Å, almost precisely the average distance between the centers of the subunits in hemoglobin.

10.10 Structural changes

One of the strengths of WAXS is its sensitivity to small structural changes induced, for instance, by ligand binding. ▶Figure 10.6 includes patterns from five forms of human hemoglobin, HbA, wherein scattering from HbCO A is compared to the other four individually. Some comparisons appear to reflect real differences among the molecular forms, whereas others do not. A quantitative measure of the difference between two WAXS patterns is required in order to determine the statistical significance of small differences between two patterns. The chi-square measure has proved to be a useful determinant of statistical significance (Rodi et al. 2007) in these comparisons. Typically, a reduced chi-square (chi-square divided by the number of degrees of freedom) $\chi_v > 1.0$ is indicative of a statistically significant difference between two scattering patterns. For the patterns in ▶Figure 10.6, the associated χ_v values are listed in ▶Table 10.1. These values indicate that the scattering from HbCO A is indistinguishable from met HbA, an oxidized form of adult human hemoglobin (liganded with a water molecule as ligand),

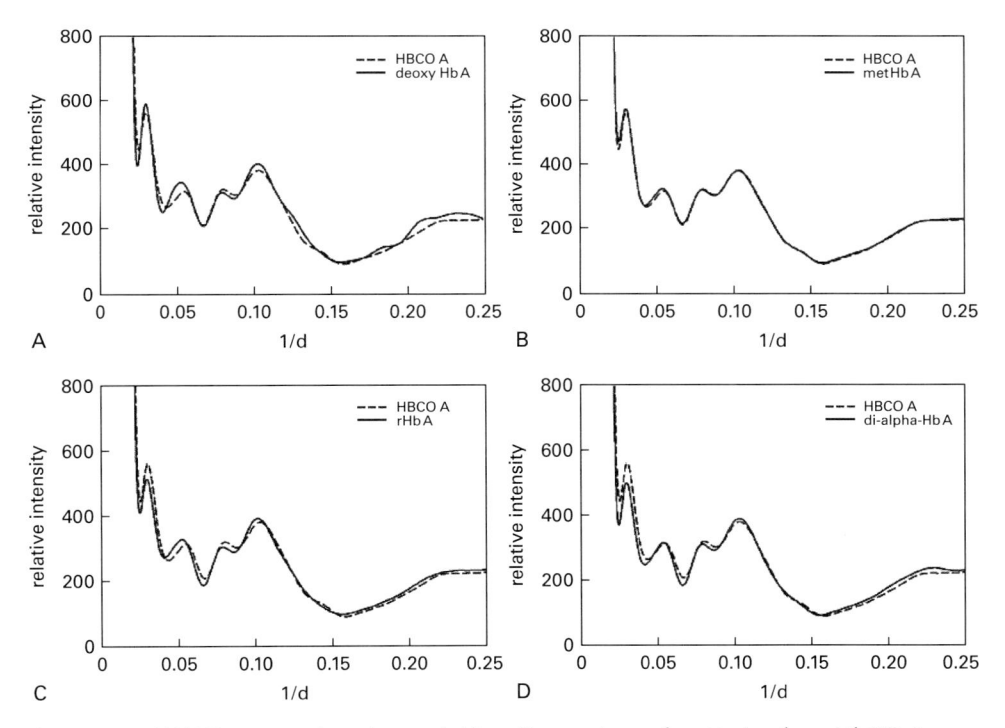

Figure 10.6: WAXS patterns from hemoglobins. Comparison of scattering from HbCO A with (A) the deoxygenated form of human hemoglobin, deoxy HbA; (B) the structurally near-identical met HbA; (C) the carbonmonoxy form of a low-affinity mutant of HbA (αV96W/βN108K), rHbA; (D) the carbonmonoxy form of di-α HbA, a hemoglobin variant in which the two α chains have been covalently linked by a single glycine residue spanning the C terminus of one α chain with the N terminus of the other. (Reproduced from Makowski et al. [2011a] with permission.)

and that all other comparisons reflect statistically significant differences between the scattering molecules.

Frequently, establishing that two patterns are statistically distinguishable is inadequate to answer the biological question that motivated the experiment. When comparing scattering from several patterns that correspond to different states or variants of the same molecule, it may be important to quantitatively categorize, cluster or systematize the differences. Qualitative descriptions of differences, such as those apparent in ▶Figure 10.6, may not fully clarify the relationships among different structures or provide information about the structural origins of the differences observed. A quantitative approach to classify and distinguish among the patterns can be carried out by a PCA (Makowski et al. 2008a) that identifies the features that most distinguish the patterns being compared and allows them to be clustered, plotted and ordered. ▶Figure 10.7 contains the results of a PCA of scattering patterns from a number of different hemoglobin molecules (including those in ▶Figure 10.6) and demonstrates the categorization based on WAXS patterns. In this plot, the projection of each pattern along the first two (most significant) parameters of the PCA are plotted. The statistically indistinguishable HbCO A and met HbA (▶Table 10.1) fall close to one another in the plot, providing a measure of the uncertainties associated with the plot – pairs of points falling closer together than these two may not be experimentally distinguishable. The transition from liganded (met or CO) to deoxy hemoglobin (R to T transition as represented by the arrows in ▶Figure 10.7) appears to generate shifts in WAXS patterns from both human and bovine hemoglobin that in the PCA analysis are of similar length and direction. This is consistent with the well-known

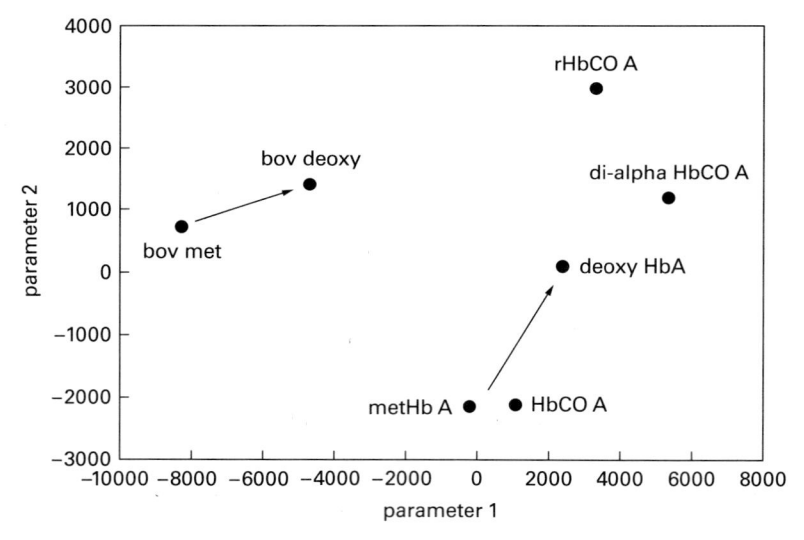

Figure 10.7: PCA. The first two parameters (most significant components) from a PCA of WAXS patterns from hemoglobins are plotted against one another. These components represent the most significant differences in features among the diffraction patterns. Patterns up and to the right correspond to structures that have lower oxygen affinity (T-state); down and to the left, higher oxygen affinity (e.g. liganded conformations; R-state). rHbCO A is the low-affinity mutant (αV96W/βN108K); and di-α HbCO A, the variant in which the two α chains have been covalently linked (Figure 10.6).

Table 10.1: Reduced chi-square values for comparison of WAXS patterns in Figure 10.7. Values less than 1 correspond to pairs of diffraction patterns that are indistinguishable statistically. Values greater than 1 correspond to pairs of diffraction patterns that are statistically distinct.

	deoxy HbA	met HbA	r-HbA	di-α HbA
HbCO A	15.2	0.81	15.2	5.6
deoxy HbA	–	35.3	7.5	6.1
met HbA	–	–	10.4	9.1
r-HbA			–	2.3

fact that the structural changes associated with ligand binding are similar for these two molecules. The di-α HbA pattern was from a hemoglobin variant in which the two α chains were covalently linked by a single glycine residue (Kroeger and Kundrot 1997) and the r-HbA pattern was from a low-affinity double mutant (αV96W/βN108K) (Maillett et al. 2008). Both were taken from the carbonmonoxy form of the molecules, but nonetheless fall to the right in ▶Figure 10.7, suggesting that their structures have more similarity with the deoxy form of HbA than the carbonmonoxy form. This observation is consistent with data that indicate both molecules exhibit low oxygen affinity (Maillett et al. 2008) and NMR data from the mutant that indicate the low oxygen affinity is associated with increased dynamics (Maillett et al. 2008) as also appears to be the case in deoxy HbA (Song et al. 2007). The conclusion motivated by the data in ▶Figure 10.7 is that all hemoglobin species studied undergo similar conformational changes when switching to a lower-affinity form.

10.11 Unfolding

Folding and unfolding of proteins in response to changes in environment remains an important topic of research. The segregation of different scale features within a WAXS pattern suggests that WAXS could provide important new information on the process of protein folding. As an example, consider the alcohol-induced unfolding of β-lactoglobulin, a process that has been studied in detail. β-lactoglobulin has been observed to transition from a native, largely β-structure, to an open α-helical structure in the presence of high concentrations of alcohols (Hirota et al. 1997; Kumar et al. 2003). WAXS patterns from β-lactoglobulin in varying amounts of ethanol are shown in ▶Figure 10.8A. The small differences between the observations in 0% ethanol and 12% ethanol are probably due to changes in contrast (differences in the electron density of the ethanol solution compared to that in the absence of ethanol). In scattering from the sample in 30% ethanol, most of the features at small angle ($1/d < 0.1$ Å^{-1}) disappear or are muted, suggesting the dissolution of ordered tertiary structure. All small-angle features of the pattern are gone in patterns from the sample in 50% ethanol. The only feature that remains almost unchanged is the strong peak at $1/d = 1/4.7$ Å^{-1}, corresponding to the strand-to-strand separation in β sheets. This provides strong evidence that much of the β-sheet structure of the protein remains intact even in the near complete absence of tertiary structure. Interestingly, although the literature suggests this structural transition involves an increase in α-helical structure, there is no increase in the intensity observed in the vicinity of $1/d = 1/10$ Å^{-1} as could be expected for α helices that are packed against one another. However, for an open structure, bereft of tertiary organization, there is no reason to expect the enhancement of intensity at this scattering angle even if α-helical content increases substantially.

The pair-distribution function can be used to track overall changes in shape of the molecule during structural transitions. ▶Figure 10.8B includes $P(r)$ for β-lactoglobulin in native conformation (0% ethanol), molten globule (30% ethanol) and unfolded state (50% ethanol). The maximum intramolecular vector is nearly identical in the molten globule and unfolded states of this molecule, although there are clear differences in the overall distribution of interatomic vector lengths observed.

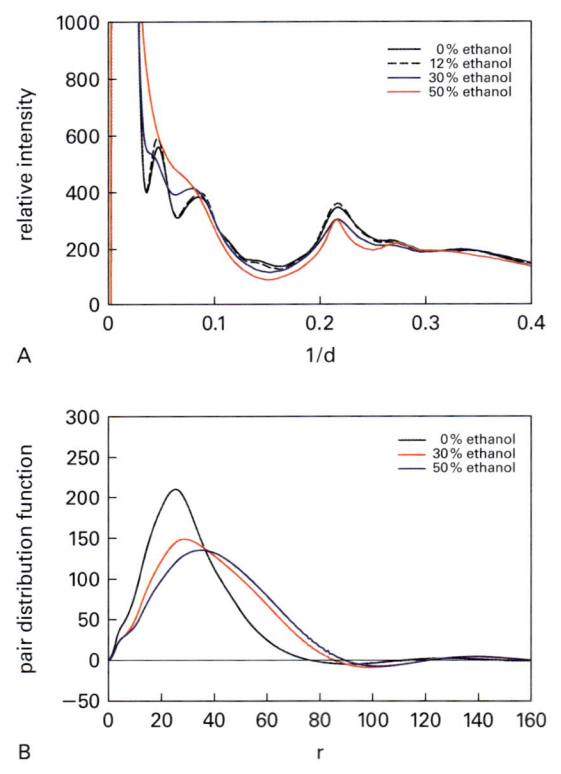

Figure 10.8: Unfolding of β-lactoglobulin in ethanol. (A) WAXS patterns from β-lactoglobulin in buffer and in 12%, 30% and 50% ethanol. Strong features that correspond to tertiary structure begin to disappear in ethanol solutions > 12% and are almost completely gone in 50% ethanol. However, the 4.7 Å peak that corresponds to β-strand separation remains strong in 50% ethanol, indicating the preservation of at least a part of the β-sheet structure, even in the near complete absence of tertiary structure. (B) Pair-distribution function from β-lactoglobulin in the absence of ethanol and in 30% ethanol (molten globule) and 50% ethanol. Ethanol induces a loosening of the packing of structural elements, leading to a larger average radius and larger maximum diameter.

10.12 Molecular modeling

The ability to calculate WAXS patterns accurately from atomic coordinates greatly enhances the utility of WAXS, making it a sensitive method of assessing the accuracy of atomic-scale models. The most accurate method for this calculation utilizes an explicit atom representation of water and results in calculation of scattered intensities that are within experimental error of observation for most rigid proteins (Grishaev et al. 2010; Park et al. 2009).

Methods for calculating solution scattering from atomic coordinates focused on SAXS for many years, and the program CRYSOL (Svergun et al. 1995) remains widely used and continues to make a major contribution to the growing use of SAXS for study of protein structure. For rapid calculation of SAXS from thousands of structures generated by molecular dynamics (MD), the FastSAXS algorithm has been introduced (Yang et al. 2009). The assumptions used in both of these approaches are less accurate at wider scattering angles, and for modeling WAXS data, a new approach was required (Bardhan et al. 2009). The reason that a simple application of the Debye formula does not work directly is (i) protein is immersed in water making it necessary to develop a representation of the water and/or the volume of water excluded by the protein, and (ii) the hydration layer of water around proteins is inhomogeneous and, on average, packed more densely than bulk water. CRYSOL uses a continuum model of water and the hydration layer that takes both of these factors into account. However, for wider-angle calculations, the inhomogeneities in the hydration layer must also be accounted for. To accomplish this, we developed an explicit atom representation of water that could be used to calculate WAXS data. This approach is implemented in the software package, XS (Park et al. 2009).

In this approach, water molecules are placed computationally around a protein out to a distance at least 7 Å from the protein surface and subject to 100 picoseconds of MD simulation during which a "snapshot" of water positions is collected every picosecond. A corresponding calculation is made for a "droplet" of water (no protein) of the same size and shape as the protein-containing droplet. The scattering patterns from these two "droplets" are then calculated, resulting in an approximation of *excess intensity'*, I_{xs}, which corresponds to the experimental difference between scattering from protein-solution-filled capillary and buffer-filled capillary,

$$I_{xs} = I_{obs} - I_{buffer} \tag{6}$$

Excess intensity is essentially identical to scattering from the protein, I_{prot} out to about 5 Å spacing, at which point scattering from buffer becomes nonnegligible. I_{prot} can be derived from the calculations using explicit atom representation of water by appropriate weighting of the scattering from buffer analogous to that shown in Equation 1. Several investigators are in the process of implementing programs for calculation of I_{xs} from atomic coordinates.

▶Figure 10.9 compares the intensity calculated from atomic coordinates to that observed from ubiquitin. The correspondence is excellent, but at several places in the pattern, the observed features appear slightly muted compared to those in the calculated intensity. This is due to structural fluctuations of the protein in solution. The calculated pattern is that of a rigid protein. The discrepancy between calculated and observed is largest in the vicinity of the β-strand 1/4.7 Å^{-1} peak, suggesting that the strand separation

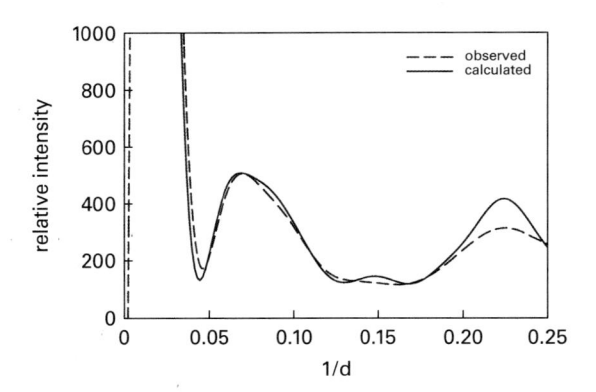

Figure 10.9: WAXS patterns from ubiquitin observed (dashed line) and calculated (solid line) from an atomic coordinate set using an explicit atom representation of water. The close coincidence of the two curves out to $1/d \sim 0.2$ Å^{-1} indicates that the atomic model accurately reflects the average structure of ubiquitin in solution. The discrepancy at $1/d \sim 0.22$ Å^{-1} – in which features of the computed pattern are muted in the observed pattern – suggests that ubiquitin undergoes a modest amount of structural fluctuations in solution.

is subject to some fluctuation in solution. Fluctuations of as little as 2%–3% of the interatomic vector length are adequate to account for the discrepancies seen in ▶Figure 10.9 (see Section 10.13 and Makowski et al. 2011b).

10.13 Modeling of structural fluctuations

Proteins fluctuate in aqueous solution, and those fluctuations impact WAXS data (Makowski et al. 2011b). Flexibility of proteins often prevents high-resolution visualization of flexible loops by crystallography. SAXS/WAXS may provide a means of obtaining information about flexible loops absent from crystal structures (Tsutakawa et al. 2007). Modeling the effect of fluctuations on a WAXS pattern is challenging. MD calculations designed to generate an unbiased representation of the ensemble of solution structures of a protein are computationally intensive and generate many coordinate sets that must subsequently be used to compute WAXS patterns – a daunting challenge. Nevertheless, a number of coarse-grained approaches (Bernado et al. 2007; Yang et al. 2010) have been put forward for predicting SAXS data, and they have enjoyed substantial success.

A simpler representation of protein fluctuations is often sufficient to characterize an ensemble in which fluctuations occur around a single consensus structural conformation. A formalism that makes it possible to predict the effect of those fluctuations on WAXS data has been developed (Makowski et al. 2011b). This method, called vector length convolution, models the fluctuations by replacing each interatomic vector length in the protein by a distribution of vector lengths, where the width of the distribution varies as a function of the length of the vector. The effect of these fluctuations on a WAXS pattern is determined by first calculating the pair-distribution function, $P(r)$, as in Equation 5, convoluting $P(r)$ by a Gaussian of width $\sigma(r)$ and then computing the reverse transform. In general, the half-width of the Gaussian is given by $\sigma(r) = a\, r^n$, where a and n are parameters that characterize the ensemble.

▶Figure 10.10 provides an example of this approach when applied to the effect of a single site mutation, L50E, on the structure and dynamics of ubiquitin. Leucine 50 occupies the hydrophobic core of the protein, and replacing it by a charged residue destabilizes the protein. The effect of this mutation on the WAXS pattern from ubiquitin is shown in ▶Figure 10.10. The features, although maintaining their original positions, are substantially suppressed in scattering from the mutant. Vector length convolution provides an approach to estimating the magnitude of fluctuations that will give rise to these changes in the scattering. The difference between WAXS patterns from wild type (WT) and L50E is well fit using the parameters $a = 0.7$ and $n = 0.5$. These parameters indicate that interatomic vectors of ~ 10 Å length in the WT protein vary in length by more than 20% (i.e. > +/–2 Å) in the mutant. Secondary structural elements (α helices and β sheets) tend to pack roughly 10 Å apart. So these numbers indicate that adjacent secondary structures are moving over 2 Å relative to one another in the mutant, enough to badly disrupt intramolecular interfaces within the protein. Comparison of observed intensity with that calculated from vector length convolution in ▶Figure 10.10 indicates reasonable correspondence between calculated and observed. It also highlights the limitations of modeling complex structural fluctuations of a protein with a two-parameter model. Whereas the 1/4.7 Å peak is still quite clear in the observed pattern – indicating the preservation of a substantial proportion of the β sheet in the mutant structure – the peak is essentially absent in the calculated intensity. Nevertheless, the agreement out to ~ 5 Å spacing provides confidence that the average magnitude of the fluctuations has been approximated well, and the fluctuations of adjacent secondary structures are of a magnitude reflected in the quantitative parameters derived from the model.

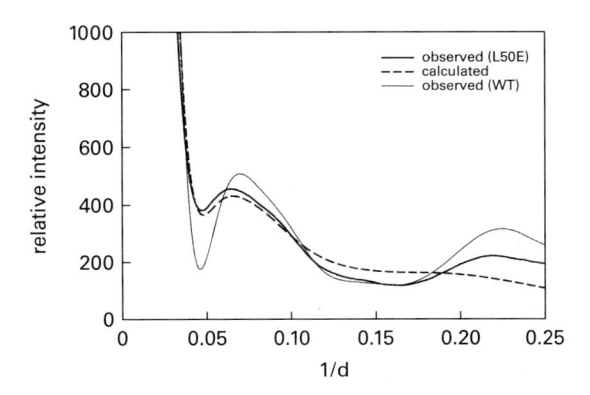

Figure 10.10: WAXS patterns observed from WT ubiquitin (thin solid line) and ubiquitin with a single substitution of a hydrophobic core residue by a charged residue L50E (solid line) are compared to that calculated from an atomic coordinate set and then subject to vector length convolution as a model for the structural fluctuations intrinsic to the mutant. The excellent correspondence between the data from L50E and that calculated demonstrates that the model can successfully represent the range of structural fluctuations present.

10.14 Outlook

WAXS is a relatively new technique with virtually all activity occurring during the past 15 years. In May 2011, a PubMed search for WAXS resulted in 220 hits, and for SAXS, about eight times that many. Presently, the technique is undergoing rapid growth with many new groups entering the field. The availability of an increasing number of WAXS beamlines at synchrotrons across the world will help fuel expansion of the technique, as will availability of software to carry out interpretation and quantitative analysis of the data. The examples provided here reflect the diversity of phenomena that can be studied using WAXS. Next generation WAXS will benefit from additional contributions of advanced computing and novel properties of X-ray beams generated by new X-ray sources.

Areas of research that will expand rapidly in the near future include (i) time-resolved studies, (ii) studies of ensembles, (iii) characterization of reaction intermediates, (iv) functional screening of drug candidates, and (v) structural studies of membrane proteins.

(i) Collection of time-resolved (TR-WAXS) data has been demonstrated for phototriggered processes (e.g. Andersson et al. 2009; Cammarata et al. 2008; Cho et al. 2010) and for conformational changes triggered by other processes – such as protein-ligand interactions. Use of microfludics for rapid mixing is an intrinsically slow process, but investigators are exploring ways to use light to trigger a wide set of protein processes that are not intrinsically light sensitive using an approach that has been referred to as optogenetics (Möglich and Moffat 2010). Combining TR-WAXS with optogenetic approaches and atomic resolution structural models will result in the eventual construction of experimentally validated molecular movies for watching molecular processes triggered by a wide variety of stimuli.

(ii) Studies of changes in molecular ensemble in response to regulatory interactions (e.g. Yang et al. 2010) should provide new insights into the role of both abundant and minor conformations in molecular signaling pathways and other processes. Signal transduction is central to many human diseases including cancer. Detailed molecular characterizations of every step in these processes will contribute substantially to design of novel therapeutic strategies.

(iii) Structural intermediates defy characterization because, in many cases, they never exist in solution in the absence of other molecular species. Data from proteins in a large number of environments, such as during titration of a ligand, will be used with dimensionality reduction methods to identify and characterize scattering profiles from these intermediates. Data from enzymes undergoing catalytic cycling will provide the basis for structural characterization of intermediates in the catalytic process.

(iv) Molecular interactions that alter structure can be readily discriminated from ones that generate little if any structural or functional change. Affinity assays detect both. WAXS can be staged in a moderate throughput manner to identify those resulting in the largest structural change (Rodi et al. 2007). Since an interaction that alters structure is also likely to modulate function, this provides a generic surrogate to functional screening of ligands for those that alter function of a protein.

(v) Membrane proteins are a particular challenge for X-ray solution scattering, but there is some evidence that new approaches may soon make these studies practical. Given the relative dearth of structural information about membrane proteins, establishing protocols for their study by WAXS could have substantial impact.

Solution studies of membrane proteins requires the protein to be solubilized in either detergent micelles, lipid vesicles or lipid nanodiscs, and the scattering from lipids or detergents is very strong at spacings of 30–40 Å ($1/d \sim 1/30$–$1/40$ Å^{-1}) due to the contrast between electron dense hydrophilic head groups and electron poor aliphatic chains. When exposed to a change in conditions (e.g. mixing with ligand, dilution, change in pH or temperature), it is difficult to unambiguously assign changes in scattered intensity to either protein or lipid. Subtraction of scattering from protein-free micelles, vesicles or nanodiscs is not practical because in the absence of protein, their structures may be different; cross-terms between the lipids and protein represent an additional challenge. Neutron scattering is often used for membrane systems because of the ease with which contrast can be changed through manipulation of the hydrogen-deuterium (H/D) ratio of the solvent, but this strategy is not readily practical with X-rays. Nevertheless, the introduction of nanodiscs (Denisov et al. 2004; Nath et al. 2007) to provide well-defined, homogeneous packages for solubilization of membrane proteins offers substantial promise for WAXS studies. TR-WAXS is also likely to be important for the study of membrane proteins, particularly in systems where changes in structure due to transient stimuli can be unambiguously assigned to protein (Andersson et al. 2009). The importance of membrane proteins in all aspects of biology is motivating substantial efforts to apply SAXS and WAXS to their study.

The confluence of high-brilliance X-ray sources with explosive growth in computing infrastructure and capabilities is providing a foundation for very rapid development of solution scattering studies of proteins and other macromolecules in the future.

References

Andersson, M., Malmerberg, E., Westenhoff, S., Katona, G., Cammarata, M., Wohrl, A.B., et al. (2009). Structural dynamics of light-driven proton pumps. Structure *17*, 1265–1275.

Bardhan, J.P., Park, S. and Makowski, L. (2009). SoftWAXS: a computational tool for modeling wide-angle X-ray solution scattering from biomolecules. J Appl Crystallogr. *42*, 932–943.

Bernado, P., Mylonas, E., Petoukhov, M.V., Blackledge, M. and Svergun, D.I. (2007). Structural characterization of flexible proteins using small-angle X-ray scattering. J Am Chem Soc *129*, 5656–5664.

Cammarata, M., Levantino, M., Schotte, F., Anfinrud, P.A., Ewald, F., Choi, J., et al. (2008). Tracking the structural dynamics of proteins in solution using time-resolved wide-angle X-ray scattering. Nat Methods *5*, 881–886.

Chacon, P., Moran, F., Diaz, J.F., Pantos, E., and Andreu, J.M. (1998). Low-resolution structures of proteins in solution retrieved from X-ray scattering with a genetic algorithm. Biophys J *74*, 2760–2775.

Cho, H.S., Dashdorj, N., Schotte, F., Graber, T., Henning, R., and Anfinrud, P. (2010). Protein structural dynamics in solution unveiled via 100-ps time-resolved x-ray scattering. Proc Natl Acad Sci *107*, 7281–7286.

Denisov, I.G., Grinkova, Y.V., Lazarides, A.A., and Sligar, S.G. (2004). Directed self-assembly of monodisperse phospholipid bilayer nanodiscs with controlled size. J Am Chem Soc *126*, 3477–3487.

Doniach, S. (2001). Changes in biomolecular conformation seen by small angle x-ray scattering. Chem Rev 101, 1763–1778.

Eliezer, D. (2009). Biophysical characterization of intrinsically disordered proteins. Curr Opin Struct Biol *19*, 23–30.

Fischetti, R., Stepanov, S., Rosenbaum, G., Barrea, R., Black, E., Gore, D., et al. (2004a). The BioCAT undulator beamline 18ID: a facility for biological non-crystalline diffraction and X-ray absorption spectroscopy at the Advanced Photon Source. J Synchrotron Radiat *11*, 399–405.

Fischetti, R.F., Rodi, D.J., Gore, D.B., and Makowski, L. (2004b). Wide angle x-ray solution scattering as a probe of ligand-induced conformational changes in proteins. Chem Biol *11*, 1431–1443.

Fischetti, R.F., Rodi, D.J., Mirza, A., Kondrashkina, E., Irving, T., and Makowski, L. (2003). Wide angle X-ray scattering of proteins: effect of beam exposure on protein integrity. J Synch Res *10*, 398–404.

Grishaev, A., Guo, L., Irving, T., and Bax, A. (2010). Improved fitting of solution X-ray scattering data to macromolecular structures and structural ensembles by explicit water modeling. J Am Chem Soc *132*, 15484–15486.

Hammersley, A.P. (1997). FIT2D: An introduction and overview. ESRF Internal Report, Grenoble, France, ESRF97HA02T.

Hammersley, A.P. (1998). FIT2D V9.129 reference manual V3.1. ESRF Internal Report, Grenoble, France, ESRF98HA01T.

Hammersley, A.P., Svensson, S.O., Hanfland, M., Fitch, A.N., and Hausermann, D. (1996). Two-dimensional detector software: from real detector to idealised image or two-theta scan. High Pressure Res *14*, 235–248.

Hirai, M., Iwase, H., Hayakawa, T., and Inoue, K. (2002). Structural hierarchy of several proteins observed by wide-angle solution scattering. J Synchrotron Radiat *9*, 202–205.

Hirai, M., Koizumi, M., Hayakawa, T., Takahashi, H., Abe, S. Hirai, H., et al. (2004). Hierarchical map of protein unfolding and refolding at thermal equilibrium revealed by wide-angle X-ray scattering. Biochemistry *43*, 9036–9049.

Hirota, N., Mizuno, K., and Goto, Y. (1997). Cooperative a-helix formation of P-lactoglobulin and melittin induced by hexafluoroisopropanol. Protein Sci *6*, 416–421.

Hura, G.L., Menon, A.L., Hammel, M., Rambo, R.P., Poole, F.L., Tsutakawa, S.E., et al. (2009). Robust, high-throughput solution structural analyses by small angle X-ray scattering (SAXS). Nat Methods *6*, 606–612.

Kroeger, K.S., and Kundrot, C.E. (1997). Structures of a hemoglobin-based blood substitute: insights into the function of allosteric proteins. Structure *5*, 227–237.

Kumar, S., Modig, K., and Halle, B. (2003). Trifluoroethanol-induced $\beta - > \alpha$ transition in β-lactoglobulin: hydration and cosolvent binding studied by ^2H, ^{17}O, and ^{19}F magnetic relaxation dispersion. Biochemistry *42*, 13708–13716.

Luzzati, V., and Tardieu, A. (1980). Recent developments in solution x-ray scattering. Annu Rev Biophys Bioeng *9*, 1–29.

Maillett, D.H., Simplaceanu, V., Shen, T.J., Ho, N.T., Olson, J.S., and Ho, C. (2008). Interfacial and distal-heme pocket mutations exhibit additive effects on the structure and function of hemoglobin. Biochemistry *47*, 10551–10563.

Makowski, L. (2010). Characterization of proteins with wide-angle X-ray solution scattering (WAXS). J Struct Funct Genomics *11*, 9–19.

Makowski, L., Bardhan, J., Fischetti, R., Gore, D., Lal, S.J., Mandava, S., et al. (2011a). WAXS studies of the structural diversity of hemoglobin in solution. J Mol Biol *408*, 909–921. PMID: 21420976.

Makowski, L., Gore, D., Mandava, S., Minh, D., Park, S., Rodi, D.J., et al. (2011b). X-ray solution scattering studies of the structural diversity intrinsic to protein ensembles. Biopolymers *95*, 531–542. PMID:21462170.

Makowski, L., Rodi, D.J., Mandava, S., Devrapahli, S., and Fischetti, R.F. (2008a). Characterization of protein fold using wide angle X-ray solution scattering. J Mol Biol *383*, 731–744.

Makowski, L., Rodi, D.J., Mandava, S., Minh, D.,Gore, D. and Fischetti, R.F. (2008b). Molecular crowding inhibits intramolecular breathing motions in proteins. J Mol Biol *375*, 529–546.

Möglich, A., and Moffat, K. (2010). Engineered photoreceptors as novel optogenetic tools. Photochem Photobiol Sci *9*, 1286–1300.

Nath, A., Atkins, W.M., and Sligar, S.G. (2007). Applications of phospholipid bilayer nanodiscs in the study of membranes and membrane proteins. Biochemistry *46*, 2059–2069.

Park, S., Bardhan, J.P., Roux, B., and Makowski, L. (2009). Simulated X-ray scattering of protein solutions using explicit-solvent molecular dynamics. J Chem Phys *130*, 134114.

Petoukhov, M.V., and Svergun, D.I. (2007). Analysis of X-ray and neutron scattering from biomacromolecular solutions. Curr Opin Struct Biol *17*, 562–571.

Putnam, C.D., Hammel, M., Hura, G.L., and Tainer, J.A. (2007). X-ray solution scattering (SAXS) combined with crystallography and computation: defining accurate macromolecular structures, conformations and assemblies in solution. Q Rev Biophys *40*, 191–285.

Rodi, D.J., Mandava, S., Gore, D.B., Makowski, L., and Fischetti, R.F. (2007). Detection of functional ligand binding events using synchrotron X-ray scattering. J Biomol Screen *12*, 994–998.

Song, X-J., Yuan, Y., Simplaceanu, V., Sahu, S., Ho, N.T., and Ho, C. (2007). A comparative NMR study of the polypeptide backbone dynamics of hemoglobin in the deoxy and carbonmonoxy forms. Biochemistry *46*, 6795–6803.

Svergun, D., Barberato, C., and Koch, M.H.J. (1995). CRYSOL-a program to evaluate x-ray solution scattering of biological macromolecules from atomic coordinates. J Appl Crystallogr *28*, 768–773.

Svergun, D.I. (1992). Determination of the regularization parameter in indirect-transform methods using perceptual criteria. J Appl Crystallogr *25*, 405–503.

Svergun, D.I. (1999). Restoring low resolution structure of biological macromolecules from solution scattering using simulated annealing. Biophys J 2879–2886.

Svergun, D.I., Petoukhov, M.V., and Koch, M.H.J. (2001). Determination of domain structure of proteins from X-ray solution scattering. Biophys J *80*, 2946–2953.

Tiede, D.M., Zhang, R., and Seifert, S. (2002). Protein conformations explored by difference high-angle solution X-ray scattering: oxidation state and temperature dependent changes in cytochrome c. Biochemistry *41*, 6605–6614.

Tsutakawa, S.E., Hura, G.L., Frankel, K.A., Cooper, P.K., and Tainer, J.A. (2007). Structural analysis of flexible proteins in solution by small angle X-ray scattering combined with crystallography. J Struct Biol *158*, 214–223.

Walther, D., Cohen, F.E., and Doniach, S. (2000). Reconstruction of low resolution three-dimensional density maps from one-dimensional small angle X-ray scattering data for biomolecules in solution. J Appl Crystallogr *33*, 350–363.

Yang, S., Blachowicz, L., Makowski, L., and Roux, B. (2010). Multidomain assembled states of Hck tyrosine kinase in solution. Proc Natl Acad Sci *107*, 15757–15762.

Yang, S., Park, S., Makowski, L., and Roux, B. (2009). A rapid coarse residue-based computational method for X-ray solution scattering characterization of protein folds and multiple conformational states of large protein complexes. Biophys J *96*, 4449–4463.

11 Where purity matters: recombinant versus synthetic peptides in beta amyloid formation

Toni Vagt and Rudi Glockshuber

11.1 Amyloid fibrils possess a defined quaternary structure

There are at least 20 different diseases in humans associated with the abnormal aggregation of an endogenous protein into amyloid fibrils (Chiti and Dobson 2006). Besides type 2 diabetes, the presumably best-known examples are neurodegenerative diseases including Alzheimer's disease (AD), Parkinson's disease, Huntington's disease and transmissible spongiform encephalopathy (Creutzfeld-Jakob disease [CJD]). A common feature of all amyloid diseases is the accumulation of amyloid fibrils that constitute the end product of misfolding and aggregation of the respective endogenous polypeptide. Importantly, and in contrast to the formation of unspecific protein aggregates in the form of "inclusion bodies" that often accumulate under heat-stress conditions or during recombinant protein production in bacteria (Allen et al. 1992; Hartley and Kane 1988; Marston 1986), the formation of amyloid fibrils is an ordered process that leads to insoluble aggregates with a *regular* quaternary structure. During the last decades, numerous X-ray diffraction studies, electron microscopic and biophysical experiments as well as biochemical data demonstrated that the common structural principle in all amyloid fibrils is the formation of an extended, intermolecular β-sheet structure in which identical segments of the respective polypeptide are aligned parallel to the fibril axis (▶Figure 11.1) (Greenwald and Riek 2010). As the main interactions stabilizing the amyloid fibril structure are intermolecular β-sheet hydrogen bonds between adjacent strands, the ability of forming amyloid fibrils resides in any polypeptide chain and is, in principle, independent of primary structure (Fändrich et al. 2001). This is evident from the incorporation of polypeptide segments into the β-sheet structure of amyloid fibrils that adopt α-helical structure in the native, biologically active tertiary structure of the respective protein. Examples are the increased β-sheet and decreased α-helical content in human prions relative to the benign, cellular prion protein (Pan et al. 1993), and the observation that even myoglobin, an all-α-helical protein, can form β-sheet-rich amyloid fibrils under the certain conditions (Fändrich et al. 2001).

The reasons why certain polypeptides have an increased tendency of forming amyloid fibrils are not fully understood (Belli et al. 2011; Thompson et al. 2006). Besides structural requirements such as steric and energetic compatibility of side chains in β strands aligned parallel to the fibril axis and the stability of the respective protein against unfolding and/or structural rearrangements, kinetic barriers for the different steps in the amyloid formation pathway also play a crucial role. The commonly accepted kinetic scheme of amyloid fibril formation is a nucleated polymerization reaction, in which monomers sequentially associate to a homo-oligomer of critical size that acts as seed for further fibril growth, during which monomers successively associate with the growing fibril ends (Jarrett and Lansbury 1993; Prusiner 1991). This mechanism is characterized by a lag phase in the kinetics of spontaneous fibril formation starting with monomers,

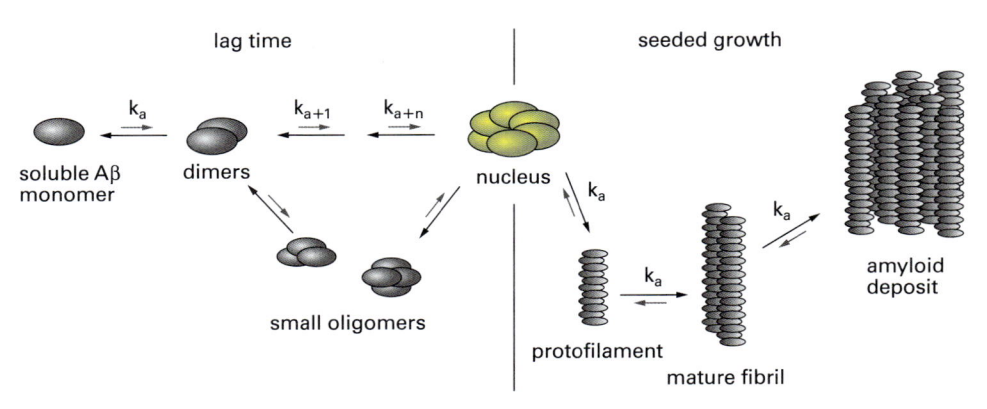

Figure 11.1: Reaction scheme of the nucleated polymerization of amyloid fibrils, illustrated with the amyloid β peptide (Aβ) as an example. The reaction mechanism is characterized by a rate-limiting assembly of monomers via oligomeric species to a stable nucleus, followed by a rapid fibril growth phase in which protofilaments are formed that assemble to mature fibrils, which eventually aggregate to amyloid deposits in vivo.

as the formation of the seeds is generally the rate-limiting step of fibril formation. Recent results show that a quantitative description of the kinetics of fibril formation in vitro also requires the inclusion of an additional reaction step, namely, the fragmentation of growing fibrils that simultaneously increases the concentration of growing ends during the aggregation reaction (Knowles et al. 2009). Together, amyloid fibril formation has many similarities with the growth of crystals, in which structurally identical molecules are arranged in a regular, three-dimensional lattice.

Another important, structural aspect of amyloid fribril formation is the ability of amyloidogenic polypeptides to form different types of amyloid fibrils that slightly differ in their quaternary structure, a phenomenon that is again analogous to crystal growth where different crystal forms can be formed even under identical conditions. Probably the most prominent example in this context is the occurrence of "strains" in prion diseases that are associated with a defined, specific set of disease symptoms (Chien et al. 2004; Osherovich and Weissman 2002). In the context of the "protein only hypothesis," which postulates that the prion, the infectious agent of transmissible spongiform encephalopathies, is an ordered aggregate of the "scrapie" form of the prion protein (PrP^{Sc}) with increased β-sheet content relative to the monomeric, cellular form (PrP^C), it is now well established that different forms of human prion diseases (i.e. different forms of sporadic and familial CJD) are linked with different aggregation forms of PrP that differ, for example, in the fractions of the incorporated glycoforms of PrP and the size of the protease resistant core of the PrP^{Sc} aggregates (Glockshuber 2005). This means that only one type of prion amyloid is present in an affected individual, and only this amyloid form is propagated in a newly infected individual after transmission. Another example of fibrillar heterogeneity is the in vitro formation of amyloid beta fibrils, the potential causative agent of Alzheimer's disease (see Section 11.2). Recent high-resolution electron microscopy (EM) studies showed that up to 12 different forms of beta amyloid fibrils with slightly different quaternary structures can be identified in a single in vitro aggregation reaction of the Aβ, which obviously have very similar stability and aggregation kinetics. While the mechanisms underlying this heterogeneity of the fibrils are unknown, it is well conceivable that it arises from differences in the side chain conformations of β-sheet residues aligned parallel to the fibril axis (▶Figure 11.2). Thus, different disease symptoms in AD may also be related to the fraction of different types of fibrils in an affected individual.

Besides high-resolution X-ray structures of fibrils of small peptides, only very few three-dimensional structures of amyloid fibrils at atomic resolution are available to date (Luhrs et al. 2005). It appears that the main difficulty for structure determination (for example by solid-state nuclear magnetic resonance [NMR] spectroscopy), is indeed the structural heterogeneity of fibril preparations.

11.2 The importance of purity for reproducible kinetics of amyloid fibril formation in vitro: the Aβ as an example

The Aβ is a natural proteolytic degradation product of the human amyloid precursor protein (APP) and the main component of amyloid plaques accumulating in the brain of AD patients (Masters et al. 1985). There is a growing body of evidence that accumulation of Aβ amyloid is the cause rather than a consequence of AD (Finder and Glockshuber 2007; Hardy and Selkoe 2002; Jakob-Roetne and Jacobsen 2009). Thus, there is also of fundamental interest to elucidate its mechanism of fibril formation through in vitro aggregation experiments. The two main Aβ species in humans are the 40- and 42-residue peptides Aβ1–40 and Aβ1–42, which have identical N termini and only differ by the two additional C-terminal residues in Aβ1–42 (▶Figure 11.3). The molar Aβ1–40/Aβ1–42 ratio in humans is about 10:1 (Irie et al. 2005) and is a consequence of the substrate specificity of γ-secretase, the protease which generates the Aβ C terminus in vivo and preferably cleaves after Aβ residue 40. Although Aβ1–40 is the predominant Aβ species, Aβ1–42 is most likely the pathophysiologically relevant Aβ form, as the majority of familial Alzheimer's disease (FAD) forms is associated with mutations in either APP or γ-secretase that lead to an increase in the fraction of Aβ1–42 relative to

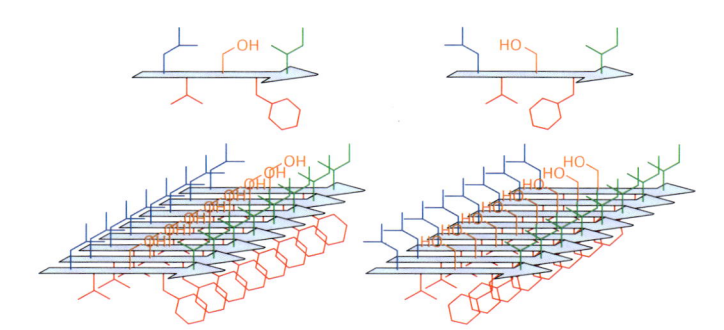

Figure 11.2: Fibrils with different quaternary structure may be formed by a single amyloidogenic peptide. The scheme illustrates one of several possibilities of how fibrils with different quaternary structure can be formed by the same peptide. Both fibrils only differ in the conformation of the side chains that are aligned parallel to the fibril axis (bottom). The conformation of the corresponding monomer in the context of the fibril is shown at the top.

Aβ1–40 (Duff et al. 1996; Finder and Glockshuber 2007; Scheuner et al. 1996; Suzuki et al. 1994; Wong et al. 2002), and as Aβ1–42 has strong cytotoxic effects on neuronal cell cultures compared to Aβ1–40 (Kuperstein et al. 2010). In addition, there are also FAD cases with mutations in the Aβ sequence itself (▶Figure 11.3).

Until today, almost all studies on the *in vitro* aggregation of different Aβ forms have been performed with chemically synthesized peptides and are mainly focused on the aggregation mechanism of Aβ1–40, which is much simpler to handle in solution compared to the more hydrophobic and less soluble Aβ1–42 peptide. Unfortunately, the research on the *in vitro* aggregation of Aβ is characterized by numerous contradictory results from different laboratories, the fact that results often cannot be reproduced by others and, even when experiments are performed within the same laboratory and Aβ preparations from the same supplier are used, by significant batch-to-batch variations between different Aβ preparations (Dobeli et al. 1995; Soto et al. 1995; Zagorski et al. 1999). As outlined further on in this section, there is evidence that these discrepancies result from different amounts of impurities in synthetic Aβ preparations that cannot be avoided due to the technical limitations of peptide synthesis.

▶Figure 11.1 shows a general reaction scheme of Aβ aggregation that agrees with most experimental data and follows the generally accepted nucleation-polymerization model. Starting with Aβ monomers, it assumes the formation of different oligomeric species until a nucleus of sufficient size and stability is formed that then acts as seed for fibril growth via transient accumulation of smaller protofibrils and protofilaments, followed by association of protofilaments to mature fibrils. While the structures of oligomeric Aβ fibril precursors are unknown, different models have been proposed for the structure of the Aβ monomer within the fibril (Fändrich et al. 2011). Although all models share the general features of amyloid fibrils (i.e. the parallel orientation of monomers parallel to the fibril axis), they differ in the exact positions of β-strand regions

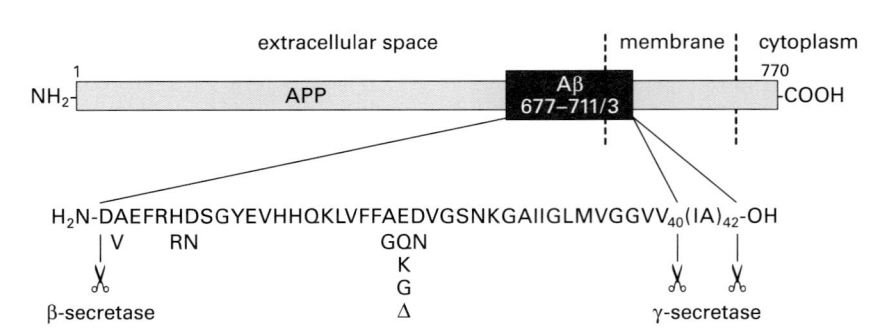

Figure 11.3: Primary structure of the Aβ forms Aβ1–40 and Aβ1–42 (amino acid numbering according to Aβ). Both Aβ1–40 and Aβ1–42 are natural degradation products of the APP, comprising APP residues 672–711 and 672–713, respectively. Residues 1–28 of Aβ correspond to the last residues of the extracellular APP domain, and Aβ residues 29–40/42 correspond to the N-terminal part of the APP transmembrane domain. The N terminus of Aβ1–40 and Aβ1–42 results from cleavage with β-secretase after APP residue 671, the C termini are generated by γ-secretase cleavage after APP residues 711 and 713, respectively. Single amino acid replacements in the Aβ sequence that are associated with FAD cases are indicated below the Aβ sequence.

and their relative orientations, and in the number of monomers per cross-section in protofibrils and fibrils. Recent high-resolution EM data on Aβ1–40 fibrils suggest that fibrils of Aβ1–40 contain four, and fibrils of Aβ1–42 two, monomers per cross-section, respectively (Sachse et al. 2008; Schmidt et al. 2009). In the case of Aβ1–40, up to 12 different classes of fibrils can be distinguished via transmission electron microscopy and atomic force microscopy (Fändrich et al. 2011). This structural heterogeneity appears to be the main reason for the fact that it is still notoriously difficult to obtain homogeneous fibrils for structure determination at atomic resolution. Toward this goal, solid-state NMR spectroscopy appears to be the most promising method that is presently available (Böckmann and Meier 2010).

Fibrils of Aβ1–40 and Aβ1–42 formed *in vitro* are generally longer than 1 μm (Fändrich et al. 2011) and consequently consist of several thousand Aβ monomers with identical tertiary structure that are assembled into a regular three-dimensional array in the mature fibril. It automatically follows that highly pure preparations of Aβ peptide are required for obtaining reproducible *in vitro* aggregation kinetics. This is because small amounts of impurities of monomeric Aβ derivatives in the range of ~1% that are typically observed in synthetic Aβ preparations may bind to growing fibril ends and prevent further fibril growth by blocking further incorporation of authentic Aβ monomers (▶Figure 11.4). If one assumes that the concentration of growing fibril ends due to fibril fragmentation does significantly increase during an in vitro aggregation reaction, a strong deceleration of fibril growth is expected through such impurities, as nonauthentic Aβ variants would first need to dissociate form the growing ends to allow further fibril growth. In other words, even minute amounts of impurities with similar covalent structure may be sufficient to decelerate or even inhibit the fibril formation reaction.

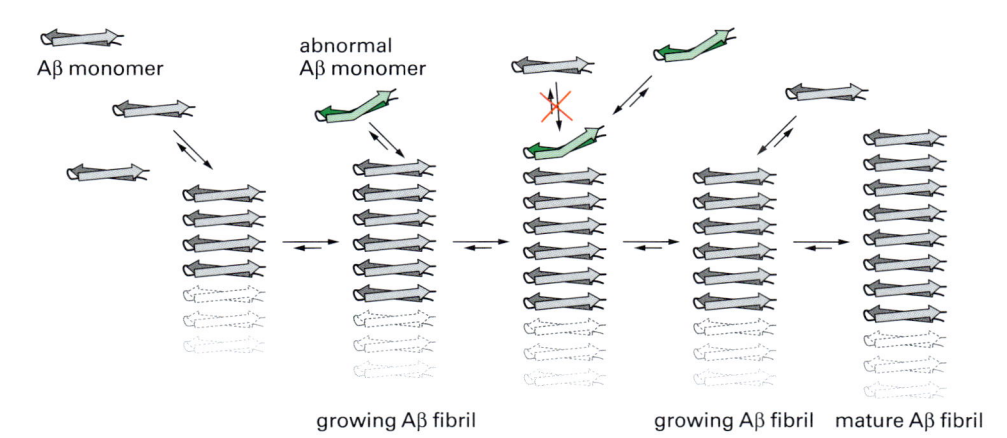

Figure 11.4: Even small amounts of impurities in synthetic Aβ peptides that possess a covalent structure similar to authentic Aβ can significantly retard fibril growth. An abnormal Aβ monomer (e.g. synthetic Aβ containing a single D-histidine residue, depicted in green) that becomes incorporated at the growing fibril end may prevent further incorporation of authentic Aβ, so that further fibril growth is only possible after dissociation of the abnormal Aβ from the growing end. It follows that even very small quantities of abnormal Aβ relative to authentic Aβ can have a strong inhibitory effect on fibril growth.

To circumvent these difficulties, we recently developed a highly efficient protocol for recombinant production of Aβ1–40, Aβ1–42 and variants thereof in *Escherichia coli* (Finder et al. 2010). We fused the corresponding Aβ sequences to the C terminus of the polypeptide repeat sequence H$_6$GS(NANP)$_{19}$, a previously established Aβ fusion tag for high-level production of Aβ in the *E. coli* cytoplasm (Luhrs et al. 2005), and introduced the cleavage sequence of the Tobacco Etch Virus (TEV) protease at the fusion site (▶Figure 11.5A). As TEV protease is insensitive with respect to the N-terminal residue of the C-terminal cleavage product (Kapust et al. 2002), we could efficiently enrich the H$_6$-tagged fusion protein via metal chelate affinity chromatography and obtain authentic, recombinant Aβ after TEV protease cleavage and reversed-phase HPLC purification in quantities of 20 milligrams per liter of bacterial culture. Compared to previously reported methods for recombinant Aβ production, the Aβ peptide was obtained without impurities according to analytical HPLC analysis and mass spectrometry in very high yields (other strategies for the recombinant production of Aβ peptides are reviewed in Finder et al. 2010 and in Long et al. 2011).

We next performed a systematic comparison of the aggregation kinetics of recombinant Aβ1–42 with different preparations of commercially available, synthetic Aβ1–42 preparations. In all cases, fibril formation of synthetic Aβ1–42 (monitored by the increase in fluorescence of the fibril-specific dye thioflavin T) proceeded slower than the same reaction with recombinant Aβ1–42 (▶Figure 11.5B and 11.5D) (Finder et al. 2010). Even after further purification of the synthetic peptides to HPLC homogeneity, they still aggregated significantly slower than recombinant Aβ1–42 and formed less regular fibrils according to negative-stain EM. A racemization analysis of the purest synthetic preparations then revealed a D-histidine content of 1.01%. As there are three histidine residues in the Aβ sequence (▶Figure 11.3), this means that about 3% of the molecules in HPLC-pure, synthetic Aβ1–42 contained a D-histidine, and these impurities cannot be removed by conventional, reversed-phase HPLC. In 9-fluorenylmethoxycarbonyl (FMOC)–based peptide synthesis, racemization in the range of 1%–2% during the coupling of histidine and cysteine is still a technically unsolved problem (Jones 2002). If Aβ peptides with a single D-histidine residue indeed acted as fibril formation inhibitors according to the mechanism outlined in ▶Figure 11.4, it follows that recombinant, racemic-free Aβ peptides should be the material of choice for all future *in vitro* studies on the mechanism of Aβ aggregation. This is underlined by the fact that recombinant Aβ1–42 always showed significantly higher toxicity toward rat primary neuronal cultures compared to synthetic peptides (Finder et al. 2010). Our assumption is that different amounts of impurities in synthetic Aβ preparations, which cannot be removed by conventional chromatographic techniques also provides a very plausible explanation for the contradictory results for synthetic Aβ preparations obtained by different groups. An additional advantage of using recombinant instead of synthetic Aβ peptides is that Aβ variants (e.g. with single amino acid replacements linked to FAD), are readily accessible.

11.3 Future challenges for the characterization of fibrillar structures

Regarding the structural biology of Aβ amyloid fibrils, the biggest challenge at present is certainly the elucidation of the three-dimensional structure of Aβ fibrils at atomic resolution. To date, the most promising approach toward this goal is the structural analysis of

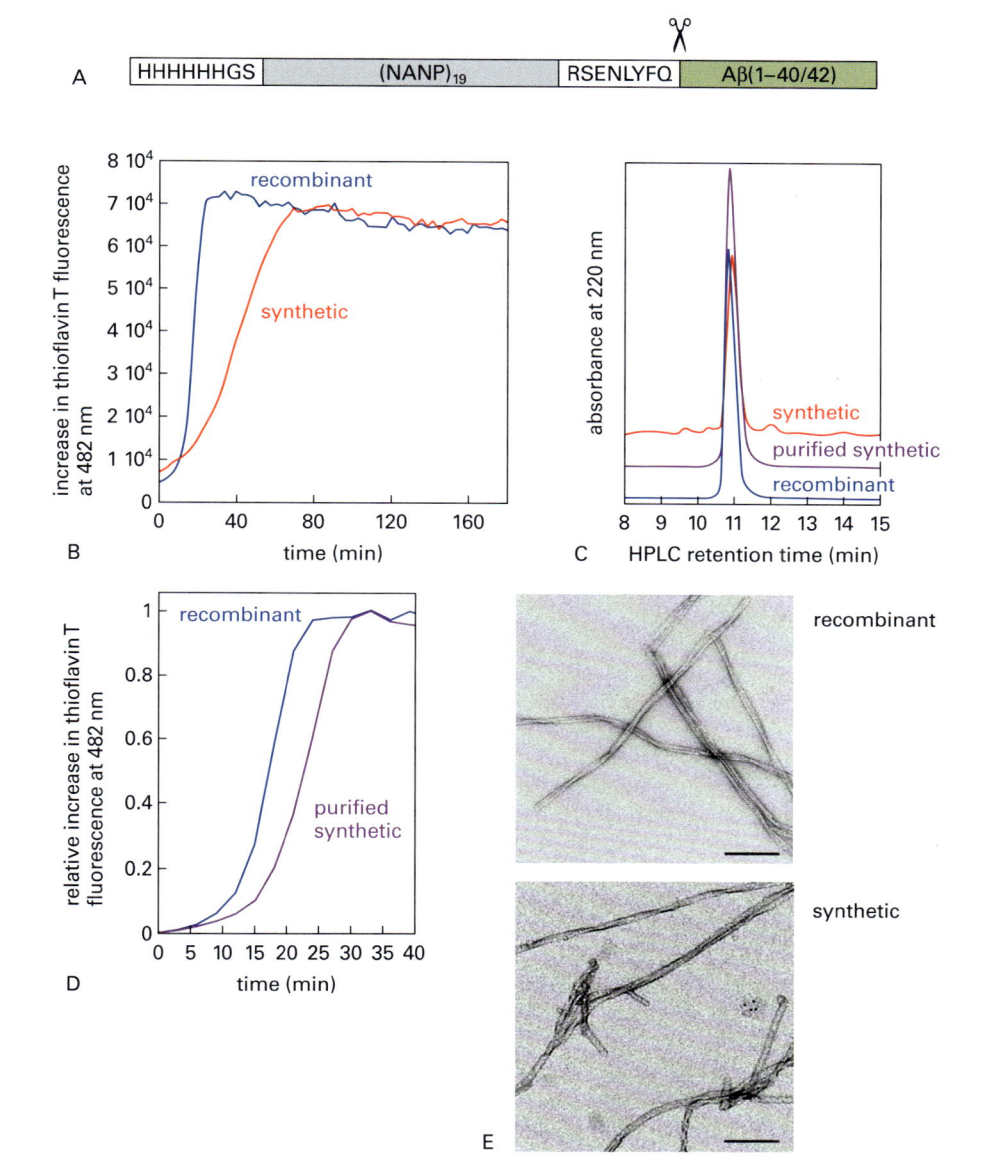

Figure 11.5: Recombinant, racemic-free Aβ1–42 aggregates faster than HPLC-purified, synthetic Aβ1–42. (A) Recombinant Aβ1–42 was produced in *E. coli* as fusion protein in which the hexahistidine-tagged repeat sequence (NANP)$_{19}$ was fused to the N terminus of Aβ via a TEV protease recognition sequence. Cleavage with TEV protease (scissors) yielded authentic, recombinant Aβ1–42 that can be purified to homogeneity by reversed-phase HPLC. (B) Recombinant Aβ1–42 aggregates faster than synthetic Aβ1–42 (indicated purity: 97.1%). Fibril formation at pH 7.4 and 37°C (total monomer concentration: 7.4 μM) was followed by the increase in the fluorescence at 482 nm of the amyloid-specific dye thioflavin T. (C) Analytical HPLC runs of recombinant (blue), synthetic (red) and further HPLC-purified, synthetic Aβ1–42 (pink). (D) Even after further purification to apparent HPLC-homogeneity, synthetic Aβ1–42 still aggregates slower than recombinant Aβ1–42 (reaction conditions as described in [B]; adapted from Finder et al. [2010]). (E) EM pictures of fibrils formed by recombinant and synthetic Aβ1–42. The scale bar represents 100 nm.

Aβ amyloid fibrils by solid-state NMR spectroscopy, for which fibrils of isotope-labeled Aβ are required that are, in practice, only accessible via recombinant Aβ produced in bacteria. In addition, the previously described heterogeneity of synthetic Aβ1–40 fibrils may be reduced or even avoided with the purer, recombinant Aβ preparations. Regarding the tendency of Aβ to simultaneously form different fibril types under the same experimental conditions, another important task for the future is the establishment of criteria according to which different types of Aβ fibrils can be distinguished. Possible biochemical parameters that are specific to different fibril forms could, for example, be the number of fibrillar thioflavin T binding sites per Aβ monomer; the half-life of fibril dissociation in the presence of denaturants; extreme pH values or heat; the resistance of the fibrils against different proteases; the preference for different, amyloid-specific dyes and the seeding activity of fragmented fibrils. We propose that recombinant instead of synthetic Aβ should be used in future studies on Aβ aggregation.

References

Allen, S.P., Polazzi, J.O., Gierse, J.K., and Easton, A.M. (1992). Two novel heat shock genes encoding proteins produced in response to heterologous protein expression in *Escherichia coli*. J Bacteriol *174*, 6938–6947.

Belli, M., Ramazzotti, M., and Chiti, F. (2011). Prediction of amyloid aggregation in vivo. EMBO Rep. 12, 657–663.

Böckmann, A., and Meier, B.H. (2010). Prions – en route from structural models to structures. Prion *4*, 72–79.

Chien, P., Weissman, S.J., and DePace, A.H. (2004). Emerging principles of conformation-based prion inheritance. Annu Rev Biochem *73*, 617–656.

Chiti, F., and Dobson, C.M. (2006). Protein misfolding, functional amyloid, and human disease. Annu Rev Biochem *75*, 333–366.

Dobeli, H., Draeger, N., Huber, G., Jakob, P., Schmidt, D., Seilheimer, B., et al. (1995). A biotechnological method provides access to aggregation competent monomeric Alzheimer's 1–42 residue amyloid peptide. Nature Biotechnology *13*, 988–993.

Duff, K., Eckman, C., Zehr, C., Yu, X., Prada, C.M., Perez-tur, J., et al. (1996). Increased amyloid-beta 42(43) in brains of mice expressing mutant presenilin 1. Nature *383*, 710–713.

Fändrich, M., Fletcher, M.A., and Dobson, C.M. (2001). Amyloid fibrils from muscle myoglobin. Nature *410*, 165–166.

Fändrich, M., Schmidt, M., and Grigorieff, N. (2011). Recent progress in understanding Alzheimer's β-amyloid structures. Trends Biochem Sci *36*, 338–345.

Finder, V.H., and Glockshuber, R. (2007). Amyloid-β aggregation. Neurodegenerative Dis *4*, 13–27.

Finder, V.H., Vodopivec, I., Nitsch, R., and Glockshuber, R. (2010). The recombinant amyloid-β peptide Aβ1–42 aggregates faster and is more toxic than synthetic Aβ1–42. J Mol Biol *396*, 9–18.

Glockshuber, R. (2005). Biochemistry and structural biology of mammalian prion disease. In: Protein folding handbook, part II, J. Buchner and T. Kiefhaber, eds (Weinheim, Germany: WILEY-VCH Verlag GmbH & Co. KGaA), pp. 1114–1143.

Greenwald, J., and Riek, R. (2010). Biology of amyloid: structure, function, and regulation. Structure *18*, 1244–1260.

Hardy, J., and Selkoe, D.J. (2002). The amyloid hypothesis of Alzheimer's disease: progress and problems on the road to therapeutics. Science *297*, 353–356.

Hartley, D.L., and Kane, J.F. (1988). Properties of inclusion bodies from recombinant *Escherichia coli*. Biochem Soc Trans *16*, 101–102.

Irie, K., Murukami, K., Masuda, Y., Morimoto, A., Ohigashi, H., Ohashi, R., et al. (2005). Structure of β-amyloid fibrils and its relevance to their neurotoxicity: implications for the pathogenesis of Alzheimer's disease. J Biosci Bioeng 99, 437–447.

Jakob-Roetne, R., and Jacobsen, H. (2009). Alzheimer's disease: from pathology to therapeutic approaches. Angew Chem Int Ed Engl 48, 3030–3059.

Jarrett, J.T., and Lansbury, P.T., Jr. (1993). Seeding "one dimensional cristallization" of amyloid: a pathogenic mechanism in Alzheimer's disease and scrapie? Cell 73, 1055–1058.

Jones, J.H. (2002). Synthesis of peptides and peptidomimetics. In: Houben-Weyl. Methods of organic chemistry, M. Goodman, A. Felix, L. Moroder, and C. Toniolo, eds (Stuttgart, Germany: Thieme Stuttgart), E22a, pp. 334–346.

Kapust, R.B., Toszer, J., Copeland, T.D., and Waugh, D.S. (2002). The P1' specificity of tobacco etch virus protease. Biochem Biophys Res Commun 294, 949–955.

Knowles, T.P.J., Waudby, C.A., Devlin, G.L., Cohen, S.I.A., Aguzzi, A., Vendruscolo, M., et al. (2009). An analytical solution to the kinetics of breakable filament assembly. Science 326, 1533–1537.

Kuperstein, I., Broersen, K., Benilova, I., Rozenski, J., Jonckheere, W., Debulpaep, M., et al. (2010). Neurotoxicity of Alzheimer's disease Aβ peptides is induced by small changes in the $A\beta_{42}$ to $A\beta_{40}$ ratio. EMBO J 29, 3408–3420.

Long, F., Cho, W., and Ishii, Y. (2011). Expression and purification of ^{15}N- and ^{13}C-isotope labeled 40-residue human Alzheimer's' β-amyloid peptide for NMR-based structural analysis. Protein Expr Purif. 79, 16–24.

Luhrs, T., Ritter, C., Adrian, M., Riek-Loher, D., Bohrmann, B., Dobeli, H., et al. (2005). 3D structure of Alzheimer's beta amyloid-beta(1–42) fibrils. Proc Natl Acad Sci U S A 102, 17342-17347.

Marston, F.A.O. (1986). The purification of eukaryotic polypeptides synthesized in *Escherichia coli*. Biochem J 240, 1–12.

Masters, C.L., Simms, G., Weinman, N.A., Multhaupt, G., McDonald, B.L., and Beyreuther, K. (1985). Amyloid plaque core protein in Alzheimer disease and Down syndrome. Proc Natl Acad Sci U S A 82, 4245–4249.

Osherovich, L.Z., and Weissman, J.S. (2004). The utility of prions. Dev Cell 2, 143–151.

Pan, K.M., Baldwin, M., Nguyen, J., Gasset, M., Serban, A., Groth, D., et al. (1993). Conversion of alpha-helices into beta-sheets features in the formation of the scrapie prion proteins. Proc Natl Acad Sci U S A 90, 10962–10966.

Prusiner, S.B. (1991). Molecular biology of prion diseases. Science 252, 1515–1522.

Sachse, C., Fändrich, M., and Grigorieff, N. (2008). Paired β-sheet structure of an Aβ(1–40) amyloid fibril revealed by electron microscopy. Proc Natl Acad Sci U S A 105, 7462–7466.

Scheuner, D., Eckman, C., Jensen, M., Song, X., Citron, M., Suzuki, N., et al. (1996). Secreted amyloid beta-protein similar to that in the senile plaques of Alzheimer's disease is increased *in vivo* by presenilin 1 and 2 and APP mutations linked to familial Alzheimer's disease. Nat Med 2, 864–870.

Schmidt, M., Sachse, C., Richter, W., Xu, C., Fändrich, M., and Grigorieff, N. (2009). Comparison of Alzheimer Aβ(1–40) and Aβ(1–42) amyloid fibrils reveals similar protofibril structures. Proc Natl Acad Sci U S A 106, 19813–19818.

Soto, C., Castano, E.M., Kumar, R.A., Beavis, R.C., and Frangione, B. (1995). Fibrillogenesis of synthetic amyloid-beta peptides is dependent on their initial secondary structure. Neurosci Lett 200, 105–108.

Suzuki, N., Cheung, T.T., Cai, X.D., Odaka, A., Otvos, L., Jr., Eckman, C., et al. (1994). An increased percentage of long amyloid beta protein secreted by familial amyloid beta protein precursor (beta APP717) mutants. Science 264, 1336–1340.

Thompson, M.J., Sievers, S.A., Karanicolas, J., Ivanova, M.I., Baker, D., and Eisenberg, D. (2006). The 3D profile method for identifying fibril-forming segments of proteins. Proc Natl Acad Sci U S A *103*, 4074–4078.

Wong, P.C., Cai, H., Borchelt, D.R., and Price, D.L. (2002). Genetically engineered mouse models of neurodegenerative disease. Nat Neurosci *5*, 633–639.

Zagorski, M.G., Yang, J., Shao, H., Ma, K., Zeng, H., and Hong, A. (1999). Methodological and chemical factors affecting amyloid beta peptide amyloidogenicity. Methods Enzymol *309*, 189–204.

12 Chemical modification of proteins in living cells

Ariele Hanek and Ivan R. Corrêa Jr.

12.1 Introduction

The ability to characterize functioning proteins in living cells is paramount to improving our knowledge of complex biological networks. In recent years, a number of chemical and biochemical methods have been developed to selectively detect and manipulate proteins in their native environment (Giepmans et al. 2006; Marks and Nolan 2006). Since the discovery, and subsequent development, of green fluorescent protein (GFP), an ever-growing number of researchers have used fluorescent proteins (FPs) to label and study proteins of interest. FPs have been used to detect reporter gene expression, follow protein trafficking, and study many dynamic biochemical signals in living cells and whole animals (Davidson and Campbell 2009; Day and Davidson 2009). FPs have many benefits including a wide range of proven protocols, a large body of work published using FPs and, most importantly, the fact that the labels are generated within the cell, therefore eliminating the need for exogenous labeling agents, fixation, or permeabilization procedures (Shaner et al. 2005). Nonetheless, FP-based approaches face limitations. Although a wide range of FP variants that span the visible spectrum have been developed, the optical properties of the available monomeric versions of FPs still present limitations such as broad absorption spectra and lower quantum yields, particularly in the red and near-infrared spectral region. More importantly, FPs are notably less bright or photostable than many synthetic fluorescent dyes, making them difficult to employ in certain applications such as time-resolved experiments and single-molecule microscopy. Additional considerations include the size of FPs (~27 kDa), which may interfere with the function of the target protein, and FP oligomerization, which may lead to mistrafficking or aggregation of the target protein (Campbell et al. 2002).

Many of the disadvantages of FPs can be overcome by using a small-molecule probe to chemically label a protein of interest. Several approaches have been developed to study proteins in living systems (Sletten and Bertozzi 2009). Classic bioconjugation relies on the reaction of reporter probes with the side chains of the canonical amino acids. For instance, lysine side chains can be readily modified with *N*-hydroxysucciminidyl esters, sulfonyl chorides or iso(thio)cyanates, while cysteine side chains can be modified with maleimides, iodoacetamides and disulfides. These methods have been widely used for in vitro conjugation of proteins with biotin, fluorophores, and solid surfaces. However, these classic chemical bioconjugation methods lack specificity and are often incompatible with the chemical complexity of living systems. Therefore, more recent efforts have focused on developing bioorthogonal chemical reactions that are compatible with physiological conditions and inert to the plethora of functionalities present in cellular systems.

An emerging technique for live cell imaging and proteomics applications utilizes specific recognition sequences to recruit chemical probes for in situ labeling of cellular proteins (Hinner and Johnsson 2010; Johnsson and Johnsson 2007; O'Hare et al. 2007). It differs from conventional chemical labeling approaches by the fact that small-molecule probes bind to a peptide or protein tag that is fused to the protein being studied, or target protein. The binding of the small molecule occurs either via a self-labeling protein/peptide tag or by enzymatic ligation. This approach combines the convenience and specificity of genetically encoded systems with the versatility of small-molecule probes. There are two steps to using this system: cloning and expression of the target protein as the peptide/protein tag fusion, and labeling of the tagged fusion with the reactive substrate of choice. The unique combination of high spatial and temporal resolution, nondestructive compatibility with living cells and organisms and molecular specificity render this approach suitable for a broad range of applications from in vivo imaging to drug discovery processes.

The first part of this chapter is dedicated to reviewing the more readily used methods for specific labeling of proteins with chemical probes. This is followed by a section dedicated to recent applications such as pulse-chase experiments, multicolor protein imaging, and superresolution microscopy. The last part of the chapter contains experimental protocols and technical notes for SNAP-tag labeling. Experimental procedures for other labeling techniques are referenced throughout the chapter. The labeling methods using intein-mediated self-splicing (Ghosh et al. 2011), metal-based chelation to peptide tags (Sletten and Bertozzi 2009), and tRNA/aminoacyl-tRNA synthetase-mediated incorporation of unnatural amino acids (Liu and Schultz 2010) have recently been reviewed elsewhere.

12.2 Site-specific labeling of proteins with chemical probes

The covalent and noncovalent binding of chemical probes to peptide or protein tags fused to a target protein enables the incorporation of chemical or optical properties to the target proteins, thereby providing the ability to track, manipulate and interrogate the protein. The success of any labeling strategy lies in the ability to confer the desired chemical or optical properties necessary to study the target protein. Fusion of the target protein to FPs only allows monitoring via fluorescence and predetermines which wavelengths can be used. Use of a genetically encoded peptide or protein tag, and subsequent reaction with a small-molecule probe offers several advantages. Small-molecule probes offer greater diversity as a ligand can be derivatized with other biomolecules, fluorophores, or any other chemical functionality. Additionally, protein and peptide tags, unlike classic protein bioconjugation methods, have been optimized to have a high labeling efficiency and specificity, ensuring the desired chemical or optical properties are conferred only to the tagged protein. Finally, the same tag can be used with a variety of small-molecule probes without the need to reclone the target protein into a new fusion. Thus, once generated, the same fusion protein can be modified with fluorescent dyes or affinity ligands, used for direct in-gel detection or Western-blot assays or immobilized on solid surfaces for purification or pull-down experiments. Herein we discuss the most prominent approaches for site-specific labeling of fusion protein with chemical probes: the tetracysteine tag, SNAP- and CLIP-tags, HaloTag, trimethoprim (TMP)–tag, acyl carrier protein (ACP)–tag, biotin and lipoic acid ligases and sortase.

12.2.1 Self-labeling proteins and peptides

Protein and peptide tags can be broadly divided into categories, self-labeling and enzyme-mediated labeling. In both cases, the target protein, or protein of interest (POI),

12.2 Site-specific labeling of proteins with chemical probes | 199

is expressed as a fusion protein with the peptide or protein tag (▶Figure 12.1). Self-labeling tags interact directly with the chemical probe leading to a covalent or noncovalent bond. For enzyme-mediated labeling, the chemical probe is attached to the tag via an enzyme-mediated process. There are benefits and drawbacks for each system.

12.2.1.1 Tetracysteine and tetraserine tags

The tetracysteine and tetraserine tags are genetically encoded peptide tags that selectively react with fluorogenic biarsenical and bisboronic acid substrates, respectively. The most important aspect of these tagging families is their relative small size, which

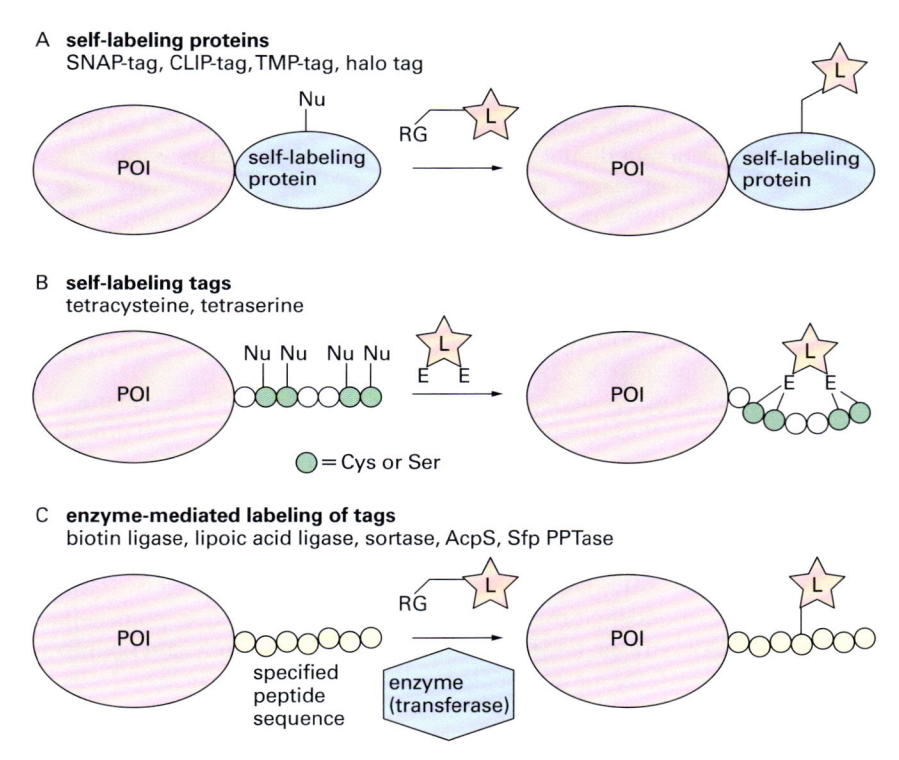

Figure 12.1: General methods for specific labeling of proteins. (A) The target protein, or POI, is expressed as a fusion with a self-labeling protein tag (SNAP-tag, CLIP-tag, TMP-tag and HaloTag). After expression, a reporter probe coupled to a substrate for the self-labeling tag (benzylguanine, benzylcytosine, TMP or haloalkane, respectively) is added. The self-labeling tag binds the substrate and links the probe to the protein fusion. (B) The target protein is expressed with a peptide tag (tetracysteine or tetraserine). Subsequent incubation with a biarsenical or boronic acid probe, respectively, covalently attaches the probe to the peptide tag. (C) In enzyme-mediated labeling (biotin ligase, lipoic acid ligase, Sortase A, AcpS or Sfp 4′-phosphopantetheinyl transferases [PPTases]), the target protein is expressed with a tag that is recognized by an enzyme. Addition of the enzyme and the corresponding enzyme substrate (biotin, alkanoic acid, Gly$_5$, coenzyme A [CoA], respectively) carrying a chemical probe leads to the transfer of the substrate to the tag.

potentially minimizes the interference with the functions of the target protein fused to them. The tetracysteine tag, pioneered by Tsien and coworkers (Griffin et al. 1998), is the first and, so far, the most widespread of the tag-mediated protein-labeling systems. It is based on the selective recognition by biarsenical fluorogenic substrates (e.g. FlAsH and ReAsH) of a minimal tetracysteine sequence CCXXCC situated at a hairpin structure, where XX has been optimized as ProGly. A unique feature of the biarsenical substrates is the generation of an intense fluorescence response only after binding to the tetracysteine motif, therefore greatly reducing background fluorescence of any unbound species. The availability of cell membrane-permeant substrates, such as FlAsH and ReAsH, permits the labeling of tetracysteine-tagged intracellular proteins (Adams and Tsien 2008; Hoffmann et al. 2010), further extending the usefulness of this tag. However, the specificity of these substrates is less than ideal due to the presence of naturally occurring cysteine-rich sequences in proteins other than the target protein and, more importantly, to nonspecific hydrophobic interactions with the fluorogenic substrates, making the detection of proteins expressed at low levels or of smaller temporal pool sizes difficult. Background labeling can be reduced by use of high concentrations of a competing dithiol reagent. However, this increases washing requirements to remove the dithiol reagent prior to imaging. Finally, the biarsenicals are potentially cytotoxic, precluding the use of the tetracysteine-labeling in live animals. Despite these limitations, this method has been successful in many cellular applications, including studies of acetylcholine (Ziegler et al. 2011) and G protein-coupled receptor (GPCR) activation (Hoffmann et al. 2005), visualization of mRNA translation (Rodriguez et al. 2006), amyloid formation (Roberti et al. 2007) and detection of protein-protein interactions (Luedtke et al. 2007).

The tetraserine tag developed by Schepartz and coworkers is similar to the tetracysteine tag but uses a fluorogenic rhodamine-derived bisboronic acid (RhoBo) that binds peptides or proteins containing the SSPGSS motif with nanomolar affinity (Halo et al. 2009). The intracellular labeling of tetraserine containing proteins has been demonstrated using the RhoBo probe in HeLa cells. A tetraserine tag combines the power of a short peptide tag, such as tetracysteine, without the requirement for cytotoxic biarsenical compounds. Although the bisboronic acids are preferred over biarsenicals for this reason, the abundance of SSXXSS-like sequences in endogenous cellular proteins makes labeling of only the target protein difficult. Further study and optimization of this technique are necessary for widescale use of the tetraserine tag.

12.2.1.2 SNAP- and CLIP-tags

Among the most versatile self-labeling fusion tags is the SNAP-tag®, a stable and monomeric 20 kDa engineered mutant of the human repair protein O^6-alkylguanine-DNA alkyltransferase (hAGT) developed by Johnsson and coworkers (Keppler et al. 2003, 2004). The SNAP-tag covalently reacts with O^6-benzylguanine (BG) and O^6-benzyl-4-chloropyrimidine (CP) derivatives bearing a chemical or optical probe. During the labeling reaction, a stable thioether bond is formed between the active site cysteine (Cys 145) and the BG substrate, in which the label appended at the *para*-position of the benzyl group is irreversibly transferred to the SNAP-tag. Each tag reacts once with a single substrate molecule, in a well-defined mechanism, with a predictable stoichiometry and rapid kinetics, irrespective of the attached target protein. The SNAP-tag protein-labeling system enables the specific, covalent attachment of virtually any molecule to a protein of interest. SNAP-tag fusion proteins can be labeled with fluorescent dyes, affinity ligands or other binding moieties (Kindermann et al. 2004); used for selective cross-linking of interacting protein partners (Gautier et al. 2009; Lemercier et al. 2007) and immobilized on solid surfaces (Engin et al. 2010; Iversen et al. 2008; Kindermann et al. 2003). Selective labeling of membrane and intracellular targets can be achieved by appropriate choice of cell-permeable and cell-impermeable substrates, respectively. Owing to the covalent and quantitative nature of the labeling reaction, the SNAP-tag is well suited for the analysis and quantification of the fused target protein using in-gel fluorescence scanning.

The CLIP-tag™ is a modified version of the SNAP-tag, which was engineered to react with O^2-benzylcytosine (BC) rather than BG derivatives (Gautier et al. 2008). Although CLIP-tag labeling is slower than SNAP-tag, the two tags have been used in conjugation for orthogonal and complementary labeling of two proteins simultaneously in the same cell. The concurrent labeling of SNAP and CLIP fusion proteins has been employed to visualize different generations of two different proteins in one sample via double pulse-chase experiments (Gautier et al. 2008), to sense the concentration of cell metabolites via FRET change (Brun et al. 2009) to explore protein-protein interactions (Gautier et al. 2009) as well as to analyze cell surface protein GPCR heteromeric complexes (Doumazane et al. 2011).

Use of self-labeling proteins, such as SNAP-tag and CLIP-tag, as well as the HaloTag and TMP-tag discussed in Section 12.2.1.3 and 12.2.1.4, respectively, is limited to target proteins whose function is not altered by fusion to the protein tag. There is a trade-off between the size of the label (6 residues vs. ~20 kDa) and the specificity of labeling as well as the diversity of chemical probes available. Thus choosing a self-labeling peptide tag or self-labeling protein tag discussed in Section 12.2.1.2 depends on the target protein to be studied and the experiment used. The self-labeling protein tags discussed in this section and in subsequent sections may not offer a significant size advantage over FPs, but they offer a wide range of fluorophores that are brighter and more photostable, diverse selection of chemical probes and overcome issues with FP oligomerization.

12.2.1.3 HaloTag

The HaloTag protein-labeling approach is based on an engineered 33 kDa haloalkane dehalogenase (DhaA) from *Rhodococcus rhodochrous* developed by Promega (Los et al. 2008). This mutated haloalkane dehalogenase permits the covalent tethering of chemical probes containing a chloride-terminated ligand to cell surface or intracellular HaloTag fusion proteins. Similar to other self-labeling protein systems, HaloTag substrates can be derivatized with a diverse array of chemical probes (e.g. fluorophores, biotin, or resins) enabling the system to be used in a variety of labeling experiments. The HaloTag has been successfully employed for in vivo applications, such as cell imaging (Lee et al. 2010), protein interaction studies (Urh et al. 2008) and protein purification (Ohana et al. 2011). A recent study exploited this protein-labeling technology for live animal cancer imaging (Kosaka et al. 2009). In this study, a SHIN3 tumor cell line overexpressing a HaloTag receptor was intraperitoneally implanted in mice and labeled with exogenously injected fluorophore-conjugated HaloTag substrates. The HaloTag has also been used in combination with superresolution techniques to elucidate subcellular distribution of β1-integrin-HaloTag fusion in HeLa cells (Schroder et al. 2009).

Although the utility of HaloTag system has been demonstrated in a number of biological settings, the intrinsic reactivity of haloalkane substrates toward other nucleophiles present in the cell can be problematic and leads to nonspecific labeling. For example, Gautier and coworkers have reported that the incubation of six different cell lines (HEK293T, CHO, BHK, HeLa CCL2, HeLa MZ and HT29) with HaloTag diAcFAM ligand led to the labeling of an unknown 28 kDa protein labeled at 10–30 pmol/mg (Gautier et al. 2008), demonstrating the promiscuity of the substrate. More recently, it was reported that HaloTag diAcFAM-ligand stained the exogenous protein GSTO1 (glutathione S-transferase omega 1) in NIH/3T3 live cells, which were not expressing HaloTag (Son et al. 2010).

12.2.1.4 TMP- and A-TMP-tags

The TMP-tag is a fusion protein-labeling system developed by Cornish and coworkers based on the *Escherichia coli* dihydrofolate reductase (eDHFR) (Miller et al. 2005). In this system, an intracellular protein of interest is fused to the receptor domain of eDHFR (18 kD), which specifically binds TMP conjugates through noncovalent interactions. The TMP-tag can be labeled with cell-permeable TMP substrates with no apparent cross-reactivity or toxicity in mammalian cells. Recently, a covalent version of the TMP-tag (A-TMP-tag) was obtained by a proximity-induced reaction of an acrylamide electrophile installed on the TMP substrate and a mutated Cys residue (L28C) positioned just outside of the eDHFR binding pocket (Gallagher et al. 2009). The A-TMP-tag provides a more enduring modification of the tagged proteins for single-molecule tracking, pulse-chase labeling and other applications. The A-TMP-tag offers the ability to choose between a covalent or noncovalent interaction with the probe based solely on choosing a TMP derivative with or without the acrylamide electrophile, allowing the user increased flexibility for removing a label. The TMP-tag has been shown to specifically label intracellular proteins in living with a variety of small-molecule fluorophores, quantum dots, and lanthanide chelates for several applications, including protein-protein interactions (Rajapakse et al. 2010), two-photon imaging (Gallagher et al. 2010) and superresolution microscopy (Wombacher et al. 2010). Thus far, labeling of cell surface proteins has not been demonstrated with TMP-tag or A-TMP-tag.

12.2.2 Enzyme-mediated labeling

In addition to the fusion protein and the chemical probe required for self-labeling tags, enzyme-mediated labeling strategies require the addition of a third component, an enzyme, which attaches the chemical probe to the fusion protein. Limitations of the enzymatic approaches include the added level of complexity, narrow range of chemical probes due to enzyme-substrate specificity and, often times, restriction to cell surface proteins. Intracellular labeling is difficult because the chemical probe and enzyme both need to gain entry to the cell for labeling to occur. Despite these limitations, enzyme-mediated labeling offers a key advantage for the study of target proteins that are sensitive to protein fusions. Enzyme-mediated labeling combines the high specificity of self-labeling proteins with the small size of peptide tags.

12.2.2.1 ACP- and MCP-tags

ACP-tag (8 kDa) and MCP-tag (8 kDa) are small-protein tags based on the ACP from *E. coli* (George et al. 2004; Yin et al. 2006). In contrast to the self-labeling approaches, the presence of an additional enzyme is required for the formation of a covalent link between the ACP-tag or MCP-tag and their substrates, which are derivatives of CoA) (Sunbul and Yin 2009). In the labeling reaction, the substituted 4-phosphopantetheininyl group of CoA, carrying either fluorescence or affinity labels, is covalently attached to a conserved serine residue of the ACP-tag by the PPTases AcpS from *E. coli* or Sfp from *Bacillus subtilis*. The MCP-tag contains two mutations (D36T and D39G) relative to the ACP-tag and can only be labeled by Sfp. In the presence of the appropriate PPTases, ACP-tag and MCP-tag allow sequential labeling of two different proteins in the same living cells. Because CoA conjugates are cell impermeable, this approach is limited to labeling proteins in cell lysates or on the cell surface. More recently, Yin and coworkers used phage-displayed libraries to identify S6 and A1, two 12-residue-length peptide tags that are efficient substrates for site-specific protein-labeling catalyzed by Sfp and AcpS, respectively (Zhou et al. 2007). These tags are suitable for labeling cell membrane proteins (e.g. ion channels) that may be sensitive to larger tags. In addition to live cell imaging of cell surface proteins (Kropf et al. 2008; Vivero-Pol et al. 2005), applications of Sfp and AcpS labeling include functional protein microarrays (Wong et al. 2008) and formation of bioactive hydrogels for tissue engineering (Mosiewicz et al. 2010).

12.2.2.2 Biotin and lipoic acid ligases

Enzyme-mediated site-specific modification of proteins has also been accomplished by Ting and coworkers using biotin and lipoic acid ligases. Biotin ligase (BirA) from *E. coli* recognizes a specific peptide sequence (15 amino acid residues) genetically fused to a protein of interest and catalyzes the covalent attachment of a biotin probe to this acceptor peptide (sometimes termed AP). Similarly to Sfp and AcpS, BirA is of bacterial origin, and therefore mammalian proteins tagged with the BirA recognition motif can be selectively biotinylated in the presence of mammalian biotin ligases (Chen et al. 2005; Howarth and Ting 2008). However, in contrast to Sfp and AcpS, which are highly promiscuous and can tolerate a wide range of biophysical probes appended to their natural substrate, BirA only accepts biotin or biotin isosteres (e.g. ketobiotin) substrates. Once transferred by BirA to the protein of interest, the functionalized biotin isostere can serve as a handle to introduce labels in a second step. For example the transferred ketobiotin can react with hydrazine or amino-oxy compounds carrying a fluorescent probe. More recently, Ting and coworkers discovered that yeast and *Pyrococcus horikoshii* biotin ligases can transfer alkyne and azide biotin analogues to tagged proteins, allowing bioorthogonal detection under physiological conditions using 1,3-dipolar cycloadditions (Slavoff et al. 2008). Recent work targeting quantum dots to surface proteins (Howarth et al. 2005) and detecting protein-protein interactions (Fernandez-Suarez et al. 2008; Thyagarajan and Ting 2010) has demonstrated the potential of this system. Nevertheless, the presence of other biotinylated proteins in both prokaryotic and eukaryotic cells and the fact that fluorescent probes cannot be transferred to the targeted protein in a single step currently limits the utility of this labeling technique, especially for live cell imaging of intracellular proteins.

Lipoic acid ligase (LplA) catalyzes the ATP-dependent covalent ligation of lipoic acid to specific proteins involved in oxidative metabolism. The LplA from *E. coli* was initially adapted by Ting and coworkers to specifically attach an alkyl azide onto an engineered LplA acceptor peptide (LAP) (Fernandez-Suarez et al. 2007). The azido group functions as a chemical handle and, following the labeling, can be selectively derivatized with cyclooctyne probes. LplA has been utilized in conjunction with BirA to simultaneously label two different protein receptors in the same cell. Further mutation of LplA permitted the direct transfer of a photo-cross-linker onto LAP-fused proteins. This system was then employed for site-specific mapping of protein-protein interactions (Baruah et al. 2008). Very recently, a new, mutant LplA capable of recognizing a cell-permeable substrate derived from the 7-hydroxycoumarin was identified. It has been shown that this mutant can be expressed in mammalian cells and catalyze the covalent conjugation of the fluorescent substrate to a transposable 13-amino-acid peptide (Jin et al. 2011; Uttamapinant et al. 2010). The method, termed PRIME (probe incorporation mediated by enzymes), has already been successfully utilized for the site-specific protein labeling within the cytosol and the nucleus. Compared to other protein-labeling methods, PRIME offers a unique combination of a small-peptide tag, high-labeling specificity and compatibility with the intracellular labeling of mammalian living cells; however, further reengineering of LplA is necessary to allow the enzymatic introduction of fluorophores other than the 7-hydroxycoumarin.

12.2.2.3 Sortase

The sortase-mediated transpeptidation, also called Sortagging, is another chemoenzymatic approach for site-specific labeling of proteins and peptides (Popp et al. 2007; Tanaka et al. 2008). The bacterial Sortase A (SrtA) enzyme catalyzes a transpeptidation reaction in which a synthetic N-terminal polyglycine peptide carrying a chemical label can be specifically installed onto proteins equipped with an LPXTG peptide motif. More recently, a dual-labeling system using two sortase enzymes with unique specificities from *Staphylococcus aureus* and *Streptococcus pyogenes* was exploited for site-specific labeling the two ends of a doubly tagged protein with two different chemical probes (Antos et al. 2009). SrtA tolerates a diverse range of functionalized polyglycine substrates, and there is no apparent limitation to the size of the reporter probe, thereby eliminating the need for a two-step labeling protocol. However, the Sortagging system is restricted to the labeling of proteins in vitro and on the surface of living cells. Recent applications include the PEGylation of recombinant cytokines (Popp et al. 2011) and the installation of lipids and fatty acids on an LPETG-tagged GFP protein (Antos et al. 2008).

12.2.3 Summary

All the methods presented in Sections 12.2.1 and 12.2.2 share the same starting point – genetically encoding a "tag" to the target protein to yield a fusion protein. The fusion protein is then specifically labeled with chemical probes providing a high degree of specificity and temporal control of labeling. Additionally, these techniques offer the advantage of using small-molecule probes, which can confer more diverse functionality than FPs. The exact method used depends on the target protein and downstream application. A summary of the techniques discussed, their substrates, recognition sequences and commercial availability are presented in ▶Table 12.1.

Table 12.1: Summary of labeling strategies

Method	Tag	Tag size (aa)	Substrate derivative	Enzyme used	Covalent attachment	Commercial availability
Self-labeling						
Tetraserine	SSPGSS	6–10	Bisboronic acid		Yes	
Tetracysteine	CCPGC	6–12	Biarsenical		Yes	Invitrogen (Lumio™)
SNAP-tag®	hAGT	182	Benzylguanine		Yes	New England Biolabs
CLIP-tag™	hAGT	182	Benzylcytosine		Yes	New England Biolabs
HaloTag®	DhaA	296	Chloroalkane		Yes	Promega
TMP-tag	eDHFR	157	TMP		No	Active Motif (LigandLink™)
A-TMP-tag	eDHFR:L28C	157	A-TMP		Yes	
Enzyme-mediated labeling						
ACP-tag	ACP	77	CoA	AcpS or Sfp	Yes	New England Biolabs
MCP-tag	ACP:D36T/D39G	77	CoA	Sfp	Yes	New England Biolabs
A1	GDSLDMLEWSLM	12	CoA	AcpS	Yes	
S6	GDSLSWLLRLLN	12	CoA	Sfp	Yes	
Biotin ligase	GLNIFEAQKIEH	15	Biotin, ketobiotin	BirA	Yes	
Lipoic acid ligase	GFEIDKVWYDLDA	12–17	Lipoic acid	LplA	Yes	
PRIME	GFEIDKVWYDLDA	13	7-hydroxycoumarin	LplA:W37V	Yes	
SorTag	LPXTG	6	Gly$_5$	Sortase A	Yes	

12.3 Selecting an appropriate labeling technique

No labeling method is perfect for every imagined experiment. Each method has its own advantages and disadvantages. A successful experiment matches the labeling technique to the research question and to the biological requirements of the experiment. ▶Table 12.2 evaluates each of the labeling methods discussed according to location of the target protein and application.

The nature of the target protein will partially dictate which labeling methods can be used. An important consideration is the *location of the target protein*. If the target protein is expressed intracellularly, be sure to pick a labeling technique appropriate for intracellular labeling. Also ensure the environment (oxidative, reducing) is appropriate for the technique. For example, the tetracysteine tag works best in a relatively reducing environment, such as the cytosol. What is the *expression level* of the target protein? Several labeling methods can label endogenous components to a small degree. For these methods, labeling of the target protein competes with labeling of the endogenous components; thus a higher level of expression of the target protein is needed to make the experiment successful. For target proteins with low expression, steer toward methods and chemical probes with low background labeling. Lastly, think about *sensitivity to heterologous fusion*. The smaller peptide tags are preferable for target proteins that are sensitive to heterologous fusion. Proteins that are sensitive to heterologous fusion but are abundantly expressed can take advantage of the tetracysteine tag. If a low level of background is paramount and a protein tag inhibits function, consider one of the enzyme-mediated strategies.

In addition to the target protein, consider the nature of the experiment. Which *cell line* will you use? There is variability in performance for each labeling technique across cell lines. A quick check of the literature should inform you as to which cell lines are commonly used with which methods. Most importantly, determine what *question you hope to answer* with your experiment. Do you need multicolor capability? Are you doing a pulse-chase experiment or superresolution microscopy? Do you need long-term labeling? Do you need to use multiple chemical probes with different functionality for several experiments? Be sure the labeling method you choose is compatible with the experiment you wish to conduct. Use ▶Table 12.2 to guide you in your selection of a labeling technique based on your experimental needs.

12.4 Live cell applications

The advent of small-molecule chemical probes and new labeling strategies suitable for use in live cells and animals directly impacts our ability to study the proteins in living systems. In the following subsections, we discuss recent applications of these labeling strategies. Specific examples are cited for SNAP-tag; however, other labeling methods may also be applicable. ▶Table 12.3 summarizes which labeling techniques are suitable for a given application.

12.4.1 Pulse-chase labeling

Studying protein trafficking, turnover, and internalization is greatly improved by the ability to differentially label protein cohorts synthesized at different times. The

12.4 Live cell applications

Table 12.2: Evaluation of labeling methods in mammalian cells. (-) indicates a method cannot be used; (+) indicates a publication using the method; (•) indicates the application is possible but there is not a publication; (++) indicates the method is superior to others in the column with (+).

	Cellsurface labeling	Intracellular labeling	Low background	Number of steps	Free choice of labels	Localization	Trafficking/ Internalization	Pulse Chase	Superresolution microscopy	Examples
Tetraserine	•	-	-	1	-	•	-	-	-	Halo et al. (2009)
Tetracysteine	+	+	-	1	+	+	+	++	-	Martin et al. (2005), Gaietta et al. (2002)
SNAP-tag	+	+	++	1	++	+	+	++	+	Klein et al. (2011), Gautier et al. (2008), Dellagiacoma et al. (2010), Milenkovic et al. (2009)
CLIP-tag	+	+	++	1	++	+	•	++	•	Gautier et al. (2008)
HaloTag	+	+	++	1	++	+	+	++	+	Kardon et al. (2009), Yamaguchi et al. (2009), Schroder et al. (2009)
TMP-tag	-	+	++	1	++	+	-	-	+	Calloway et al. (2007), Wombacher et al. (2010)
A-TMP-tag	-	+	+	1	+	+	•	•	•	Gallagher et al. (2009)
ACP-tag	+	-	++	1	++	+	+	++	•	George et al. (2004), Vivero-Pol et al. (2005), Kropf et al. (2008)
Biotin ligase	+	-	+	1–2[a]	+	+	+	+[b]	•	Thyagarajan and Ting (2010), Howarth et al. (2005)

(Continued)

Table 12.2: Evaluation of labeling methods in mammalian cells. (-) indicates a method cannot be used; (+) indicates a publication using the method; (•) indicates the application is possible but there is not a publication; (++) indicates the method is superior to others in the column with (+). *(Continued)*

	Cellsurface labeling	Intracellular labeling	Low background	Number of steps	Free choice of labels	Localization	Trafficking/ Internalization	Pulse Chase	Superresolution microscopy	Examples
Lipoic acid ligase	+	+	++	2	+	+	-	-	•	Fernandez-Suarez (2007)
PRIME	+	+	++	1	-	+	•	-	-	Uttamapinant et al. (2010)
SorTag	+	-	++	1	++	+	•	•	•	Popp et al. (2007)

[a] Number of steps depends on label used.
[b] Pulse-chase experiments were done with one-step labeling.

Table 12.3: Typical conditions for specific labeling with chemical probes

	Enzyme concentration	Label concentration	Time	Reference
Self-labeling methods				
Tetracysteine	N/A	1–10 µM	30–90 min	Invitrogen website
SNAP-tag	N/A	1–5 µM	30 min	NEB website
CLIP-tag	N/A	1–5 µM	30–60 min	NEB website
HaloTag	N/A	1–10 µM	15 min	Promega website
TMP-tag	N/A	1–2 µM	10–60 min	Calloway et al. (2007)
A-TMP-tag	N/A	1 µM	10 min	Gallagher et al. (2009)
Enzyme-mediated labeling methods				
Sfp	2 µM	1 µM	20 min	Zhou et al. (2007)
AcpS	2 µM	1 µM	30 min	Zhou et al. (2007)
Biotin ligase	0.3 µM	10 µM	15–60 min	Howarth and Ting (2008)
Lipoic acid ligase	10 µM	350 µM	60 min	Fernandez-Suarez et al. (2007)
Sortase	200 µM	500 µM	10–30 min	Popp et al. (2007)

labeling technologies discussed in this chapter allow the user temporal control over labeling, whereas genetically encoded labels, such as FPs, do not. The added temporal control enables direct assessment of the fate of nascent proteins and the determination of their turnover rates. Pulse-chase and quench-pulse-chase labeling allow for selective labeling of mature protein fusions or newly synthesized protein fusions, respectively. In a pulse-chase experiment, cells are exposed to a chemical probe long enough to ensure complete labeling of the fusion protein. Excess probe is removed by washing, which serves the dual purpose of decreasing background in subsequent steps and ensuring that any newly synthesized protein fusion is not labeled. The fate of the labeled species can then be followed without competing signals from nascent protein expression. To study only newly synthesized protein, a block, or quench step, is first applied. Here, a nonfluorescent substrate is added to the media so that any already synthesized fusion protein reacts with the nonfluorescent substrate. After washing away excess nonfluorescent substrate, a fluorescent substrate is added to the media. Since any "old" fusion protein is previously blocked with the nonfluorescent substrate, only new fusion protein will be labeled. Researchers have used the temporal control to investigate such diverse topics as the fate of mature and nascent proteins involved in centromere assembly during mitosis (Jansen et al. 2007) or trafficking of Na,K-ATPases (Alves et al. 2010; Farr et al. 2009).

12.4.2 Superresolution microscopy

The diffraction limit of the visible spectrum prevents nanoscale resolution of cellular structures or dynamic movements with a conventional light microscope. Recently, many research groups have developed superresolution microscopy techniques (e.g. photoactivated localization microscopy [PALM], stochastic optical reconstruction microscopy [STORM], and stimulated emission depletion microscopy [STED]) that allow for nanometer resolution. The success of these techniques is highly linked to the suitability of the fluorophore label (Fernandez-Suarez and Ting 2008). Photoactivatable or photoswitchable probes are able to undergo photoinduced conversions between dark (nonfluorescing) and light (fluorescing) states. Advancing beyond the diffraction limit requires either limiting the area of active fluorophores at a given point in time (STED) or limiting the number of fluorophores active over a larger area (PALM, STORM). The point spread function (PSF) of the fluorophores is fit to a mathematical model (such as a Gaussian distribution), and the fit is used to calculate the center of the emission. The precision with which the PSF can be fit depends on the photostability and brightness of the fluorophore. Small-molecule probes can be custom designed to meet the biophysical requirements for these superresolution techniques and are able to emit over 10-fold more photons than photoswitchable FPs, thereby increasing the resolution of the images. Small-molecule probes have been successfully used in combination with tagging techniques to image cytoskeletal and cell membrane proteins (STED) with 40 nm resolution (Hein et al. 2010), microtubules (STORM) with 25 nm resolution (Dellagiacoma et al. 2010) and histone H2B proteins (direct stochastic optical reconstruction microscopy [dSTORM]) with 20 nm resolution (Klein et al. 2011).

12.4.3 Protein-protein interactions

Protein-protein interactions are at the center of virtually all biological processes, including the formation of macrostructures, cell signaling, regulation and metabolic pathways. The identification of protein-protein interactions continues to be one of the main challenges in functional proteomics (Wells and McClendon 2007). Fluorescence resonance energy transfer (FRET) is one of the most exploited strategies for studying protein-protein interactions in living cells. FRET occurs by nonradiative transfer of energy between two chromophores in close proximity (1–10 nm) and good spectral overlap between fluorescence donors and acceptors. FRET measurements between FPs have been widely used to study the dynamics of proteins interactions in cells (Frommer et al. 2009). However, the low signal-to-noise ratio, broad absorption/emission spectra and variation in brightness of FPs limit the number of FRET partners that can be employed in FP-based FRET experiments. Furthermore, the analysis of interactions among protein receptors at the surface of living cells is complicated by background fluorescence due to the accumulation of FPs in intracellular compartments. To overcome these issues, researchers have developed time-resolved FRET approaches that combine the advantages of long lifetime lanthanoid cryptates with site-specifically labeling technologies, and they have employed these approaches to study the oligomeric state of several membrane proteins, including γ-aminobutyric (GABA) (Maurel et al. 2008) and muscarinic (Alvarez-Curto et al. 2010) receptors.

A different approach for the investigation of protein interactions in living cells and in cell extracts takes advantage of the ability to label fusion proteins with cross-linking probes. In this approach, interacting proteins are genetically fused to orthogonal tags, such as SNAP-tag or CLIP-tag, and covalently cross-linked using selective homobifunctional or heterobifunctional substrates. After cross-linking, the trapped protein complexes are analyzed by Western blot or in-gel fluorescence scanning. In contrast to chemical or light-induced cross-linking methods, which suffer from poor selectivity, tag-based approaches can be used to specifically detect interactions of proteins expressed at levels as low as 1 pmol/mg of total protein. Moreover, tag-based approaches allow the detection of weak or transient interactions that can be difficult to detect by conventional cross-linking approaches (Gautier et al. 2009; Lemercier et al. 2007).

12.4.4 Other Applications

In addition to the aforementioned applications, site-specific labeling techniques can be used as biochemical tools to sense the intracellular concentration of cell metabolites with spatial and temporal resolution. Use of a fusion tag enables precise targeting of molecular biosensors to specific subcellular compartments in living systems. Recent examples include fluorescent sensors for intracellular Zn^{2+} (Tomat et al. 2008), Ca^{2+} (Bannwarth et al. 2009; Kamiya and Johnsson 2010), O_2 (Geissbuehler et al. 2010) and H_2O_2 (Srikun et al. 2010).

Site-specific labeling systems have also been successfully employed to study protein function using chromophore-assisted laser inactivation (CALI). CALI is a powerful tool for functional analysis of a protein of interest in live cells (Jacobson et al. 2008). In contrast to gene knockout or RNA interference, which suffer from genetic compensation and low time resolution, CALI allows rapid inactivation of tagged proteins in living cells with high spatial and temporal control. Successful CALI experiments depend on generation of a reactive oxygen species (ROS) and subsequent contact between the ROS and POI. Genetically encoded photosensitizers, such as eGFP and Killerred, are inefficient primarily due to the low quantum yield of ROS. Tag-based CALI approaches are advantageous because the photosensitizer can be designed specifically to generate ROS in high quantum yield, thereby improving the efficiency of protein depletion and allowing the role of the target protein in a given cellular pathway to be assessed (Keppler and Ellenberg 2009).

12.5 Technical protocols for SNAP-tag labeling

The experimental protocols and technical notes for SNAP-tag labeling are based on the instructions for cellular labeling provided by New England Biolabs (http://www.neb.com). For cell culture and transfection methods, refer to established protocols. A video-article illustrating the fluorescent labeling of COS-7 expressing SNAP-tag fusion proteins was recently published in the *Journal of Visualized Experiments* (Provost and Sun 2010).

12.5.1 Standard protocol for labeling in living cells

12.5.1.1 Expression of SNAP-tag fusion protein constructs:

1. Clone the gene of interest upstream or downstream of the SNAP-tag coding sequence, making a fusion to the N or C terminus of the SNAP-tag, respectively. Plasmids for stable or transient expression of SNAP-tag protein fusions in mammalian cells and SNAP-tag fluorescent substrates are available from New England Biolabs.
2. Seed cells of choice in culture chambers and incubate in appropriate medium (e.g. complete F-12K medium) at 37°C, 5% CO_2 overnight (16–24 hours). Cell density should appear approximately 50%–60% confluent.
3. Transfect cells in culture with the SNAP-tag plasmid using standard transfection methods (NEB's TransPass D2 in combination with TransPass V or Roche's FuGENE® 6 have been used for both transient and stable transfections). For stable expression, begin selecting mammalian cultures in 600–1,200 µg/mL G418 (geneticin) 24–48 hours after transfection. The optimal selection conditions will vary with the cell line, thus establishing a kill curve for each cell line to determine optimal selection conditions is recommended. After 8–12 days of continuous selection, stable colonies will become visible.
4. Once stable colonies are visible, incubate cells another 16–24 hours at 37°C, 5% CO_2 to allow expression of the fusion protein. The appropriate time for adequate protein expression should be empirically determined.

12.5.1.2 Labeling reaction of SNAP-tag fusion proteins:

1. Dissolve one 50 nmol vial of SNAP-tag substrate (or one 30 nmol vial of SNAP-Cell TMR-Star) in 50 µL of DMSO to make a 1 mM stock solution of SNAP-tag substrate (or 0.6 mM for SNAP-Cell TMR-Star). Vortex 1–5 minutes to ensure full dissolution of the substrate. Immediately before use, prepare labeling medium by diluting the

SNAP-tag substrate stock solution 1:200 in complete medium. Mix thoroughly by pipetting. The final concentration of the substrate is 5 µM (or 3 µM for SNAP-Cell TMR-Star).
2. Carefully remove growth medium from transfected cells by vacuum suction. Wash cells twice with complete F-12K and add the labeling medium. Incubate cells at 37pC, 5% CO_2 for 30 minutes.
3. While cells are incubating, make a dilute Hoechst 33342 solution for nuclear staining by adding 1 µL of 16.2 mM solution Hoechst 33342 to 10 mL of complete F-12K.
4. Remove the labeling medium and add Hoechst solution. Incubate cells at 37°C, 5% CO_2 for 2–3 minutes.
5. Wash cells three times with complete F-12K.
6. *For cell surface labels,* proceed to imaging. *For intracellular labels* incubate at 37°C, 5% CO_2 for 30 minutes to allow any unincorporated fluorophore to diffuse out of the cells. Replace the medium and proceed to imaging.
7. Image cells using a fluorescent microscope with an appropriate filter set.

12.5.2 Technical notes

12.5.2.1 Cloning and expression

The SNAP-tag vector has cleavage sites for easy C- or N- terminal cloning. The linker between the protein of interest and SNAP-tag should be limited to ~10 residues to reduce nonspecific protease cleavage of long, unstructured peptides. Labeled SNAP-tag protein has been detected in mammalian cells over 48 hours indicating stability of the signal over this period. The length of time over which the signal from labeled SNAP-tag fusion protein can be detected depends upon the stability and turnover of the fusion protein, which, in turn, depends upon the protein of interest.

12.5.2.2 Substrates

Labeling concentrations of ≤5 µM are recommended for cellular labeling. However, no significant toxicity has been noted in proliferation or viability assays over 2 hours when substrate concentrations of 20 µM are used. Excess substrate is washed away from cells after the incubation. Nonpermeable substrates can be used for labeling SNAP-tag fusion proteins specifically on the surface of living cells. Unlike FP approaches, only mature proteins localized on the cell surface – and not proteins in the secretory or recycling pathways – will be labeled. Intracellular labeling of SNAP-tag fusion proteins can be accomplished using cell-permeable SNAP-tag substrates. Cell-permeable SNAP-tag substrates are available across the visible spectrum. Currently the far-red and near-infrared emitting SNAP-tag substrates are impermeable to mammalian cells. The labeling of intracellular protein targets with cell-impermeable substrates has been achieved via microinjection of the substrates into cells (Keppler et al. 2006) or after permeabilization of the cell's plasma membrane using fixation procedures. More recently, a cell-loading technique based on temporary membrane disruption by mechanical friction with glass beads was utilized to create direct access of impermeant SNAP-tag substrates to the cell's cytoplasm (Maurel et al. 2010). A protocol for coinjection of the capped mRNA of different SNAP-tagged proteins and SNAP-tag nonpermeable substrates has been described for live imaging and cell tracking of Zebrafish embryos (Campos et al. 2011).

12.5.2.3 Fixation

Labeling of SNAP-tag fusion proteins can be carried out prior to or after fixation. In contrast to some FP variants, labeled SNAP-tag fusion proteins can be fixed with standard fixation methods, such as formaldehyde or methanol, without significant loss of signal. After fixation, SNAP-tag fusion proteins can also be processed for immunofluorescence using anti-SNAP antibodies. Cells expressing SNAP-tag can be fixed prior to labeling; however, the fixative used and washing steps may need to be optimized to attain good results.

12.5.2.4 Blocking

The labeling of nontransfected or mock-transfected cells is usually sufficient for a negative control. In those cases where it is desirable to generate a further negative control by blocking the SNAP-tag activity in cells expressing SNAP-tag fusion protein, a nonfluorescent SNAP-tag substrate SNAP-Cell Block (bromothenylpteridine, BTP) can be used. SNAP-Cell Block can also be used in pulse-chase experiments to block the SNAP-tag reactivity during the chase between two pulse-labeling steps (Jansen et al. 2007) or in quench-pulse-chase strategies to saturate all previously synthesized SNAP-tag proteins and permit the fluorescent labeling of only newly synthesized protein pools (Farr et al. 2009). In addition to the cell-permeable SNAP-Cell Block, a nonpermeable nonfluorescent substrate SNAP-Surface Block is available for selectively blocking cell surface SNAP-tag fusion proteins.

12.5.2.5 Troubleshooting

Optimal substrate concentrations and reaction times range from 1 to 10 µM and 15 to 60 minutes, respectively, depending on the experimental conditions and expression levels of the SNAP-tag fusion protein. Best results are usually obtained at concentrations between 1 and 5 µM substrate and 30 minutes of reaction time. Increasing substrate concentration and reaction time usually results in higher background and does not necessarily increase the signal to background ratio. Nonspecific background can be reduced by the presence of 10% fetal calf serum or 1% BSA in the labeling solution. If the labeled fusion protein cannot be detected, it is most likely due to lack of expression. It is advisable to verify the transfection method to confirm that the cells contain the fusion gene of interest. If this is confirmed, check for expression of the SNAP-tag fusion protein. Ideally, this is performed with an antibody for the protein of interest. If no antibody against the fusion partner exists, an anti-SNAP-tag antibody is commercially available. Alternatively, a fluorescent substrate, such as SNAP-Vista Green, can be used to confirm the presence of SNAP-tag fusion in cell extracts following SDS-PAGE, without the need for Western blotting.

If labeling signal is weak, it may be due to insufficient exposure of the fusion protein to the substrate. Try increasing the concentration of SNAP-tag substrate and/or the incubation time. Another possibility is that the protein may be poorly expressed or rapidly turned over. The turnover rates of the SNAP-tag fusion protein will vary depending on the fusion partner (half-life values may range from a few minutes to several hours). If the fluorescence signal decreases rapidly, it may be due to instability of the fusion protein. When the protein turnover is fast or the fusion protein has limited stability, analyzing the cells under the microscope immediately after labeling is recommended. In some cases, the signal may be stabilized by fixing the cells. To visualize proteins with fast turnover rates or low thermal stability, SNAP-tag fusion proteins can be labeled at lower temperatures (for instance at 4°C or 16°C). Photobleaching may be reduced by using commercially available anti-fade reagents.

References

Adams, S.R., and Tsien, R.Y. (2008). Preparation of the membrane-permeant biarsenicals FlAsH-EDT2 and ReAsH-EDT2 for fluorescent labeling of tetracysteine-tagged proteins. Nat Protoc *3*, 1527–1534.

Alvarez-Curto, E., Ward, R.J., Pediani, J.D., and Milligan, G. (2010). Ligand regulation of the quaternary organization of cell surface M3 muscarinic acetylcholine receptors analyzed by fluorescence resonance energy transfer (FRET) imaging and homogeneous time-resolved FRET. J Biol Chem *285*, 23318–23330.

Alves, D.S., Farr, G.A., Seo-Mayer, P., and Caplan, M.J. (2010). AS160 associates with the Na+,K+-ATPase and mediates the adenosine monophosphate-stimulated protein kinase-dependent regulation of sodium pump surface expression. Mol Biol Cell *21*, 4400–4408.

Antos, J.M., Chew, G.L., Guimaraes, C.P., Yoder, N.C., Grotenbreg, G.M., Popp, M.W., et al. (2009). Site-specific N- and C-terminal labeling of a single polypeptide using sortases of different specificity. J Am Chem Soc *131*, 10800–10801.

Antos, J.M., Miller, G.M., Grotenbreg, G.M., and Ploegh, H.L. (2008). Lipid modification of proteins through sortase-catalyzed transpeptidation. J Am Chem Soc *130*, 16338–16343.

Bannwarth, M., Correa, I.R., Sztretye, M., Pouvreau, S., Fellay, C., Aebischer, A., et al. (2009). Indo-1 derivatives for local calcium sensing. ACS Chem Biol *4*, 179–190.

Baruah, H., Puthenveetil, S., Choi, Y.A., Shah, S., and Ting, A.Y. (2008). An engineered aryl azide ligase for site-specific mapping of protein-protein interactions through photo-cross-linking. Angew Chem Int Ed Engl *47*, 7018–7021.

Brun, M.A., Tan, K.T., Nakata, E., Hinner, M.J., and Johnsson, K. (2009). Semisynthetic fluorescent sensor proteins based on self-labeling protein tags. J Am Chem Soc *131*, 5873–5884.

Calloway, N.T., Choob, M., Sanz, A., Sheetz, M.P., Miller, L.W., and Cornish, V.W. (2007). Optimized fluorescent trimethoprim derivatives for in vivo protein labeling. Chembiochem *8*, 767–774.

Campbell, R.E., Tour, O., Palmer, A.E., Steinbach, P.A., Baird, G.S., Zacharias, D.A., et al. (2002). A monomeric red fluorescent protein. Proc Natl Acad Sci U S A *99*, 7877–7882.

Campos, C., Kamiya, M., Banala, S., Johnsson, K., and Gonzalez-Gaitan, M. (2011). Labelling cell structures and tracking cell lineage in zebrafish using SNAP-tag. Dev Dyn *240*, 820–827.

Chen, I., Howarth, M., Lin, W., and Ting, A.Y. (2005). Site-specific labeling of cell surface proteins with biophysical probes using biotin ligase. Nat Methods *2*, 99–104.

Davidson, M.W., and Campbell, R.E. (2009). Engineered fluorescent proteins: innovations and applications. Nat Methods *6*, 713–717.

Day, R.N., and Davidson, M.W. (2009). The fluorescent protein palette: tools for cellular imaging. Chem Soc Rev *38*, 2887–2921.

Dellagiacoma, C., Lukinavicius, G., Bocchio, N., Banala, S., Geissbuhler, S., Marki, I., et al. (2010). Targeted photoswitchable probe for nanoscopy of biological structures. ChemBioChem *11*, 1361–1363.

Doumazane, E., Scholler, P., Zwier, J.M., Eric, T., Rondard, P., and Pin, J.P. (2011). A new approach to analyze cell surface protein complexes reveals specific heterodimeric metabotropic glutamate receptors. FASEB J *25*, 66–77.

References

Engin, S., Trouillet, V., Franz, C.M., Welle, A., Bruns, M., and Wedlich, D. (2010). Benzylguanine thiol self-assembled monolayers for the immobilization of SNAP-tag proteins on microcontact-printed surface structures. Langmuir 26, 6097–6101.

Farr, G.A., Hull, M., Mellman, I., and Caplan, M.J. (2009). Membrane proteins follow multiple pathways to the basolateral cell surface in polarized epithelial cells. J Cell Biol 186, 269–282.

Fernandez-Suarez, M., Baruah, H., Martinez-Hernandez, L., Xie, K.T., Baskin, J.M., Bertozzi, C.R., et al. (2007). Redirecting lipoic acid ligase for cell surface protein labeling with small-molecule probes. Nat Biotechnol 25, 1483–1487.

Fernandez-Suarez, M., Chen, T.S., and Ting, A.Y. (2008). Protein-protein interaction detection in vitro and in cells by proximity biotinylation. J Am Chem Soc 130, 9251–9253.

Fernandez-Suarez, M., and Ting, A.Y. (2008). Fluorescent probes for super-resolution imaging in living cells. Nat Rev Mol Cell Biol 9, 929–943.

Frommer, W.B., Davidson, M.W., and Campbell, R.E. (2009). Genetically encoded biosensors based on engineered fluorescent proteins. Chem Soc Rev 38, 2833–2841.

Gaietta, G., Deerinck, T.J., Adams, S.R., Bouwer, J., Tour, O., Laird, D.W., Sosinsky, G.E., Tsien, R.Y., and Ellisman, M.H. (2002). Multicolor and electron microscopic imaging of connexin trafficking. Science 296, 503–507.

Gallagher, S.S., Jing, C., Peterka, D.S., Konate, M., Wombacher, R., Kaufman, L.J., et al. (2010). A trimethoprim-based chemical tag for live cell two-photon imaging. ChemBioChem 11, 782–784.

Gallagher, S.S., Sable, J.E., Sheetz, M.P., and Cornish, V.W. (2009). An in vivo covalent TMP-tag based on proximity-induced reactivity. ACS Chem Biol 4, 547–556.

Gautier, A., Juillerat, A., Heinis, C., Correa, I.R., Jr., Kindermann, M., Beaufils, F., and Johnsson, K. (2008). An engineered protein tag for multiprotein labeling in living cells. Chem Biol 15, 128–136.

Gautier, A., Nakata, E., Lukinavicius, G., Tan, K.T., and Johnsson, K. (2009). Selective cross-linking of interacting proteins using self-labeling tags. J Am Chem Soc 131, 17954–17962.

Geissbuehler, M., Spielmann, T., Formey, A., Marki, I., Leutenegger, M., Hinz, B., et al. (2010). Triplet imaging of oxygen consumption during the contraction of a single smooth muscle cell (A7r5). Biophys J 98, 339–349.

George, N., Pick, H., Vogel, H., Johnsson, N., and Johnsson, K. (2004). Specific labeling of cell surface proteins with chemically diverse compounds. J Am Chem Soc 126, 8896–8897.

Ghosh, I., Considine, N., Maunus, E., Sun, L., Zhang, A., Buswell, J., et al. (2011). Site-specific protein labeling by intein-mediated protein ligation. Methods Mol Biol 705, 87–107.

Giepmans, B.N., Adams, S.R., Ellisman, M.H., and Tsien, R.Y. (2006). The fluorescent toolbox for assessing protein location and function. Science 312, 217–224.

Griffin, B.A., Adams, S.R., and Tsien, R.Y. (1998). Specific covalent labeling of recombinant protein molecules inside live cells. Science 281, 269–272.

Halo, T.L., Appelbaum, J., Hobert, E.M., Balkin, D.M., and Schepartz, A. (2009). Selective recognition of protein tetraserine motifs with a cell-permeable, pro-fluorescent bis-boronic acid. J Am Chem Soc 131, 438–439.

Hein, B., Willig, K.I., Wurm, C.A., Westphal, V., Jakobs, S., and Hell, S.W. (2010). Stimulated emission depletion nanoscopy of living cells using SNAP-tag fusion proteins. Biophys J 98, 158–163.

Hinner, M.J., and Johnsson, K. (2010). How to obtain labeled proteins and what to do with them. Curr Opin Biotechnol 21, 766–776.

Hoffmann, C., Gaietta, G., Bunemann, M., Adams, S.R., Oberdorff-Maass, S., Behr, B., et al. (2005). A FlAsH-based FRET approach to determine G protein-coupled receptor activation in living cells. Nat Methods 2, 171–176.

Hoffmann, C., Gaietta, G., Zurn, A., Adams, S.R., Terrillon, S., Ellisman, M.H., et al. (2010). Fluorescent labeling of tetracysteine-tagged proteins in intact cells. Nat Protoc *5*, 1666–1677.

Howarth, M., Takao, K., Hayashi, Y., and Ting, A.Y. (2005). Targeting quantum dots to surface proteins in living cells with biotin ligase. Proc Natl Acad Sci U S A *102*, 7583–7588.

Howarth, M., and Ting, A.Y. (2008). Imaging proteins in live mammalian cells with biotin ligase and monovalent streptavidin. Nat Protoc *3*, 534–545.

Iversen, L., Cherouati, N., Berthing, T., Stamou, D., and Martinez, K.L. (2008). Templated protein assembly on micro-contact-printed surface patterns. Use of the SNAP-tag protein functionality. Langmuir *24*, 6375–6381.

Jacobson, K., Rajfur, Z., Vitriol, E., and Hahn, K. (2008). Chromophore-assisted laser inactivation in cell biology. Trends Cell Biol *18*, 443–450.

Jansen, L.E., Black, B.E., Foltz, D.R., and Cleveland, D.W. (2007). Propagation of centromeric chromatin requires exit from mitosis. J Cell Biol *176*, 795–805.

Jin, X., Uttamapinant, C., and Ting, A.Y. (2011). Synthesis of 7-aminocoumarin by Buchwald-Hartwig cross coupling for specific protein labeling in living cells. ChemBioChem *12*, 65–70.

Johnsson, N., and Johnsson, K. (2007). Chemical tools for biomolecular imaging. ACS Chem Biol *2*, 31–38.

Kamiya, M., and Johnsson, K. (2010). Localizable and highly sensitive calcium indicator based on a BODIPY fluorophore. Anal Chem *82*, 6472–6479.

Kardon, J.R., Reck-Peterson, S.L., and Vale, R.D. (2009). Regulation of the processivity and intracellular localization of Saccharomyces cerevisiae dynein by dynactin. Proc Natl Acad Sci USA *106*, 5669–5674.

Keppler, A., Arrivoli, C., Sironi, L., and Ellenberg, J. (2006). Fluorophores for live cell imaging of AGT fusion proteins across the visible spectrum. Biotechniques *41*, 167–170, 172, 174–165.

Keppler, A., and Ellenberg, J. (2009). Chromophore-assisted laser inactivation of alpha- and gamma-tubulin SNAP-tag fusion proteins inside living cells. ACS Chem Biol *4*, 127–138.

Keppler, A., Gendreizig, S., Gronemeyer, T., Pick, H., Vogel, H., and Johnsson, K. (2003). A general method for the covalent labeling of fusion proteins with small molecules in vivo. Nat Biotechnol *21*, 86–89.

Keppler, A., Pick, H., Arrivoli, C., Vogel, H., and Johnsson, K. (2004). Labeling of fusion proteins with synthetic fluorophores in live cells. Proc Natl Acad Sci U S A *101*, 9955–9959.

Kindermann, M., George, N., Johnsson, N., and Johnsson, K. (2003). Covalent and selective immobilization of fusion proteins. J Am Chem Soc *125*, 7810–7811.

Kindermann, M., Sielaff, I., and Johnsson, K. (2004). Synthesis and characterization of bifunctional probes for the specific labeling of fusion proteins. Bioorg Med Chem Lett *14*, 2725–2728.

Klein, T., Loschberger, A., Proppert, S., Wolter, S., van de Linde, S., and Sauer, M. (2011). Live-cell dSTORM with SNAP-tag fusion proteins. Nat Methods *8*, 7–9.

Kosaka, N., Ogawa, M., Choyke, P.L., Karassina, N., Corona, C., McDougall, M., et al. (2009). In vivo stable tumor-specific painting in various colors using dehalogenase-based protein-tag fluorescent ligands. Bioconjug Chem *20*, 1367–1374.

Kropf, M., Rey, G., Glauser, L., Kulangara, K., Johnsson, K., and Hirling, H. (2008). Subunit-specific surface mobility of differentially labeled AMPA receptor subunits. Eur J Cell Biol *87*, 763–778.

Lee, H.L., Lord, S.J., Iwanaga, S., Zhan, K., Xie, H., Williams, J.C., et al. (2010). Superresolution imaging of targeted proteins in fixed and living cells using photoactivatable organic fluorophores. J Am Chem Soc *132*, 15099–15101.

Lemercier, G., Gendreizig, S., Kindermann, M., and Johnsson, K. (2007). Inducing and sensing protein–protein interactions in living cells by selective cross-linking. Angew Chem Int Ed Engl *46*, 4281–4284.

Liu, C.C., and Schultz, P.G. (2010). Adding new chemistries to the genetic code. Annu Rev Biochem *79*, 413–444.

Los, G.V., Encell, L.P., McDougall, M.G., Hartzell, D.D., Karassina, N., Zimprich, C., et al. (2008). HaloTag: a novel protein labeling technology for cell imaging and protein analysis. ACS Chem Biol *3*, 373–382.

Luedtke, N.W., Dexter, R.J., Fried, D.B., and Schepartz, A. (2007). Surveying polypeptide and protein domain conformation and association with FlAsH and ReAsH. Nat Chem Biol *3*, 779–784.

Marks, K.M., and Nolan, G.P. (2006). Chemical labeling strategies for cell biology. Nat Methods *3*, 591–596.

Martin, B.R., Giepmans, B.N., Adams, S.R., and Tsien, R.Y. (2005). Mammalian cell-based optimization of the biarsenical-binding tetracysteine motif for improved fluorescence and affinity. Nat Biotechnol *23*, 1308–1314.

Maurel, D., Banala, S., Laroche, T., and Johnsson, K. (2010). Photoactivatable and photoconvertible fluorescent probes for protein labeling. ACS Chem Biol *5*, 507–516.

Maurel, D., Comps-Agrar, L., Brock, C., Rives, M.L., Bourrier, E., Ayoub, M.A., et al. (2008). Cell-surface protein-protein interaction analysis with time-resolved FRET and snap-tag technologies: application to GPCR oligomerization. Nat Methods *5*, 561–567.

Milenkovic, L., Scott, M.P., and Rohatgi, R. (2009). Lateral transport of Smoothened from the plasma membrane to the membrane of the cilium. J Cell Biol *187*, 365–374.

Miller, L.W., Cai, Y., Sheetz, M.P., and Cornish, V.W. (2005). In vivo protein labeling with trimethoprim conjugates: a flexible chemical tag. Nat Methods *2*, 255–257.

Mosiewicz, K.A., Johnsson, K., and Lutolf, M.P. (2010). Phosphopantetheinyl transferase-catalyzed formation of bioactive hydrogels for tissue engineering. J Am Chem Soc *132*, 5972–5974.

Ohana, R.F., Hurst, R., Vidugiriene, J., Slater, M.R., Wood, K.V., and Urh, M. (2011). HaloTag-based purification of functional human kinases from mammalian cells. Protein Expr Purif *76*, 154–164.

O'Hare, H.M., Johnsson, K., and Gautier, A. (2007). Chemical probes shed light on protein function. Curr Opin Struct Biol *17*, 488–494.

Popp, M.W., Antos, J.M., Grotenbreg, G.M., Spooner, E., and Ploegh, H.L. (2007). Sortagging: a versatile method for protein labeling. Nat Chem Biol *3*, 707–708.

Popp, M.W., Dougan, S.K., Chuang, T.Y., Spooner, E., and Ploegh, H.L. (2011). Sortase-catalyzed transformations that improve the properties of cytokines. Proc Natl Acad Sci U S A *108*, 3169–3174.

Provost, C.R., and Sun, L. (2010). Fluorescent labeling of COS-7 expressing SNAP-tag fusion proteins for live cell imaging. J Vis Exp *39*, e1876.

Rajapakse, H.E., Gahlaut, N., Mohandessi, S., Yu, D., Turner, J.R., and Miller, L.W. (2010). Time-resolved luminescence resonance energy transfer imaging of protein-protein interactions in living cells. Proc Natl Acad Sci U S A *107*, 13582–13587.

Roberti, M.J., Bertoncini, C.W., Klement, R., Jares-Erijman, E.A., and Jovin, T.M. (2007). Fluorescence imaging of amyloid formation in living cells by a functional, tetracysteine-tagged alpha-synuclein. Nat Methods *4*, 345–351.

Rodriguez, A.J., Shenoy, S.M., Singer, R.H., and Condeelis, J. (2006). Visualization of mRNA translation in living cells. J Cell Biol *175*, 67–76.

Schroder, J., Benink, H., Dyba, M., and Los, G.V. (2009). In vivo labeling method using a genetic construct for nanoscale resolution microscopy. Biophys J *96*, L01–03.

Shaner, N.C., Steinbach, P.A., and Tsien, R.Y. (2005). A guide to choosing fluorescent proteins. Nat Methods 2, 905–909.

Slavoff, S.A., Chen, I., Choi, Y.A., and Ting, A.Y. (2008). Expanding the substrate tolerance of biotin ligase through exploration of enzymes from diverse species. J Am Chem Soc 130, 1160–1162.

Sletten, E.M., and Bertozzi, C.R. (2009). Bioorthogonal chemistry: fishing for selectivity in a sea of functionality. Angew Chem Int Ed Engl 48, 6974–6998.

Son, J., Lee, J.J., Lee, J.S., Schuller, A., and Chang, Y.T. (2010). Isozyme-specific fluorescent inhibitor of glutathione s-transferase omega 1. ACS Chem Biol 5, 449–453.

Srikun, D., Albers, A.E., Nam, C.I., Iavarone, A.T., and Chang, C.J. (2010). Organelle-targetable fluorescent probes for imaging hydrogen peroxide in living cells via SNAP-Tag protein labeling. J Am Chem Soc 132, 4455–4465.

Sunbul, M., and Yin, J. (2009). Site specific protein labeling by enzymatic posttranslational modification. Org Biomol Chem 7, 3361–3371.

Tanaka, T., Yamamoto, T., Tsukiji, S., and Nagamune, T. (2008). Site-specific protein modification on living cells catalyzed by Sortase. ChemBioChem 9, 802–807.

Thyagarajan, A., and Ting, A.Y. (2010). Imaging activity-dependent regulation of neurexin-neuroligin interactions using trans-synaptic enzymatic biotinylation. Cell 143, 456–469.

Tomat, E., Nolan, E.M., Jaworski, J., and Lippard, S.J. (2008). Organelle-specific zinc detection using zinpyr-labeled fusion proteins in live cells. J Am Chem Soc 130, 15776–15777.

Urh, M., Hartzell, D., Mendez, J., Klaubert, D.H., and Wood, K. (2008). Methods for detection of protein-protein and protein-DNA interactions using HaloTag. Methods Mol Biol 421, 191–209.

Uttamapinant, C., White, K.A., Baruah, H., Thompson, S., Fernandez-Suarez, M., Puthenveetil, S., et al. (2010). A fluorophore ligase for site-specific protein labeling inside living cells. Proc Natl Acad Sci U S A 107, 10914–10919.

Vivero-Pol, L., George, N., Krumm, H., Johnsson, K., and Johnsson, N. (2005). Multicolor imaging of cell surface proteins. J Am Chem Soc 127, 12770–12771.

Wells, J.A., and McClendon, C.L. (2007). Reaching for high-hanging fruit in drug discovery at protein-protein interfaces. Nature 450, 1001–1009.

Wombacher, R., Heidbreder, M., van de Linde, S., Sheetz, M.P., Heilemann, M., Cornish, V.W., et al. (2010). Live-cell super-resolution imaging with trimethoprim conjugates. Nat Methods 7, 717–719.

Wong, L.S., Thirlway, J., and Micklefield, J. (2008). Direct site-selective covalent protein immobilization catalyzed by a phosphopantetheinyl transferase. J Am Chem Soc 130, 12456–12464.

Yamaguchi, K., Inoue, S., Ohara, O., and Nagase, T. (2009). Pulse-chase experiment for the analysis of protein stability in cultured mammalian cells by covalent fluorescent labeling of fusion proteins. Methods Mol Biol 577, 121–131.

Yin, J., Lin, A.J., Golan, D.E., and Walsh, C.T. (2006). Site-specific protein labeling by Sfp phosphopantetheinyl transferase. Nat Protoc 1, 280–285.

Zhou, Z., Cironi, P., Lin, A.J., Xu, Y., Hrvatin, S., Golan, D.E., et al. (2007). Genetically encoded short peptide tags for orthogonal protein labeling by Sfp and AcpS phosphopantetheinyl transferases. ACS Chem Biol 2, 337–346.

Ziegler, N., Batz, J., Zabel, U., Lohse, M.J., and Hoffmann, C. (2011). FRET-based sensors for the human M1-, M3-, and M5-acetylcholine receptors. Bioorg Med Chem 19, 1048–1054.

13 Proteomics of human bronchoalveolar lavage fluid: discovery of biomarkers of chronic obstructive pulmonary disease (COPD) with difference gel electrophoresis (DIGE) and mass spectrometry (MS)

Francesco Giorgianni, Valentina Mileo, Li Chen, Dominic M. Desiderio, and Sarka Beranova-Giorgianni

13.1 Introduction

13.1.1 Proteomics and biomarker discovery

A proteome represents the complete repertoire of proteins present in a biological system at a particular time and under a particular set of conditions. Proteomics is a rapidly expanding area of scientific inquiry that focuses on examination of proteomes on a global scale. In that respect, proteomics is a component of systems biology, which centers on discovery-based characterization of the elements of a biological system in a comprehensive fashion (Smith and Figeys 2006). It is believed that through this comprehensive description, new, emergent properties of the system are revealed; these properties cannot be captured by targeted studies of individual components (such as single proteins).

A promising direction of proteomics research is toward discovery and development of novel biomarkers for diagnosis and staging of a disease, to monitor disease progression and evaluate therapeutic effects. Biomarker discovery with proteomics relies on comparative, quantitative examination of the proteomes in sets of control versus perturbed (i.e. diseased) systems to pinpoint a group of proteins whose abundances differ between the two sample sets. These differentially expressed proteins are expected to arise from disease-related perturbations of molecular networks, and hence to reflect the disease status of the organism. (Analogous concepts apply, e.g. for biomarkers intended to monitor response to therapy.) A related idea that has been formulated posits that a combination of multiple biomarkers more accurately represents a fingerprint of disease and thus provides diagnostic specificity not achievable with a single biomarker. Identification of such multiprotein fingerprints can best be achieved through global-scale proteomic examinations.

An important consideration in biomarker research is the selection of the biological system to study. The ultimate goal of these research efforts is translation of findings and their application to diagnostic tests in clinical settings. From this standpoint, it is desirable to focus on examination of serum, which is the biospecimen obtainable through least invasive means. On the other hand, the serum proteome reflects the systemic status of an organism, which results from exceedingly complex contributions of many different processes that occur throughout the body. Efforts to search and locate markers of a particular disease among all the serum proteins, the vast majority of which are unrelated to the disease phenotype, might suffer from "needle-in-the-haystack" limitations. One possible means to address these limitations is to interrogate in the discovery phase the proteomes in specimens procured directly from the diseased organ or from sites proximal to the diseased organ. Thus, differentially expressed proteins relevant to disease mechanisms will be easier to identify. The candidate biomarkers would be evaluated for diagnostic purposes in subsequent studies in the biomarker development pipeline.

In the context of lung diseases, the clinical specimens suitable for proteome interrogation include lung tissue, bronchoalveolar lavage fluid (BAL) (Plymoth et al. 2003; Wattiez and Falmagne 2005), induced sputum and exhaled breath condensate. BAL fluid is obtained with washout of the epithelial lining of the lung with a saline solution via a fiberoptic bronchoscope. The procedure to obtain BAL is invasive but generally well tolerated. Proteins in BAL might be of diverse origin: they might be released locally by epithelial and/or inflammatory cells, or they might be leaked from circulation. Overall, BAL fluid is considered to reflect the pulmonary protein expression in a more accurate way compared to sputum or exhaled breath condensate. Therefore, the BAL proteome is an important target for comparative profiling in biomarker research.

13.1.2 COPD

COPD is defined as a "preventable and treatable disease, whose pulmonary component is characterized by airflow limitation that is not fully reversible. The airflow limitation is usually progressive and associated with an abnormal inflammatory response of the lung to noxious particles or gases" (Rabe et al. 2007). COPD is currently the fifth leading cause of death worldwide and is projected by the World Health Organization to become the third leading cause of death by 2020.

The inflammatory reaction in COPD can be quite different, depending on the site of action. In the small airways, a cellular inflammatory response causes an increase in tissue and a consequent decrease in the luminal area for airflow; in the parenchyma, the inflammation leads to loss of alveolar tissue and to the development of emphysema. In addition to airflow limitation, COPD might include extrapulmonary effects that might contribute to the severity of the disease. It is recognized that COPD is, in fact, an umbrella term for a complex, multifactorial syndrome that encompasses multiple pathological elements that might be manifested with variable degree of severity (Barnes and Celli 2009; Punturieri et al. 2007). The most important symptoms of COPD are breathlessness, excessive sputum production and chronic cough that often impact the functional capacity of patients and their ability to perform normal activities. Overall, besides a serious risk of early death, COPD patients often experience a dramatic decline in their quality of life, with limited mobility, frequent hospitalizations and need for continuous medication and oxygen therapy.

The severity of COPD is classified based on spirometric testing of airflow limitation and ranges from mild COPD (Stage I) to very severe COPD (Stage IV). The management of COPD includes four components: (a) assessment and monitoring of the disease, (b) reduction of risk factors, (c) management of stable COPD and (d) management of disease exacerbations (Rabe et al. 2007). Current pharmacotherapy options only treat symptoms, and none of the existing medications reverses the gradual decline in lung function that is characteristic for COPD. There is a great need for novel biomarkers to assess susceptibility to COPD, for early and accurate evaluation of disease severity, for selection of optimum therapeutic regimens and for monitoring of response to therapy.

Current knowledge of the molecular underpinnings of COPD highlights the complex nature of the multiple, interconnected factors and processes: inflammation, protease/antiprotease imbalance, oxidative stress, apoptosis and others (MacNee 2005, 2007; Owen 2008; Zhan and Desiderio 2011). Each process contributes a number of alterations in specific proteins to produce the overall pathologic phenotype. In order to effectively capture the molecular perturbations associated with COPD, investigation of proteins in human BAL with proteomics methods constitutes a powerful approach to obtain new insights into the molecular mechanisms of COPD, and to identify accurate biomarkers for early detection and improved management of COPD (Barnes et al. 2006; Chen et al. 2010).

13.1.3 Proteomics technologies for COPD biomarker discovery

From the bioanalytical standpoint, analyses of proteomes in human biological fluids are a highly challenging undertaking. This challenge stems from the complexity of the proteomes: the large number of proteins, vastly different protein abundance levels, post-translational protein modifications and diverse physicochemical properties of different proteins (Mallick and Kuster 2010). Major advancements in three areas have enabled the expansion of proteome research. These advancements include separation technologies (electrophoresis, chromatography), MS and bioinformatics. The proteomics research community recognizes that there is no single technology to comprehensively probe complex proteomes. Instead, to meet the challenge of proteome analyses, diverse combinations of powerful analytical methods and technologies have been developed and applied in proteomics research. Traditionally, separation of highly complex protein mixtures has been accomplished with two-dimensional polyacrylamide gel electrophoresis (2D-PAGE) (Friedman et al. 2009; Gorg et al. 2009). In conjunction with MS/bioinformatics for downstream identification of proteins of interest (POI), 2D-PAGE has been the key technology in gel-based workflows used in many proteomics applications. A modified 2D-PAGE separation methodology that has been developed for multiplexed quantitative proteomics is termed DIGE (Minden et al. 2009; Timms and Cramer 2008).

13.1.4 DIGE

Two-dimensional gel electrophoresis is a protein separation technique that combines isoelectric focusing (IEF) and sodium dodecyl sulfate–polyacrylamide gel electrophoresis (SDS-PAGE). A number of major advances in 2D-PAGE have been made over the years (Friedman et al. 2009; Gorg et al. 2009). DIGE represents a significant improvement over conventional 2D-PAGE because it overcomes issues associated with gel-to-gel variability encountered in comparative proteomics. The critical aspect of DIGE is the incorporation of labeling with spectrally resolvable fluorescent dyes (Cy2, Cy3, and Cy5), which are used to derivatize proteins in different samples prior to 2D-PAGE. Labeling enables multiplexed analysis of different protein mixtures on the same gel, because the fluorescence signals of each individual sample can be recorded separately at excitation/emission wavelengths specific for each dye. A reference sample (pooled internal standard), which is typically included in experimental design, facilitates spot matching between gels and allows normalization of spot abundances for accurate quantification. Thus, in DIGE, gel-to-gel variability in spot patterns is minimized, and the number of gels needed per experiment is reduced. The introduction of DIGE in the 2D-PAGE proteomic workflow allows a more accurate determination of differentially expressed proteins, especially in samples characterized by large biological variability, such as BAL fluid proteomes.

Two different dye chemistries are available: CyDye Fluor saturation dyes and CyDye Fluor minimal dyes. The CyDye saturation dyes are used for special applications involving scarce samples. The CyDye minimal dyes, which are used for regular applications, have an N-hydroxysuccinimidyl-ester reactive group that covalently binds to lysine residue of a protein via formation of an amide linkage. The CyDye minimal dyes possess a single positive charge; when coupled to a protein, the charge carried by the CyDye replaces the single positive charge that lysine residues carry at neutral or acidic pH. In this way, the isoelectric point of the protein does not significantly change upon labeling. In the minimal labeling strategy, ca. 1%–2% of lysine residues are labeled; each labeled protein carries only a single dye label and is visualized as a single spot. The CyDye labeling gives DIGE the advantages of standardization, high accuracy, reproducibility and high sensitivity (down to 25 pg of a single protein). Moreover, a linear dynamic range for protein abundance change of up to five orders of magnitude (10^5) is provided.

The CyDye minimal dye strategy allows multiplexing of up to three separate samples on the same 2D gel. The preferred experimental design for DIGE includes a pooled internal standard, which is composed of equal amounts of all biological samples within the experiment. This pooled internal standard is labeled with one of the dyes, usually Cy2, and it is then coelectrophoresed on every single gel within the same experiment. Cy3 and Cy5 are used to label experimental (control and diseased) samples in a randomized design in order to avoid bias and systematic errors. The three labeled samples are mixed together and separated simultaneously in a single 2D-PAGE gel. In the first dimension, the proteins are subjected to IEF that resolves proteins based on their charge. In the second dimension, the proteins are separated according to their molecular weights by SDS-PAGE. Because CyDyes are resolvable through fluorescence, the final gel can be scanned at different wavelengths to collect quantitative images of protein patterns from the three respective samples. Quantitative comparison of protein abundances across multiple gels is performed using specialized software (DeCyder), and proteins with statistically significant alterations in relative expression levels are located. For identification of these proteins, it is recommended to use preparative gels produced with higher protein loads and poststaining. These preparative gels can be matched to the analytical gels, and the spots of interest can be picked. In our experiments, however, the low protein amount per sample did not permit inclusion of preparative gels, and spots of interest were excised from the DIGE analytical gels.

13.2 Application of DIGE platform to COPD biomarker discovery

In this section, application of DIGE to discover COPD biomarkers is presented, and methods for BAL proteome analysis are described in detail. The study focused on comparative analyses of human BAL from a group of control smoker subjects versus COPD patients. A total of 26 subjects from each group were included in the analyses.

To probe the molecular composition of the BAL specimens in a global-scale, quantitative manner, a combination of DIGE, state-of-the-art MS and advanced bioinformatics tools was used. The general bioanalytical strategy is outlined in ▶Figure 13.1. The strategy encompasses four steps: (1) processing of BAL samples, (2) CyDye labeling and 2D-PAGE separation, (3) computer-assisted differential expression analysis and

13.2 Application of DIGE platform to COPD biomarker discovery

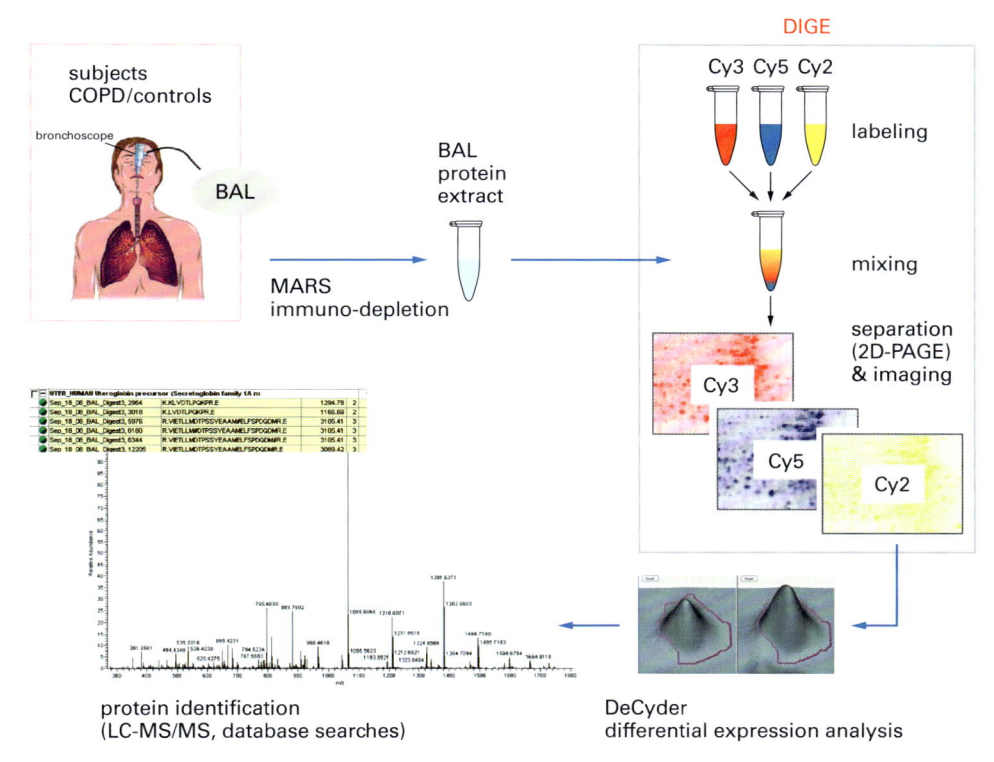

Figure 13.1: Outline of the bioanalytical strategy for comparative proteomics of human BAL. The workflow encompasses processing of BAL samples (desalting, depletion of selected high-abundance proteins), DIGE (labeling of proteins with CyDyes, protein separation with 2D-PAGE, gel imaging), computer-assisted differential expression analysis and identification of differentially expressed proteins by MS and searches of a protein sequence database.

(4) protein identification with liquid chromatography–tandem mass spectrometry (LC-MS/MS) and database searches.

13.2.1 Processing of BAL samples

Robust methods for sample cleanup and protein extraction are a critical component of any proteomics experiment. The objective is to obtain a protein mixture suitable for two-dimensional gel electrophoresis. Human BAL has specific characteristics that determine the nature of the sample-processing steps. First, because BAL proteins are diluted in a large volume of saline, sample concentration and removal of salts are needed. Second, the BAL proteome contains high amounts of plasma-derived proteins that interfere with detection of biologically relevant low-level biomarker proteins. Therefore, to effectively mine proteomes in BAL and other complex biological fluids, high-abundance proteins are removed.

13.2.1.1 Initial cleanup and desalting

The original volumes of the human BAL samples used in the study varied between 4 and 11 mL. The samples were thawed, transferred to 14-mL polypropylene tubes and centrifuged to remove cell debris. If needed, the sample volume was reduced to approximately 4 mL in a vacuum centrifuge. The samples were desalted by ultrafiltration with spin concentrators (M_W cutoff of 5,000 Da [Agilent]). The samples were placed in the concentrators and were centrifuged (25 min; 5,000 × g; 4°C). Each sample volume was reduced to ca. 100–200 µL of retentate. After the first concentration step, water (4 mL) was added to the retentate, and the concentration step was repeated. This process was repeated for a total of three times. The final retentate (ca. 100 µL) was collected, placed in 1.5 mL Eppendorf tubes and dried in a vacuum centrifuge.

13.2.1.2 Removal of high-abundance proteins

A group of six high-abundance plasma proteins that could mask lower-abundance, high-interest biomarkers were depleted with immunoaffinity chromatography. The Hu-6 Multiple Affinity Removal System (MARS) spin cartridge (Agilent) was used for this purpose. This MARS column removes albumin, transferrin, haptoglobin, antitrypsin, IgG and IgA. Prior to MARS chromatography, the dried samples were dissolved in 400 µL of MARS Buffer A. Each sample was centrifuged (9,000 × g). The supernatants were filtered through a 0.2 µm spin filter (part of the MARS kit).

After filtration, the protein concentration in MARS Buffer A was measured. The protein content in each sample was determined with the Micro BCA Protein Assay (Pierce). Bovine serum albumin (BSA) was used as a protein standard. The absorbance readings at 562 nm from the BSA standards were fitted to a second-degree polynomial equation, and the parameters derived from the equation were used to calculate the concentration of each BAL sample from the respective absorbance readings at 562 nm. It is important to note that the protein content in the individual BAL samples varied to a large degree. In the set of BAL samples used in the study, the protein content before MARS depletion generally ranged from ca. 250 µg to more than 1,400 µg. However, it has to be noted that several samples contained much lower protein amounts and were not processed further. This variability in protein content must be taken into account in study design; additional BAL samples might have to be processed to obtain sufficient number of biological replicates.

MARS depletion was performed according to the manufacturer's procedure. (BAL samples that contained more than 1 mg of protein were processed in two aliquots). Briefly, BAL samples in MARS Buffer A were applied to the MARS spin cartridge, and the flow-through containing the depleted proteome was collected. The column was washed with two 400 µL aliquots of Buffer A. The eluates were combined with the original flow-through to produce ca. 1.2 mL of depleted sample. To regenerate the column, the bound proteins were eluted with MARS Buffer B. The depleted samples were desalted by ultrafiltration as described in Section 13.2.1.1. For desalting, the concentration-dilution step was repeated for a total of five times. It is critical to remove all salts from the samples because residual salts affect isoelectric focusing. After desalting, protein concentration was determined with the Micro Protein Assay kit. Samples with a sufficient amount of protein (minimum of 75 µg; Section 13.2.2) were used for DIGE. Aliquots of the samples corresponding to 75 µg of protein were dried in a vacuum centrifuge and redissolved in DIGE labeling buffer.

13.2.2 CyDye labeling of MARS-depleted BAL samples

To allow multiplexed separation of control and COPD samples, CyDye DIGE Fluor dyes were introduced according to the minimal labeling chemistry. The CyDye Fluor minimal dyes possess a reactive group that forms an amide bond with the amino group of lysine residues. In the minimal labeling strategy, 50 µg of protein is labeled, and the dye/protein ratio (400 pmol of dye/ 50 µg of protein) is chosen so that only a small portion of the total protein in the sample is labeled, and that each protein carries only one dye label. An internal standard (a pool of all samples within the experiment) is included to enable normalization of protein abundances and matching of protein patterns.

In the biomarker discovery study, which included analyses of 52 samples (26 controls and 26 COPD), a total of three experiments were performed, each with 8–9 controls and 8–9 COPDs. One of the three comparative proteomics experiments that is described here involved 18 samples – 9 controls and 9 COPDs. The experimental design included labeling of the control and COPD samples with Cy3 or Cy5 dyes in a randomized fashion, to minimize bias and systematic errors. An internal standard obtained by pooling of aliquots of all 18 samples was used; this standard was labeled with Cy2. The typical setup of the BAL DIGE experiment is shown in ▶Table 13.1. Each final protein solution to be subjected to isoelectric focusing contained 50 µg of Cy3-labeled protein (control or COPD; ▶Table 13.1), 50 µg of Cy5-labeled protein (control or COPD) and 50 µg of Cy2-labeled pooled standard. The final solution volume was 350 µL; this specific volume is required for proper rehydration of the immobilized pH gradient (IPG) strip used for IEF.

For CyDye labeling, each dried BAL protein sample (75 µg of protein) was redissolved in 37.5 µL of labeling buffer that contained 30 mM Tris, 7 M urea, 2 M thiourea and 4% (w/v) CHAPS detergent. The pH of the labeling buffer was adjusted to 8.5 for the labeling reaction. An aliquot of 12.5 µL was taken from each of the 18 samples, and all aliquots were pooled to make the internal standard (225 µL, 450 µg of protein). The CyDyes (5 nmol CyDye DIGE Fluor minimal dye labeling kit; GE Healthcare) were reconstituted in fresh anhydrous dimethylformamide (DMF) and diluted to 200 pmol/µL working solution with DMF. Two microliters of the CyDye working solution were added to each reconstituted protein sample. Labeling was allowed to proceed for 60 min at 4°C in the dark. The reaction was terminated with lysine stop solution (1 µL of 10 mM solution); the mixtures were incubated for 10 min in the dark. After labeling, each sample was mixed with 25 µL of rehydration buffer (7 M urea, 2 M thiourea, 2% [w/v] CHAPS). Each trio of samples (▶Table 13.1) was combined in a single tube and diluted with 200 µL of modified rehydration buffer that contained dithiothreitol (DTT, final concentration 20 mM) and IPG buffer 4–7 L (final concentration 2% v/v). No Bromophenol Blue dye was used in the rehydration buffer.

13.2.3 Two-dimensional gel electrophoresis

The labeled BAL proteins were separated with 2D-PAGE; a total of nine gels per experiment was used, as outlined in ▶Table 13.1.

13.2.3.1 Isoelectric focusing

In the first dimension of 2D-PAGE, the CyDye labeled BAL proteins were separated with IEF (18 cm long IPG strips; linear pH gradient from pH 4 to pH 7 [GE Healthcare]). The

Tab. 13.1: Experimental setup for comparative proteomics of human BAL

Gel no.	1	2	3	4	5	6	7	8	9
Cy3	139 (control)	148 (COPD)	163 (control)	135 (COPD)	115 (control)	175 (COPD)	155 (control)	170 (COPD)	200 (control)
Cy5	137 (COPD)	147 (control)	153 (COPD)	210 (control)	176 (COPD)	120 (control)	172 (COPD)	190 (control)	164 (COPD)
Cy2	Pooled std.	Pooled std.	Pooled std.	Pooled std.	Pooled std.	Pooled std.	Pooled std.	Pooled std.	Pooled std.
IPG strip number	88286	88285	88284	88283	88282	88281	85897	85896	85895

IPG strips were rehydrated with the sample solutions (350 µL; overnight incubation in a reswelling tray). The IEF was performed with the Multiphor II IEF unit (GE Healthcare). A 5-step voltage program was applied: 0–100 V (gradient over 1 min); 100 V (fixed for 120 min); 100–500 V (gradient over 1 min); 500–3,500 V (gradient over 90 min); 3,500 V (fixed for 8 h). The maximum current limit was set at 2 mA.

13.2.3.2 SDS-PAGE

The BAL proteins were separated in the second dimension with SDS-PAGE in large-format (20 × 24 cm) gradient gels of 8%–16% T, cast between low-fluorescence glass plates. The gels were cast in-house with a DALT gradient maker (GE Healthcare) and used within 2 weeks after preparation. The SDS-PAGE separations were performed with a Protean Plus Dodeca electrophoresis cell (Bio-Rad).

The SDS-PAGE was performed immediately after IEF. Prior to SDS-PAGE, the proteins in the IPG strips were reduced with DTT and alkylated with iodoacetamide. For these reduction/alkylation steps, the IPG strips were equilibrated for 15 min in a solution of urea (6 M), SDS (2%, w/v), Tris.HCl (0.375 M, pH 8.8), glycerol (25% v/v) and DTT (0.1 M), followed by a second 15 min incubation in an analogous solution that contained iodoacetamide (0.1 M) instead of DTT. The IPG strips were loaded onto the SDS-PAGE gels and secured with an agarose overlay containing Bromophenol Blue dye. It is important to avoid bubbles in the agarose between the IPG strip and the SDS-PAGE gel. The gel cassettes were placed in the Dodeca cell filled with 25 L of electrophoresis buffer (pH 8.3) of Tris base (25 mM), glycine (192 mM) and SDS (0.1%, w/v). Electrophoresis of the nine gels was performed simultaneously at 200 V (constant voltage) for 12 h. For some of the gels, additional time was needed to complete the SDS-PAGE separation.

13.2.4 Image acquisition

The DIGE gels were scanned with a Typhoon 9410 Variable Mode Imager (GE Healthcare). Because of the low-fluorescence characteristics of the plates, the gels were imaged while still sandwiched between the plates. Emission filter and laser settings for each CyDye were chosen as recommended by GE Healthcare, and were: 520 BP 40 emission filter/Blue2 (488) for Cy2; 580 BP 30 emission filter/Green (532) for Cy3; 670 BP 30 emission filter/Red (633) for Cy5. The gels were first prescanned at low resolution (1,000 µm) to adjust the photomultiplier tube (PMT) voltage settings to optimize sensitivity. The Cy3 and Cy5 channels were adjusted, depending on the spot intensity and number of saturated spots for the individual gels. The gel images were first loaded into the ImageQuant software and inspected; the maximum pixel values were calculated for selected large spots. The PMT values were optimized to avoid spot saturation. For quantitative analysis, the final images were acquired at high resolution (100 µm).

Examples of 2D gels with protein spot patterns from control and COPD BAL are shown in ▶Figure 13.2A, B. These gels depict proteins in the isoelectric point range of 4–7, with molecular weights between ca. 15,000 and 250,000 Da. It is evident that the BAL-processing protocol yielded clean samples, without interfering substances. Although DIGE compensates for gel-to-gel differences in spot quality, it is still important to obtain well-focused, high-quality protein patterns for effective pattern comparison and for confident protein identification. The gels shown in ▶Figure 13.2A, B contain several

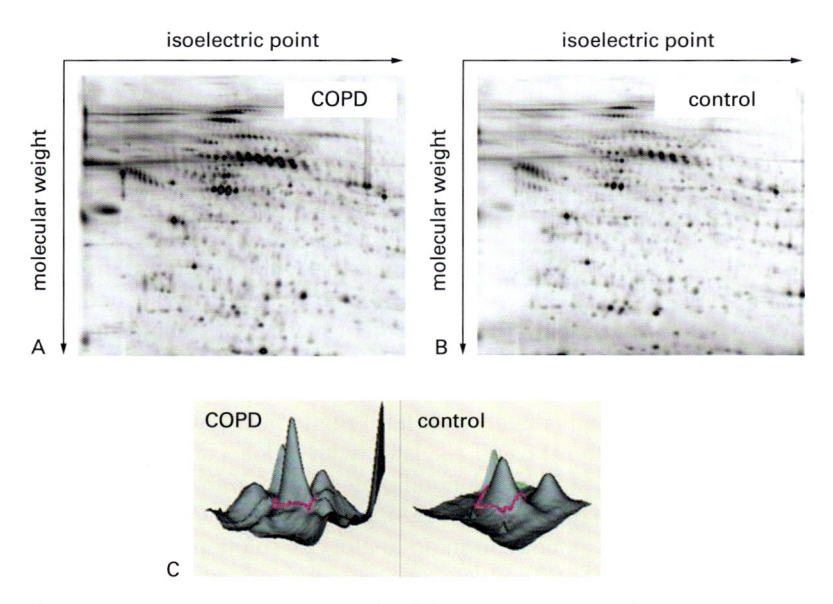

228 | 13 Proteomics of human bronchoalveolar lavage fluid

Figure 13.2: Representative 2D gels of the BAL proteomes from COPD (A) and control (B) subject. Separation of proteins was performed with 18 cm IPG strips with a pH range 4–7 and with large-format SDS-PAGE gels of 8%–16% T. (C) Example of output from DeCyder differential expression analysis.

hundreds of well-resolved spots; some of the spots, in particular those that appear as spot "trains," are known to arise from multiple isoforms of the same protein.

13.2.5 Differential expression analysis

The images were loaded onto the DeCyder 2D software (version 7.0, GE Healthcare) and cropped. For spot detection, background subtraction and normalization of a set of images from the same gel, the images were analyzed with differential in-gel analysis (DIA) module with selection criteria on spot slopes and spot volumes that minimize the inclusion of nonprotein spots (e.g. dust); this process is crucial to optimize comparative image analysis. The DIA-generated gel images were loaded into the biological variation analysis (BVA) module for comparative analyses of images across multiple gels to identify protein spots with differential expression between the control versus COPD proteomes. The t-test statistics were used to calculate the statistical significance of the differentially expressed spots.

An example of the graphical output from the DeCyder analysis, which shows differential display data for one of the protein spots, is depicted in ▶Figure 13.2C. The DeCyder outputs were inspected manually to confirm the accuracy of spot detection and the quality of the overall spot appearance. In total, 34 spots that exhibited statistically significant abundance differences in control versus COPD were located, and the proteins in these spots were successfully identified with methods described in Section 13.2.6.

13.2.6 Protein identification

The differentially expressed proteins were identified with LC-MS/MS and searches of the SwissProt protein sequence database. Because of low amounts of BAL proteins, it was not possible to produce a separate "pick" gel with higher protein load; therefore, analytical gels were used to procure protein for identification. The proteins were identified via LC-MS/MS analyses of corresponding peptide digests. The tandem mass spectrometry (MS/MS) data contained series of product ions generated from gas-phase dissociation of corresponding peptide precursor ions. Product ions that were diagnostic of the amino acid sequence of the peptides were used for database searches.

13.2.6.1 Spot visualization

The DIGE gels from which spots were to be excised were stained with a modified silver stain protocol compatible with subsequent mass spectrometric analysis (Shevchenko et al. 1996). The spots for excision were located by careful manual comparison of the spot patterns in DIGE images with those in the poststained gels. To obtain the spots for protein identification, the gels were placed onto a light box, and the gel plugs that contained POI were excised manually with a pipette tip. Excessive handling of gels should be avoided to minimize chance of gel breakage and the possibility of keratin contamination. Robotics for spot picking and protein digestion are commercially available.

13.2.6.2 Protein digestion

The gel plugs were destained with a solution of sodium thiosulfate/potassium ferricyanide (Gharahdaghi et al. 1999). After washing with water, the gel plugs were incubated for 20 min a 200 mM ammonium bicarbonate buffer. After incubation, plugs were washed with water and dehydrated with three aliquots of acetonitrile. The dehydrated gel pieces were dried in a vacuum centrifuge. For digestion, 20 µL of trypsin digestion solution was added to the gel pieces; after reswelling, the mixtures were incubated overnight at 37°C. (The trypsin digestion solution contained PROMEGA sequencing-grade trypsin at a concentration of 16.7 ng/µL in 50 mM ammonium bicarbonate buffer, pH 8.) After digestion, the digest solutions were collected in separate test tubes. To extract residual peptides, the gel pieces were incubated in an extraction solution of 60% acetonitrile/35% water/5% trifluoroacetic acid; the samples were sonicated in this solution for 20 min. The extracts were combined with the original digest mixtures, and the samples were dried in a vacuum centrifuge. Prior to LC-MS/MS analysis, the digests were desalted and concentrated with ZipTip C18 microcolumns (Millipore).

13.2.6.3 MS

The LC-MS/MS analyses were performed with an LTQ quadrupole linear ion trap mass spectrometer (Thermo Scientific) interfaced with a nanoflow liquid chromatography system (Famos/Ultimate from Dionex). The samples were injected manually through a six-port valve. The LC separations were carried out with a fused silica capillary column/spray needle (75 µm I.D., 10 cm packing length; New Objective) packed in-house with MAGIC C18 reversed-phase particles (5 µm, 200 Å; Michrom Bioresources). The peptides were eluted with a linear gradient of 0–90% B over 30 min, at a flow rate of

300 nL/min. Mobile phase A was 98% water/2% methanol/0.05% formic acid; mobile phase B was 10% water/90% methanol/0.05% formic acid (Giorgianni et al. 2004).

The peptides that eluted from the LC were ionized with nano-electrospray and analyzed with the linear ion trap mass spectrometer. The objective of the mass spectrometric measurements was to determine the masses of the intact peptide ions and, most importantly, the sequence-diagnostic product-ion patterns. Therefore, for each peptide, MS and MS/MS data were acquired in the data-dependent mode. This acquisition mode is used to analyze mixtures of unknown composition such as peptide digests. In data-dependent acquisition, the instrument performs cycles of MS and MS/MS measurements: in MS, the masses of the precursor peptide(s) that elute from LC at the particular time are measured, and that information is used to set the parameters for a subsequent series of MS/MS measurements, which are carried out following the MS scan. Many MS-MS/MS cycles are performed during a single LC-MS/MS analysis. To identify differentially expressed BAL proteins, the LTQ mass spectrometer cycled through the acquisition of a full-scan MS spectrum, followed by seven MS/MS scans of the most abundant ions from the MS scan.

13.2.6.4 Protein database searches

The LC-MS/MS data set obtained for each protein digest was used to search the SwissProt protein sequence database. The search engine TurboSequest (part of the Bioworks 3.2 software package) was used to perform the searches. As is customary in proteomics, this search program uses minimally processed MS/MS data, and no manual interpretation is needed. In these Sequest-based database searches, the experimentally measured mass of a precursor peptide was used to retrieve from the database peptide sequences whose theoretical masses matched the measured mass. Next, the theoretical product-ion (MS/MS) patterns generated in silico were compared with experimental MS/MS data for each candidate peptide (Sadygov et al. 2004). The presence of carbamidomethyl-cysteine and/or oxidized methionine residues were considered as possible modifications. The degree of correlation between the theoretical and experimental patterns was reflected chiefly as the correlation score (Xcorr). The database search outputs were filtered with the following criteria: Xcorr cutoff values of 1.5, 2.0 and 3.5 for singly, doubly and triply charged precursor ions, respectively; and a protein probability cutoff of 1×10^{-3}. In selected cases, the Sequest-based results were confirmed with a different search algorithm (Mascot). All search results were inspected manually to evaluate the quality of the associated MS/MS data. The majority of the differentially expressed proteins were identified based on the MS/MS data for ≥2 peptides. An example of a typical MS/MS spectrum obtained in the BAL protein-identification step is shown in ▶Figure 13.3. This spectrum displays a near-complete series of sequence-determining C-terminal product ions of the *y* series. The *y* series is supplemented by a number of N-terminal product ions from the *b* series. As indicated in the inset of ▶Figure 13.3, these ions arise from cleavages along the peptide backbone. This information (i.e. the *m/z* values of the product ions, together with the measured mass of the precursor ion) was used to pinpoint through a database search the peptide sequence AYYHLLEQVAPK as the top match. This sequence belongs to the protein Plastin-2. Six other peptides that mapped

13.2 Application of DIGE platform to COPD biomarker discovery

to human Plastin-2 were also retrieved in the search based on their product-ion patterns. It should be noted that typically a portion of the protein sequence is mapped by the identified peptides. Isoelectric point and molecular weight of the intact protein can be estimated from the position of the corresponding spot on 2D gel.

Taken together, application of DIGE in combination with LC-MS/MS produced a panel of 34 putative biomarkers of COPD. Twenty-five proteins were found up-regulated in COPD, and 9 proteins were down-regulated in COPD. The identities of the putative biomarkers are summarized in ▶Table 13.2 (up-regulated proteins) and ▶Table 13.3 (down-regulated proteins). These differentially expressed proteins are functionally diverse and include proteases and other enzymes, surfactant proteins, proteins involved in cell structure and other functional categories. To assess their clinical utility, the panel of candidate biomarkers will be further evaluated in future targeted studies of larger groups of BAL samples.

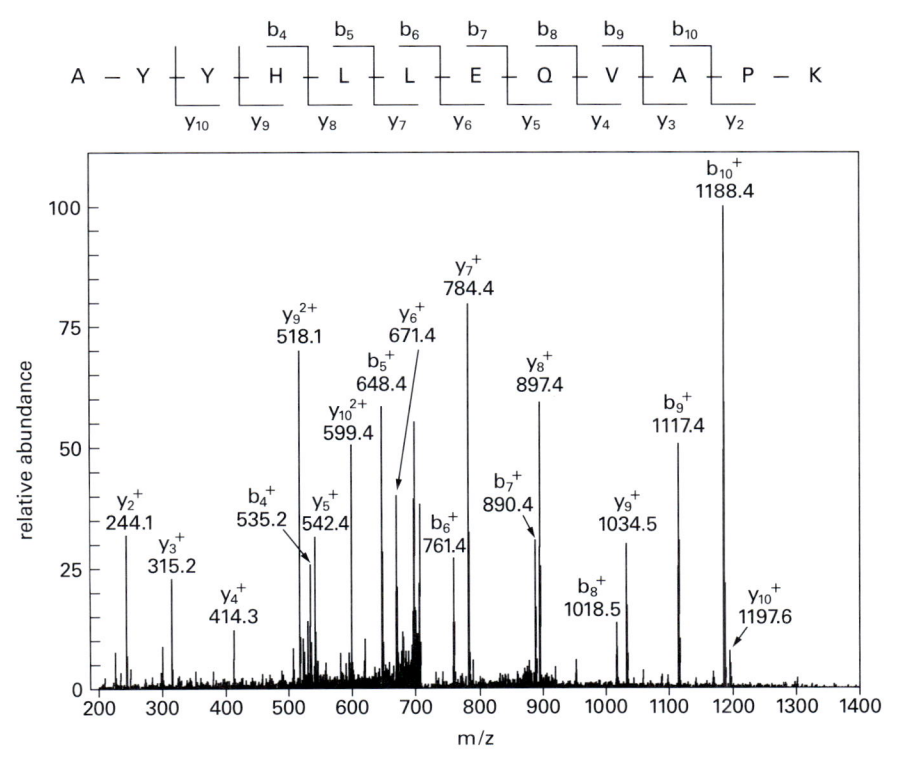

Figure 13.3: Example of a peptide MS/MS spectrum obtained in the LC-MS/MS analysis of one of the differentially expressed BAL proteins. The spectrum contains series of y and b product ions diagnostic of the amino acid sequence of the peptide. Based on these data, the peptide AYYHLLEQVAPK was identified that belongs to Plastin-2.

Tab. 13.2: Differentially expressed proteins up-regulated in COPD

Protein name	SwissProt accession code
Retinal dehydrogenase 1	P00352
Alpha-1-antichymotrypsin	P01011
Alpha-2-macroglobulin	P01023
Complement C3	P01024
HLA class II histocompatibility antigen	P01903
Apolipoprotein A-I	P02647
Alpha-1-acid glycoprotein 1	P02763
Alpha-2-HS-glycoprotein	P02765
Transthyretin	P02766
Serotransferrin	P02787
Heat shock beta-1	P04792
Mitochondrial aldehyde dehydrogenase	P05091
Tropomyosin alpha-3 chain	P06753
Cathepsin D	P07339
Cathepsin L1	P07711
Vimentin	P08670
Fructose-1,6-bisphosphatase 1	P09467
Leukotriene A-4 hydrolase	P09960
Plastin-2	P13796
Alpha-1-acid glycoprotein 2	P19652
Rab GDP dissociation inhibitor alpha	P31150
Lysozyme C	P61626
Tropomyosin alpha-4 chain	P67936
Calmodulin-like protein 5	Q9NZT1
Dipeptidyl-peptidase 2	Q9UHL4

Tab. 13.3: Differentially expressed proteins down-regulated in COPD

Protein name	Swissprot accession code
Lactotransferrin	P02788
Alpha-amylase 1	P04745
Pulmonary surfactant-associated protein B	P07988
Calreticulin	P27797
Phosphatidylethanolamine-binding protein 1	P30086
Caspase-14	P31944
Neutrophil gelatinase-associated lipocalin	P80188
Scavenger receptor cysteine-rich type 1 protein M130	Q86VB7
Cathepsin Z	Q9UBR2

13.3 Outlook

Implementation of DIGE has significantly improved the capabilities of 2D-PAGE for comparative proteomics of complex samples. DIGE allows accurate quantification of protein expression changes, and the platform has been widely used in a variety of proteomics applications, including biomarker discovery. Alternative 1D-gel and gel-free approaches that have been developed in recent years provide researchers with a variety of choices in proteomics workflows (Fang et al. 2010; Giorgianni et al. 2003; Piersma et al. 2010). A unique benefit offered by 2D-PAGE and DIGE is the availability of information on intact proteins. This information, which cannot be obtained in gel-free workflows that include analyses at the peptide level, may be critically important, for example, for quantification of changes in specific protein isoforms, including post-translationally modified proteins. DIGE will likely remain an attractive component of proteomics technologies, and it will continue to play an important role in comparative proteomics research.

References

Barnes, P.J., and Celli, B.R. (2009). Systemic manifestations and comorbidities of COPD. Eur Respir J *33*, 1165–1185.

Barnes, P.J., Chowdhury, B., Kharitonov, S.A., Magnussen, H., Page, C.P., Postma, D., et al. (2006). Pulmonary biomarkers in chronic obstructive pulmonary disease. Am J Respir Crit Care Med *174*, 6–14.

Chen, H., Wang, D., Bai, C., and Wang, X. (2010). Proteomics-based biomarkers in chronic obstructive pulmonary disease. J Proteome Res *9*, 2798–2808.

Fang, Y., Robinson, D.P., and Foster, L.J. (2010). Quantitative analysis of proteome coverage and recovery rates for upstream fractionation methods in proteomics. J Proteome Res *9*, 1902–1912.

Friedman, D.B., Hoving, S., and Westermeier, R. (2009). Isoelectric focusing and two-dimensional gel electrophoresis. Methods Enzymol *463*, 515–540.

Gharahdaghi, F., Weinberg, C.R., Meagher, D.A., Imai, B.S., and Mische, S.M. (1999). Mass spectrometric identification of proteins from silver-stained polyacrylamide gel: a method for the removal of silver ions to enhance sensitivity. Electrophoresis *20*, 601–605.

Giorgianni, F., Cappiello, A., Beranova-Giorgianni, S., Palma, P., Trufelli, H., and Desiderio, D.M. (2004). LC-MS/MS analysis of peptides with methanol as organic modifier: improved limits of detection. Anal Chem *76*, 7028–7038.

Giorgianni, F., Desiderio, D.M., and Beranova-Giorgianni, S. (2003). Proteome analysis using isoelectric focusing in immobilized pH gradient gels followed by mass spectrometry. Electrophoresis *24*, 253–259.

Gorg, A., Drews, O., Luck, C., Weiland, F., and Weiss, W. (2009). 2-DE with IPGs. Electrophoresis *30 Suppl 1*, S122–S132.

MacNee, W. (2005). Pulmonary and systemic oxidant/antioxidant imbalance in chronic obstructive pulmonary disease. Proc Am Thorac Soc *2*, 50–60.

MacNee, W. (2007). Pathogenesis of chronic obstructive pulmonary disease. Clin Chest Med *28*, 479–513, v.

Mallick, P., and Kuster, B. (2010). Proteomics: a pragmatic perspective. Nat Biotechnol *28*, 695–709.

Minden, J.S., Dowd, S.R., Meyer, H.E., and Stuhler, K. (2009). Difference gel electrophoresis. Electrophoresis *30 Suppl 1*, S156–S161.

Owen, C.A. (2008). Roles for proteinases in the pathogenesis of chronic obstructive pulmonary disease. Int J Chron Obstruct Pulm Dis *3*, 253–268.

Piersma, S.R., Fiedler, U., Span, S., Lingnau, A., Pham, T.V., Hoffmann, S., et al. (2010). Workflow comparison for label-free, quantitative secretome proteomics for cancer biomarker discovery: method evaluation, differential analysis, and verification in serum. J Proteome Res *9*, 1913–1922.

Plymoth, A., Lofdahl, C.G., Ekberg-Jansson, A., Dahlback, M., Lindberg, H., Fehniger, T.E., et al. (2003). Human bronchoalveolar lavage: biofluid analysis with special emphasis on sample preparation. Proteomics *3*, 962–972.

Punturieri, A., Croxton, T.L., Weinmann, G., and Kiley, J.P. (2007). The changing face of COPD. Am Fam Physician *75*, 315–316.

Rabe, K.F., Hurd, S., Anzueto, A., Barnes, P.J., Buist, S.A., Calverley, P., et al. (2007). Global strategy for the diagnosis, management, and prevention of chronic obstructive pulmonary disease: GOLD executive summary. Am J Respir Crit Care Med *176*, 532–555.

Sadygov, R.G., Cociorva, D., and Yates, J.R., III. (2004). Large-scale database searching using tandem mass spectra: looking up the answer in the back of the book. Nat Methods *1*, 195–202.

Shevchenko, A., Wilm, M., Vorm, O., and Mann, M. (1996). Mass spectrometric sequencing of proteins silver-stained polyacrylamide gels. Anal Chem *68*, 850–858.

Smith, J.C., and Figeys, D. (2006). Proteomics technology in systems biology. Mol Biosyst *2*, 364–370.

Timms, J.F., and Cramer, R. (2008). Difference gel electrophoresis. Proteomics *8*, 4886–4897.

Wattiez, R., and Falmagne, P. (2005). Proteomics of bronchoalveolar lavage fluid. J Chromatogr B Anal Technol Biomed Life Sci *815*, 169–178.

Zhan, X., and Desiderio, D.M. (2011). Nitroproteins identified in human ex-smoker bronchoalveolar lavage fluid. Aging Dis *2*, 100–115.

14 Proteomic analysis of Duchenne muscular dystrophy (DMD)

Kunigunde Stephani-Kosin, Hisashi Hirano, and Roza Maria Kamp

14.1 Introduction

DMD is a genetic disorder, which causes irreversible loss of muscle mass and consequently leads to early death. The dystrophin gene (Hoffman et al. 1987) contains 2.6 million base pairs and is the largest gene of the human DNA (Roberts et al. 1993). Many nonreversible changes of this gene are responsible for the development of muscular dystrophies (Davies and Nowak 2006). In 1861, Duchenne described muscle dystrophy, which is observed in 1 of 3,500 boys born and which is caused by absence of the dystrophin protein, due to mutations of the corresponding gene. The first symptoms of the disease occur by the age of 3 years. The children have problems with sitting, walking and running. The progressive weakness of skeletal muscle and accumulation of connective and fat tissue causes such a high loss of mobility that the children are wheel-chair dependent by the age of 12. In the last phase of the disease, the patients have problems also with eating, breathing and myocardial insufficiency (Speer and Oexle 2002). The average life span of DMD patients is about 20 years.

The diagnosis of DMD starts usually with blood analysis and measurement of serum creatine kinase, which is increased up to ten times compared to healthy patients. Electromyography is used to test muscle activity, but molecular or protein diagnostic methods like the detection of gene mutations or Western blot analysis for dystrophin or immunohistochemical analyses are the main diagnostic procedures.

Although the established diagnostic methods are very good, reliable therapies are not yet available.

The known, but not very efficient and applicable therapies can be classified as follows (Grunwald 2009; Sakamoto et al. 2002; Zhou et al. 2006):

- molecular therapies, including substitution of the dystrophin gene, correction of the mutation or up-regulation of dystrophin-related proteins;
- cell-based strategies, including stem-cell therapy or myoblast transfer;
- pharmacological treatment, including up-regulation of utrophin, calpain, proteasome, serine protease inhibiton, cyclosporin or deflazacort treatment and blocking of the calcium channel.

Because causal therapy is not possible, it is important to develop new methods to reduce or stop progression of this disease. Some therapeutic treatments are already in use, such as cortisol treatment, but the side effects are very strong. For this reason, we applied proteome analysis to search for proteins that are involved in the development of DMD. DMD is a fast-progressing disease in comparison with Becker muscular dystrophy, where progression of the disease is much milder. We decided to use human myoblasts (precursor cells) for the investigation of the DMD disease because human cells have never been applied for proteome in studies of DMD. In previous tests, mice or dogs were used for animal experiments, but big discrepancies between both were observed. In the case of dog experiments, the progression of the DMD was comparable with humans, but mouse experiments showed a quite different disease process. We compared cells of four healthy and two diseased donors, and we also studied the progression of the DMD disease in patients between the ages of 3 months and 6 years. Such investigations allowed us to identify not only of alterations in protein expression, but also the relative quantitation of the expression level. The results of proteome analysis gave new and valuable insights into molecular processes in DMD cells and present a basis for further studies of cellular signal transduction, which is disturbed in DMD patient cells. The analyzed, differently expressed proteins serve as new biomarkers and will be used for further therapeutic investigations in studies with human cells.

14.2 Materials

14.2.1 Cells

The primary human skeletal muscle cells (SkMCs) were derived from quadriceps biopsies at the Charité Berlin (von Moers), according to the declaration of Helsinki. The DMD diagnostics were examined by histological staining of dystrophin at the protein level. Dystrophin-negative myoblasts were separated from other cell types using cell-specific antigens at the MTCC Munich. The purified myoblasts were cryopreserved for transport, then cultured for two passages and frozen again in liquid nitrogen. ▶Table 14.1, further on in this chapter, shows the primary myoblasts that were used.

14.2.2 Cell culturing

For the cell experiments, the following items were used:

Skeletal muscle cell growth medium (SMCGM), supplement mix (SM), trypsin neutralization solution (TNS), trypsin/EDTA solution (0.04%/0.03%), freezing medium and CryoSFM were purchased from PromoCell. The antibiotics used, including penicillin/streptomycin (10,000 µg/mL), gentamycin (10 mg/mL), and Dulbecco's PBS (10x) were from Biochrom AG.

Cell incubator (Function line) and clean bench (HeraSafe) were from Heraeus; Cell culture flasks, T25 from TPP; Cryo-tubes, 2 mL from Corning; Microscope, Axiovert 40 C and Microscope, Axiovert 40 CLF from Zeiss; and Camera, Jenoptik ProgRes C10^{plus} from Jenaoptik.

14.2.3 Protein extraction and analysis

For protein extraction, the ProteoExtract Complete Mammalian Proteome Extraction kit was from Calbiochem (Merck). For the Bradford protein concentration assay, the RotiQuant solution from Roth was used. Trypsin gold mass spectrometry (MS) grade was from Promega.

All buffer chemicals were from Sigma and Roth. The 2D-PAGE electrophoresis equipment was from BioRad. Data processing softwares used for comparison of 2D-electropherograms were from Progenesis Linear Dynamics.

14.3 Methods

14.3.1 Cell culture of human, primary myoblasts

The myoblasts were cultured in SMCGM/SM under sterile conditions in a cell incubator at 37°C and water-saturated atmosphere containing 5% carbon dioxide. For dystrophin-positive cells, penicillin/streptomycin (100 µg/mL) was used in order to avoid bacterial infections. The dystrophin-negative cells were treated with penicillin/streptomycin only for a short time. T25 polyethylene/polystyrene flasks (TPP) were used for cultivation. The exchange of media was performed two times a week. Cells with a confluence of 90% were split in 1:2 ratio.

Trypsin/EDTA solution (50 µL/cm^2) was applied for 1–3 min for detaching the cells. TNS was added to inhibit the trypsin activity (100 µL/cm^2). The suspension was transferred in a 15 mL Corning tube and was centrifuged in EconoSpin (Sorvall) at 200 × g for 5 min. The supernatant was discarded, and the cell pellet was washed using PBS buffer and then centrifuged again. The supernatant was removed, the cell pellet was resuspended in 8 mL SMCGM/SM and split in equal volumes in T25 culture flasks (TPP). The cells were controlled and documented with the Axiovert 40 CLF microscope (Zeiss) with integrated ProgRes C10plus camera (Jenoptik).

For cryoconservation of the adherent cells, a T25 culture flask was treated with 500 µL of the trypsin solution. The trypsin digestion was stopped after the detachment of cells by adding 1 mL TNS . After washing, the cell pellet was resuspended in 1mL freezing media CryoSFM. The cell-solution was transferred into a cryotube, placed first on ice, to later at –20°C for 30 min and finally at –80°C for 1 day. For longer storage, the cells we kept in liquid nitrogen.

The frozen cells were fast thawed in warm water, until about 2/3 of the cell solution was liquid. Then the thawed part was transferred into T25 culture flask, and 9 mL SMCGM was added. In the final step, the culture flask was incubated for 6 h at 37°C in the cell incubator for attachment of the cells to the culture surface. After 6 h, the medium was exchanged.

For protein extraction from the cells, the growth medium was removed carefully without disturbing the cell monolayer. The cell layer was washed one time with PBS before the attached cells were treated with trypsin solution. The reaction was stopped with TNS, and the cells were collected by centrifugation at 150 × g for 10 min. Then the cells were washed three times using 4 mL wash buffer from the ProteoExtract Kit® for Complete Mammalian Proteome (Calbiochem). The cell pellet was stored at –80°C until protein extraction.

14.3.2 Protein extraction

Proteins were extracted using the ProteoExtract® kit for Complete Mammalian Proteome from Calbiochem.

The frozen cell pellet of a T25 culture flask with 90% confluence was thawed on ice. The pellet was resuspended in 15 µL resuspension buffer. To avoid protein degradation, proteinase- and phosphates-protease-inhibitor solutions in a ratio of 1:10 (v/v) and EDTA (1 mM) were added. Additionally, 78.5 µL of extraction buffer and 5 µL reducing agent were added. The solution was turned several times. The viscosity of the solution increased. For the reduction of the viscosity caused by DNA, 1.5 µL of a 1:10

benzonase solution was added. The solution was incubated at room temperature under shaking for solubilization of all proteins.

The insoluble components were removed after centrifugation at 25,000 × g and 10°C for 30 min. The supernatant containing the cell extract was then transferred into a fresh tube. This extract was ready to use for isoelectric focussing (IEF).

14.3.3 Determination of protein concentrations

To determine the concentrations of all soluble, extracted proteins, the RotiQuant Bradford Protein Assay was used. The assay was performed in a microtiter plate scale. For protein detection 200 µL of a 1:5 dilution of RotiQuant reagent (Roth) and MilliQ water and 50 µL of a suitable dilution of protein extract were added to each well. The absorbance was measured at 595 nm with the Infinite F200 (TECAN) photometer, and protein concentration was calculated using a standard curve of bovine serum albumin (BSA).

14.3.4 Two-dimensional gel electrophoresis

The 2D-PAGE electrophoresis equipment was from BioRad. For the first dimension in the Protean IEF cell, 17 cm immobilized pH gradient (IPG) strips were used. About 500 µg in a 330 µL sample was applied. This amount is sufficient to easily visualize a large number of spots using Coomassie Blue staining. For silver staining, about 50 µg proteins is sufficient.

- The sample was pipetted in the middle of a tray of the same size as the IPG strips. It is important to avoid any bubbles, which can disturb the electrophoresis.
- The Ready-Strip IPGs (stored at −20°C) were set aside in the tray.
- To prevent drying up, the strips were overlayed with mineral oil and rehydrated for 6 h at 50 V.
- Electrophoresis was performed for 16–18 h, 50 µA/stripe and 50,000 Vh.
- After ID IEF electrophoresis, the stripes were removed from the tray and equilibrated in buffer 1 for 10 min and in buffer 2 again for 10 min. Buffers were always prepared freshly. Buffer 1 was used for denaturation of proteins and consisted of 6 M urea, 2% SDS, 0.375 M Tris-HCl, pH 8.8, 20% glycerol, 2% DTT (w/v). Buffer 2 was used for carboxymethylation of cysteins to prevent disulfide formation and consisted of the same reagents as buffer 1, but instead of DTT, 135 mM iodacetamide was used.
- After incubation, the strips were transferred on the 15% SDS-PAGE, by facing the plastic foil toward the glass plate.
- For fixing of the IPGs, the strips were overlayed by using preheated low-melting agarose solution, mixed with one drop of Bromophenol Blue.
- The second dimension was run for 30 min at 100 V and for about 5 h at 200 V until the Bromophenol Blue marker had traveled out.

14.3.5 Staining of proteins

Polyacrylamide gels were stained with Roti-Blue (Roth). The gels were incubated overnight in the staining solution on a rotary shaker and destained for 30 min in MilliQ water.

14.3.6 Preparation of peptides for MS

After separation of proteins using 2D-PAGE, enzymatic "in-gel" digestion was necessary. It is impossible to elute direct proteins from the gel, because of very low yields. After "in-gel" digestion, the elution of the peptides is very efficient and allows further mass spectrometric analysis for identification and characterization of proteins.

14.3.7 Trypsin digestion

Protein spots were cut from the gel and stored in 20% methanol/5% acetic acid. Before cleavage, the storage solution was discarded and the gels were destained three times with 60% acetonitril/50 mM NH_4HCO_3. The supernatant was discarded, and the gel was dried in an Eppendorf tube for 20 min at room temperature. For digestion, 5 µL of trypsin (1µg/µL in 50 mM NH_4HCO_3) was added and the tube was kept on the ice for 10 min. After this time, 20 µL of 50 mM NH_4HCO_3 was added and the digestion was performed overnight. Before MS analysis, the supernatant containing the peptides had to be desalted using ZIP-TIP C18 to remove all disturbing reagents.

14.3.8 Desalting and preparation of peptides for MALDI

To remove all disturbing reagents from the protein sample before further MS analyses, ZIP-TIP (C18) was used.

- For removing all contaminants, the ZIP-TIP was washed 10 times with 10 µL 80% acetonitril/0.2% TFA in a 500 µL Eppendorf tube (pipetted 10 times with 10 µL up and down).
- For equilibration, 10 µL of 0.1% TFA in a 500 µL tube was pipetted up and down.
- For binding of peptides 10 µL of the tryptic digest was pipetted 20 times slowly up and down.
- To remove disturbing reagents and salts, the ZIP-TIP was washed 20 times using 10 µL 0.1% TFA (pipetted up and down).
- For elution I of peptides, 1 µL of elution solution was pipetted 10 times up and down.

Elution solution contained 3.5 mL 70% acetonitrile, 5 µL of 0.1% TFA, 20 mg of CHCA-4-hydroxycinamic acid.

- For elution II of peptides, 1 µL of elution solution was pipetted 10 times up and down.
- After elution, the desalted peptides were applied directly on a MALDI target.

14.3.9 MALDI MS

For recognition of digested proteins, MALDI-TOF/TOF MS (Applied Biosystems 4800) was used. The identification of proteins was performed using the Mascot search engine (Matrix science). The parameters used were as follows: peptides tolerance 300 ppm, peptide charge 1+, data format Mascot generic, MS/MS tolerance 0.3 Da.

14.4 Results and Discussion

The proteome analysis was applied for the study of DMD. But all methods described may generally also be applicable to other known diseases. The procedures of such investigations are always identical, as shown in ▶Figure 14.1. Differences are necessary in the cultivation of cells, depending on their source. For the extraction of proteins, different methods may be applied, for example, in the use of commercially available kits, which allow the most simple and reproducible procedures with high recovery of proteins.

14.4.1 Muscle cell culture

For the subtractive, comparative analysis of DMD-diseased and control cells, we used cells from human myoblast cell culture obtained from donors of different ages (▶Table 14.1). The muscle cells were derived from quadriceps biopsies;
 dys-, dystrophin-negative DMD-diseased cells; dys+, control, dystrophin-positive cells.

For the study of DMD progression, different age-dependent myoblasts were used: 3 months, 1 year, 2 years, 4 years, 6 years and 9 years.

The DMD myoblasts have in comparison with the control cells reduced metabolism, and they differ in their morphology (▶Figure 14.2). The duplication rate of control cells was 5 days, but DMD myoblasts showed lower rates, depending on the age of the donor; usually it took 10 days or longer. For this reason, it was very difficult to culture enough material necessary for further analysis using 2D-PAGE. The control cells,

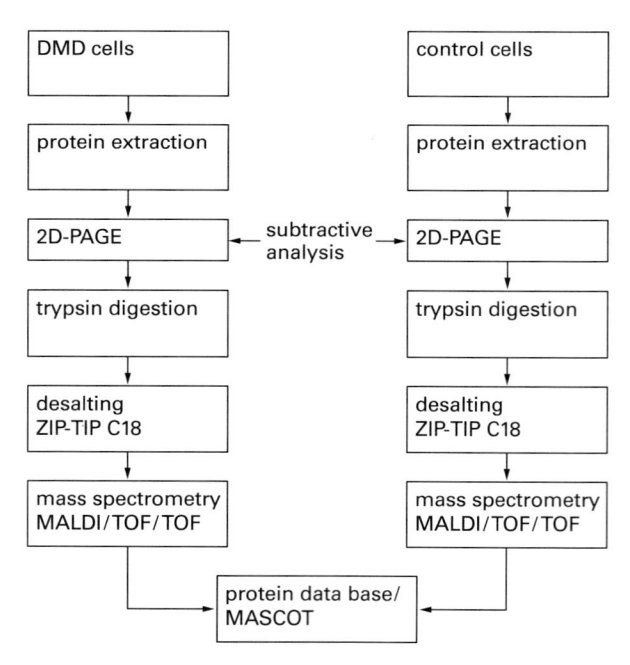

Figure 14.1: Scheme for the strategy to identify and characterize proteins involved in various diseases.

Table 14.1: The myoblasts used in the study

Patient	Dystrophin	Biopsy
1	dys-	3 months
2	dys+	1 year
3	dys+	2 years
4	dys+	4 years
5	dys-	6 years
6	dys +	9 years

healthy myoblasts, were elongated and well shaped, and they showed regular spreading in the culture medium. In the DMD cell culture, obviously more dead cells were present than in the control group. The diseased cells showed big star-shaped cytoplasm and an irregular shape. The number of the control cells in T75 flask was usually 3×10^6 cells and for DMD samples only 5×10^5 cells.

14.4.2 Two-dimensional gel electrophoresis

The best-known method for separation of thousands of proteins in the pmol or fmol range is 2D-PAGE. Two different techniques are available. It is possible to use ampholytes for regeneration of a pH gradient or IPG strips with already performed gradients. IPG was developed by Angelika Görg from TU Munich. Because of better reproducibility, we decided to use IPG strips for the first dimension and 15% SDS-PAGE for the second dimension. The disadvantage of this method is the high amount of sample that is necessary. It is always required to decide which parameters are more important. If not enough material is available, the ampholyte technique with inductive gradients is more efficient and is ten times more sensitive. But much easier, especially for beginners, is the IPG technique. For a good separation of proteins, it is important to use large gels.

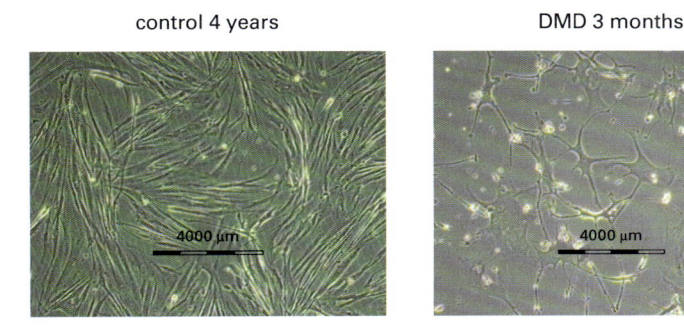

Figure 14.2: Cell cultures of primary diseased (3 months) and control cells (4 years), 100× magnification.

The possibilities are to choose between 7 cm, 17 cm and 40 cm strips. Because limited amounts of DMD myoblasts were available, 17 cm strips were chosen, though larger gels would have been advantageous. The 2D separation of proteins was optimized using 7 cm IPG strips and was scaled up to 17 cm strips. The applied pH gradient was 3–10, and the amount of proteins used per one 17 cm strip was about 500 µg. The protein concentration was determined using Bradford assay.

14.4.2.1 Two-dimensional gel electrophoresis of control and DMD samples

For separation of DMD and control proteomes, 2D electrophoresis with IPGs between pH 3–10 was used. The gels were stained using Coomassie Blue. To analyze all proteins, ruthenium, silver and phosphostain were used. Each staining allowed visualization of additional proteins.

Our results show that the proteomes of control cells differ according to the age of the donor. For this reason, we used DMD and control samples of about the same age in order to compare diseased and control cells (▶Table 14.2). This grouping allowed a comparison not only of dys+ and dys- samples, but also an identification of changes in proteomes with progression of the disease.

▶Figure 14.3 shows all performed 2D gels of diseased and control samples, which were applied for the further subtractive analysis. Usually a master gel from three gels of the same sample was applied for a further differential analysis. For comparison of proteomes derived from diseased and control cells, the software Progenesis from Non-linear Dynamics was used. It became possible to choose the level of expression of proteins (e.g. 1.5- or 2-fold). Also, the manual correction of false comparable spots was performed. All applied 2D electrophoreses showed good reproducibility, and the comparison of the 3-month DMD cells and the 1-year-old control cells was especially easy and efficient.

14.4.2.2 MS

▶Figure 14.4 shows all protein spots, which were differently expressed in all analyzed samples and which were identified and characterized using MALDI mass spectrometric analysis.

The marked spots were cut out and digested with high-quality trypsin overnight, and then the next day desalted using ZIP-TIP and applied for MALDI-TOF/TOF (Applied Biosystems 4800). The analysis of achieved mass spectra and peak lists was performed using Mascot search engine (Matrix Sciences).

Table 14.2: Groups of 2D-PAGEs used for the differential analysis

Group	DMD	Control
1	3 months	1 year
2	6 years	9 years
3	6 years	4 years
4	6 years	2 years

14.4 Results and Discussion | 243

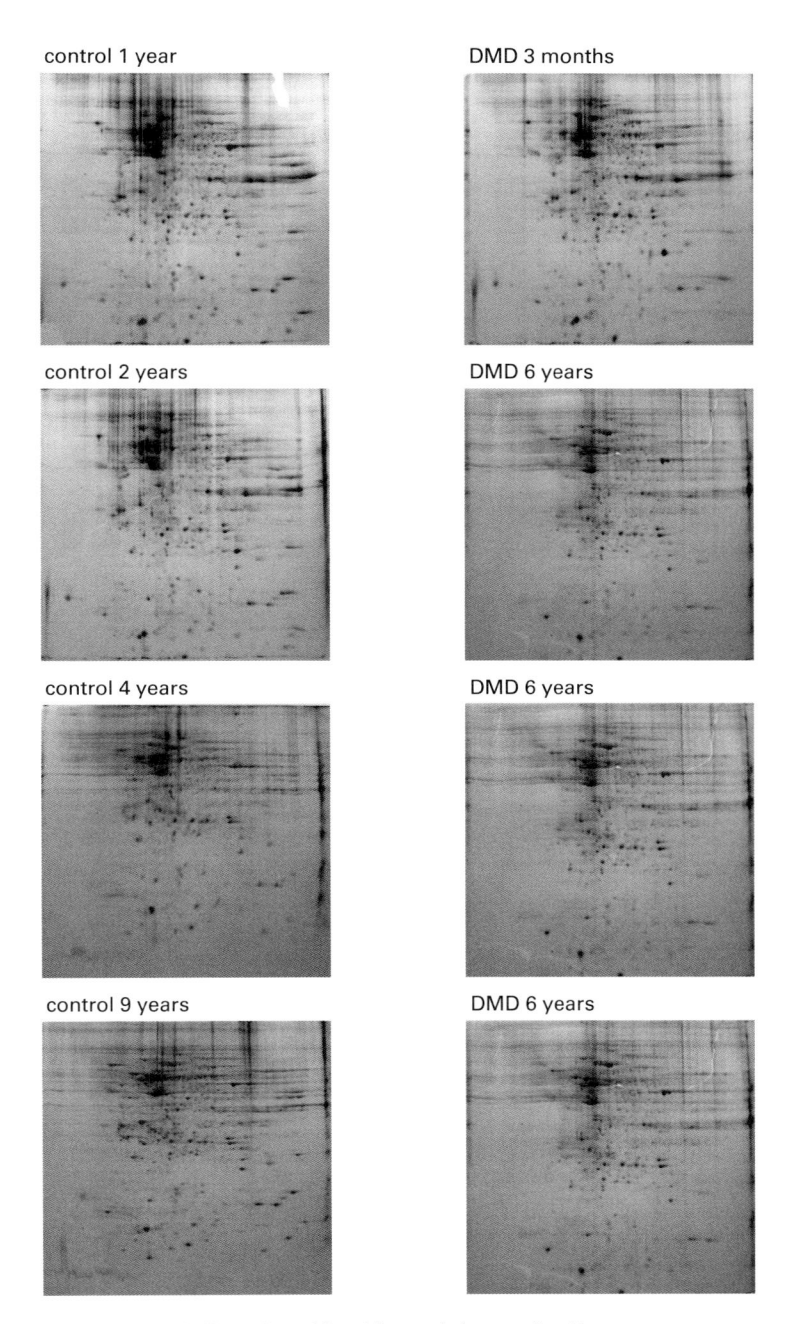

Figure 14.3: Master gels of all analyzed healthy and diseased cells.

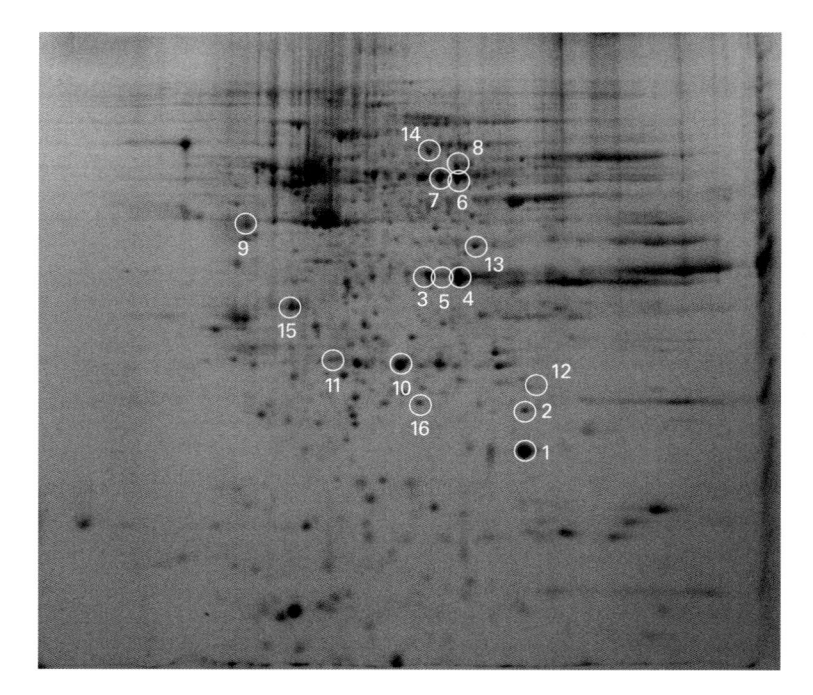

244 | 14 Proteomic analysis of Duchenne muscular dystrophy (DMD)

Figure 14.4: Comparison of all protein spots expressed differently in all analyzed samples.

14.4.2.3 Identified proteins

▶Table 14.3 shows all identified and differentially expressed proteins in DMD and control cells. Most of the identified proteins were overexpressed in the diseased cells. Very strongly overexpressed was the protein in spot 1, which was identified as the alpha-crystallin B chain. This protein is a small heat-shock protein and is present in skeleton, heart, lungs, kidney, nerves and especially in eyes. Heat-shock proteins are involved in different cellular processes, for example, protection against oxidative stress, cell growth, cell proliferation or differentiation (Welsh and Gaestel 1998; Guo-Thürmann 2008). The overexpression can be caused by oxidative stress or cell toxic substances, which lead to aggregation of proteins (Horwitz 2003; Launay et al. 2006; van den Ijssel et al. 1999; Vayssier and Polla 1998).

The highest overexpression of alpha-crystallin was found in the 3-month-old patient (▶Figure 14.5). Which function alpha-crystallin plays in signal transduction is not yet identified and needs further investigation. Another heat-shock protein identified is beta-1 in spots 10 and 11, which is involved in stress resistance and actin organization. It is known that many diseased cells show higher levels of heat-shock proteins.

In contrast, we found very low concentrations of underexpressed proteins in the diseased cells, like SOD, which is known as an antioxidant that destroys toxic radicals (▶Figure 14.5). An underexpression of SOD could be related to the very fast progression of DMD and consequently to death of the cells.

Another antioxidative, differently expressed protein is the thioredoxin-dependent peroxidase reductase, which is highly overexpressed. The higher concentration of this

14.4 Results and Discussion

Table 14.3: All identified and differentially expressed proteins in DMD and control cells

Spot	Protein name	M_W	Matched peptides	Score	Function	Up-/down-regulation
1	Alpha-crystallin B chain	20,146	10	681	Small heat-shock glycosylated protein	↑
2	Superoxide dismutase	24,707	5	365	Destroys radicals, antioxidant	↓
3	Annexin A1	38,690	10	1,296	Calcium-binding protein, promotes membrane fusion	↑
4	Annexin A1	38,690	10	1,261	Calcium-binding protein, promotes membrane fusion	↑
5	Annexin A1	38,690	8	1,056	Calcium-binding protein, promotes membrane fusion	↑
5	LIM and SH3 domain protein	29,698	5	433	Regulation of dynamic actin-based cytoskeletal activities	↑
6	Retinol dehydrogenase 1	54,827	10	876	Retinol oxidation for the synthesis of retinoic acid	↑
7	Retinol dehydrogenase 1	54,827	7	564	Retinol oxidation for the synthesis of retinoic acid	↑
8	Retinol dehydrogenase 1	54,827	10	817	Retinol oxidation for the synthesis of retinoic acid	↑
9	Reticulocalbin-1 precursor	38,866	7	564	Regulates calcium-dependent activities in ER	↑
10	Heat-shock protein beta-1	22,768	10	713	Involved in stress resistance and actin organization	↑
11	Heat-shock protein beta-1	22,768	9	484	Involved in stress resistance and actin organization	↑

(Continued)

Table 14.3: All identified and differentially expressed proteins in DMD and control cells (*Continued*)

Spot	Protein name	M_W	Matched peptides	Score	Function	Up-/down-regulation
12	Peroxiredoxin-1	22,096	6	287	Antioxidant enzymes, controls peroxide levels, mediates signal transduction	↑
13	Calreticulin	48,112	3	311	Calcium-binding chaperone promoting folding	↑
14	Dihydropyrimidinase-related protein 2	62,255	7	732	Signaling and remodeling of the cytoskeleton.	↑
15	Annexin 5	35,937	10	852	Anticoagulant in blood cascade	↓
16	Thioredoxin-dependent peroxidase	27,675	4	537	Involved in redox regulation of the cell, reduces peroxide	↑

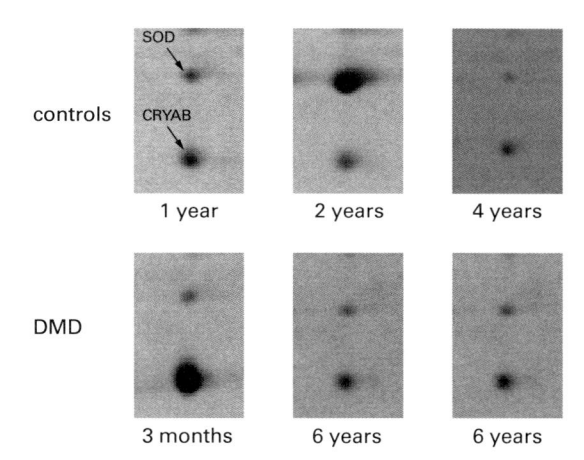

Figure 14.5: Comparison of differently expressed crystallin B (CRYAB) and superoxide dismutase (SOD) in patients of variable age.

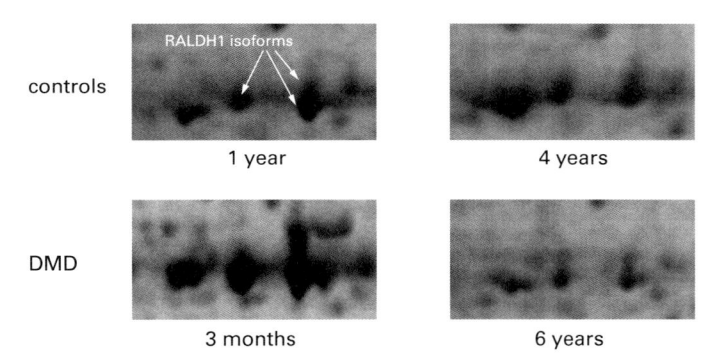

Figure 14.6: Comparison of retinal dehydrogenase in DMD and control cells in patients of different age.

protein corresponds likely to the very high concentration of alpha-crystallin in the cell and could be necessary to balance the action of alpha-crystallins and to prevent their aggregation.

The subtractive proteome analysis of DMD also allowed the analysis of some calcium-binding proteins (e.g. annexin A1, A5, reticulocalbin-1 and calreticulin). It is typical in DMD patients that intracellular calcium is elevated and becomes toxic. The high concentration of calcium is related to a permeable cell membrane, due to the absence of dystrophin. For this reason, not only calcium, but also a high concentration of creatin kinase is a typical biomarker for DMD. Only annexin 5 was underexpressed.

Interestingly, retinal dehydrogenase was found overexpressed in the DMD young cells (▶Figure 14.6). This enzyme is responsible for oxidation of retinal to retinoic acid, which is involved in the metabolism of fat. The high concentration of retinal dehydrogenase could be responsible for the loss of muscle cell mass and the formation of adipocytes. The cell culture experiments with different concentrations of retinoic acid are in progress.

14.5 Conclusion

The results show that proteome analysis can be applied to studies of DMD. The same procedure may be used for other diseases to elucidate new biomarkers to follow the disease process and its progression. We identified several proteins mostly overexpressed in DMD and in control cells. They are involved in different biological functions within the cells (e.g. calcium binding, antioxidative action, signal transduction and metabolism). The obtained data may allow further investigations on human cells with respect to developing new therapeutic strategies.

References

Davies, K.E., and Nowak, K.J. (2006). Molecular mechanisms of muscular dystrophies: old and new players. Nat Rev Mol Cell Biol 7, 762–73.

Grunwald, S. (2009). Identification and characterisation of muscular dystrophy Duchenne modifying genes and signal transduction pathways. Dissertation, Humboldt University, Berlin.

Guo-Thürmann, J. (2008). Proteomik von Muskelzellen. Diplomarbeit, Technische Fachhochschule, Berlin.

Hoffman, E.P., Brown, R.H., and Kunkel, L.M. (1987). Dystrophin: the protein product of the Duchenne muscular dystrophy locus. Cell *24;51*, 919–28.

Horwitz, J. (2003). Alpha-crystallin. Exp Eye Res *76*,145–153.

Launay, N., Goudeau, B., Kato, K., Vicart, P., and Lilienbaum, A. (2006). Cell signalling pathways to alpha B-crystallin following stresses of the cytoskeleton. Exp Cell Res *312*, 3570–3584.

Roberts, R.G., Coffey, A.J., Bobrow, M., and Bentley, D.R. (1993). Exon structure of the human dystrophin gene. Genomics *16*, 536–538.

Sakamoto, M., Yuasa, K., Yoshimura, M., Yokota, T., Ikemoto, T.S., Suzuki, M., et al. (2002). Micro-dystrophin cDNA ameliorates dystrophic phenotypes when introduced into mdx mice as a transgene. Biochem Biophys Res Commun *293*,1265–1272.

Speer, A., and Oexle, K. (2002). Muskeldystrophien. Handbuch der Molekularen Medizin. (Berlin: Springer-Verlag).

van den Ijssel, P., Norman, D.G., and Quinlan, R.A. (1999). Molecular chaperones: small heat shock proteins in the limelight. Curr Biol *9*,103–105.

Vayssier, M., and Polla, B.S. (1998). Heat shock proteins chaperoning life and death. Cell Stress Chaperones *3*, 221–227.

Welsh, M.J., and Gaestel, M. (1998). Small heat-shock protein family: function in health and disease. Ann N Y Acad Sci 851, *28*–35.

Zhou, W., Scott, S.A., Shelton, S.B., and Crutcher, K.A. (2006). Cathepsin D-mediated proteolysis of apolipoprotein E: possible role in Alzheimer's disease. Neuroscience 143, 689–701.

15 Target-oriented peptide arrays in a palliative approach to cystic fibrosis (CF)

Prisca Boisguerin, Dean R. Madden, and Rudolf Volkmer

15.1 Introduction

Structural analysis of functional protein complexes suggests at least two classes of protein-protein interactions (Jones and Thornton 1996; Ma et al. 2003; Reineke 2009). In the first class, which reflects the majority of protein-protein interactions, the complementary surfaces of the interacting partners are both extensive. In these cases, each binding surface is assembled from discontinuous sequence elements that are brought together upon protein folding. It is a great challenge to map such discontinuous binding sites (Reineke et al. 1998, 1999); however, the concept of hot spots (Bogan and Thorn 1998; Clackson and Wells 1995) shows the feasibility in principle of interfering with interactions that involve extensive surfaces. The second class comprises asymmetric interactions, in which a protein interaction domain (PID) docks with a short peptide sequence on the partner protein. In contrast to discontinuous binding sites, the binding determinants of a PID may be mapped to short linear motifs matching the sequence of the ligand peptide. The importance of PIDs, also called "small recognition domains," in the formation of protein complexes and cell signaling was first demonstrated in the late 1980s and early 1990s, and the role of PIDs in fundamental cellular functions such as transcription, scaffolding, endocytosis and proteostasis has since been confirmed (Cesareni et al. 2005; Pawson and Nash 2003).

Since the early 1990s, biological library techniques such as phage display (Scott and Smith 1990; Smith 1985), yeast two-hybrid (Fields and Song 1989) and pull-down assays (affinity chromatography) in combination with mass spectrometry (Gavin et al. 2002; Ho et al. 2002) have been the predominant tools for characterizing both classes of cellular protein-protein interactions. In the case of PID-interaction networks, phage display has become one of the major techniques for exploring highly diverse combinatorial peptide libraries (Tong et al. 2002). Additionally, bioinformatics and computational tools were developed to identify modular domains and their cognate ligands (Linding et al. 2005). Nowadays, several databases are freely available, such as MINT (http://mint.bio.uniroma2.it/mint), a public repository for molecular interactions reported in peer-reviewed journals, and the SMART database (http://smart.embl-heidelberg.de).

In comparison, array technologies, especially protein arrays, arrived relatively late in the field of protein-protein interactions (MacBeath and Schreiber 2000; Zhu et al. 2001) due to critical factors such as native folding stability or functionality. Peptides, in contrast, are easier to handle and retain many of the same function characteristics they exhibit in proteins. The fact that PIDs recognize short linear peptides perfectly corresponds to the scope of peptide array technologies and peptide libraries. Thus, peptide arrays are suitable to support proteomic research, particularly in the case of PID recognition events: revealing the binding specificity of small recognition domains, or screening for their cellular interaction partners or developing selective PID inhibitors.

Two techniques for the preparation of peptide arrays were published almost simultaneously: Frank presented the SPOT-synthesis technique (Frank 1992), while Fodor and coworkers (Fodor et al. 1991) reported the concept of light-directed, spatially addressable chemical synthesis. Both techniques represented milestones in the advancement of peptide array technologies. However, the majority of peptide arrays reported to date have been produced using the SPOT synthesis concept. This is due to the fact that SPOT synthesis is a very simple but extremely robust method for the highly parallel synthesis of peptides on planar surfaces. The method itself has been reviewed several times (e.g. Frank 1992; Reineke et al. 2005; Volkmer 2009), and protocols with useful modification have been developed (Hilpert et al. 2007; Wenschuh et al. 2000). Over a period of nearly 20 years, the SPOT synthesis technique, or for short SPOT technology, has become a widespread and essential tool in biology and biochemistry, with a literature base of more than 400 original, peer-reviewed papers. Nowadays, we are in the comfortable situation that a diverse collection of SPOT technology methods is available, permitting the application of peptide arrays to therapeutic targets, as described in Section 15.5 in detail for the case of CF.

15.2 PDZ domains

The interaction domains that mediate protein-peptide interactions generally cluster into a small number of families (e.g. PDZ, SH2, SH3, WW), each of which recognizes characteristic linear sequence motifs. These domains are conserved, functionally autonomous protein sequence modules, which can behave as independent genetic elements within genomes (Pawson and Nash 2003). The analysis of the resulting interaction networks ideally requires a quantification of domain specificity and selectivity toward all possible ligands with physiologically relevant affinity.

PDZ domains were discovered as sequence repeats apparent in the three proteins postsynaptic density 95 (PSD-95), discs large (Dlg) and zonula occludens-1 (ZO-1), and they are one of the most commonly found PIDs in organisms from bacteria to humans. They were originally referred to as DHR (discs large homology repeat) domains or GLGF repeats (after the four-residue GLGF sequence interacting with the carboxy-group of the PDZ domain ligands) (Cho et al. 1992; Kim et al. 1995; Woods and Bryant 1993). Shortly thereafter, the acronym PDZ (from the initial letters of PSD-95, Dlg and ZO-1) was proposed and adopted by the scientific community (Kennedy 1995). PDZ domains occur in one or multiple copies, primarily in cytoplasmic proteins. PDZ domain-containing proteins can be classified into three principal families according to their modular organization. The membrane-associated guanylate kinases (MAGUKs), including the proteins PSD-95, Dlg and ZO-1, which contain 1 or 3 PDZ domains, 1 SH3 domain and 1 guanylate kinase domain (GuK), make up the first family. The second family comprises proteins consisting largely of PDZ domains. The number of PDZ domains can vary from 2 to more than 10, as in the human multiple PDZ domain protein (MPDZ), which contains 13. The third family encompasses multidomain proteins without a GuK domain (Nourry et al. 2003).

PDZ domains are ~90 residues β-barrel structures flanked by α helices (Morais Cabral et al. 1996), and they typically recognize specific amino acids at the C-terminal end of peptide motifs or target proteins. In addition, some PDZs can interact with internal peptide motifs of target proteins, with another PDZ domain (Fan and Zhang 2002;

Hung and Sheng 2002; Nourry et al. 2003) or even with lipids (Zimmermann 2006; Zimmermann et al. 2002). The phenomenon of binding to the peptide C terminus is known as the classical or canonical binding mode. Peptide-ligand binding takes place in an elongated groove on the PDZ surface, in which the peptide adopts an extended conformation that forms an anti-pallelel β-sheet interaction with the second β strand (βB) and also binds to the second helix (αB). As shown in ▶Figure 15.1 for the third PDZ domain of the synaptic protein PSD-95 (Doyle et al. 1996), the free carboxylate group at the end of the peptide ligand forms hydrogen bonds with a carboxylate-binding loop that precedes βB and that contains the (R/K)xxxGψGψ motif, where ψ represents an aliphatic amino acid. The N and C termini of the PDZ domain are close to each other on the opposite side of the domain relative to the peptide-binding groove, an arrangement common to protein interaction modules.

The first studies with PDZ domains rapidly clarified which C-terminal residues were crucially required for protein interaction. Deletion and most mutations of the C-terminal residue drastically reduced binding affinity, as did many mutations at position −2 (position 0 [P^0] refers to the C-terminal residue, preceding residues are numbered as P^{-1}, P^{-2}, and so on). Songyang et al. (1997) confirmed these findings using a degenerate peptide library and described four major classes of PDZ domains: class I (S/T-x-Φ), class II (Φ-x-Φ), class III ([+]-x-Φ) and class IV (D-x-V), where x is any amino acid, Φ is a hydrophobic residue (V, I, L, A, G,W, C, M, F) and [+] is a basic, hydrophilic residue (H, R, K). The residues at P^0 and P^{-2} of the canonical peptide ligands have been described as playing the most critical role for the specificity and the affinity of the interactions.

However, several PDZ-mediated interactions do not conform to these four major canonical types of recognition, and new classes have been introduced to account for

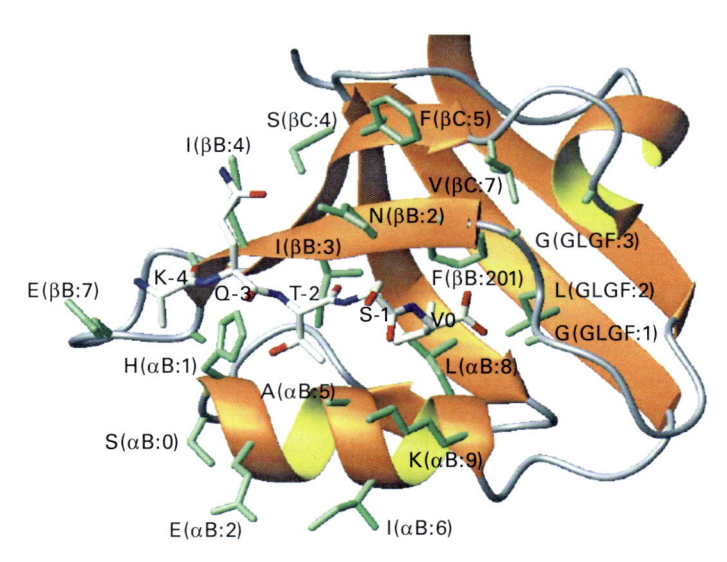

Figure 15.1: Ribbon model of the third PSD-95 PDZ Domain. Crystal structure (PDB: 1BE9) of the third PDZ domain from the synaptic protein PSD-95 in complex with the C-terminal peptide (sequence KQTSV$_{COOH}$) derived from cysteine-rich interactor of PDZ three (CRIPT) (Doyle et al. 1996).

them (Bezprozvanny and Maximov 2001). Consequently, it has been also shown that the classification based on the chemical properties of residues at the P⁰ and P⁻² positions within these C-terminal targets fail to predict the specificity of binding (Kurakin et al. 2007; Vaccaro and Dente 2002).

15.3 CF

CF is the most common fatal genetic disorder in populations of European descent (Welsh 1995). An autosomal recessive disease, CF is caused by loss-of-function mutations in a single protein, the cystic fibrosis transmembrane conductance regulator (CFTR) (Kerem et al. 1989; Riordan et al. 1989; Rommens et al. 1989), an epithelial ion channel that serves as a critical regulator of fluid and ion homeostasis in a variety of tissues (Anderson et al. 1991; Bear 1991). As a result, CF affects multiple organs (Boat and Cheng 1989; Rowe et al. 2005). Historically, it was identified with excessive levels of sweat chloride ("salty sweat"). In the gut, CFTR deficiency can lead to meconium ileus, a neonatal intestinal obstruction that is fatal if not treated. In the pancreas, loss of CFTR activity is associated with retention of digestive enzymes and associated deficits in nutrient absorption. Prompt clearance of meconium ileus and enzyme therapy represented critical advances in extending the life span of CF patients. However, the multiorgan nature of CF causes additional pathologies, including diabetes and liver disease, as well as congenital male infertility. Currently, the predominant cause of patient morbidity and mortality is the development of chronic airway infection and inflammation (Rowe et al. 2005). CFTR activity is required for airway mucociliary clearance, and CF patients are unable to efficiently clear inhaled pathogens from the lung. Compounding the problem, many opportunistic pathogens develop high levels of drug resistance in the lung, particularly associated with the formation of biofilms – clusters of microbes embedded in a viscous matrix (Cos et al. 2010). As pathogens establish persistent, drug-resistant colonies, they also contribute to inflammatory responses that further compromise lung function (Davies and Bilton 2009; Ziady and Davis 2011).

Over the past several decades, therapeutic advances have substantially extended the life span of CF patients, so that the median age of survival now exceeds 35 years (Cystic Fibrosis Foundation Patient Registry 2010). However, most advances to date have been due to the improved treatment of symptoms, rather than correction of the underlying molecular defect of CF – the loss of CFTR function in epithelial tissues. More than 1,800 mutations have been identified in CF patients (http://genet.sickkids.on.ca/app), but a single mutation, loss of a phenylalanine at position 508 (ΔF508-CFTR), accounts for 70% of alleles (Kerem et al. 1989). As a result, half of CF patients are homozygous for ΔF508-CFTR, and another 40% carry one copy. Furthermore, although substantially impaired, ΔF508-CFTR does retain some channel activity (Dalemans et al. 1991; Denning et al. 1992; Drumm et al. 1991), and strategies that rescue the mutant protein could thus potentially address the basic defect of CF in these patients.

There are three distinct functional deficits associated with the ΔF508 mutation. ΔF508-CFTR protein folds inefficiently, leading to accumulation in the endoplasmic reticulum (ER) and ER-associated degradation (Lukacs et al. 1994; Ward and Kopito 1994; Ward et al. 1995). Furthermore, protein that does fold and reach the plasma membrane exhibits reduced channel activity (Dalemans et al. 1991; Denning et al. 1992; Drumm et al. 1991). Finally, mature protein is also subject to accelerated degradation (Lukacs et al. 1993, 1994; Swiatecka-Urban et al. 2005). Extensive screening efforts have identified small-molecule compounds that serve either as "correctors" of the folding defect and/or as "potentiators" of channel activity, addressing the first two of these defects (Sloane and Rowe 2010). However, until recently, comparatively little activity has been directed at the identification of "stabilizers," compounds that could specifically extend the half-life of the rescued protein. Potential targets for stabilizing interventions include components of the peripheral quality-control, ubiquitination/deubiquitination and endocytic pathways that control the uptake, recycling and degradation of mature CFTR at the apical membrane (Bomberger et al. 2009; Cheng et al. 2002; Cholon et al. 2010; Favia et al. 2010; Okiyoneda et al. 2010; Ye et al. 2010).

15.4 Role of PDZ domains in CFTR trafficking

The CFTR C-terminal tail ($DTRL_{COOH}$) conforms to the binding motif for class I PDZ domains, and deletion of the PDZ-binding sequence has been reported to drive diverse effects on CFTR function and localization (Benharouga et al. 2003; Ostedgaard et al. 2003; Swiatecka-Urban et al. 2002, 2005). Correspondingly, a number of PDZ-containing proteins have been identified among the trafficking and scaffolding proteins that bind CFTR. These include NHERF1 (Na^+/H^+ exchanger regulatory factor 1, also known as EBP50), NHERF2 (also known as E3KARP) and CAL (CFTR-associated ligand, also known as PIST, GOPC and FIG) (Guggino and Stanton 2006; Li and Naren 2005). NHERF1 contains two PDZ domains, is localized in the apical compartment and is believed to tether CFTR to cytoskeletal actin filaments via interaction with the adaptor protein ezrin (Short et al. 1998). Although less well studied, NHERF2 is highly homologous to NHERF1 and seems to have similar functions (Hall et al. 1998; Li et al. 2002). Both of these PDZ proteins can mediate CFTR dimerization or multimerization and can couple CFTR with kinases and membrane receptors that regulate its channel activity. In contrast, a third CFTR-binding PDZ protein, CAL, is mainly localized in the Golgi compartment, and particularly in the trans-Golgi network (Cheng et al. 2004). CAL contains two N-terminal coiled-coil domains, which promote dimerization, followed by a single PDZ domain. The PDZ domain specifically binds the C-terminal end of CFTR (Cheng et al. 2002; Cushing et al. 2008). Consistent with the differences in their localization, the different CFTR-binding proteins exhibit countervailing effects on CFTR. Overexpression of CAL dramatically reduces the half-life and net expression level of CFTR at the cell surface via lysosome-mediated degradation in a dose-dependent manner (Cheng et al. 2002, 2004). This effect can be blocked by mutation of the CAL PDZ domain, and RNAi-mediated knockdown of CAL significantly increases the apical expression of ΔF508-CFTR (Wolde et al. 2007). CAL's effect can be reversed by the concomitant overexpression of NHERF1 (Cheng et al. 2002, 2004), and overexpression of NHERF1 by itself has been shown to increase apical levels of ΔF508-CFTR (Guerra et al. 2005). These data suggest that the two PDZ proteins compete for binding to the CFTR C terminus and mediate antagonistic effects: NHERF1 favors recycling to the plasma membrane, whereas CAL favors degradation.

Taken together, these facts suggest that development of CAL-selective inhibitors using peptide array screenings could rescue CFTR at the plasma membrane. Selective CAL inhibitors also represent a potentially novel class of CFTR modulators, allowing us to investigate the possibility of cooperative rescue in conjunction with "correctors" of the primary folding defect of the most common CF-associated allele, ΔF508-CFTR.

15.5 Target-oriented peptide arrays

In general, synthetic peptide arrays can be utilized either for qualitative exploratory investigations or for quantitative analytical studies, and the initial choices of array content and assay design often depend on which kind of analysis is required. Exploratory investigations may involve either simple or comparative functional assays. In a simple functional assay, the noncovalent binding interactions between a single sample and a set of immobilized peptides (probes) are analyzed in parallel to obtain relative peptide-binding preferences for the sample. The sample is usually a purified protein, which is challenged with relevant peptide probes collected in a peptide array. These peptides are defined by some shared characteristics, for example, the targets of a consensus on a sequence data bank or frame and length versions of a protein sequence. Such an array is synthetically generated on a cellulose membrane using SPOT technology (Frank 1992; Volkmer 2009). We use membranes with a dimension of 18.3 x 28.5 cm for the special platform of the fully automated synthesizer (MultiPep, Intavis AG). As a result, if different kinds of peptide arrays are required, they can be generated on a single membrane to ensure that they are produced under the same chemical conditions. In such cases, the membrane is cut after synthesis to yield individual arrays, which are then individually assayed. Generally, for a simple functional assay, previous knowledge of absolute binding affinity is not required, and densitometric measurements yield internally consistent qualitative data. Intraarray comparison of spot signal intensities yields a reliable ranking of peptide binding. However, it is important to realize that no quantitative information at all is available from simple functional assays; the affinity may be in the range of micromolar or even nanomolar and cannot be calculated from measured spot signal intensities. However, an impression of affinity can be achieved if binding affinities of several array peptides are known or have been determined independently by quantitative affinity measurement like surface plasmon resonance (SPR). Then spot signal intensities can be correlated to the real binding affinities, yielding a semiquantitative binding assay. Such an approach was applied to a proteomic functional assay of yeast SH3 domains interacting with putative immobilized binders of the yeast proteome (Landgraf et al. 2004).

Comparative functional assays are based on interarray comparison. For instance, two peptide arrays with identical content are challenged with two related proteins (e.g. a wild-type [WT] protein and a related mutant version). Spot signal intensities are then interarray comparable when proteins are applied at identical concentration and when arrays are of equal synthetic quality. Equal peptide array quality is achieved by applying identical chemical conditions during synthesis. Therefore, we strongly recommend taking peptide arrays from the same cellulose membrane (intramembrane arrays) because this guarantees the generation of spots with peptide densities (peptide concentration in a spot) of the same magnitude (Weiser et al. 2005). Such a comparable functional assay was demonstrated for challenging the PQBP WW domain and its Y65C missense mutant with two arrays of identical potential human WW domain binding sequences (Tapia et al. 2010). However, as mentioned in Section 15.3 for the simple functional assays, comparative functional assays are also inadequate for the direct calculation quantitative affinity values. Finally, we would like to stress that intramembrane arrays are a good choice when simple functional assays should be repeated.

Analytical assays, on the contrary, are real quantitative binding assays. Intramembrane arrays with identical content are probed in parallel with different concentrations of a protein sample. The approach is shown in ▶Figure 15.2 where uniform intramembrane peptide arrays are challenged with different concentrations of WW domains and spot signal intensities were measured. Plotting spot signal intensities against the applied protein concentrations resulted sigmoid graphs from which EC_{50} (half effective concentration) values could be calculated for each array peptide (▶Figure 15.2B). As a good laboratory practice the peptide arrays should contain repeated sequences in order to obtain reliable error bars of the calculated EC50 values (Weiser et al. 2005).

However, one should keep in mind that the obtained EC_{50} values are calculated under the following assumptions: immobilized peptides are all of the same quality, peptide concentrations are identical at each spot, peptide accessibility is the same for each spot, protein tags do not influence protein-peptide binding and the measured protein concentration is the effective assay concentration. In practice, however, this will never be the case, and, strictly speaking, the obtained quantitative EC_{50} values are reflections of the true dissociation constants (K_D), depending on knowledge of the quality and concentration of peptide in each spot, as well as the effective concentration of the protein. High throughput screening, however, generally does not accommodate evaluation of these factors, and as a result, determination of K_D values is based on single experiments in practice. Nevertheless, especially for proteomic approaches, EC_{50} values are extremely helpful to create a quantitative protein interaction network. This has been shown for a human proteomic set of SH2 and PTB domains with phosphorylated peptides from ErbB

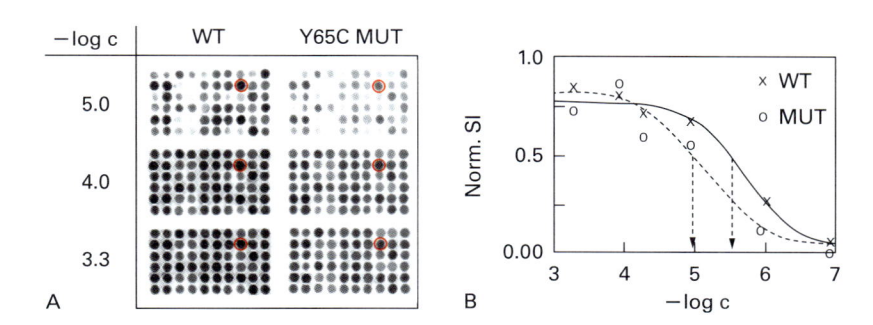

Figure 15.2: The principles of an analytical assay based on peptide arrays. (A) Uniform peptide arrays were synthesized in multiple copies on one cellulose membrane. Afterward the membrane is cut to yield individual arrays, which are then individually assayed. Exemplarily six peptide arrays are shown each displaying the same set of WW domain binding peptides formerly recovered from a simple functional peptide array. Left column: Arrays were assayed with different concentrations of the WT PQBP1 WW domain. Right column: Arrays were probed with different concentrations of the PQBP1 Y65C WW domain mutant. The spot signal intensities are measured and as shown spot signal intensities are proportional to the concentration of both WW domains. (B) Exemplarily, calculation of binding affinity is shown for a peptide depict by the red circle. The obtained spot signal intensities (SI, here normalized) are plotted against domain concentrations and sigmoid graphs for the WT (solid line) as well as for the Y65C mutant domain (dashed line) are resulted. At the inflection point, the binding affinity could be calculated as EC_{50} (half effective concentration) values. As shown by the curve shift of the Y65C mutant the assigned binding affinity is significantly reduced (higher EC_{50} value) compared to the WT domain.

receptor tyrosine kinases (Jones et al. 2006). The resulting network was able to describe the selectivity of recruitment of proteins to the receptor at various affinity thresholds and also to demonstrate that receptor tyrosine kinases become more promiscuous when overexpressed. This and an associated publication (Gordus and MacBeath 2006) show the importance of applying different analyte concentrations to obtain saturation curves from which essential experimental parameters, such as estimates of the amount of active probes on the surface, can be calculated. Alternatively, solution binding approaches such as fluorescence polarization can be used to determine individual K_D values directly (Vouilleme et al. 2010).

PDZ proteins require a very special kind of synthetic peptide array. As mentioned in Section 15.4, PDZ domains generally recognize the C-terminal four to seven residues of their protein binding partner and require a free C terminus for ligand recognition. In other words, PDZ domains recognize short linear peptides containing a free C terminus. Unfortunately, standard SPOT-synthesized peptides lack free C termini due to their coupling to the cellulose support. The first reliable and robust SPOT synthesis concept for synthesizing inverted peptide arrays with free C termini was published in 2004 by the Volkmer lab, and the approach was recently improved (Boisguerin et al. 2004, 2007). ▶Figure 15.3 depicts the major chemical aspects of the approach. In contrast to the standard SPOT synthesis protocol, synthesis of inverted peptides was performed on a cellulose membrane carrying a stable N-functionalized anchor (Licha et al. 2000), which retained the inverted peptides. The inverted and N-terminally fixed peptides (step 8 in ▶Figure 15.3) display a free C terminus resulting from reversal of the peptide orientation, achieved by successive thioether-cyclization/ester-cleavage (▶Figure 15.3F, G).

▶Figure 15.4 shows the key SPOT synthesis steps of the procedure actually used to prepare an inverted peptide array. Fortunately, the 3-brompropyesters (Fmoc-aa-OPBr) (2) used in the first approach (▶Figure 15.3) can be replaced by the commercially available 4-hydroxymethylphenoxyacetic acid (HMPA) linker (▶Figure 15.4). This permits convenient acidic ester bond hydrolysis while performing the linearization step.

15.6 An engineered peptide inhibitor of CAL extends the half-life of ΔF508-CFTR

To initiate the engineering of CAL-specific inhibitors, we decided to exploit the advantages of inverted peptide arrays for exploring PDZ binding preferences. Arrays encoding a human C-terminal peptide library (6223HumLib) were synthesized and probed in a qualitative binding assay with each of the five PDZ domains (CALP, N1P1, N1P2, N2P1 and N2P2). The 80 best binding sequences (highest signal intensities) of the respective incubations were analyzed using the WebLogo algorithm, revealing clear C-terminal binding motifs (▶Figure 15.5A).

By comparing all binding motifs resulting from the HumLib and substitutional analyses (SubAnas), we could clearly determine a preference for isoleucine at position 0 for CALP, which is not observed for the NHERF PDZ domains (▶Figure 15.5B). This illustrates the general principle of obtaining a selective ligand by focusing on determinants that engage the target but not the countertarget domains. Although the four C-terminal amino acids mostly contribute to high affinity PDZ-ligand interactions, positions upstream of the C-terminal core may further enhance selectivity. Combinatorial libraries (CombLib) are an appropriate tool to determine upstream positions in a pairwise manner.

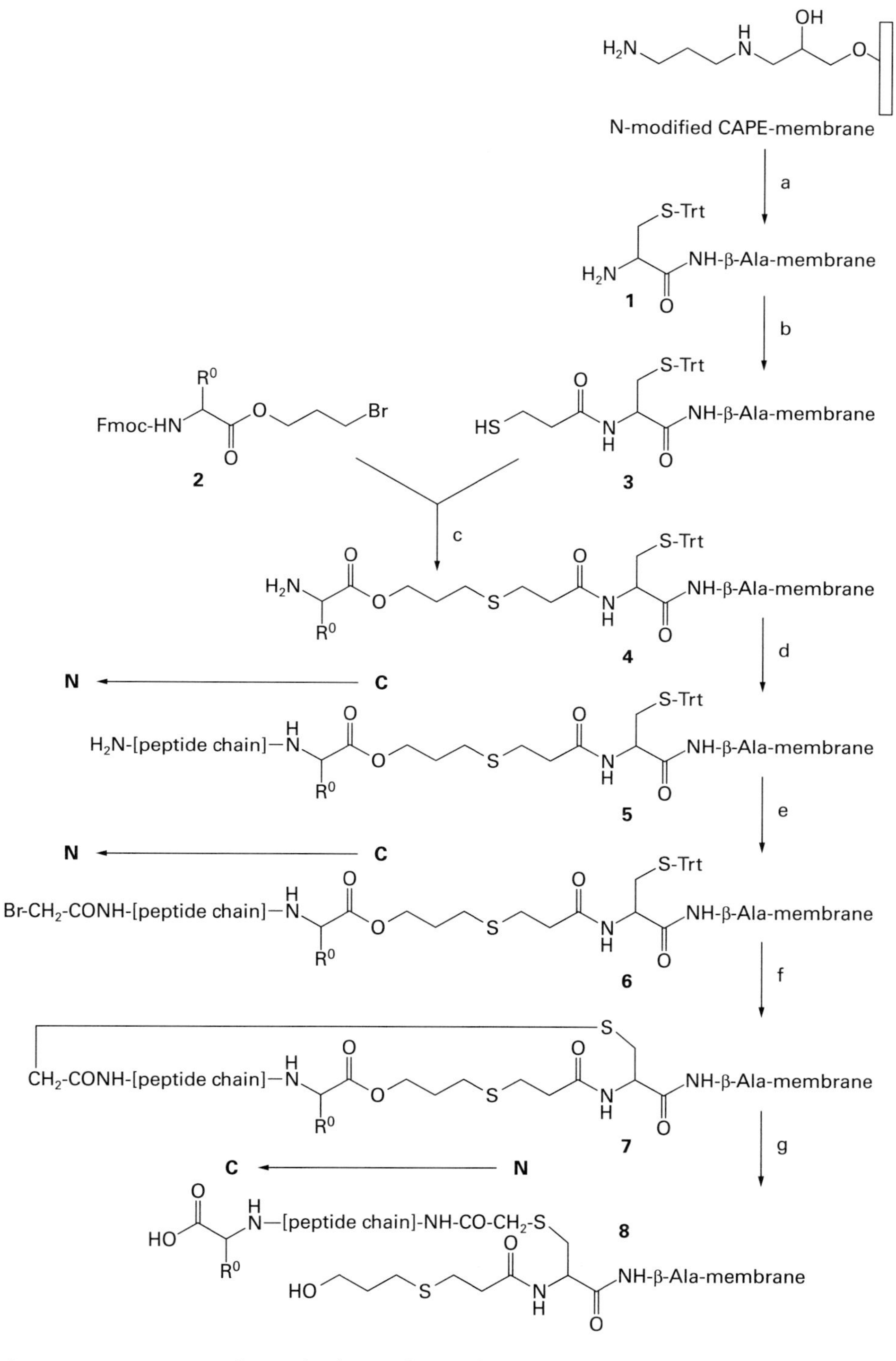

Figure 15.3: Reaction scheme for the synthesis of inverted peptides on cellulose membranes. Key compounds in the synthesis of inverted peptide arrays are Fmoc-amino acid
(Continued)

Figure 15.3: (*Continued*)

3-brompropyesters (Fmoc-aa-OPBr) (2), the membrane-bound mercaptopropionyl cysteine adduct (3), the matrix-bound amino acid ester derivative (4) and the cyclic peptide (7). Reaction conditions: (A) Fmoc-β-alanine-Opfp in DMSO, then piperidine (spacer-formation), followed by Fmoc-cysteine-(Trt)-Opfp in N-methylpyrolidone (NMP) and again followed by piperidine; (B) Mmt-S-CH$_2$-CH$_2$-COOH in DMF preactivated with ethyl 1,2-dihydro-2-ethoxyquinoline-1-carboxylate (EEDQ) and, subsequently, dichloroacetic acid, TFA, triisobutylsilane (TIBS), dichloromethane (DCM) for Mmt-cleavage; (C) aqueous solution of Cs$_2$CO$_3$, then a Fmoc-amino acid 3-bromopropyl ester in DMF and piperidine; (D) peptide synthesis using the standard SPOT synthesis protocol; (E) Fmoc-β-alanine-Opfp in NMP, then piperidine and 2,4-dinitrophenyl-bromoacetate in NMP; (F) TFA, TIBS, DCM and, subsequently, aqueous solution of Cs$_2$CO$_3$; (G) saturated aqueous solution of Li$_2$CO$_3$.

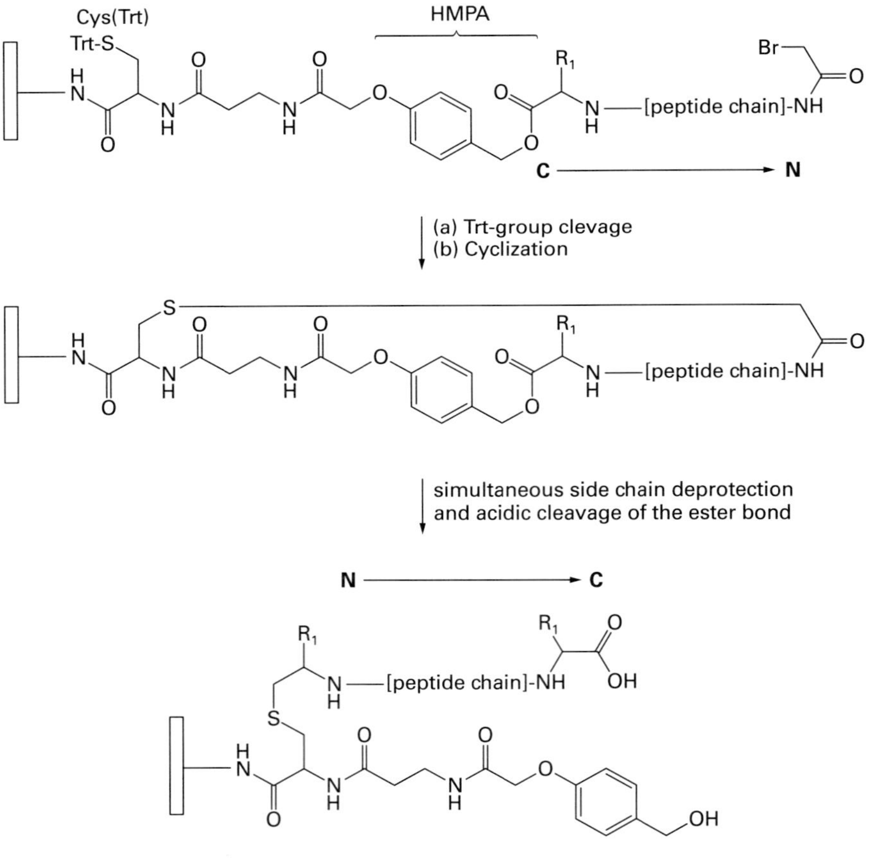

Figure 15.4: The actual approach for the synthesis of inverted peptides on cellulose membranes. First row: The sequence β-ala-cys(Trt)-β-ala forms the anchor molecule, and the cleavage site is formed by HMPA and the C-terminal amino acid. Peptide completion is carried out via standard SPOT synthesis conditions. As shown in Figure 14.3, bromoacetic acid is attached as the cyclization moiety. Second row: After selective cleavage of the Trt-group, cyclization occurs on treatment with cesium carbonate. Third row: Hydrolysis and side-chain deprotection under acid conditions occurs.

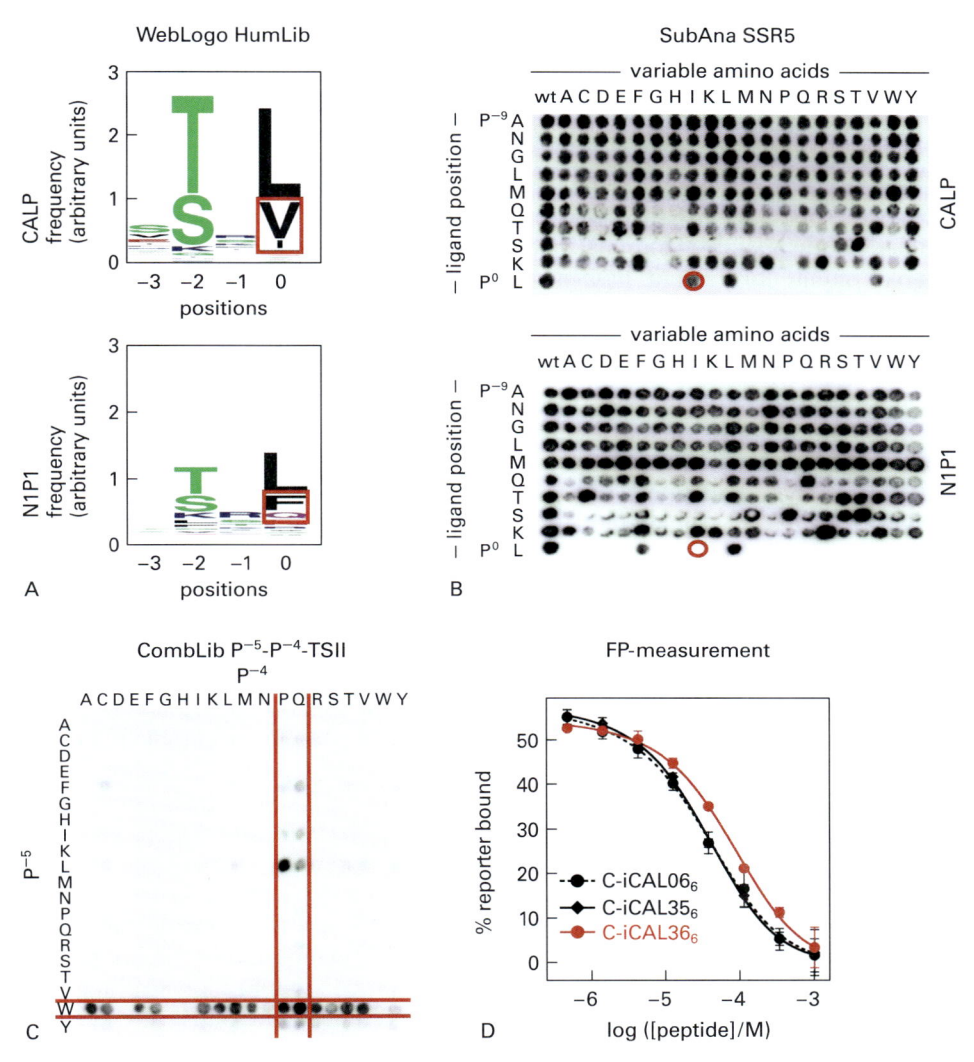

15.6 An engineered peptide inhibitor of CAL extends | 259

Figure 15.5: Consensus motif of two PDZ domains involved in CFTR binding. (A) 6223-HumLibs were incubated with the CALP and N1P1. The four most frequent C-terminal amino acids of the 80 best binding peptides were plotted using WebLogo. Amino acid abundance is defined by the letter size. (B) Substitutional analyses (SubAna) of the C-terminal SSR5 sequence demonstrate the interaction with CALP and N1P1. (C) To determine the CALP preference at position −4 and −5, we synthesized a CombLib. The recognition pattern shows a clear preference regarding the combination of glutamine and trytophane as well as proline and trytophane for the position −4 and −5. (D) The three peptides iCAL06$_6$, iCAL35$_6$ and iCAL36$_6$ were measured with the five different PDZ domains. By using W/P or W/Q at position −5/−4, respectively, we were able to enhance the selectivity of CALP. Especially the W/P combination abolished the binding affinity to the NHERF PDZ domains (K_i >5,000 μM, data not shown). Data were fit to single-site binding models. Values shown are mean ± SD, $N = 3$.

For example, a CombLib focused on P^{-5}/P^{-4} identified the combinations W/P and W/Q, both of which were found to improve the affinity and selectivity of the ligand interaction with CALP (▶Figure 15.5C). Every single step of peptide array screening technologies was validated by Fluorescence Polarisation (FP) displacement assays in a parallel manner (▶Figure 15.5D).

The resulting iCAL36 peptide inhibitor achieved a remarkable degree of selectivity among PDZ domains that share a common target and primary binding motif. When modified to include an N-terminal fluorescein moiety, F*-iCAL36 exhibits single micromolar affinity for the CAL PDZ domain. In contrast, its affinity for each NHERF1 and NHERF2 domain was millimolar or weaker, corresponding to a >750-fold drop relative to CAL (Vouilleme et al. 2010). The inhibitor selectivity is further enhanced when compared to the inverted affinity profile of the CFTR C terminus – the target of displacement. As shown previously, CFTR binds the NHERF PDZ domains with generally submicromolar affinity, whereas it binds CAL only weakly, with a K_d value exceeding 100 µM (Cushing et al. 2008). As a result, the destabilizing CAL:CFTR interaction is a "soft" target, whereas the functionally favorable NHERF:CFTR interactions represented "hardened" targets that should more resistant to inhibition.

Protein-trafficking interactions and pathways can depend critically both on cell type and on the functional state of a given cell line. In order to assess the ability of iCAL36 to rescue ΔF508-CFTR function in a physiologically relevant context, we therefore decided to probe the effect of inhibitors in polarized epithelial cells that are derived from the airway of a CF patient and that express ΔF508-CFTR (CFBE+ΔF cells) (Bruscia et al. 2002; Li et al. 2006). Fluorescence microscopy demonstrated peptide uptake in polarized CFBE+ΔF cells treated with BioPORTER reagent in combination with fluorescein-labeled iCAL36 inhibitor or scrambled control peptides. Peptide-loaded CFBE+ΔF cells were mounted in an Ussing chamber, and short-circuit currents (I_{sc}) monitored to reveal CFTR chloride-channel activity. Consistent with the proposed role of CAL in postendocytic CFTR trafficking, incubation with iCAL36 produced a significant increase in CFTR-specific I_{sc} compared to incubation with the control peptide (Cushing et al. 2010). CAL knockdown increased the apical-membrane half-life of ΔF508-CFTR, and blockade of protein synthesis with cycloheximide did not prevent the iCAL36-mediated increase, confirming a role for protein stabilization (Cushing et al. 2010). Furthermore, the magnitude of the effect was comparable to that seen with the first-generation corrector corr-4a (Pedemonte et al. 2005), and simultaneous treatment with both corr-4a and iCAL36 led to an additive effect, suggesting that correctors and stabilizers act via nonredundant pathways (Cushing et al. 2010).

15.7 Methods

Detailed methods have been previously described (Boisguerin et al. 2004, 2007; Cushing et al. 2008, 2010; Vouilleme et al. 2010) and are reproduced in this section to provide a comprehensive overview.

15.7.1 PDZ domain and peptide motif nomenclature

Homologous residue positions are numbered according to a general PDZ domain numbering scheme (based on Songyang et al. 1997) and modified to account for the PDZ domain secondary structure elements (SSE) consensus. The residue number (e.g. αB:8) is composed of a secondary structure prefix (αB = second conserved alpha helix) and a number indicating the position of this residue within the conserved SSE (8 = 8th residue). Highly conserved sequence motifs outside SSEs are named separately (e.g. GLGF:2 indicates the L residue within the GLGF-loop).

All peptide sequence motifs are given according to a modified Seefeld Convention 2001 nomenclature (Aasland et al. 2002): x denoting any amino acid; B, a mixture of 17 L-amino acids (without C/M/W); Φ, hydrophobic amino acids; ψ, aliphatic amino acids; [+], positively charged amino acids. All amino acids are given in the one-letter code. Within a target sequence pattern, the different amino acids of one position are ordered due to their frequency of occurrence.

15.7.2 Improved method of inverted peptides

The peptides were synthesized on N-modified CAPE-membranes (Bhargava et al. 2002) and prepared by a MultiPep SPOT-robot (INTAVIS Bioanalytical Instruments AG, Cologne, Germany). Array design was performed with the aid of the in-house software LISA 1.83. The synthesis started with spot definition by a standard protocol (Frank 1992), followed by the coupling of a solution of Fmoc-cysteine-(Trt)-Opfp (0.3 M) in NMP and Fmoc-alanine-Opfp (2x coupling, 15 min reaction each). After Fmoc cleavage with piperidine in DMF (20%), HMPA dissolved in DMF (0.6 M solution) and activated with EEDQ (1.1 equiv) was added, and samples were directly spotted on the membrane (4x coupling, 15 min reaction each). The membrane was acetylated with acetic anhydride (20%) in DMF; washed with DMF (5x3 min), ethanol (2x3 min) and diethyl ether (2x3 min); then finally air dried. Solutions of Fmoc-amino acid-OH (0.4 M) activated with 1,1'-carbonyldiimidazole (CDI, 3 equiv) (Ay et al. 2007) in DMF were spotted on the membrane (4x coupling, 15 min reaction time each). Proline, tyrosine and glutamine were activated with 1,1'-carbonyldi(1,2,4-triazole) (CDT) (Ay et al. 2007). The membrane was acetylated to quench the hydroxyl groups of the HMPA linker with acetic anhydride (20%) and diisopropylethylamine (DIPEA) (10%) in DMF; washed with DMF (5x3 min), ethanol (2x3 min) and diethyl ether (2x3 min); then finally air dried. The Fmoc group was removed from the spots, and the sequences of the peptides were completed by the standard SPOT synthesis protocol (Frank 1992) followed by an N-terminal tag with β-alanine. For standard SPOT synthesis, Fmoc-aa-Opfp were used with the following side-chain protection: E-, D-(OtBu); C,- S-, T-, Y-(tBu); K-, W-(Boc); N-, Q-, H-(Trt); R-(Pbf).

For thioether cyclization, all peptides were N-acylated with bromoacetic acid 2,4-dinitrophenyl ester in NMP (1 M), double coupling, 15 min reaction time each. The membrane was washed with DMF (3x3 min) and DCM (3x3 min) and dried.

To enable cyclization, the trityl side-chain protecting group of the cysteine was cleaved with TFA (7%), H_2O (2%) in DCM (1x5 min) followed by TFA (7%), TIBS (3 %), H_2O (2%) in DCM (3x5 min, each step). The membrane was washed with DCM (3x3 min), DMF (2x3 min) and DMF (2x3 min). The peptides were cyclized overnight by treatment with aqueous Cs_2CO_3/H_2O/DMF (1:1:2). The membrane was washed with DMF (2x3 min), H_2O (2x3 min), EtOH (2x3 min) and diethyl ether (2x3 min) and air-dried.

Hydrolysis and side-chain deprotection were achieved through one treatment with TFA (60%), TIBS (3%) and H_2O (2%) in DCM for 2.5 h without shaking, followed by washing steps (DCM 3x3 min, DMF 3x3 min, EtOH 3x3 min, diethyl ether 2x3 min), followed by TFA (90%), TIBS (3%) and H$_2$O (2%) in DCM for 30 min without shaking. The membrane was washed with DCM (3x3 min), DMF (3x3 min), EtOH (2x3 min), phosphate buffer (pH 7.4, 2x3 min), H$_2$O (2x3 min), EtOH (2x3 min) and diethyl ether (2x3 min) and air-dried.

15.7.3 Introduction of the different library types

Library definition: 6223HumLib = 6223 C termini (11 mers) of human proteins; SubAna = substitutional analyses (each residue of the ligand was substituted by 20 L-amino acids); CombLib = combinatorial library of the type B1-B2-(x)n (B1 and B2 are the positions which will be permuted and x represents sequence specific amino acids that were held constant).

15.7.4 PDZ domain expression and purification

NHERF1 (SwissProt accession O14745; N1P1; residues 1 139; N1P2; residues 133–234) and NHERF2 (SwissProt accession Q15599; N2P1; residues 1 151; N2P2 residues 143–230) constructs were subcloned as described previously (Wolde et al. 2007). The vectors encoding the WT CAL PDZ domain (SwissProt accession Q9HD26-2; CALP; residues 278–362) have also been described previously (Cushing et al. 2008). All constructs were verified by DNA sequencing at the Dartmouth Molecular Biology Core Facility. Bacteria transformed with individual PDZ constructs were induced with Isopropy-ß-D-1-thiogalacto-pyranoside (IPTG), harvested and lysed, and the lysates were clarified as previously described (Wolde et al. 2007), except that NHERF1 and NHERF2 PDZ domains were expressed in Rosetta 2 (DE3) cells (Novagen). Immobilized metal-affinity and size-exclusion chromatographic (SEC) purification of the PDZ domains of NHERF1, NHERF2 and CAL was performed essentially as previously described for CALP (Cushing et al. 2008; Wolde et al. 2007). Proteins were concentrated using Amicon Ultra-15 5000 MWCO (NHERF1 and NHERF2 PDZ domains) or YM-3 concentrators (CAL PDZ domain constructs) (Millipore). Following concentration, proteins were dialyzed into storage buffer: 25–50 mM NaH2PO4, pH 7.4 (N1P1, N1P2 and N2P1), 25 mM Tris, pH 7.5 (N2P2), or 25 mM Tris, pH 8.5 (CAL PDZ).

15.7.5 Binding studies of cellulose-bound peptides

The peptide libraries were prewashed once with EtOH (1x10 min), with Tris-buffered saline (TBS buffer), pH 8.0 (3x10 min), and then blocked for 4 h with blocking buffer (blocking reagent [Sigma-Genosys, Cambridge, MA, USA] in TBS buffer, pH 8.0, containing 5% sucrose). The membranes were incubated with polyhistidine-tagged PDZ domains (10–20 µg/mL) in blocking buffer overnight at 4°C. His-tagged PDZ domains were detected using a mouse anti-polyHis antibody (Sigma; 1:2,600 in blocking buffer, 2 h at RT) followed by horseradish-peroxidase conjugated anti-mouse antibody (Calbiochem; 1:2,000 in blocking buffer, 1 h at RT). To remove excess antibody the membrane was washed with TBS buffer pH 8.0 (3x10 min). Uptilight HRP blot chemiluminescent substrate (Uptima) was applied for detection using a LumiImager® (Boehringer Mannheim GmbH, Mannheim, Germany). The signal intensities were recorded as Biochemical Light Units (BLU) using the LumiAnalystTM software.

15.7.6 Fluorescence polarization

A single stock solution was prepared in FP buffer containing fixed concentrations of both fluorescently labeled reporter peptide and protein. This mixture was allowed to equilibrate for 20–60 min at RT. Unlabeled competitor peptide was dissolved and serially diluted in DMSO (Fluka). Each serial dilution was aliquoted at 1/20 final volume, to which was added 19/20 volume of the protein:reporter mixture. The final reporter peptide concentration was 30 nM and the final protein concentration was 0.25 K_D–3.0 K_D, depending on the measurement. Plates were mixed by vibration, centrifuged and allowed to incubate for an additional 15 min at 24°C in the microplate reader before measurement. Fluorescence polarization was determined at an excitation wavelength of 485 nm and an emission wavelength of 525 nm.

For NHERF1 PDZ domains, the reporter peptide used was F^*-CFTR$_6$ (F^*-VQDTRL), and for NHERF2 domains, it was F^*-CFTR$_{10}$ (F^*-TEEEVQDTRL). For CALP, it was either F^*-iCAL36 (F^*-ANSRWPTSII) or F^*-C-SSR5 (F^*-C-ANGLMQTSKL). For weakly interacting peptides, K_i values were estimated as follows: theoretical fluorescence anisotropy values were calculated for inhibitor K_i based on the known reporter:protein fluorescence anisotropy (FP$_{PL}$), free reporter anisotropy (FP$_L$) and the known K_D of the reporter:protein complex. The estimated K_i value was increased until the theoretical fluorescence anisotropy value (with assumed equal variance) at the 1 mM concentration significantly increased ($P < 0.05$) above experimental fluorescence anisotropy value.

15.7.7 Cell culture condition for differentiated CF cells

CFBE41o- cells engineered to express ΔF508-CFTR from a cytomegalovirus promoter (CFBE+ΔF cells) were generously provided by J.P. Clancy (University of Alabama, Birmingham). Cells were cultured as described (Cushing et al. 2010; Wolde et al. 2007) at 37°C in 5% CO_2. Following polarization, complete medium was replaced with minimal essential medium (MEM) for >24 h prior to experimentation.

15.7.8 Ussing chamber experiments

For cells treated with the corrector compound, 3 μM corr-4A or DMSO (control) was applied. Cells were then incubated for 24 h before experiments. Peptide inhibitors were dissolved in DMSO and diluted in PBS to 1.25 mM, and then 100 μL of diluted peptide was added to lyophilized BioPorter reagent (Sigma). Some 150 μL MEM was added to reaction tubes, and the resulting 500 μM peptide solution was applied to filters. Peptide complexes were incubated with cells for 3.5 h and washed to remove noninternalized peptide. Cells were maintained at 37°C throughout treatments. The DMSO concentration did not exceed 0.06% in peptide-only treatments or 0.14% in combined treatments with corr-4a. Short circuit current (I_{SC}) measurements were performed as described (Wolde et al. 2007), except that all measurements were performed at 37°C. CFTR-specific chloride efflux was computed as the current changed (ΔI_{SC}) following application of CFTR$_{inh}$-172. Resistances were monitored throughout each experiment to ensure monolayer integrity.

15.8 Outlook

We started our collaboration by identifying PID players involved in postmaturational CFTR trafficking and regulation. As described in chapter 4, a number of PDZ domains influence the apical-membrane lifetime and functional activity of both WT- and ΔF508-CFTR. While PDZ domains of the NHERF proteins increase CFTR half-life, interactions with the PDZ domain of CAL appear to shunt the chloride channels into a lysosomal degradation pathway. Our aim was thus to inhibit the CAL PDZ domain with a peptide showing high CAL-binding specificity and affinity, hypothesizing that such an approach could rescue ΔF508-CFTR that has reached the plasma membrane. Fortunately, we were able to identify a CAL-selective peptide inhibitor (iCAL36) using a highly iterative peptide-array screening approach and thus to provide proof of principle for the ability of CAL inhibitors to rescue ΔF508-CFTR at the apical membrane of airway epithelial cells.

For further studies, it will be important to test, and if necessary enhance, the proteolytic stability of the iCAL36 inhibitor. In addition, it is important to remember that CFTR trafficking is regulated largely by intracellular interactions, including those with CAL, and that CAL-specific stabilizers must therefore cross the cell membrane. Reports of successful intracellular targeting of protein-protein interactions by a peptidic drug are rare, and the optimization of iCAL bioavailability is therefore a major focus of ongoing studies. In particular, the use of known and novel cell-penetrating sequences offers the prospect of CAL inhibitor peptides with inherent membrane permeability.

References

Aasland, R., Abrams, C., Ampe, C., Ball, L.J., Bedford, M.T., Cesareni, G., et al. (2002). Normalization of nomenclature for peptide motifs as ligands of modular protein domains. FEBS Lett *513*, 141–144.

Anderson, M.P., Gregory, R.J., Thompson, S., Souza, D.W., Paul, S., Mulligan, R.C., et al. (1991). Demonstration that CFTR is a chloride channel by alteration of its anion selectivity. Science *253*, 202–205.

Ay, B., Volkmer, R., and Boisguerin, P. (2007). Synthesis of cleavable peptides with authentic C-termini: an application for fully automated SPOT synthesis. Tetrahedron Lett, *48*, 361–364.

Bear, C.E. (1991). Regulation of ion conductance in human skin fibroblasts. Adv Exp Med Biol *290*, 273–283, discussion 283–285.

Benharouga, M., Sharma, M., So, J., Haardt, M., Drzymala, L., Popov, M., et al. (2003). The role of the C terminus and Na+/H+ exchanger regulatory factor in the functional expression of cystic fibrosis transmembrane conductance regulator in nonpolarized cells and epithelia. J Biol Chem *278*, 22079–22089.

Bezprozvanny, I., and Maximov, A. (2001). Classification of PDZ domains. FEBS Lett *509*, 457–462.

Bhargava, S., Licha, K., Knaute, T., Ebert, B., Becker, A., Grotzinger, C., et al. (2002). A complete substitutional analysis of VIP for better tumor imaging properties. J Mol Recognit *15*, 145–153.

Boat, T.F., and Cheng, P.W. (1989). Epithelial cell dysfunction in cystic fibrosis: implications for airways disease. Acta Paediatr Scand Suppl *363*, 25–29; discussion 29–30.

Bogan, A.A., and Thorn, K.S. (1998). Anatomy of hot spots in protein interfaces. J Mol Biol *280*, 1–9.

Boisguerin, P., Ay, B., Radziwill, G., Fritz, R.D., Moelling, K., and Volkmer, R. (2007). Characterization of a putative phosphorylation switch: adaptation of SPOT synthesis to analyze PDZ domain regulation mechanisms. ChemBioChem *8*, 2302–2307.

Boisguerin, P., Leben, R., Ay, B., Radziwill, G., Moelling, K., Dong, L., et al. (2004). An improved method for the synthesis of cellulose membrane-bound peptides with free C termini is useful for PDZ domain binding studies. Chem Biol *11*, 449–459.

Bomberger, J.M., Barnaby, R.L., and Stanton, B.A. (2009). The deubiquitinating enzyme USP10 regulates the post-endocytic sorting of cystic fibrosis transmembrane conductance regulator in airway epithelial cells. J Biol Chem *284*, 18778–18789.

Bruscia, E., Sangiuolo, F., Sinibaldi, P., Goncz, K.K., Novelli, G., and Gruenert, D.C. (2002). Isolation of CF cell lines corrected at DeltaF508-CFTR locus by SFHR-mediated targeting. Gene Ther *9*, 683–685.

Cesareni, G., Gimona, M., Sudol, M., and Yaffe, M. (2005). Modular protein domains. (Weinheim, Germany: Wiley-VCH).

Cheng, J., Moyer, B.D., Milewski, M., Loffing, J., Ikeda, M., Mickle, J.E., et al. (2002). A Golgi-associated PDZ domain protein modulates cystic fibrosis transmembrane regulator plasma membrane expression. J Biol Chem *277*, 3520–3529.

Cheng, J., Wang, H., and Guggino, W.B. (2004). Modulation of mature cystic fibrosis transmembrane regulator protein by the PDZ domain protein CAL. J Biol Chem *279*, 1892–1898.

Cho, K.O., Hunt, C.A., and Kennedy, M.B. (1992). The rat brain postsynaptic density fraction contains a homolog of the *Drosophila* discs-large tumor suppressor protein. Neuron *9*, 929–942.

Cholon, D.M., O'Neal, W.K., Randell, S.H., Riordan, J.R., and Gentzsch, M. (2010). Modulation of endocytic trafficking and apical stability of CFTR in primary human airway epithelial cultures. Am J Physiol Lung Cell Mol Physiol *298*, L304–314.

Clackson, T., and Wells, J.A. (1995). A hot spot of binding energy in a hormone-receptor interface. Science *267*, 383–386.

Cos, P., Tote, K., Horemans, T., and Maes, L. (2010). Biofilms: an extra hurdle for effective antimicrobial therapy. Curr Pharm Des *16*, 2279–2295.

Cushing, P.R., Fellows, A., Villone, D., Boisguerin, P., and Madden, D.R. (2008). The relative binding affinities of PDZ partners for CFTR: a biochemical basis for efficient endocytic recycling. Biochemistry *47*, 10084–10098.

Cushing, P.R., Vouilleme, L., Pellegrini, M., Boisguerin, P., and Madden, D.R. (2010). A stabilizing influence: CAL PDZ inhibition extends the half-life of DeltaF508-CFTR. Angew Chem Int Ed Engl *49*, 9907–9911.

Cystic Fibrosis Foundation Patient Registry. (2010). Annaul data report 2009. (Bethesda, MD: Cystic Fibrosis Foundation).

Dalemans, W., Barbry, P., Champigny, G., Jallat, S., Dott, K., Dreyer, D., et al. (1991). Altered chloride ion channel kinetics associated with the delta F508 cystic fibrosis mutation. Nature *354*, 526–528.

Davies, J.C., and Bilton, D. (2009). Bugs, biofilms, and resistance in cystic fibrosis. Respir Care *54*, 628–640.

Denning, G.M., Anderson, M.P., Amara, J.F., Marshall, J., Smith, A.E., and Welsh, M.J. (1992). Processing of mutant cystic fibrosis transmembrane conductance regulator is temperature-sensitive. Nature *358*, 761–764.

Doyle, D.A., Lee, A., Lewis, J., Kim, E., Sheng, M., and MacKinnon, R. (1996). Crystal structures of a complexed and peptide-free membrane protein-binding domain: molecular basis of peptide recognition by PDZ. Cell *85*, 1067–1076.

Drumm, M.L., Wilkinson, D.J., Smit, L.S., Worrell, R.T., Strong, T.V., Frizzell, R.A., et al. (1991). Chloride conductance expressed by delta F508 and other mutant CFTRs in *Xenopus* oocytes. Science *254*, 1797–1799.

Fan, J.S., and Zhang, M. (2002). Signaling complex organization by PDZ domain proteins. Neurosignals *11*, 315–321.

Favia, M., Guerra, L., Fanelli, T., Cardone, R.A., Monterisi, S., Di Sole, F., et al. (2010). Na+/H+ exchanger regulatory factor 1 overexpression-dependent increase of cytoskeleton organization is fundamental in the rescue of F508del cystic fibrosis transmembrane conductance regulator in human airway CFBE41o- cells. Mol Biol Cell *21*, 73–86.

Fields, S., and Song, O. (1989). A novel genetic system to detect protein-protein interactions. Nature *340*, 245–246.

Fodor, S.P., Read, J.L., Pirrung, M.C., Stryer, L., Lu, A.T., and Solas, D. (1991). Light-directed, spatially addressable parallel chemical synthesis. Science *251*, 767–773.

Frank, R. (1992). Spot synthesis an easy technique for positionally addressable, parallel chemical synthesis on a membrane support. Tetrahedron *48*, 9217–9232.

Gavin, A.C., Bosche, M., Krause, R., Grandi, P., Marzioch, M., Bauer, A., et al. (2002). Functional organization of the yeast proteome by systematic analysis of protein complexes. Nature *415*, 141–147.

Gordus, A., and MacBeath, G. (2006). Circumventing the problems caused by protein diversity in microarrays: implications for protein interaction networks. J Am Chem Soc *128*, 13668–13669.

Guerra, L., Fanelli, T., Favia, M., Riccardi, S.M., Busco, G., Cardone, R.A., et al. (2005). Na+/H+ exchanger regulatory factor isoform 1 overexpression modulates cystic fibrosis transmembrane conductance regulator (CFTR) expression and activity in human airway 16HBE14o- cells and rescues DeltaF508 CFTR functional expression in cystic fibrosis cells. J Biol Chem *280*, 40925–40933.

Guggino, W.B., and Stanton, B.A. (2006). New insights into cystic fibrosis: molecular switches that regulate CFTR. Nat Rev Mol Cell Biol *7*, 426–436.

Hall, R.A., Premont, R.T., Chow, C.W., Blitzer, J.T., Pitcher, J.A., Claing, A., et al. (1998). The beta2-adrenergic receptor interacts with the Na+/H+-exchanger regulatory factor to control Na+/H+ exchange. Nature *392*, 626–630.

Hilpert, K., Winkler, D.F., and Hancock, R.E. (2007). Peptide arrays on cellulose support: SPOT synthesis, a time and cost efficient method for synthesis of large numbers of peptides in a parallel and addressable fashion. Nat Protoc *2*, 1333–1349.

Ho, Y., Gruhler, A., Heilbut, A., Bader, G.D., Moore, L., Adams, S.L., et al. (2002). Systematic identification of protein complexes in *Saccharomyces cerevisiae* by mass spectrometry. Nature *415*, 180–183.

Hung, A.Y., and Sheng, M. (2002). PDZ domains: structural modules for protein complex assembly. J Biol Chem *277*, 5699–5702.

Jones, R.B., Gordus, A., Krall, J.A., and MacBeath, G. (2006). A quantitative protein interaction network for the ErbB receptors using protein microarrays. Nature *439*, 168–174.

Jones, S., and Thornton, J.M. (1996). Principles of protein-protein interactions. Proc Natl Acad Sci U S A *93*, 13–20.

Kennedy, M.B. (1995). Origin of PDZ (DHR, GLGF) domains. Trends Biochem Sci *20*, 350.

Kerem, B., Rommens, J.M., Buchanan, J.A., Markiewicz, D., Cox, T.K., Chakravarti, A., et al. (1989). Identification of the cystic fibrosis gene: genetic analysis. Science *245*, 1073–1080.

Kim, E., Niethammer, M., Rothschild, A., Jan, Y.N., and Sheng, M. (1995). Clustering of Shaker-type K+ channels by interaction with a family of membrane-associated guanylate kinases. Nature *378*, 85–88.

Kurakin, A., Swistowski, A., Wu, S.C., and Bredesen, D.E. (2007). The PDZ domain as a complex adaptive system. PLoS One *2*, e953.

Landgraf, C., Panni, S., Montecchi-Palazzi, L., Castagnoli, L., Schneider-Mergener, J., Volkmer-Engert, R., et al. (2004). Protein interaction networks by proteome peptide scanning. PLoS Biol *2*, E14.

Li, C., and Naren, A.P. (2005). Macromolecular complexes of cystic fibrosis transmembrane conductance regulator and its interacting partners. Pharmacol Ther *108*, 208–223.

Li, N., Pei, P., Bu, D.F., He, B., and Wang, G.F. (2006). A novel CFTR mutation found in a Chinese patient with cystic fibrosis. Chin Med J (Engl) *119*, 103–109.

Li, Y., Li, J., Straight, S.W., and Kershaw, D.B. (2002). PDZ domain-mediated interaction of rabbit podocalyxin and Na(+)/H(+) exchange regulatory factor-2. Am J Physiol Renal Physiol *282*, F1129–1139.

Licha, K., Bhargava, S., Rheinländer, C., Becker, A., Schneider-Mergener, J., and Volkmer-Engert, R. (2000). Highly parallel nano-synthesis of cleavable peptide-dye conjugates on cellulose membranes. Tetrahedon Lett *41*, 1711–1715.

Linding, R., Letunic, I., Gibson, T.J., and Bork, P. (2005). Computational analysis of modular protein architectures. In: Modular protein domains, G. Cesareni, M. Gimona, M. Sudol and M. Yaffe, eds (Weihnheim, Germany: Wiley-VCH), pp. 439–476.

Lukacs, G.L., Chang, X.B., Bear, C., Kartner, N., Mohamed, A., Riordan, J.R., et al. (1993). The delta F508 mutation decreases the stability of cystic fibrosis transmembrane conductance regulator in the plasma membrane. Determination of functional half-lives on transfected cells. J Biol Chem *268*, 21592–21598.

Lukacs, G.L., Mohamed, A., Kartner, N., Chang, X.B., Riordan, J.R., and Grinstein, S. (1994). Conformational maturation of CFTR but not its mutant counterpart (delta F508) occurs in the endoplasmic reticulum and requires ATP. EMBO J *13*, 6076–6086.

Ma, B., Elkayam, T., Wolfson, H., and Nussinov, R. (2003). Protein-protein interactions: structurally conserved residues distinguish between binding sites and exposed protein surfaces. Proc Natl Acad Sci U S A *100*, 5772–5777.

MacBeath, G., and Schreiber, S.L. (2000). Printing proteins as microarrays for high-throughput function determination. Science *289*, 1760–1763.

Morais Cabral, J.H., Petosa, C., Sutcliffe, M.J., Raza, S., Byron, O., Poy, F., et al. (1996). Crystal structure of a PDZ domain. Nature *382*, 649–652.

Nourry, C., Grant, S.G., and Borg, J.P. (2003). PDZ domain proteins: plug and play! Sci 179, RE7, 1-12.

Okiyoneda, T., Barriere, H., Bagdany, M., Rabeh, W.M., Du, K., Hohfeld, J., et al. (2010). Peripheral protein quality control removes unfolded CFTR from the plasma membrane. Science *329*, 805–810.

Ostedgaard, L.S., Randak, C., Rokhlina, T., Karp, P., Vermeer, D., Ashbourne Excoffon, K.J., et al. (2003). Effects of C-terminal deletions on cystic fibrosis transmembrane conductance regulator function in cystic fibrosis airway epithelia. Proc Natl Acad Sci U S A *100*, 1937–1942.

Pawson, T., and Nash, P. (2003). Assembly of cell regulatory systems through protein interaction domains. Science *300*, 445–452.

Pedemonte, N., Lukacs, G.L., Du, K., Caci, E., Zegarra-Moran, O., Galietta, L.J., et al. (2005). Small-molecule correctors of defective DeltaF508-CFTR cellular processing identified by high-throughput screening. J Clin Invest *115*, 2564–2571.

Reineke, U. (2009). Antibody epitope mapping using de novo generated synthetic peptide libraries. Methods Mol Biol *524*, 203–211.

Reineke, U., Kramer, A., and Schneider-Mergener, J. (1999). Antigen sequence- and library-based mapping of linear and discontinuous protein-protein-interaction sites by spot synthesis. Curr Top Microbiol Immunol *243*, 23–36.

Reineke, U., Sabat, R., Volk, H.D., and Schneider-Mergener, J. (1998). Mapping of the interleukin-10/interleukin-10 receptor combining site. Protein Sci *7*, 951–960.

Reineke, U., Schneider-Mergener, J., and Schutkowski, M. (2005). Peptide arrays in proteomics and drug discovery. In: BioMEMS and Biomedical Nanotechnology, M. Ozkan and M. Heller, eds (New York: Springer), pp. 161–282.

Riordan, J.R., Rommens, J.M., Kerem, B., Alon, N., Rozmahel, R., Grzelczak, Z., et al. (1989). Identification of the cystic fibrosis gene: cloning and characterization of complementary DNA. Science 245, 1066–1073.

Rommens, J.M., Iannuzzi, M.C., Kerem, B., Drumm, M.L., Melmer, G., Dean, M., et al. (1989). Identification of the cystic fibrosis gene: chromosome walking and jumping. Science 245, 1059–1065.

Rowe, S.M., Miller, S., and Sorscher, E.J. (2005). Cystic fibrosis. N Engl J Med 352, 1992–2001.

Scott, J.K., and Smith, G.P. (1990). Searching for peptide ligands with an epitope library. Science 249, 386–390.

Short, D.B., Trotter, K.W., Reczek, D., Kreda, S.M., Bretscher, A., Boucher, R.C., et al. (1998). An apical PDZ protein anchors the cystic fibrosis transmembrane conductance regulator to the cytoskeleton. J Biol Chem 273, 19797–19801.

Sloane, P.A., and Rowe, S.M. (2010). Cystic fibrosis transmembrane conductance regulator protein repair as a therapeutic strategy in cystic fibrosis. Curr Opin Pulm Med 16, 591–597.

Smith, G.P. (1985). Filamentous fusion phage: novel expression vectors that display cloned antigens on the virion surface. Science 228, 1315–1317.

Songyang, Z., Fanning, A.S., Fu, C., Xu, J., Marfatia, S.M., Chishti, A.H., et al. (1997). Recognition of unique carboxyl-terminal motifs by distinct PDZ domains. Science 275, 73–77.

Swiatecka-Urban, A., Brown, A., Moreau-Marquis, S., Renuka, J., Coutermarsh, B., Barnaby, R., et al. (2005). The short apical membrane half-life of rescued {Delta}F508-cystic fibrosis transmembrane conductance regulator (CFTR) results from accelerated endocytosis of {Delta}F508-CFTR in polarized human airway epithelial cells. J Biol Chem 280, 36762–36772.

Swiatecka-Urban, A., Duhaime, M., Coutermarsh, B., Karlson, K.H., Collawn, J., Milewski, M., et al. (2002).PDZ domain interaction controls the endocytic recycling of the cystic fibrosis transmembrane conductance regulator. J Biol Chem 277, 40099–40105.

Tapia, V.E., Nicolaescu, E., McDonald, C.B., Musi, V., Oka, T., Inayoshi, Y., et al. (2010). Y65C missense mutation in the WW domain of the Golabi-Ito-Hall syndrome protein PQBP1 affects its binding activity and deregulates pre-mRNA splicing. J Biol Chem 285, 19391–19401.

Tong, A.H., Drees, B., Nardelli, G., Bader, G.D., Brannetti, B., Castagnoli, L., et al. (2002). A combined experimental and computational strategy to define protein interaction networks for peptide recognition modules. Science 295, 321–324.

Vaccaro, P., and Dente, L. (2002). PDZ domains: troubles in classification. FEBS Lett 512, 345–349.

Volkmer, R. (2009). Synthesis and application of peptide arrays: quo vadis SPOT technology. ChemBioChem 10, 1431–1442.

Vouilleme, L., Cushing, P.R., Volkmer, R., Madden, D.R., and Boisguerin, P. (2010). Engineering peptide inhibitors to overcome PDZ binding promiscuity. Angew Chem Int Ed Engl 49, 9912–9916.

Ward, C.L., and Kopito, R.R. (1994). Intracellular turnover of cystic fibrosis transmembrane conductance regulator. Inefficient processing and rapid degradation of wild-type and mutant proteins. J Biol Chem 269, 25710–25718.

Ward, C.L., Omura, S., and Kopito, R.R. (1995). Degradation of CFTR by the ubiquitin-proteasome pathway. Cell 83, 121–127.

Weiser, A.A., Or-Guil, M., Tapia, V., Leichsenring, A., Schuchhardt, J., Frommel, C., et al. (2005). SPOT synthesis: reliability of array-based measurement of peptide binding affinity. Anal Biochem 342, 300–311.

Welsh, M.J. (1995). The development of gene transfer for cystic fibrosis. Adv Intern Med *40*, 429–444.

Wenschuh, H., Gausepohl, H., Germeroth, L., Ulbricht, M., Matuschewski, H., Kramer, A., et al. (2000). Positionally addressable parallel synthesis on continuous membranes. In: Combinatorial chemistry: a practical approach, H. Fenniri, ed. (Oxford: Oxford University Press), pp. 95–116.

Wolde, M., Fellows, A., Cheng, J., Kivenson, A., Coutermarsh, B., Talebian, L., et al. (2007). Targeting CAL as a negative regulator of DeltaF508-CFTR cell-surface expression: an RNA interference and structure-based mutagenetic approach. J Biol Chem *282*, 8099–8109.

Woods, D.F., and Bryant, P.J. (1993). ZO-1, DlgA and PSD-95/SAP90: homologous proteins in tight, septate and synaptic cell junctions. Mech Dev *44*, 85–89.

Ye, L., Knapp, J.M., Sangwung, P., Fettinger, J.C., Verkman, A.S., and Kurth, M.J. (2010). Pyrazolylthiazole as DeltaF508-cystic fibrosis transmembrane conductance regulator correctors with improved hydrophilicity compared to bithiazoles. J Med Chem *53*, 3772–3781.

Zhu, H., Bilgin, M., Bangham, R., Hall, D., Casamayor, A., Bertone, P., et al. (2001). Global analysis of protein activities using proteome chips. Science *293*, 2101–2105.

Ziady, A.G., and Davis, P.B. (2011). Methods for evaluating inflammation in cystic fibrosis. Methods Mol Biol *742*, 51–76.

Zimmermann, P. (2006). The prevalence and significance of PDZ domain-phosphoinositide interactions. Biochim Biophys Acta *1761*, 947–956.

Zimmermann, P., Meerschaert, K., Reekmans, G., Leenaerts, I., Small, J.V., Vandekerckhove, J., et al. (2002). PIP(2)-PDZ domain binding controls the association of syntenin with the plasma membrane. Mol Cell *9*, 1215–1225.

16 Probing protein dynamics in vivo using backbone cyclization: bacterial acyl carrier protein as a case study

Gerrit Volkmann, Peter W. Murphy, Elden E. Rowland, John E. Cronan Jr., and David M. Byers

16.1 Introduction

16.1.1 Studying protein structures and dynamics in living cells

Proteins are inherently flexible and structurally dynamic biomolecules. The ability to switch between different conformational states is often a pivotal feature for such diverse cellular processes as protein-protein and protein-ligand interactions, rebuilding of active sites in enzyme catalysis and changing the molecular architecture of entire cells. Much of our current understanding of protein dynamics is derived from biophysical analyses including spectroscopic techniques such as nuclear magnetic resonance (NMR), circular dichroism (CD), electron paramagnetic resonance (EPR) and fluorescence spectroscopy, as well as mass spectrometry (MS), molecular dynamics simulations and X-ray crystallography. Although these powerful techniques can provide detailed molecular information with atomic precision, they usually lack the capacity to look beyond the artificial experimental setting in which they take place. In other words, the data obtained from the measurements performed in vitro or in silico cannot be faithfully extrapolated to the natural cellular environment of a protein (i.e. to the in vivo situation). The difficulty lies in the tremendous complexity of the interior of a living cell, which makes it almost impossible to isolate a single protein species from thousands of other biomolecules and obtain meaningful data about its structural makeup.

In order to provide such physiologically relevant data on protein flexibility and conformational changes, several promising techniques have recently emerged. Volker Dötsch and coworkers showed in 2001 that isotopically labeled proteins are amenable to NMR spectroscopy when overexpressed in the cytoplasm of *Escherichia coli* grown in medium containing stable isotopes (Serber et al. 2001a, 2001b). Since then, in-cell NMR has also been developed for spectra acquisition of proteins in *Xenopus laevis* oocytes (Bodart et al. 2008; Sakai et al. 2006; Selenko et al. 2006) and cultured mammalian cells (Inomata et al. 2009; Ogino et al. 2009; Serber et al. 2006). In-cell NMR has provided interesting structural insights of proteins that would otherwise have been impossible to deduce from in vitro experiments. For example, the transcriptional coregulator FlgM from *Salmonella typhimurium* is an intrinsically disordered protein (IDP), which lacks stabilized secondary structural elements when NMR is performed on purified protein (Daughdrill et al. 1998). Strikingly, in the highly crowded cytoplasm of *E. coli*, the FlgM C terminus becomes structured while the N terminus remains disordered (Dedmon et al. 2002). The folding stability of ubiquitin inside living mammalian cells was found to be decreased by 20-fold (Inomata et al. 2009), which was assumed to be a necessary feature for interaction with the various protein partners in the cytoplasm. In-cell NMR can also be used to probe protein-protein and protein-small molecule interactions in living cells (Burz and Shekhtman 2008; Xie et al. 2009). A common pitfall of in-cell NMR is that larger globular proteins remain largely invisible in the cytoplasm of living cells while only disordered proteins give high-resolution spectra (Barnes et al. 2011; Bertini et al. 2011; Li et al. 2008). Efforts are currently underway to overcome this obstacle (e.g. by incorporating ^{19}F isotopes into the protein of interest during expression) (Li et al. 2010).

Another approach to study conformational changes and dynamics of proteins in vivo is to determine intramolecular Förster resonance energy transfer (FRET) between two fluorescent probes linked to the protein of interest by spectroscopic readout using fluorescence microscopy (for a review, see Chamberlain and Hahn 2000). This approach has enabled visualization of myosin II phosphorylation (Post et al. 1995), led to the development of a calmodulin-based intracellular sensor for Ca^{2+} (Miyawaki et al. 1997; Romoser et al. 1997), and helped delineate the structural organization of voltage-gated calcium channels (Kobrinsky et al. 2009). With the advent of spectroscopic measurements of single molecules in living cells (Yang 2010), this field holds great promise for future studies of protein dynamics in vivo. Lastly, chemical labeling strategies that rely on the dynamic accessibility of certain amino acid side chains can further be employed to follow conformational changes on living cells (Azim-Zadeh et al. 2007).

In this chapter, we describe a novel, interdisciplinary methodology to probe protein dynamics in vivo, combining protein engineering, biophysical analyses and functional assays in live cells. At the heart of the strategy lies protein backbone cyclization, which will be described in the following section.

16.1.2 Cyclization of polypeptides

16.1.2.1 Cyclic peptides and proteins found in nature

Ribosomally synthesized polypeptides have long been thought to be linear macromolecules with a clearly defined beginning and end, represented by the amino(N) and carboxy(C) terminus, respectively. It wasn't until the 1990s that this central idea of protein architecture had to be revisited when several polypeptides with a seamless backbone were discovered (i.e. peptides and proteins containing a peptide bond between the amino-group of the first amino acid and the carboxyl-group of the last amino acid). These circular or cyclic polypeptides include the θ-defensins in certain mammals, the phallocidins and amatoxins of fungi, the cyclotides found in plants, as well as prokaryotic bacteriocins and cyanobactins (Cascales and Craik 2010). Certain cyclic peptides are also produced by nonribosomal peptide synthetases (Schwarzer et al. 2003). A common feature of the cyclic polypeptides is their exceptional thermodynamic stability, which is due to the cyclic backbone, and is in some cases enhanced by intramolecular disulfide bonds or other cross-links. Most of the circular polypeptides found to date exert some form of activity against other organisms or viruses with the exception of the pilin proteins from *Agrobacterium tumefaciens* and *E. coli*, which play a role in transfer of genetic information during bacterial conjugation (Kalkum et al. 2004). Another interesting group of proteins that present a circular arrangement in their structures are the lasso peptides (Rosengren and Craik 2009). In contrast to the backbone-cyclized proteins, the lasso peptides are characterized by a macrolactam ring, which forms between the N terminus and the carboxy group of an internal Glu or Asp residue, and through which the C-terminal tail is threaded (Nar et al. 2010).

The maturation of functional, cyclic pilin proteins is a multistep process (reviewed by Kalkum et al. 2004), wherein a primary translation product is proteolytically processed to the final pilin sequence. Intriguingly, the last protease (named TraF), which truncates the C terminus by four amino acids, is also responsible for the subsequent cyclization step. The biosynthesis of cyclotides also proceeds through larger precursor proteins by proteolytic processing events, where in the final step an asparaginyl endopeptidase, much like TraF, both breaks and forms a peptide bond (Saska et al. 2007). This dual functional character of proteases has since been referred to as "protease-catalyzed protein splicing" (Saska and Craik 2008). While the biosynthesis of most other natural cyclic polypeptides synthesized on ribosomes remains to be elucidated, the fact that they are usually part of a larger precursor protein indicates that similar protease-catalyzed events may take place (Cascales and Craik 2010). The lasso peptides are presumably all converted from a precursor protein by the concerted action of (i) an enzyme that activates the internal carboxyl-functionalized amino acid by adenylation and (ii) a processing enzyme that both cleaves off an N-terminal portion from the precursor and forms the macrolactam ring between the new N terminus and the AMP-activated Glu or Asp residue, as seen for capistruin (Knappe et al. 2009).

16.1.2.2 Engineering artificial cyclic peptides and proteins

The exceptional stability seen in natural cyclic peptides and proteins prompted the question whether this characteristic could be conferred upon naturally linear polypeptides. To this end, several chemical and recombinant strategies to produce cyclic polypeptides have been developed in recent years. Peptides produced by solid-phase peptide synthesis (SPPS) can be converted to circular structures by native chemical ligation (NCL) (Camarero and Muir 1999; Clark et al. 2005; Deechongkit and Kelly 2002), which requires an N-terminal cysteine residue and a C-terminal thioester to form a peptide bond between the N- and C-terminal amino acids (Dawson et al. 1994). Cyclic proteins that are rich in intramolecular disulfide bonds can be produced by the so-called thia-zip reaction, a diversion of NCL that progresses from the C to the N terminus through reversible thiol-thiolactone exchanges to arrive in the final step at the chemoselective reaction between N-terminal cysteine and C-terminal thioester (Tam and Lu 1998). Cystine-rich circular proteins have also been synthesized with a combination of SPPS and protease-assisted peptide bond formation (Thongyoo et al. 2007). Peptides produced by SPPS that contain a C-terminal glycolate phenylalanyl-amide ester can efficiently be cyclized by subtiligase (Jackson et al. 1995), an artificial enzyme engineered from subtilisin BPN' for reverse proteolysis (Chang et al. 1994). A conceptionally different route to cyclization of peptides is the use of a transglutaminase enzyme; however, this requires the introduction of a Gln and Lys residue near the ends of the peptide sequence and practically affords an internally cyclized peptide leaving the N and C termini free (Touati et al. 2011).

In addition to these chemical approaches for polypeptide cyclization, recombinantly expressed proteins and peptides can also be cyclized. Purified polypeptides genetically equipped with two Gly at the N terminus and the sequence Leu-Pro-Glu-Thr-Gly-Gly at the C terminus can be circularized by sortase A (Popp et al. 2011). The cyanocysteine-assisted protein ligation reaction (Ishihama et al. 1999) can also be harnessed to achieve in vitro protein cyclization (Takahashi et al. 2007). Proteins can further be recombinantly equipped with an N-terminal Cys and a C-terminal thioester for NCL (Xu and Evans 2001). Recombinant cyclic polypeptides can also be produced during their expression in vivo, which obviates the need to purify the protein prior to the cyclization reaction. Camarero et al. (2001a) have devised a technique to achieve polypeptide cyclization by NCL in living cells, where the N-terminal Cys residue is generated by action of an endogenous methionine aminopeptidase, and the C-terminal thioester bond is afforded through the N-S acyl rearrangement of a protein-splicing element. This technique, however, suffers from the fact that the thiol group of the N-terminal Cys and the thioester bond have no inherent affinity for each other, hence the proximity of the two functional groups required for the NCL reaction relies solely on correct folding of the protein to be cyclized (Kimura et al. 2006).

The most effective way to cyclize polypeptides in vivo is by means of split inteins. Split inteins are protein-splicing elements that catalyze peptide bond formation by protein *trans*-splicing (Wu et al. 1998). The N- and C-terminal fragments of a split intein, hereafter referred to as I_N and I_C, respectively, associate with nano- or low micromolar affinity (Ludwig et al. 2008; Shi and Muir 2005), which elegantly overcomes the drawback of NCL-mediated *in vivo* cyclization. It is for this reason that the split intein approach, pioneered by Scott et al (1999), has become the most widely used method for in vivo polypeptide cyclization, and is often referred to as split intein-mediated circular ligation of peptides and proteins (SICLOPPS).

SICLOPPS has been extensively used for the production of cyclic peptide libraries to identify potential drug leads (Cheng et al. 2007; Deschuyteneer et al. 2010; Horswill et al. 2004; Kinsella et al. 2002; Kritzer et al. 2009; Naumann et al. 2008; Nilsson et al. 2005; Tavassoli and Benkovic 2005; Tavassoli et al. 2008), and detailed protocols for this powerful and pharmaceutically relevant technique have been made available (Abel-Santos et al. 2003; Tavassoli and Benkovic 2007). In addition to peptides, numerous proteins have also been cyclized using the SICLOPPS approach, including bacterial dihydrofolate reductase (Scott et al. 1999) and maltose binding protein (Evans et al. 2000), jellyfish green fluorescent protein (Iwai et al. 2001; Zhao et al. 2010), the IIA^{Glc} and IICB^{Glc} subunits from a bacterial glucose transporter (Siebold and Erni 2002), the N-terminal domain of a bacterial DnaB helicase (Williams et al. 2002), the β-lactamase antibiotic resistance protein (Kwon et al. 2008), firefly luciferase (Kanno et al. 2007), as well as a complex of zinc finger LIM-only protein 4 (LMO4) and a domain of its binding partner ldb1 (Jeffries et al. 2006). Many of these reports, along with cyclization studies using the NCL approach, have provided valuable insights into the effects of protein cyclization on protein function, stability, structure and folding kinetics in vitro (Camarero et al. 2001b; Hofmann et al. 2002; Iwai and Plückthun 1999; Iwai et al. 2001; Jeffries et al. 2006; Scott et al. 1999; Siebold and Erni 2002; Watt et al. 2007; Williams et al. 2002, 2005; Zhao et al. 2010); however, complementary detailed studies in living cells were not undertaken.

16.1.3 Motivation for the presented methodology

Given the potential of split intein-mediated protein cyclization to investigate protein structure-function relationships, we reasoned that intracellular protein cyclization might provide a novel opportunity to probe protein dynamics in vivo. We chose acyl carrier protein (ACP) from the marine bioluminescent bacterium *Vibrio harveyi* as our target protein, because it has been extensively characterized both biochemically and structurally (Byers and Gong 2007; Chan and Vogel 2010; Chan et al. 2010). ACP is an essential protein that must reversibly interact with at least 30 partner enzymes in the synthesis of fatty acids and other specialized bacterial products (Butland et al. 2005). Like all other ACPs of known structure, folded *Vha* ACP exhibits a characteristic "four-helix bundle" (Chan et al. 2010), helices II-IV of which likely provide a hydrophobic pocket that can expand to enclose fatty acyl chains covalently attached to the phosphopantetheine moiety of holo-ACP. ACPs are also highly dynamic and flexible proteins, properties believed to be essential for their function as they must undergo significant conformational changes to alternately sequester and present their acyl chains to the active sites of partner enzymes (Byers and Gong 2007; Chan and Vogel 2010).

Vha ACP poses a particularly interesting model system to explore the relationship between conformational dynamics and protein structure and function. *Vha* ACP is very acidic (pI ~ 4.0) and exhibits many features of natively unfolded proteins, including anomalous SDS-PAGE mobility (de la Roche et al. 1997) and lack of higher order structure at neutral pH (Flaman et al. 2001). Folding of *Vha* ACP can be induced upon charge neutralization (e.g. by lowering pH, increasing salt concentration, mutagenic replacement of acidic residues or binding of divalent cations to helix II), attachment of fatty acyl groups or interaction with enzyme partners (Gong et al. 2007, 2008). The overall scope of our investigation encompassed the following question: how does constraining the N and C-termini of ACP impact its conformation, dynamics, stability and function, both in vitro and in vivo? The following part of the chapter describes in detail our efforts to probe the essentiality of the ACP dynamic structure for its function in vivo using intracellular split intein-mediated protein cyclization.

16.2 In vivo protein cyclization, biophysical analyses and functional assays

16.2.1 Plasmid construction

For producing a backbone-cyclized protein using intramolecular protein *trans*-splicing, an expression vector is required that contains a circularly permuted split intein (Scott et al. 1999), and preferably unique cleavage sites for restriction endonucleases, which allow cloning of a target protein between the split intein sequences. To this end, we have constructed the plasmid pTI$_C$(NS)I$_N$H, which harbors the C- and N-terminal sequences of the protein splicing domain of the *Synechocystis* sp. PCC6803 GyrB (*Ssp* GyrB) intein, and unique NheI and SapI restriction sites between the split intein fragment-encoding sequences for insertion of a target protein sequence (▶Figure 16.1A).

During PCR of the intein-encoding sequences using plasmid pMSG (Appleby et al. 2009) as a template, a SacI-site was appended to the 5'-end of the C-intein (I$_C$) sequence, and a hexahistidine(His$_6$-tag)-encoding sequence followed by a stop codon and a PstI-site was appended to the 3'-end of the N-intein (I$_N$) sequence. The I$_C$ and I$_N$ sequences were subsequently fused by anneal-extension PCR, during which a codon for the first natural C-extein residue (Ser+1) was introduced after the I$_C$ sequence, followed by recognition sites for NheI and SapI (▶Figure 16.1A). The advantage of using the NheI-site downstream of the Ser+1 codon is that it provides the codon for the second natural C-extein residue (Ala+2), which has proven to be pivotal for efficient protein

16 Probing protein dynamics in vivo using backbone cyclization

A

end of C-intein			Nhel			SapI			start of N-intein		
CAT	AAC	AGC	G|CT	AGC	ATA	TGC	TCT	TCC	|TGT	TTT	TCT
GTA	TTG	TCG	CGA	TC|G	TAT	ACG	AGA	AGG	ACA|	AAA	AGA
His	Asn	Ser	Ala	Ser					Cys	Phe	Ser
		+1	+2						1	2	3

B

Nhel		target gene without start and stop codon					SapI		
G|CT	AGC	NNN	NNN	NNN	NNN	GGT	|TGT	AGA	AGA	GC
CGA	TC|G	NNN	NNN	NNN	NNN	CCA	ACA|	TCT	TCT	CG
Ala	Ser					Gly	Cys			
+2						−1	1			

C

end of C-intein			Nhel		target gene without start and stop codon				start of N-intein			
CAT	AAC	AGC	G|CT	AGC	NNN	NNN	NNN	NNN	GGT	TGT	TTT	TCT
GTA	TTG	TCG	CGA	TC|G	NNN	NNN	NNN	NNN	CCA	ACA	AAA	AGA
His	Asn	Ser	Ala	Ser					Gly	Cys	Phe	Ser
		+1	+2						−1	1	2	3

Figure 16.1: Plasmid construction for split intein-mediated protein cyclization. (A) The multiple cloning site within vector pTI$_C$(NS)I$_N$H, which allows cloning of a target gene between the NheI- and SapI-sites. The end and beginning of the split intein sequences are indicated along with selected amino acid residues. (B) Sequence of a PCR product to be cloned into pTI$_C$(NS)I$_N$H. (C) Restriction-based cloning of the PCR product from B into vector from A yields the cyclization plasmid (backbone omitted).

splicing of the *Ssp* GyrB split intein (G. Volkmann, unpublished observation), as well as the codon for Ser, which is the first amino acid of *Vha* ACP subsequent to removal of the starting methionine by methionine endopeptidase. The advantage of using the SapI-site is that it allows one to customize the sticky-end overhangs since SapI does not cleave within its recognition sequence (Xu et al. 1998); in our case, we chose the overhang to be the codon for the first residue of I$_N$ (Cys1).

The complete PCR product was then digested with SacI and PstI and cloned into a similarly digested expression vector pT, yielding pTI$_C$(NS)I$_N$H (▶ Figure 16.1A). Vector pT is a modified version of pTWIN1 (New England Biolabs), in which the IMPACT™ open reading frame was replaced by consecutive cloning sites for NdeI, SacI, BglII and HindIII. Recombinant protein expression in pT is governed by the T7 promoter, so an *E. coli* strain is required that contains an integrated copy of the T7 RNA polymerase gene (e.g. BL21[DE3] and derivatives). Basal expression of the recombinant gene is minimized by the plasmid-borne *lacI* gene, whose gene product (the Lac repressor) binds immediately downstream of the T7 promoter and inhibits gene expression, hence expression of the recombinant gene is inducible by isopropyl β-D-1-thiogalactopyranoside (IPTG).

The coding sequence for the *Vha* ACP-L46W mutant was PCR amplified (without start and stop codons) containing the NheI-site at the 5′-end, and the codons for Gly and Cys1 followed by the SapI-site at the 3′-end, respectively (▶Figure 16.1B). We chose the L46W mutant of *Vha* ACP because the lone tryptophan residue can serve as an indicator for the folding-unfolding process of ACP (Gong et al. 2008). Cloning of the NheI plus SapI-digested PCR product into similarly digested pTI$_C$(NS)I$_N$H yielded plasmid pTCYC-L46W (▶Figure 16.1C).

Theoretically, the *Ssp* GyrB split intein-mediated in vivo cyclization of ACP-L46W should result in (a) the liberation of the split intein fragments from the preL46W protein and (b) a cyclic *Vha* ACP-L46W protein (hereafter referred to as cycL46W), whose N and C termini are connected by a three-residue linker (Gly-Ser-Ala) (▶Figure 16.2A). Given that the three-dimensional structure of bacterial ACP is highly conserved (Byers and Gong 2007), and the distance between the termini of folded *Vha* ACP is 8 Å or less (Chan et al. 2010), this linker should be just long enough not to cause steric constraint on the overall structure. Longer linkers encoding affinity tags for protein purification can also be incorporated, if desired. As the migration behavior of cyclic proteins during protein gel electrophoresis cannot be accurately predicted, we also constructed a plasmid, pTPRE-L46W-mut, where two catalytically important residues of the split intein (Cys at the beginning and Asn at the end) were mutated to Ala in order to prevent protein *trans*-splicing (▶Figure 16.2A), and in turn clearly define the migration of unspliced, linear precursor protein during sodium dodecyl sulfate–polyacrylamide gel electrophoresis (SDS-PAGE). Finally, it is important to consider that the split intein fragments may undergo nonproductive cleavage reactions (N- and C-terminal cleavage, respectively), which would also lead to their liberation from the preL46W precursor protein expressed from pTCYC-L46W but without cyclization of *Vha* ACP-L46W. The wild type, linear form of *Vha* ACP migrates at ~20 kDa in SDS-PAGE (de la Roche et al. 1997).

16.2.2 In vivo protein cyclization

The expression plasmids were transformed into BL21(DE3)pLysS cells. Cells were grown in LB medium supplemented with 50 µg/mL ampicillin and 34 µg/mL chloramphenicol to an OD$_{600}$ of 0.6, expression of the recombinant genes was induced with 0.8 mM IPTG, and cells were allowed to grow for 18 h at room temperature (control cultures were grown similarly without the addition of IPTG). Cells were then harvested, lysed in standard SDS-PAGE sample buffer containing 30 mM dithiothreitol (DTT) as reducing agent and boiled for 5 min prior to loading on a 12.5% SDS-PAGE gel. Gels were either stained with Coomassie Brilliant Blue G-250, or subjected to Western blotting with a primary mouse anti-His antibody and a secondary anti-mouse IgG horseradish peroxidase-conjugated antibody with visualization of His$_6$-tagged proteins by chemiluminescence on X-ray film.

The cells harboring plasmid pTPRE-L46W-mut showed a strongly induced protein species migrating at ~30 kDa, which gave a signal on the anti-His Western blot (▶Figure 16.2B), and thus represents the splicing-deficient preL46W^{mut} protein (calculated molecular weight [M$_W$]: 27247.38). In contrast, the splicing-competent pTCYC-L46W revealed the presence of two induced protein species (▶Figure 16.2B), one migrating at ~14 kDa, which reacted with anti-His antibodies and thus corresponds to the His-tagged N-terminal split intein fragment I$_N$H (calculated M$_W$: 13317.02), and the other migrating

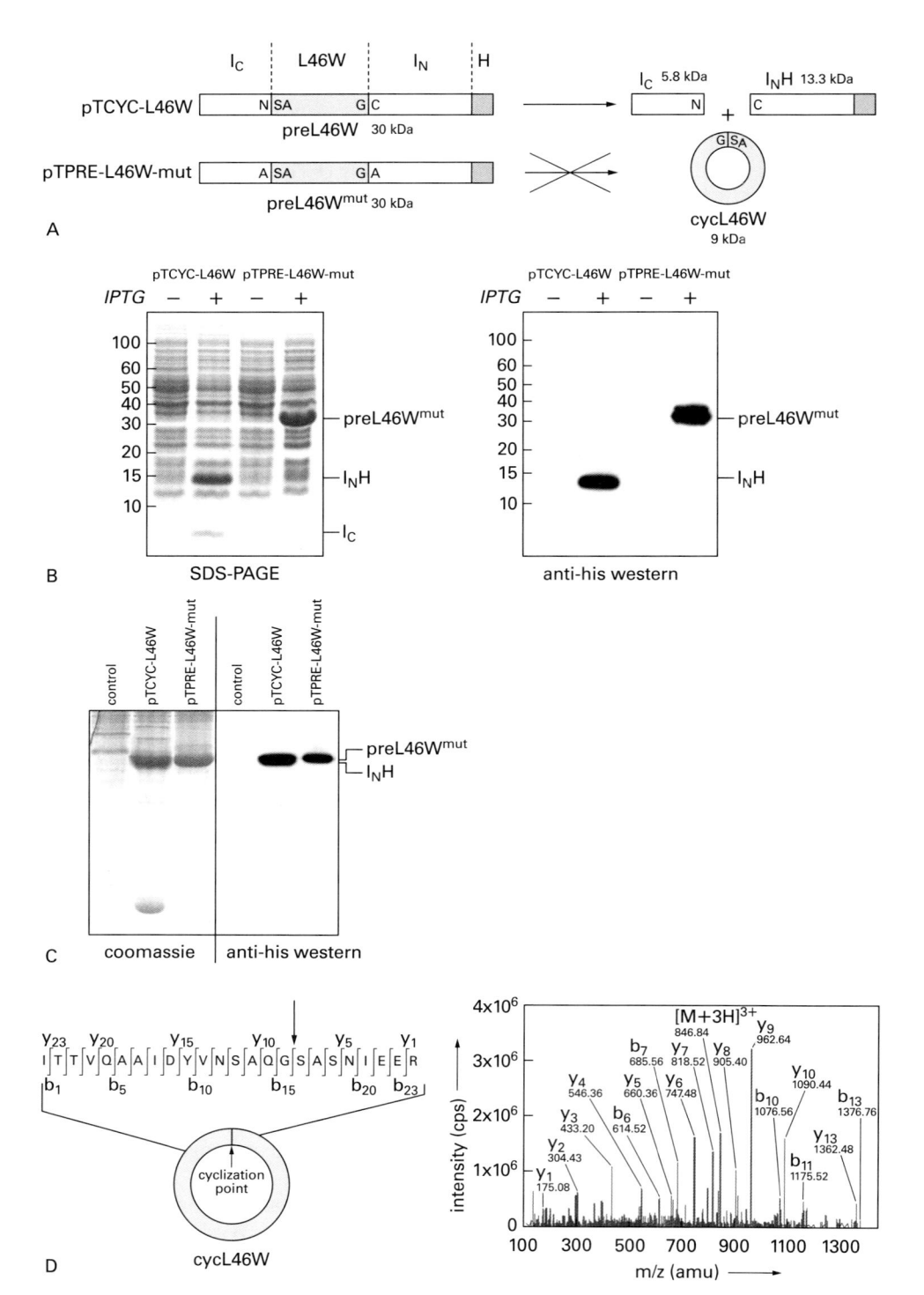

Figure 16.2: In vivo cyclization of *Vha* ACP-L46W. (A) Schematic representation of the expression constructs and expected products from split intein catalyzed backbone

(Continued)

Figure 16.2: (*Continued*)

cyclization. Amino acid residues at the ligation junction are indicated in one-letter code. (B) SDS-PAGE and Western blot analysis of expression constructs shown in A. Relevant protein species are indicated; molecular masses are given in kDa. (C) Native PAGE and Western blot of soluble extracts obtained from cultures expressing constructs shown in A. Cells in the control lane did not express ACP-related proteins. (D) Left: schematic representation of the cyclic *Vha* ACP-L46W protein (cycL46W) and the sequence of the diagnostic tryptic peptide. The cyclization point is indicated by an arrow. Right: confirmation of the diagnostic tryptic peptide ion shown to the left by tandem mass spectrometry. Figure adapted from Volkmann et al. (2010).

below 10 kDa, which was identified by MS as the C-terminal split intein fragment I_C (calculated M_W: 5353.03). These results clearly indicated that the expressed preL46W protein underwent a reaction inside the cells, liberating both I_NH and I_C from the precursor protein. The reaction appeared to be complete due to the absence of preL46W protein, which should have migrated at the same position as the preL46W^{mut} protein expressed from plasmid pTPRE-L46W-mut; however, due to the absence of a third induced protein species, it was not clear at this point whether the detected split intein fragments resulted from the anticipated protein cyclization reaction or from simultaneous N- and C-terminal cleavage. We therefore took advantage of the highly acidic nature of *Vha* ACP and its ability to migrate in native PAGE under basic conditions (pH 9.2) (Flaman et al. 2001). Soluble protein extracts were prepared and separated by native PAGE, which revealed a rapidly migrating protein species only in extracts prepared from cultures harboring plasmid pTCYC-L46W (▶Figure 16.2C). The less mobile proteins induced from pTCYC-L46W and pTPRE-L46W-mut at the top of the native gel were identified as the I_NH split intein fragment and the preL46W^{mut} protein, based on their reactivity with anti-His antibodies during Western blot analysis.

The rapidly migrating protein from the pTCYC-L46W extract was subsequently analyzed by MS. The circular nature of a protein can be accurately confirmed by proteolysis, liquid chromatography and peptide sequencing. In principle, a linear protein that presents five cleavage sites for the protease trypsin in its primary sequence should result in six peptide fragments upon complete digestion with trypsin, whereas the same protein in a circular form should give only five tryptic peptide fragments because the N- and C-terminal fragments are linked by a peptide bond. We refer to the peptide, whose sequence extends across the region of cyclization, as the diagnostic peptide (▶Figure 16.2D).

To look for the diagnostic peptide expected for the cycL46W protein, the protein band at the bottom of the native gel (▶Figure 16.2C) was excised, reduced with DTT, and acetylated with iodoacetamide. In-gel digestion of the protein was conducted with sequencing-grade trypsin (Promega) for 7.5 h at 37°C, and the tryptic peptides were resuspended in 30 µL of 5% acetonitrile/0.5% formic acid. Proteolysis products were injected onto an Onyx monolithic C18 capillary column (0.1×150 mm, Phenomenex), which was operated by a nanoflow Ultimate system (LC Packings) interfaced to the nanoflow electrospray ionization source of a hybrid triple quadrupole linear ion trap mass spectrometer (Applied Biosystems). Liquid chromatography was performed with solvent A (0.1% formic acid in 2% aqueous acetonitrile) and solvent B (0.1% formic acid in 98% aqueous acetonitrile), and peptides were eluted with a 3–30% solvent B gradient over 35 min at a flow rate of 1 µL/min. Spectra were acquired using Analyst 1.4.1 in information dependent acquisition mode.

The analysis revealed a triply charged peptide ion at m/z 846.84 (▶Figure 16.2D), whose deconvoluted mass (2,537.53 Da) was in excellent agreement with the expected mass of the diagnostic peptide (2,537.72 Da). This peptide ion was then fragmented by collision-induced dissociation, yielding an MS/MS spectrum that was in good agreement with the sequence ITTVQAAIDYVNSAQGSASNIEER (▶Figure 16.2D), which is the expected sequence of the diagnostic peptide (cyclization point underlined). Our MS data thus unequivocally confirmed the successful in vivo cyclization of Vha ACP-L46W using split intein-mediated protein trans-splicing.

16.2.3 In vitro biophysical characterization of purified cyclic ACP

16.2.3.1 Purification

Due to the absence of an affinity tag in the cycL46W protein, we took advantage of the heat stability of ACP for protein purification. First, BL21(DE3)pLysS cells harboring pTCYC-L46W were grown to mid-log phase (A_{660} = 0.5–0.6) in LB medium (containing 100 µg/mL ampicillin and 34 µg/mL chloramphenicol) at 37°C, at which point the culture was induced with 1 mM IPTG at 30°C and growth continued until the stationary phase was reached (A_{660} = 1.5–2.0). Cells were then harvested by centrifugation (4 °C, 13,000 × g, 10 min) and resuspended in phosphate-buffered saline (2 mL per 250 mL of culture). Cell lysis was performed by repeated passages through progressively smaller bore needles (16.5 ga, 18.5 ga, 21.5 ga) prior to sonication. Cellular debris was then removed by centrifugation (4°C, 27,000 × g, 20 min). The soluble protein fraction was heated for 10 min at 80°C, cooled on ice for 10 min and subjected to centrifugation (4°C, 14,000 × g, 10 min). This step takes advantage of the heat stability of ACP (presumably due to its inherent flexibility and anionic nature) and will not be possible with most proteins. The supernatant containing heat-stabile ACP was filtered (0.22 µm) and injected onto a Superose 6 10/300 GL size exclusion column (GE Healthcare) equilibrated with 10 mM MES (pH 6.0), 2 mM DTT. Fractions found to contain cycL46W were pooled and then filtered (0.22 µm) prior to Source15Q (GE Healthcare) anion exchange chromatography. Bound protein was eluted using a linear gradient (1 mL/min flow rate) from 0–1 M NaCl in the same buffer. Both chromatographic steps were performed using a Waters 650 protein chromatography system, and we found that the steps were essentially interchangeable. ▶Figure 16.3A shows the resulting chromatogram when anion exchange chromatography was performed first.

In order to accurately determine the effects of backbone cyclization on Vha ACP-L46W using various biophysical techniques, we also constructed a linear control protein (linL46W). This protein was expressed in fusion with glutathione-S-transferase, and was purified by a standard protocol (Flaman et al. 2001). The linL46W protein cleaved from glutathione-S-transferase had the exact same primary sequence as cycL46W but lacked the peptide bond between the N-terminal Ser and the C-terminal Gly residues.

When we examined the migration behavior of the purified cycL46W and linL46W proteins in SDS- and native PAGE, we observed that the cycL46W protein migrated significantly faster in both electrophoretic systems (▶Figure 16.3B). This gave a first

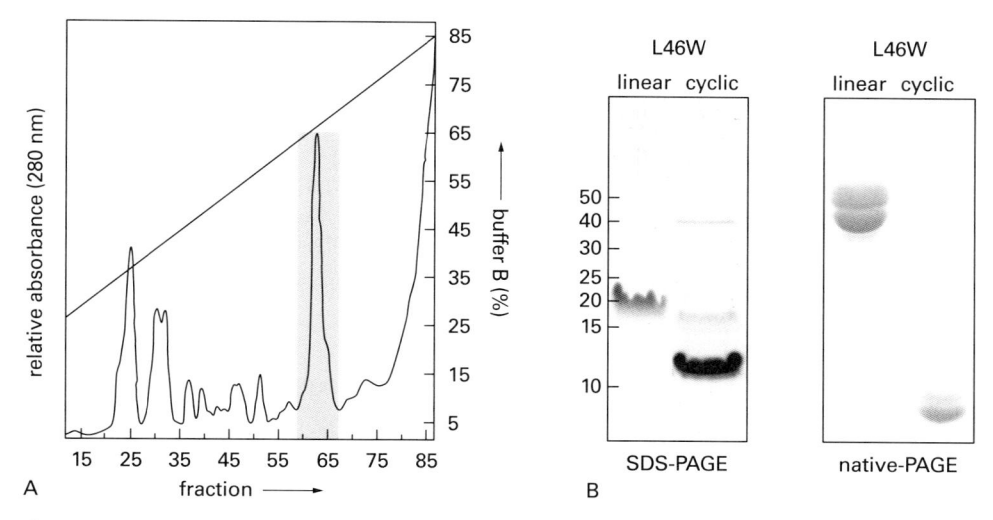

16.2 In vivo protein cyclization, biophysical analyses and functional assays | 281

Figure 16.3: Purification of cycL46W and migration behavior in protein gel electrophoresis. (A) Chromatogram from cation exchange chromatography of heated soluble extracts obtained from BL21(DE3)/pTCYC-L46W cells. The grey shaded area represents the elution peak of cycL46W. (B) SDS- and native PAGE analysis of purified linear and cyclic ACP-L46W proteins. Gels were stained with Coomassie; molecular masses are given in kDa. Figure adapted from Volkmann (2009) and Volkmann et al. (2010).

indication that the cycL46W protein exhibited a more compact structure, allowing the protein to fit through smaller pores in the polyacrylamide gels. This observation is consistent with studies of other backbone-cyclized proteins (Iwai et al. 2001; Scott et al. 1999; Williams et al. 2002) and represents a quick and easy way to assess the effect of cyclization on the overall protein structure.

16.2.3.2 CD spectroscopy

The two main secondary structural elements of proteins, α helices and β sheets, respond differently to circularly polarized light in terms of absorption. This behavior is used in CD spectroscopy to semiquantitatively evaluate the contribution of secondary structural elements to the overall protein structure. This technique thus allows to quickly determine whether changing the primary structure of a protein (e.g. by mutation or backbone cyclization) somehow affects the global protein structure.

CD spectra of linL46W and cycL46W were recorded from 190–260 nm in continuous mode with a speed of 20 nm/min using a J-810 spectropolarimeter (Jasco) with a 0.1 cm water-jacketed cell (25°C ± 0.1°C). The chosen wavelength range encompasses the region where α helices are observed using circularly polarized light (200–215 nm). The proteins were diluted to 1 µM in 10 mM sodium phosphate (pH 7.0), 0.1 mM EDTA. As noted in the introduction (Section 16.1.3), *Vha* ACP is largely unfolded at the chosen experimental pH of 7.0, but can be brought into the folded conformation by adding magnesium ions. This conformational transition is very rapid, hence spectra for both proteins were acquired before and immediately after the addition of MgSO$_4$ (10 mM final concentration), which exhibits decreased UV absorbance compared to MgCl$_2$.

As expected, CD indicated that linL46W protein is largely devoid of ordered secondary structure in the absence of MgSO$_4$, but exhibits significant α-helical content in the presence of MgSO$_4$ (▶Figure 16.4A) likely corresponding to the four-helix bundle. In contrast, cycL46W appeared to be in the folded conformation even in the absence of MgSO$_4$ (▶Figure 16.4A), indicating that cyclization of the *Vha* ACP-L46W mutant stabilized the folded four-helix bundle. Furthermore, the CD spectra of folded linL46W and cycL46W were almost identical, suggesting no detrimental effects of backbone cyclization on the global ACP structure.

16.2.3.3 Steady-state tryptophan fluorescence spectroscopy

As mentioned in Section 16.2.1, the L46W mutant of *Vha* ACP was chosen because the lone tryptophan residue has been shown to serve as an indicator of *Vha* ACP tertiary structure (Gong et al. 2008). When *Vha* ACP-L46W is unfolded (e.g. in the absence of divalent cations in vitro) the Trp residue is exposed to bulk solvent and has a peak fluorescence emission at 350 nm. If *Vha* ACP-L46W is folded (e.g. in the presence of divalent cations in vitro), the Trp is positioned in the hydrophobic core of ACP and emits at ~310 nm. This characteristic gave us the opportunity to further substantiate the results from CD spectroscopy concerning the stabilized folded conformation of cycL46W, as CD and Trp fluorescence changes have previously been shown to occur in concert in the L46W mutant (Gong et al. 2008).

All fluorescence measurements were carried out on a Photon Technology International (PTI) QuantaMaster-4CW in a 3 mm semimicro cuvette (Hellma) using the supplied cuvette holder (PTI). This assembly in turn sits in a Peltier-controlled heat/water-cooled 10 mm cuvette holder (25°C ± 0.01°C) in the main compartment of the instrument. The slit widths were set at 5 nm for both excitation and emission monochromators to achieve the best balance between spectral resolution and signal-to-noise ratio. The selection of slit widths depends on how much signal is recorded by the spectrometer and how important it is to limit excitation and emission wavelengths. An important note is that opening or closing the slits will increase and decrease, respectively, the spectral resolution and signal-to-noise of a spectrum (Lakowicz 2006).

The linL46W and cycL46W proteins were diluted to 1 μM in 10 mM sodium phosphate (pH 7.0), 0.1 mM EDTA, and were excited at 296 nm. The emission spectra were recorded from 300–450 nm in photon-counting mode with an integration time of 1 s both before and after the addition of MgSO$_4$ (10 mM final concentration). This protein concentration was chosen as Trp-46 exhibits substantial emission even at this dilute concentration. However, Trp at other positions or in other proteins may not have such a signal and may require increased protein concentrations (5–10 μM or more). Inclusion of EDTA is important to chelate any divalent cations that may be present in protein preparations. Finally, all fluorescence intensities were corrected for the Raman band from the solvent using the supplied software (Felix32™, PTI).

As expected, the fluorescence spectrum of linL46W in the absence of divalent cations (▶Figure 16.4B) exhibited peak emission of Trp-46 at 350 nm (unfolded conformation),

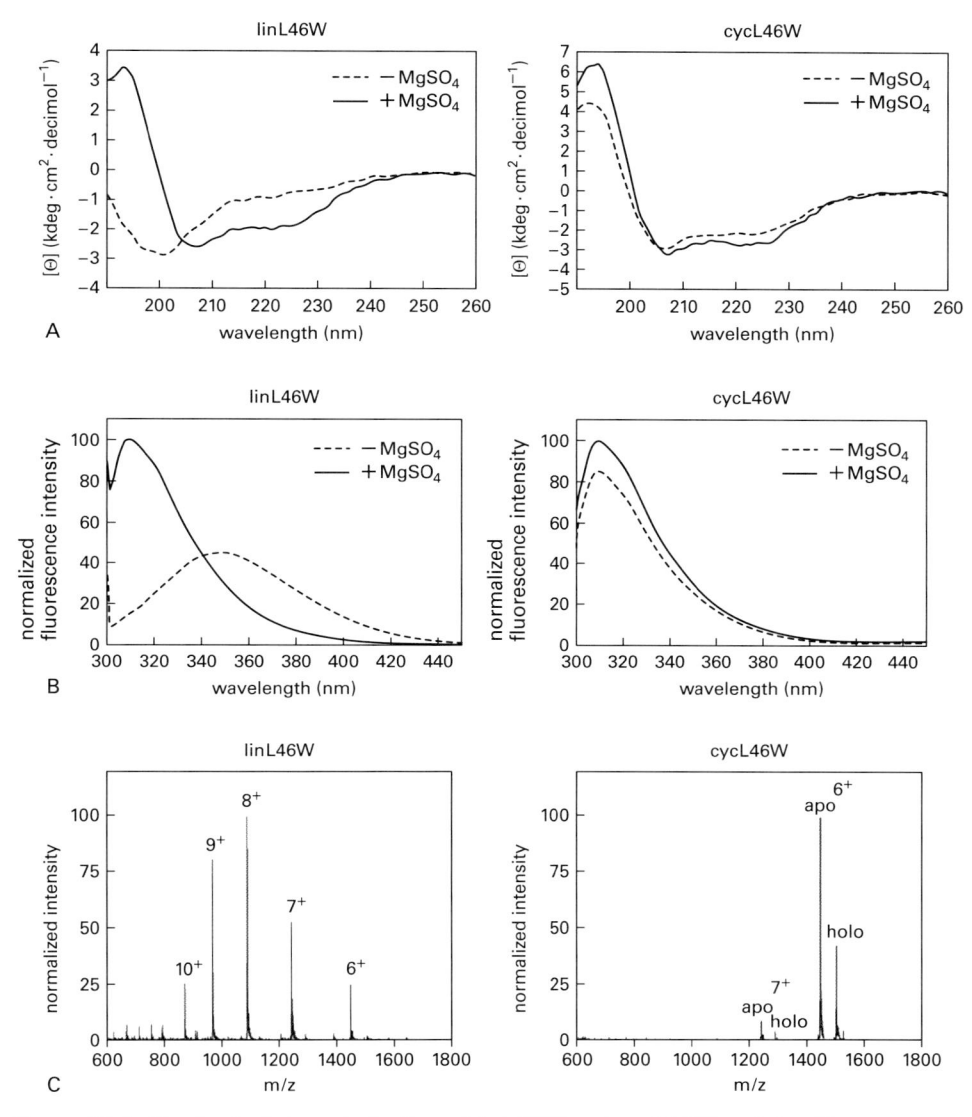

16.2 In vivo protein cyclization, biophysical analyses and functional assays | 283

Figure 16.4: Biophysical characterization of linL46W and cycL46W proteins. Shown are representative data from (A) CD spectroscopy, (B) steady-state tryptophan fluorescence spectroscopy, and (C) charge state distribution as determined by electrospray ionization mass spectrometry. Figure adapted from Volkmann et al. (2010).

whereas in the presence of MgSO$_4$ the peak was at 310 nm (folded conformation). CycL46W, on the other hand, had emission maxima of Trp-46 at 310 nm (folded conformation) both in the absence and presence of MgSO$_4$. This result was consistent with the findings from CD spectroscopy and further substantiated the assumption that cyclization of *Vha* ACP-L46W stabilized the folded conformation.

16.2.3.4 Charge state distribution

To evaluate further the effects of backbone cyclization on *Vha* ACP-L46W, we analyzed the linL46W and cycL46W proteins for their charge state distribution (CSD) using positive-mode electrospray ionization mass spectrometry. This technique is generally sensitive to the conformation of proteins (Yan et al. 2004) and has shown value in analyzing backbone-cyclized proteins (Watt et al. 2007). In general, when proteins enter the gas phase of the mass spectrometer in an unfolded state they can form a variety of differently charged ions and thus exhibit a broad CSD. Folded proteins, however, produce a narrower CSD of lower charge magnitude due to their decreased solvent-accessible surface area.

The purified proteins (5 µg in 0.1% TFA in H_2O) were desalted by centrifugation (1000 × g) over 3 µL of POROS R2 resin (Applied Biosystems) packed into a gel-loading tip with a restricted end. Prior to loading the proteins, the resin was primed with 50 µL of 70% aqueous acetonitrile/0.1% formic acid and equilibrated with 0.1% TFA in H_2O. The proteins were eluted in 70% aqueous acetonitrile/0.1% formic acid. The solvent was removed by centrifugation in a speed vac, and the proteins were resuspended in 100 µL of 10% aqueous acetonitrile/0.5% formic acid. Proteins were subjected to liquid chromatography on an Onyx monolithic C18 capillary column (0.1×50 mm, Phenomenex) prior to injection into the mass spectrometer. The flow rate was set to 3 µL/min, and the gradient was 10%–70% solvent B over 20 min (Section 16.2.2). The capillary voltage was 2.20 kV with a declustering potential of 30 V, the curtain gas and GS1 were set to 20 and 2, respectively (arbitrary units). Spectra were acquired in enhanced mass spectrum mode (400–1700 *m/z*) with an ion trap fill time of 20 ms.

The CSD of linL46W exhibited a broad pattern (ions from 10^+ to 6^+) with a maximum at 8^+, which is in agreement with the CSD of wild-type *Vha* ACP (Murphy et al. 2007) (▶Figure 16.4C). The apparent molecular mass of linL46W was 8685 Da (expected monoisotopic mass of oxidized, unmodified (i.e., apo-) linL46W is 8683 Da). In contrast, the CSD of cycL46W was narrow, with most protein ions having six positive charges (▶Figure 16.4C). We note that in the case of cycL46W we detected both apo- and holo-forms, the latter modified with 4'-phosphopantetheine at Ser-36. This modification had no effect on the CSD, as holo-cycL46W was also present mostly with six positive charges. The observed molecular mass of apo-cycL46W was 8667 Da, which is 18 Da less than the mass of apo-linL46W and is due to the loss of a water molecule during the split intein-mediated cyclization reaction.

The narrow CSD of cycL46W is consistent with data obtained by CD and Trp fluorescence spectroscopy, indicating that the backbone-cyclized *Vha* ACP-L46W protein exhibits a stabilized folded conformation. However, this technique provides additional insight into the restricted backbone dynamics of cyclized ACP, as our earlier study demonstrated that wild-type (linear) *Vha* ACP exhibits a broad (i.e. unfolded) CSD regardless of whether the protein was folded in solution prior to analysis in the mass spectrometer (Murphy et al. 2007). It is tempting to speculate that, unlike linL46W, the cycL46W protein might be incapable of rapidly unfolding in passing from the liquid to gas phase during the sub-millisecond timescale of the electrospray process (Grandori 2003).

16.2.4 In vivo complementation assays

16.2.4.1 Engineering an *E. coli* ACP-knockout strain

Our biophysical analyses strongly indicated that the cycL46W protein is present in a stabilized conformation. Because the structurally dynamic character of the ACP family has been suggested to be pivotal for their biological activity, we wanted to assess whether the cyclic protein is functional in living cells. To pursue this idea, we needed a cellular background that allowed the expression of cycL46W in the absence of endogenous ACP. To this end, we constructed an *E. coli* strain, in which the chromosomal *acpP* gene is knocked out, and expression of wild-type *E. coli* ACP (*Eco* ACP) from a plasmid strictly requires the presence of the inducer L-arabinose. Thus, when introducing a second plasmid encoding a mutant ACP (like the construct for cycL46W) that is inducible by conditions that repress expression of *Eco* ACP (e.g. by providing D-glucose and IPTG), cellular growth should indicate functionality of the mutant ACP (▶Figure 16.5A).

First, an intermediate strain, CY1878, was generated, which would later be subjected to transformation with plasmid DNA containing a synthetic version of the *Eco* ACP gene, followed by gene replacement of the chromosomal *Eco* ACP copy with a chloramphenicol resistance cassette. Using standard phage transduction techniques (Sambrook and Russell 2001), strain CY1878 was constructed by transducing the *E. coli* strain DY329 (ΔlacU169 nadA:Tn10 gal490 λcI857Δ[cro-bioA]; Yu et al. 2000) with a phage P1vir stock grown on the wild-type strain MC1061 (Casadaban et al. 1983). Transductants were selected at 30°C on glucose medium E plates in the absence of biotin and nicotinic acid. Growth on these plates indicated loss of the temperature-sensitive λ prophage from strain DY329, which was further confirmed by screening colonies for growth at 42°C. One of the temperature-resistant colonies was saved as strain CY1878. This strain shows a very low background when performing targeted gene replacements (J.E. Cronan Jr., unpublished results).

Strain CY1878 was then transformed with plasmid pCY765 with selection on agar plates containing 50 µg/mL spectinomycin. The plasmid pCY765, which contains a synthetic *Eco acpP* gene (Hill et al. 1995) under control of the arabinose-inducible P_{BAD} promoter, was constructed in two steps. First, the 1,401 bp BamHI/FspI fragment of pNRD25 (De Lay and Cronan 2006) was ligated to the 3,815 bp fragment of pBAD30 (Guzman et al. 1995) cut with the same enzymes. The resulting intermediate plasmid was then digested with FspI and ligated to the 2,067 bp HincII fragment of p34S-Sm2 (Dennis and Zylstra 1998), which harbors the *aadA* gene encoding a spectinomycin resistance determinant, to give pCY765 (insertion into the FspI site inactivated the ampicillin resistance gene of the intermediate plasmid).

In the final steps, the chromosomal copy of the wild-type *Eco acpP* gene in CY1878/pCY765 was replaced with an antibiotic resistance cassette against chloramphenicol (Δ*acpP:cat*). For this, a phage P1vir stock was prepared by growing the phage on strain NRD62, which already presents Δ*acpP:cat* on the chromosome and was obtained by λ Red recombinase-mediated gene replacement (De Lay and Cronan 2006). Transductants were selected on agar plates containing 10 µg/mL chloramphenicol and 0.2% L-arabinose, and one colony was saved as strain CY1861. Of note, the *acpP* gene on plasmid pCY765 is synthetic in nature and contains 28 silent nucleotide changes (Hill et al. 1995), which effectively prevented recombination between the phage-derived Δ*acpP:cat* and the *acpP* sequence on plasmid pCY765.

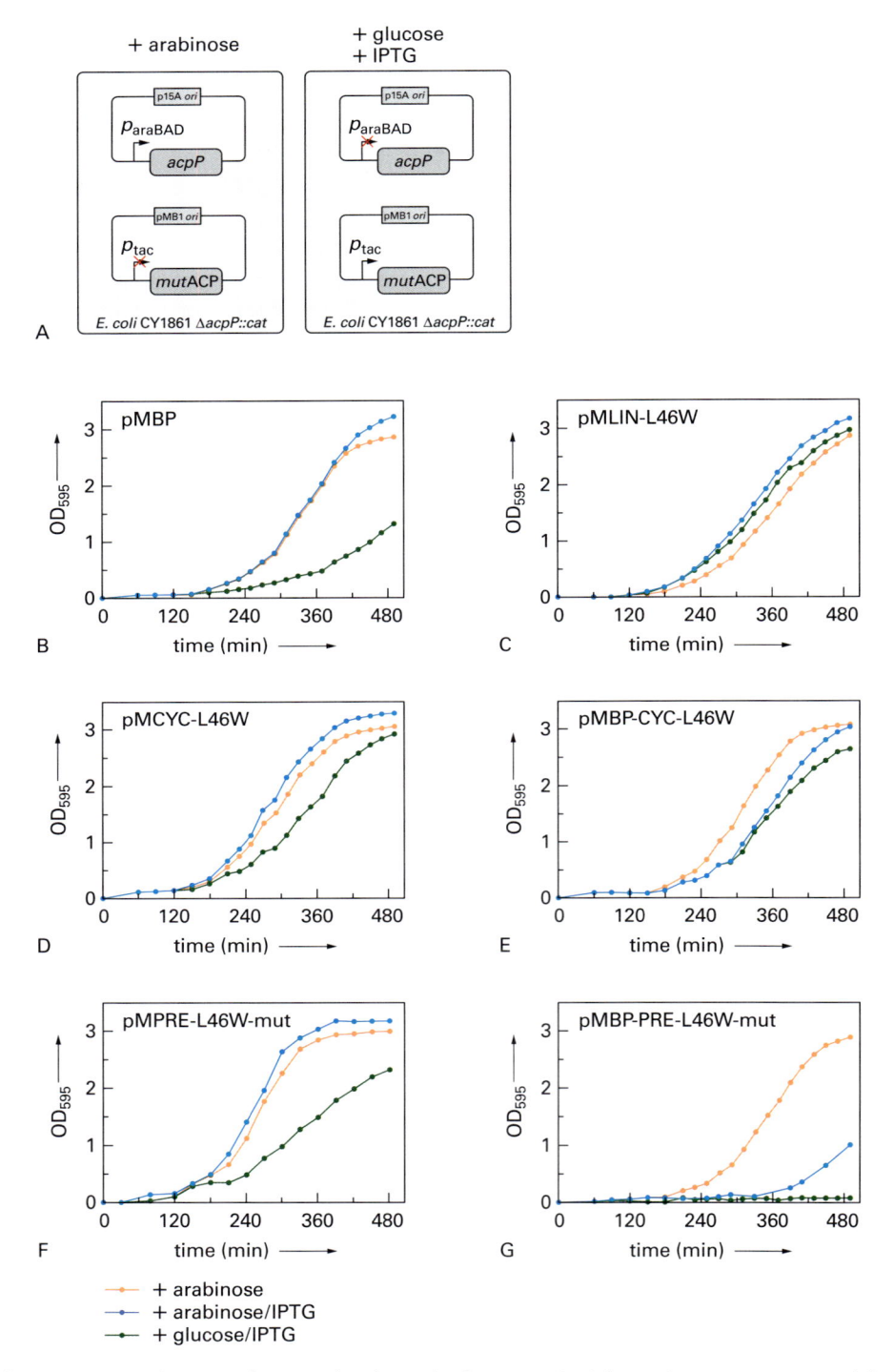

Figure 16.5: In vivo complementation in strain CY1861. (A) Schematic representation of the in vivo growth complementation experiment. *E. coli* strain CY1861 expresses wild-type *Eco* ACP only in the presence of arabinose, whereas expression of the query constructs

(Continued)

Figure 16.5: (*Continued*)

containing mutated or modified ACP constructs (mutACP, e.g. for cyclization of *Vha* ACP-L46W) is only induced in medium containing IPTG and glucose. Glucose represses expression of wild-type *Eco* ACP from the araBAD promoter. (B-G) Growth curves obtained for CY1861 transformed with the indicated query plasmids. Figure adapted from Volkmann et al. (2010).

16.2.4.2 Construction of expression vectors for in vivo complementation

To test the functionality of the cycL46W protein in strain CY1861, we first needed to reclone the cyclization open reading frame (cycORF) from pTCYC-L46W because in this plasmid, protein expression requires T7 RNA polymerase, which is not produced in CY1861. We chose to introduce the cycORF into plasmid pMAL-c2X (New England Biolabs), where recombinant protein expression is also inducible by IPTG but uses the endogenous *E. coli* transcription machinery. Moreover, this plasmid has a pMB1 origin of replication and is thus compatible with the pACYC-derived plasmid pCY765, which uses the p15A origin of replication (▶Figure 16.5A). The two plasmids also encode different antibiotic selection markers (pMAL-c2X: ampicillin, pCY765: spectinomycin).

Two sets of pMAL-c2X derived expression constructs were made: one set (pM-series) where the cycORF replaced the plasmid-borne *malE* gene encoding for *E. coli* maltose binding protein (MBP), and the other (pMBP-series) where the cycORF was cloned downstream of *malE*. The restriction sites used for the pM-series were NdeI and PstI, whereas for the pMBP-series BamHI and PstI were used. Although in the case of the pMBP-series, expression leads to an MBP-I_C-ACP (L46W)-I_NH protein, the presence of MBP upstream of I_C does not interfere with the split intein-mediated cyclization reaction (Volkmann 2009). Interestingly, we found that the expression levels for plasmids from the pM-series was about 25-fold lower than from the pMBP-series (data not shown), giving the opportunity to evaluate the effect of both high and low expression levels of the cycORFs. Control constructs for the in vivo growth assays included the cyclization-deficient PRE-L46W-mut ORF (in pM- and pMBP-series), as well as the linL46W ORF as a positive control (only pM-series), and a construct expressing only MBP as a negative control.

16.2.4.3 In vivo growth complementation assay

Strain CY1861 was made chemically competent using $CaCl_2$ (Sambrook and Russell 2001), after which the plasmids described in Section 16.2.4.2 were transformed. Transformants were grown at 37°C for 18 h on LB-agar plates supplemented with 50 µg/mL spectinomycin, 100 µg/mL ampicillin and 0.2% arabinose. From the transformation plates, three single colonies of similar size from all transformations were each resuspended in 50 µL LB medium containing 50 µg/mL spectinomycin and 100 µg/mL ampicillin (LB-SA medium). In a 96-well plate, three wells were filled with 245 µL of LB-SA medium for every colony resuspension. One well was then further supplemented with 0.2% arabinose (for expression of wild-type *Eco* ACP), the second well with 0.2% glucose (to repress the expression of *Eco* ACP) and 0.8 mM IPTG (for expression of the query *Vha* ACP-L46W proteins and the negative control, MBP), and the third well with 0.2% arabinose and 0.8 mM IPTG to assess whether expression of any of the query constructs was toxic to the cells. To the wells were pipetted 5 µL of the colony resuspensions, after which the absorbance at 595 nm (A_{595}) was recorded ($t = 0$). The plate then incubated at 37°C with vigorous shaking, and A_{595} was recorded every 20–30 min for a period of 8 h using a plate reader. The A_{595} readings were normalized to the $t = 0$ values and were multiplied by factor 4 to give the OD_{595}. We recommend not to use the outer wells of the 96-well plate during such a growth assay because pronounced evaporation occurs in these wells during the prolonged incubation period, resulting in a decrease in the culture volume and inaccurate measurements of A_{595}.

▶Figure 16.5B-G shows the growth curves from the complementation experiment. In all cases, the transformants grew well in medium containing 0.2% arabinose, and had reached the stationary phase after 6–8 h. This showed that the wild-type *Eco* ACP expressed from plasmid pCY765 rescued cell survival, as expected. The negative control construct expressing MBP in medium containing glucose and IPTG led to severely delayed growth (▶Figure 16.5B), whereas the positive control (expression of linear *Vha* ACP-L46W from pMLIN-L46W) fully complemented cell growth (▶Figure 16.5C). Expression of the cyclization constructs under both high and low levels of expression (from plasmids pMBP- and pMCYC-L46W) also resulted in growth complementation, albeit with a slight delay in reaching stationary phase when compared to the linear ACP-L46W construct (▶Figure 16.5D,E). This finding thus suggests that the backbone-cyclized *Vha* ACP-L46W protein is able to function in fatty acid synthesis in living cells despite its restricted dynamic character (Section 16.2.3). Lastly, the cyclization deficient construct pMPRE-L46W-mut was able to rescue growth at levels in between that of cyclized protein and the negative control (▶Figure 16.5F), which indicates that free protein termini of ACP are beneficial but not obligatory for its biological function. Interestingly, we observed that the cyclization-deficient construct pMBP-PRE-L46W-mut was inhibitory for cell growth even when wild-type *Eco* ACP was simultaneously expressed (▶Figure 16.5G).

16.3 Outlook

In this chapter, we have described the use of split intein-mediated protein cyclization for evaluating the impact of altered protein dynamics in living cells. Our approach necessitates that conformational changes involve the native amino and carboxy termini of a protein, which would be restricted in their flexibility upon protein cyclization. Given that a large fraction of all proteins likely have their termini in close proximity (Chiche et al. 2004), cyclizing the backbone could indeed be a novel and more generally useful avenue to probe conformational changes in proteins. This approach could also enhance crystallization success of proteins (such as ACP) that may be resistant to X-ray structure determination due to their dynamic nature. Although we have used *E. coli* as a model organism in our methodology, we believe that in vivo protein cyclization by split inteins holds great value to address similar scientific questions about protein dynamics and conformational changes in other model organisms. Fortunately, split inteins do not require any cofactors or energy equivalents for their peptide bond forming reaction, and can thus catalyze protein cyclization in eukaryotic cells and even animals (Kanno et al. 2007). A variety of techniques exist to genetically manipulate eukaryotic cells either by deleting a gene from the genome or performing transient knock-downs by silencing

RNAs, which should generally make it possible to replace a native protein by a circular variant in these more complex cells.

References

Abel-Santos, E., Scott, C.P., and Benkovic, S.J. (2003). Use of inteins for the in vivo production of stable cyclic peptide libraries in *E. coli*. Methods Mol Biol *205*, 281–294.

Appleby, J.H., Zhou, K., Volkmann, G., and Liu, X.Q. (2009). Novel split intein for trans-splicing synthetic peptide onto C terminus of protein. J Biol Chem *284*, 6194–6199.

Azim-Zadeh, O., Hillebrecht, A., Linne, U., Marahiel, M.A., Klebe, G., Lingelbach, K., et al. (2007). Use of biotin derivatives to probe conformational changes in proteins. J Biol Chem *282*, 21609–21617.

Barnes, C.O., Monteith, W.B., and Pielak, G.J. (2011). Internal and global protein motion assessed with a fusion construct and in-cell NMR spectroscopy. ChemBioChem *12*, 390–391.

Bertini, I., Felli, I.C, Gonnelli, L., Kumar, M.V.V., and Pierattelli, R. (2011). ^{13}C direct-detection biomolecular NMR spectroscopy in living cells. Angew Chem Int Ed Engl *50*, 2339–2341.

Bodart, J.F., Wieruszeski, J.M., Amniai, L., Leroy, A., Landrieu, I., Rousseau-Lescuyer, A., et al. (2008). NMR observation of Tau in *Xenopus* oocytes. J Magn Reson *192*, 252–257.

Burz, D.S., and Shekhtman, A. (2008). In-cell biochemistry using NMR spectroscopy. PLoS One *3*, e2571.

Butland, G., Peregrín-Alvarez, J.M., Li, J., Yang, W., Yang, X., Canadien, V., et al. (2005). Interaction network containing conserved and essential protein complexes in *Escherichia coli*. Nature *433*, 531–537.

Byers, D.M., and Gong, H. (2007). Acyl carrier protein: structure-function relationships in a conserved multifunctional protein family. Biochem Cell Biol *85*, 649–662.

Camarero, J.A., Fushman, D., Cowburn, D., and Muir, T.W. (2001a). Peptide chemical ligation inside living cells: in vivo generation of a circular protein domain. Bioorg Med Chem *9*, 2479–2484.

Camarero, J.A., Fushman, D., Sato, S., Giriat, I., Cowburn, D., Raleigh, D.P., et al. (2001b). Rescuing a destabilized protein fold through backbone cyclization. J Mol Biol *308*, 1045–1062.

Camarero, J.A., and Muir, T.W. (1999). Biosynthesis of a head-to-tail cyclized protein with improved biological activity. J Am Chem Soc *121*, 5597–5598.

Casadaban, M.J., Martinez-Arias, A., Shapira, S.K., and Chou, J. (1983). Beta-galactosidase gene fusions for analyzing gene expression in *Escherichia coli* and yeast. Methods Enzymol *100*, 293–308.

Cascales, L., and Craik, D.J. (2010). Naturally occurring circular proteins: distribution, biosynthesis and evolution. Org Biomol Chem *8*, 5035–5047.

Chamberlain, C., and Hahn, K.M. (2000). Watching proteins in the wild: fluorescence methods to study protein dynamics in living cells. Traffic *1*, 755–762.

Chan, D.I., Chu, B.C., Lau, C.K., Hunter, H.N., Byers, D.M., and Vogel, H.J. (2010). NMR solution structure and biophysical characterization of *Vibrio harveyi* acyl carrier protein A75H: effects of divalent metal ions. J Biol Chem *285*, 30558–30566.

Chan, D.I., and Vogel, H.J. (2010). Current understanding of fatty acid biosynthesis and the acyl carrier protein. Biochem J *430*, 1–19.

Chang, T.K., Jackson, D.Y., Burnier, J.P., and Wells, J.A. (1994). Subtiligase: a tool for semisynthesis of proteins. Proc Natl Acad Sci U S A *91*, 12544–12548.

Cheng, L., Naumann, T.A., Horswill, A.R., Hong, S.J., Venters, B.J., Tomsho, J.W., et al. (2007). Discovery of antibacterial cyclic peptides that inhibit the ClpXP protease. Protein Sci *16*, 1535–1542.

Chiche, L., Heitz, A., Gelly, J.C., Gracy, J., Chau, P.T.T., Ha, P.T., et al. (2004). Squash inhibitors: from structural motifs to macrocyclic knottins. Curr Protein Pept Sci 5, 341–349.

Clark, R.J., Fischer, H., Dempster, L., Daly, N.L., Rosengren, K.J., Nevin, S.T., et al. (2005). Engineering stable peptide toxins by means of backbone cyclization: stabilization of the alpha-conotoxin MII. Proc Natl Acad Sci U S A 102, 13767–13772.

Daughdrill, G.W., Hanely, L.J., and Dahlquist, F.W. (1998). The C-terminal half of the anti-sigma factor FlgM contains a dynamic equilibrium solution structure favoring helical conformations. Biochemistry 37, 1076–1082.

Dawson, P.E., Muir, T.W., Clark-Lewis, I., and Kent, S.B. (1994). Synthesis of proteins by native chemical ligation. Science 266, 776–779.

Dedmon, M.M., Patel, C.N., Young, G.B., and Pielak, G.J. (2002). FlgM gains structure in living cells. Proc Natl Acad Sci U S A 99, 12681–12684.

Deechongkit, S., and Kelly, J.W. (2002). The effect of backbone cyclization on the thermodynamics of beta-sheet unfolding: stability optimization of the PIN WW domain. J Am Chem Soc 124, 4980–4986.

de la Roche, M.A., Shen, Z., and Byers, D.M. (1997). Hydrodynamic properties of *Vibrio harveyi* acyl carrier protein and its fatty-acylated derivatives. Arch Biochem Biophys 34, 159–164.

De Lay, N.R., and Cronan, J.E., Jr. (2006). Gene-specific random mutagenesis of *Escherichia coli in vivo*: isolation of temperature-sensitive mutations in the acyl carrier protein of fatty acid synthesis. J Bacteriol 188, 287–296.

Dennis, J.J., and Zylstra, G.J. (1998). Improved antibiotic-resistance cassettes through restriction site elimination using *Pfu* DNA polymerase PCR. Biotechniques 25, 772–774, 776.

Deschuyteneer, G., Garcia, S., Michiels, B., Baudoux, B., Degand, H., Morsomme, P., et al. (2010). Intein-mediated cyclization of randomized peptides in the periplasm of *Escherichia coli* and their extracellular secretion. ACS Chem Biol 5, 691–700.

Evans, T.C., Jr., Martin, D., Kolly, R., Panne, D., Sun, L., Ghosh, I., et al. (2000). Protein trans-splicing and cyclization by a naturally split intein from the dnaE gene of *Synechocystis* species PCC6803. J Biol Chem 275, 9091–9094.

Flaman, A.S., Chen, J.M., Van Iderstine, S.C., and Byers, D.M. (2001). Site-directed mutagenesis of acyl carrier protein (ACP) reveals amino acid residues involved in ACP structure and acyl-ACP synthetase activity. J Biol Chem 276, 35934–35939.

Gong, H., Murphy, A., McMaster, C.R., and Byers, D.M. (2007). Neutralization of acidic residues in helix II stabilizes the folded conformation of acyl carrier protein and variably alters its function with different enzymes. J Biol Chem 282, 4494–4503.

Gong, H., Murphy, P.W., Langille, G.M., Minielly, S.J., Murphy, A., McMaster, C.R., et al. (2008). Tryptophan fluorescence reveals induced folding of *Vibrio harveyi* acyl carrier protein upon interaction with partner enzymes. Biochim Biophys Acta 1784, 1835–1843.

Grandori, R. (2003). Electrospray-ionization mass spectrometry for protein conformational studies. Curr Org Chem 7, 1589–1603.

Guzman, L.M., Belin, D., Carson, M.J., and Beckwith, J. (1995). Tight regulation, modulation, and high-level expression by vectors containing the arabinose PBAD promoter. J Bacteriol 177, 4121–4130.

Hill, R.B., MacKenzie, K.R., Flanagan, J.M., Cronan, J.E., Jr., and Prestegard, J.H. (1995). Overexpression, purification, and characterization of *Escherichia coli* acyl carrier protein and two mutant proteins. Protein Expr Purif 6, 394–400.

Hofmann, A., Iwai, H., Hess, S., Plückthun, A., and Wlodawer, A. (2002). Structure of cyclized green fluorescent protein. Acta Crystallogr D Biol Crystallogr 58, 1400–1406.

Horswill, A.R., Savinov, S.N., and Benkovic, S.J. (2004). A systematic method for identifying small-molecule modulators of protein-protein interactions. Proc Natl Acad Sci U S A 101, 15591–15596.

Inomata, K., Ohno, A., Tochio, H., Isogai, S., Tenno, T., Nakase, I., et al. (2009). High-resolution multi-dimensional NMR spectroscopy of proteins in human cells. Nature *458*, 106–109.

Ishihama, Y., Ito, O., and Oda, Y. (1999). A novel method for peptide block synthesis using unprotected peptides. Tetrahedron Lett *40*, 3415–3418.

Iwai, H., Lingel, A., and Plückthun, A. (2001). Cyclic green fluorescent protein produced in vivo using an artificially split PI-PfuI intein from *Pyrococcus furiosus*. J Biol Chem *276*, 16548–16554.

Iwai, H., and Plückthun, A. (1999). Circular beta-lactamase: stability enhancement by cyclizing the backbone. FEBS Lett *459*, 166–172.

Jackson, D.Y., Burnier, J.P., and Wells, J.A. (1995). Enzymatic cyclization of linear peptide esters using subtiligase. J Am Chem Soc *117*, 819–820.

Jeffries, C.M., Graham, S.C., Stokes, P.H., Collyer, C.A., Guss, J.M., and Matthews, J.M. (2006). Stabilization of a binary protein complex by intein-mediated cyclization. Protein Sci *15*, 2612–2618.

Kalkum, M., Eisenbrandt, R., and Lanka, E. (2004). Protein circlets as sex pilus subunits. Curr Protein Pept Sci *5*, 417–424.

Kanno, A., Yamanaka, Y., Hirano, H., Umezawa, Y., and Ozawa, T. (2007). Cyclic luciferase for real-time sensing of caspase-3 activities in living mammals. Angew Chem Int Ed Engl *46*, 7595–7599.

Kimura, R.H., Tran, A.T., and Camarero, J.A. (2006). Biosynthesis of the cyclotide Kalata B1 by using protein splicing. Angew Chem Int Ed *45*, 973–976.

Kinsella, T.M., Ohashi, C.T., Harder, A.G., Yam, G.C., Li, W., Peelle, B., et al. (2002). Retrovirally delivered random cyclic peptide libraries yield inhibitors of interleukin-4 signaling in human B cells. J Biol Chem *277*, 37512–37518.

Knappe, T.A., Linne, U., Robbel, L., and Marahiel, M.A. (2009). Insights into the biosynthesis and stability of the lasso peptide capistruin. Chem Biol *16*, 1290–1298.

Kobrinsky, E., Abrahimi, P., Duong, S.Q., Thomas, S., Harry, J.B., Patel, C., et al. (2009). Effect of $Ca_v\beta$ subunits on structural organization of $Ca_v1.2$ calcium channels. PLoS One *4*, e5587.

Kritzer, J.A., Hamamichi, S., McCaffery, J.M., Santagata, S., Naumann, T.A., Caldwell, K.A., et al. (2009). Rapid selection of cyclic peptides that reduce alpha-synuclein toxicity in yeast and animal models. Nat Chem Biol *5*, 655–663.

Kwon, J.S., Bal, J., Hwang, H.M., and Kim, J.Y. (2008). A circularly permuted beta-lactamase as a novel reporter for evaluation of protein cyclization efficiency. J Microbiol *46*, 456–461.

Lakowicz, J.R. (2006). Principles of fluorescence spectroscopy (New York: Springer Science+Business Media, LLC).

Li, C., Charlton, L.M., Lakkavaram, A., Seagle, C., Wang, G., Young, G.B., et al. (2008). Differential dynamical effects of macromolecular crowding on an intrinsically disordered protein and a globular protein: implications for in-cell NMR spectroscopy. J Am Chem Soc *130*, 6310–6311.

Li, C., Wang, G.F., Wang, Y., Creager-Allen, R., Lutz, E.A., Scronce, H., et al. (2010). Protein 19F NMR in *Escherichia coli*. J Am Chem Soc *132*, 321–327.

Ludwig, C., Schwarzer, D., and Mootz, H.D. (2008). Interaction studies and alanine scanning analysis of a semi-synthetic split intein reveal thiazoline ring formation from an intermediate of the protein splicing reaction. J Biol Chem *283*, 25264–25272.

Miyawaki, A., Llopis, J., Heim, R., McCaffery, J.M., Adams, J.A., Ikura, M., et al. (1997). Fluorescent indicators for Ca^{2+} based on green fluorescent proteins and calmodulin. Nature *388*, 882–887.

Murphy, P.W., Rowland, E.E., and Byers, D.M. (2007). Electrospray ionization mass spectra of acyl carrier protein are insensitive to its solution phase conformation. J Am Soc Mass Spectrom *18*, 1525–1532.

Nar, H., Schmid, A., Puder, C., and Potterat, O. (2010). High-resolution crystal structure of a lasso peptide. ChemMedChem 5, 1689–1692.

Naumann, T.A., Tavassoli, A., and Benkovic, S.J. (2008). Genetic selection of cyclic peptide Dam methyltransferase inhibitors. ChemBioChem 9, 194–197.

Nilsson, L.O., Louassini, M., and Abel-Santos, E. (2005). Using siclopps for the discovery of novel antimicrobial peptides and their targets. Protein Pept Lett 12, 795–799.

Ogino, S., Kubo, S., Umemoto, R., Huang, S., Nishida, N., and Shimada, I. (2009). Observation of NMR signals from proteins introduced into living mammalian cells by reversible membrane permeabilization using pore-forming toxin, streptolysin O. J Am Chem Soc 131, 10834–10835.

Popp, M.W., Dougan, S.K., Chuang, T.Y., Spooner, E., and Ploegh, H.L. (2011). Sortase-catalyzed transformations that improve the properties of cytokines. Proc Natl Acad Sci U S A 108, 3169–3174.

Post, P.L., DeBiasio, R.L., and Taylor, D.L. (1995). A fluorescent biosensor of myosin II regulatory light chain phosphorylation reports a gradient of phosphorylated myosin II in migrating cells. Mol Biol Cell 6, 1755–1768.

Romoser, V.A., Hinkle, P.M., and Persechini, A. (1997). Detection in living cells of Ca^{2+}-dependent changes in the fluorescence emission of an indicator composed of two green fluorescent protein variants linked by a calmodulin-binding sequence: a new class of fluorescent indicators. J Biol Chem 272, 13270–13274.

Rosengren, K.J., and Craik, D.J. (2009). How bugs make lassos. Chem Biol 16, 1211–1212.

Sakai, T., Tochio, H., Tenno, T., Ito, Y., Kokubo, T., Hiroaki, H., et al. (2006). In-cell NMR spectroscopy of proteins inside *Xenopus laevis* oocytes. J Biol NMR 36, 179–188.

Sambrook, J., and Russell, D. (2001). Molecular cloning: a laboratory manual. (New York: Cold Spring Harbour Press).

Saska, I., and Craik, D.J. (2008). Protease-catalyzed protein splicing: a new post-translational modification? Trend Biochem Sci 33, 363–368.

Saska, I., Gillon, A.D., Hatsugai, N., Dietzgen, R.G., Hara-Nishimura, I., Anderson, M.A., et al. (2007). An asparaginyl endopeptidase mediated in vivo protein backbone cyclization. J Biol Chem 282, 29721–29728.

Schwarzer, D., Finking, R., and Marahiel, MA. (2003). Non-ribosomal peptides: from genes to products. Nat Prod Rep 20, 275–287.

Scott, C.P., Abel-Santos, E., Wall, M., Wahnon, D.C., and Benkovic, S.J. (1999). Production of cyclic peptides and proteins in vivo. Proc Natl Acad Sci U S A 96, 13638–13643.

Selenko, P., Serber, Z., Gadea, B., Ruderman, J., and Wagner, G. (2006). Quantitative NMR analysis of the protein G B1 domain in *Xenopus laevis* egg extracts and intact oocytes. Proc Natl Acad Sci U S A 103, 11904–11909.

Serber, Z., Keatinge-Clay, A.T., Ledwidge, R., Kelly, A.E., Miller, S.M., and Dötsch, V. (2001a). High-resolution macromolecular NMR spectroscopy inside living cells. J Am Chem Soc 123, 2446–2447.

Serber, Z., Ledwidge, R., Miller, S.M., and Dötsch, V. (2001b). Evaluation of parameters critical to observing proteins inside living *Escherichia coli* by in-cell NMR spectroscopy. J Am Soc 123, 8895–8901.

Serber, Z., Selenko, P., Hänsel, R., Reckel, S., Löhr, F., Ferrel, J.E., Jr., et al. (2006). Investigating macromolecules inside cultured and injected cells by in-cell NMR spectroscopy. Nat Protoc 1, 2701–2709.

Shi, J., and Muir, T.W. (2005). Development of a tandem protein trans-splicing system based on native and engineered split inteins. J Am Chem Soc 127, 6198–6206.

Siebold, C., and Erni, B. (2002). Intein-mediated cyclization of a soluble and a membrane protein in vivo: function and stability. Biophys Chem 96, 163–171.

Takahashi, H., Arai, M., Takenawa, T., Sota, H., Xie, Q.H., and Iwakura, M. (2007). Stabilization of hyperactive dihydrofolate reductase by cyanocysteine-mediated backbone cyclization. J Biol Chem *282*, 9420–9429.

Tam, J.P., and Lu, Y.A. (1998). A biomimetic strategy in the synthesis and fragmentation of cyclic protein. Protein Sci *7*, 1583–1592.

Tavassoli, A., and Benkovic, S.J. (2005). Genetically selected cyclic-peptide inhibitors of AICAR transformylase homodimerization. Angew Chem Int Ed Engl *44*, 2760–2763.

Tavassoli, A., and Benkovic, S.J. (2007). Split intein mediated circular ligation used in the synthesis of cyclic peptide libraries in *E. coli*. Nat Protoc *2*, 1126–1133.

Tavassoli, A., Lu, Q., Gam, J., Pan, H., Benkovic, S.J., and Cohen, S.N. (2008). Inhibition of HIV budding by a genetically selected cyclic peptide targeting the Gag-TSG101 interaction. ACS Chem Biol *3*, 757–764.

Thongyoo, P., Jaulent, A.M., Tate, E.W., and Leatherbarrow, R.J. (2007). Immobilized protease-assisted synthesis of engineered cysteine-knot microproteins. ChemBioChem *8*, 1107–1109.

Touati, J., Angelini, A., Hinner, M.J., and Heinis, C. (2011). Enzymatic cyclisation of peptides with a transglutaminase. ChemBioChem *12*, 38–42.

Volkmann, G. (2009). Protein cleavage, labeling and cyclization using non-canonical split inteins. PhD thesis, Dalhousie University, Halifax, Canada.

Volkmann, G., Murphy, P.W., Rowland, E.E., Cronan, J.E., Jr., Liu, X.Q., Blouin, C., et al. (2010). Intein-mediated cyclization of bacterial acyl carrier protein stabilizes its folded conformation but does not abolish function. J Biol Chem *285*, 8605–8614.

Watt, S.J., Sheil, M.M., Beck, J.L., Prosselkov, P., Otting, G., and Dixon, N.E. (2007). Effect of protein stabilization on charge state distribution in positive- and negative-ion electrospray ionization mass spectra. J Am Soc Mass Spectrom *18*, 1605–1611.

Williams, N.K., Liepinsh, E., Watt, S.J., Prosselkov, P., Matthews, J.M., Attard, P., et al. (2005). Stabilization of native protein fold by intein-mediated covalent cyclization. J Mol Biol *346*, 1095–1108.

Williams, N.K., Prosselkov, P., Liepinsh, E., Line, I., Sharipo, A., Littler, D.R., et al. (2002). In vivo protein cyclization promoted by a circularly permuted *Synechocystis* sp. PCC6803 DnaB mini–intein. J Biol Chem *277*, 7790–7798.

Wu, H., Hu, Z., and Liu, X.Q. (1998). Protein trans-splicing by a split intein encoded in a split DnaE gene of *Synechocystis* sp. PCC6803. Proc Natl Acad Sci U S A *95*, 9226–9231.

Xie, J., Thapa, R., Reverdatto, S., Burz, D.S., and Shekhtman, A. (2009). Screening of small molecule interactor library by using in-cell NMR spectroscopy (SMILI-NMR). J Med Chem *52*, 3516–3522.

Xu, M.Q., and Evans, T.C., Jr. (2001). Intein-mediated ligation and cyclization of expressed proteins. Methods *24*, 257–277.

Xu, S.Y., Xiao, J.P., Ettwiller, L., Holden, M., Aliotta, J., Poh, C.L., et al. (1998). Cloning and expression of the ApaLI, NspI, NspHI, SacI, ScaI, and SapI restriction-modification systems in *Escherichia coli*. Mol Gen Genet *260*, 226–231.

Yan, X., Watson, J., Ho, P.S., and Deinzer, M.L. (2004). Mass spectrometric approaches using electrospray ionization charge states and hydrogen-deuterium exchange for determining protein structures and their conformational changes. Mol Cell Proteomics *3*, 10–23.

Yang, H. (2010). Progress in single-molecule spectroscopy in cells. Curr Opin Chem Biol *14*, 3–9.

Yu, D., Ellis, H.M., Lee, E.C., Jenkins, N.A., Copeland, N.G., and Court, D.L. (2000). An efficient recombination system for chromosome engineering in *Escherichia coli*. Proc Natl Acad Sci U S A *97*, 5978–5983.

Zhao, Z., Ma, X., Li, L., Zhang, W., Ping, S., Xu, M.Q., et al. (2010). Protein cyclization enhanced thermostability and exopeptidase-resistance of green fluorescent protein. J Microbiol Biotechnol *20*, 460–466.

17 The protein epitope mimetic approach to protein-protein interaction inhibitors

John A. Robinson and Kerstin Moehle

17.1 Introduction

The design of small-molecule protein epitope mimetics is now widely recognized as a promising structure/mechanism-based approach to the discovery of protein-protein interaction (PPI) inhibitors. Progress in the field has been driven, on the one hand, by the enormous increase in the knowledge base of PPIs, arising from high throughput genomic and proteomic methods (Lehne and Schlitt 2009; Stumpf et al. 2008), and on the other, by the massive growth in the 3D structural database of protein-protein complexes. The key scientific challenge is to convert this structural and functional information, through a rational molecular design process, into new PPI inhibitors with potential value in chemical biology, or in drug and vaccine discovery. Ideally, structural information on PPIs should be combined with an understanding of the mechanism(s) of PPIs, to aid the design process. In this review article, we begin with a short overview of recent mechanistic studies of PPIs. Following this, an overview is provided of methods now available for the discovery of PPI inhibitors, in particular, based on the protein epitope mimetic approach.

17.2 Mechanisms of protein-protein interactions

Protein-protein interfaces are often quite large (in the range 600 to ~2,500 $Å^2$ on each side) (Conte et al. 1999; Jones and Thornton 1996), and although they are never really flat, they do lack well-defined binding pockets of the type characteristic of enzyme active sites. It is probably worthwhile to differentiate between inhibitors targeting enzyme active sites and those directed toward PPIs. Enzymes have evolved to catalyze chemical reactions, mostly by binding tightly to the reaction transition state (TS). In other words, the required complementarity between the molecule undergoing reaction and the packing within the protein are most efficient in the enzyme-TS complex. Based upon comparisons of reaction rates in enzyme catalyzed and noncatalyzed processes, it has been estimated that enzyme-TS complexes typically have dissociation constants in the range 10^{-12} M to 10^{-20} M (average $10^{-16\pm4}$ M), which is much greater than expected from the surface areas of TSs (Zhang and Houk 2005), in particular for those enzymes that operate by noncovalent interactions with their substrate(s) (Smith et al. 2009). Many active site-directed enzyme inhibitors act as structural analogues of the TS, and in optimal cases may achieve similar levels of complementarity and affinity with the enzyme. The complementarity between the protein and TS typically includes a network of interactions extending into the protein shell surrounding the active site. This view is supported by recent studies with the enzyme purine nucleoside phosphorylase, which provided evidence for conformational collapse in the protein around a potent inhibitor in the active site due to its structural similarity to the TS (Edwards et al. 2010). Thus, with enzymes, improved packing and positive cooperativity in binding to the TS architecture may involve an extensive network of interactions, perhaps extending some way into the protein shell around the active site (Williams 2010).

For many PPIs, average dissociation constants tend to be in the micromolar to nanomolar range, not because higher affinities are impossible, but because in many cases no strong evolutionary pressure exists for further affinity maturation, as pointed out already for the case of antibody-antigen interactions (Foote and Eisen 1995). The theoretical diffusion-limited association rate for two molecules in solution is likely to be around 5×10^9 M^{-1} s^{-1} (the Einstein-Smoluchowski limit [Berg and von Hippel 1985]). Many PPIs in solution have association rates that are slower (in the range 10^4–10^7 M^{-1} s^{-1} (Harel et al. 2007; Karlsson et al. 1991; Raman et al. 1992), although electrostatic steering effects can lead to substantially faster association rates (Gabdoulline and Wade 2002; Schreiber 2002; Schreiber and Fersht 1993, 1996; Selzer and Schreiber 1999). A minimalist three-state model for protein-protein association was suggested, based on Brownian dynamics simulations, with a first rapid diffusion controlled formation of an encounter complex, in which a protein pair surrounded and trapped by a solvent cage undergoes multiple collisions and rotational reorientation during each encounter (Gabdoulline and Wade 2002; Northrup and Erickson 1992; Northrup et al. 1988; Selzer and Schreiber 2001; Tang et al. 2006). In a second rate-determining step, the two proteins rearrange to the final stereospecific bound state, which might be rate-limited by conformational transitions and desolvation steps, amongst other things (Harel et al. 2007). Assuming that a lifetime for a protein-protein complex beyond 20–30 min would (in many cases) bring little significant extra biological advantage, an effective limit for the off-rate can be fixed at 10^{-3} to 10^{-4} s^{-1}, which would in turn give a ceiling on affinity around 10^{-9} to 10^{-10} M. Enzyme inhibitor design based on mimicry of TSs may, in principle, tap-into a higher potential binding energy (average K_d $10^{-16\pm4}$ M) focused at an active site whose architecture is often complementary to that of a small-molecule ligand.

The cores of monomeric proteins are stabilized consistently by hydrophobic interactions. At some protein interfaces, hydrophobic forces play an important role (e.g. human growth hormone [hGH]-hGH-receptor [hGHR] complex]Wells 1996]), but in others mainly polar or even charge-charge interactions seem to be dominant (e.g. barnase-barstar) (Tsai et al. 1996, 1997). There appears to be no significant difference in the chemistry or geometry of individual interactions at protein interfaces compared to the cores of folded monomers. However, side-chain-to-side-chain interactions are more frequent at protein interfaces compared to protein cores (Cohen et al. 2008). A number of studies have pointed out that energetically important residues at interfaces (hot spots – ▶ Figure 17.2, further on in this chapter) are enriched particularly in tryptophan, tyrosine and arginine, while some other amino acids are found very rarely in hot-spots (Val, Leu, Ser, Thr) (Bogan and Thorn 1998; Hu et al. 2000; Ma et al. 2003; Padlan 1990). The preference for Trp, Tyr and Arg may be due, at least in part, to the ability of these side-chains to engage simultaneously in several types of favorable interactions, which in turn facilitate the creation of interaction networks at protein interfaces (see further on in this section). Recently, minimalist combinatorial libraries based on restricted amino acid usage have been used to identify synthetic protein interfaces (Kossiakoff and Koide 2008). For example, interfaces built using just two different amino acids (Tyr/Ser) in the context of antibody CDR loops are sufficient to produce antibody fragments with high affinity and specificity for protein targets (Gilbreth et al. 2008).

Alanine-scanning mutagenesis led in 1995 to the identification of so-called hot-residues located within "hot-spots" that have since been found in many protein-protein interfaces (Clackson and Wells 1995; Clackson et al. 1998; Moreira et al. 2007; Schreiber and Fersht 1995). Ala mutations that reduce affinity by >10-fold indicate a hot-residue making a disproportionately large contribution to the binding energy. Such hot residues often cluster at the center of interfaces, constitute less than half of the contact surface, and are surrounded by other residues, which by this definition, contribute relatively little in binding energy (DeLano 2002; Keskin et al. 2005; Reichmann et al. 2007). Not surprisingly, hot spots have become a major focus of interest in PPI inhibitor design (Wells and McClendon 2007). One mechanistic interpretation of hot-spots is that the peripheral residues serve as an O-ring to exclude solvent from the center, which would generate a lower effective dielectric and strengthen hydrogen-bonding and other electrostatic interactions (Bogan and Thorn 1998). Similar effects are known to be important in enzyme catalysis, where water exclusion from the active site is often important for catalytic efficiency. However, an alternative view has been proposed, where an interaction deleted by Ala mutagenesis in the periphery might be more easily compensated by a bridging water molecule, and hence causes less loss in stability (Janin 1999). From structural studies it is known that protein interfaces contain on average more bound water molecules than are found in the core of a folded protein. Modeling studies have suggested that both effects may be operative in some cases (Kortemme and Baker 2002). In one mutagenesis study, it was possible to improve the affinity of a PPI from a K_d of 23 nM to 0.15 nM through mutations in a V_H domain interacting with its protein antigen (Koide et al. 2007). In this case, a crystallographic study suggested that the mutations did not expand the antibody-antigen interface, but rather induced an extended network of interactions, and included additional hot residues at the periphery (not the center) of the epitope. These results indicate that although the O-ring architecture may occur frequently, it is not a prerequisite for high affinity PPIs.

More recently, double mutant cycle analyses with selected PPIs have suggested that interfaces between proteins are often built in a modular fashion. Double mutant cycles may reveal whether the contributions from a pair of mutated residues are additive or whether the energetic effects are coupled through neighboring residues (Albeck et al. 2000). In this way, the interface between TEM1-ß-lactamase and ß-lactamase inhibitor protein (K_D 1 nM) was shown to have a modular architecture made up of clusters of residues with strong intracluster connections and weak intercluster connections (Reichmann et al. 2005). The individual residue clusters appeared to be largely independent of each other energetically. A high degree of cooperativity within each cluster was apparent, to the point that in one case the deletion of an entire cluster had little impact on the structure of the interface, whereas single mutations within a cluster could lead to structural rearrangements of their cluster (Potapov et al. 2008). Computational analyses have also revealed that hot-spot residues may be clustered locally within tightly packed interface regions, and that within these relatively dense clusters, the residues form a dense network of interactions, suggesting that their contributions to the stability of the complex should be cooperative (Halperin et al. 2004; Keskin et al. 2005). A comparison of four different PPIs, with affinities ranging from 2 µM (CheA-CheY complex) to 0.01 pM (barnase-barstar complex) suggested that the extent of connectivity within a cluster of interactions, rather than the size of the interface per se, might be the driving force behind tight binding. The degree of connectivity was found to be much higher in the barnase-barstar complex than in the other three analyzed. The CheA-CheY interface is characterized by relatively few contacts between the proteins and only one medium-sized cluster, whereas the barnase-barstar interface has two clusters, one cluster being highly developed and forming a very elaborate network of interactions. In the barnase-barstar interaction (Schreiber and Fersht 1993), the high affinity (K_d 0.01 pM) and electrostatically driven fast rate of association (k_{on} 6×10^8 s^{-1} M^{-1}) are likely a response to the strong evolutionary pressure on the cell (*Bacillus amyloliquefaciens*) to develop a potent inhibitor (barstar) of the RNase (barnase), to keep the deadly RNase in an inactive form until it can be exported from the cell. This strong selection pressure is apparently able to overcome the inherent difficulty in constructing the high level of organization in the network of interactions seen at this protein interface (Reichmann et al. 2005). These results highlight the importance of higher network organization within protein interfaces, and not just of surface complementary principle, in accounting for binding affinity and specificity. This network or cluster view of protein-protein interactions has important implications for inhibitor design, although the focus is often on designing molecules to fill pockets ("knobs-into-holes") on the surface of proteins.

Finally, it is important to consider the key role played by water molecules in PPIs. Several surveys have shown that water is far more abundant at interfaces, than in the cores of folded proteins (Conte et al. 1999). Water-mediated polar interactions are as abundant at interfaces as direct protein–protein hydrogen bonds, and they may contribute significantly to the stability of the assembly (Ikura et al. 2004; Rodier et al. 2005). Uncertainties in the location and role of water molecules at PPIs contribute greatly to the difficulties in calculating binding affinities by computational methods (Wang et al. 2011), and in attempts to interpret thermodynamic data (binding enthalpy/entropy) in structural terms.

Binding driven by association of hydrophobic surfaces is conventionally viewed as being driven by favorable entropic contributions, due to release of surface-associated water molecules to the bulk solvent (Chandler 2005; Dunitz 1994). However, the generality of this view has recently been challenged in a simple model system comprising a spherical hydrophobic ligand binding to a solvated hemispherical cavity (Baron et al. 2010; Setny et al. 2010). According to MD simulations, the free energy of this binding process is not dominated by the direct interaction between the ligand and pocket, but by the contributions of water. Contrary to expectation, in this model system the binding is driven by enthalpy and opposed by entropy. This was explained by the release of water molecules from the pocket, and the loss of entropy as arising from the expulsion of disorganized water from the receptor cavity (i.e. cavity water is more entropic than bulk water). An interesting experimental study of protein-ligand interactions revealed just such a thermodynamic profile; an increasingly favorable enthalpy and increasingly unfavorable entropy as the hydrophobic surface area of the ligand increases (Homans 2007; Malham et al. 2005). Another recent survey of protein-ligand interactions found that for the majority of small-molecule interactions with proteins, the enthalpy change provides the largest favorable contribution to binding, and contrary to expectation the trend to enhanced affinity with greater burial of apolar surface is, in general, only weakly correlated with a favorable entropy change (Olsson et al. 2008).

17.3 Small-molecule screening approaches

High-throughput screening has often not been very successful for identifying PPI inhibitors (Arkin and Wells 2004; Fry 2006; Toogood 2002). One explanation for this poor performance might be that the large compound libraries typically used for screening, contain molecules of insufficient (or inappropriate) structural diversity. For example, relatively small, flat (sp² hybridized), aromatic or heteroaromatic compounds, might be overrepresented in libraries, because of their perceived likelihood to interact with enzyme active sites or G-coupled protein receptors (GPCRs). Mechanistic considerations, however (see Section 17.2), suggest that such molecules might not be well suited in the search for PPI inhibitors. So what types of property should an ideal screening library have? This question was addressed in a recent study in which approximately 15,000 compounds from three different sources (commercial libraries, academic diversity-oriented synthesis projects, and natural products) were shown to have quite different properties, when tested for their ability to bind a collection of unrelated proteins (Clemons et al. 2010). The study showed that increasing the content of sp³-hybridized and stereogenic centers, relative to compounds from commercial sources, improved protein binding selectivity and frequency. Not surprisingly, the compounds of highest stereochemical complexity were most often those from natural sources, although continuing improvements in organic synthesis methodology (e.g. diversity oriented synthesis) are also making such molecules increasingly accessible. The structural and stereochemical complexity of natural products relative to small drug-like molecules, and their value in the drug-discovery screening exercise, continue to be well argued (Ganesan 2008; Harvey 2008; Li and Vederas 2009). It is worth noting that many stereochemically complex natural products are macrocyclic compounds in the molecular weight range 0.5–2 kDa. Molecules of this size and shape occupy an area of molecular space that nature has obviously explored (at least to some extent), but which due to its shear size is much more difficult to explore thoroughly using diversity-oriented synthesis.

The so-called fragment-based approach is now also being applied to the discovery of PPI inhibitors (Coyne et al. 2010; Erlanson 2006; Murray and Blundell 2010). The goal here is to build drug leads by identifying small molecular fragments that bind to adjacent sites on the protein with µM or even mM affinity, and then either linking or expanding them to improve affinity and selectivity. The first practical demonstration of the approach used SAR by NMR to identify and link small fragments (200–300 Da) that bind to FK506-binding protein (Shuker et al. 1996). A related idea called "tethering" aims to discover weakly binding ligands through an intermediary disulfide tether (Erlanson et al. 2000). A native or engineered cysteine in a protein is allowed to react reversibly with a library of disulfide-containing small molecules. The cysteine-captured ligands are then identified by MS.

A more recent example of the fragment-based approach is the discovery of inhibitors of Bcl-X_L, which are important targets in anticancer therapy. The protein Bak adopts an amphipathic α helix that binds to Bcl-X_L through hydrophobic and electrostatic interactions (crystal structure, PDB 1BXL) (Sattler et al. 1997). Two small-molecule hits that target the groove-like binding site on Bcl-X_L were found using SAR-by-NMR. These were linked and the resulting ligand was optimized in several steps to give ABT-737, with subnanomolar affinity for Bcl-X_L (Oltersdorf et al. 2005). A related compound ABT-263 proved to be orally bioavailable, and advanced into clinical testing in cancer (Park et al.

2008). Interestingly, 3D structures of Bcl-X$_L$ bound to two low affinity starting fragments (PDB 1YSG), and bound to a high affinity optimized ligand (PDB 2O2N) have been reported (Bruncko et al. 2007). Upon comparison of these two structures, it is notable that the binding site on Bcl-X$_L$ has undergone considerable adaptation as the structure of the ligand has changed during optimization (▶Figure 17.1). As a result, the sites on Bcl-X$_L$ occupied by the two low affinity ligands have changed considerably when compared to the structure of Bcl-X$_L$ bound to an optimized ligand, showing that the mechanism of affinity maturation in this case does not simply involve linking two fragments in adjacent binding sites. Clearly, the protein itself adapts considerably as the ligand is changed.

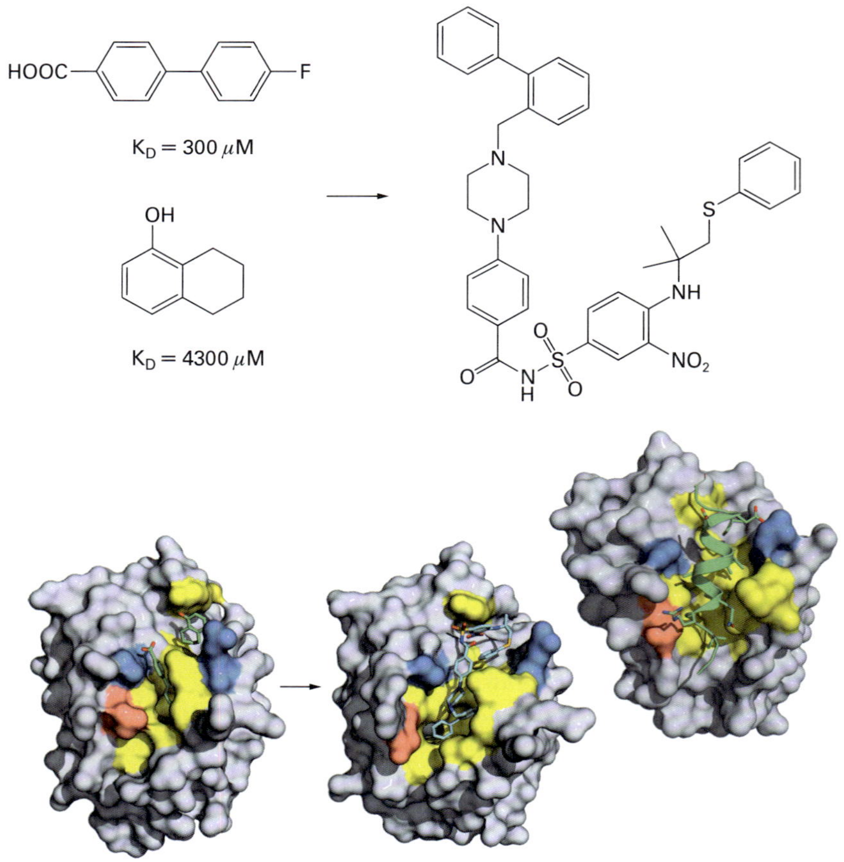

Figure 17.1: Structures of ligands bound to Bcl-x$_L$. A large conformational change in the protein Bcl-x$_L$ occurs as the ligands change from the complex with low-affinity inhibitors (left, PDB 1YSG), to that with a high affinity ligand (middle, PDB 2O2N). The protein conformation is again different when bound to a helical peptide derived from an endogenous BH3 binding partner (right, PDB 1BXL). The small-molecule ligands do not mimic the structure of the peptide ligand, but rather trap the protein in a different conformation, binding in deeper cavities with more puckered grooves.

Another example of adaptivity in a protein-binding site has been observed during the discovery and optimization of an IL2-receptor-α (IL-2Rα) antagonist. A series of acylphenylalanine derivatives was discovered in an attempt to mimic the R38–F42 loop region of IL-2, one of which was found to be a competitive inhibitor of IL-2/IL-2Rα binding with an IC_{50} of 3 µM (Tilley et al. 1997) (▶Figure 17.2). Although the acylphenylalanine derivatives were designed to complex with IL-2Rα by emulating residues R38 and F42 of the IL-2 ligand, they were instead shown to interact with the ligand IL2. From this starting point, new molecules that bind to IL-2 were optimized, with dissociation constant in the mid-nanomolar range. These molecules were assembled in a fragment-based approach, guided by X-ray structures and medicinal chemistry (Arkin et al. 2003; Braisted et al. 2003; Raimundo et al. 2004). Although the small molecules were assembled before the structure of the IL-2–IL-2Rα complex had been reported (Rickert et al. 2005), they bind close to the center of the receptor contact region on IL-2. The contact epitope for the small molecule is about half the size of that for the receptor, but the small molecule and the receptor bind to IL-2 with nearly equivalent affinities. The contact surface on IL-2 for binding to IL-2Rα is relatively flat. By contrast, the small molecule traps a conformation of IL-2 in which a groove is present for small-molecule binding, and in which a loop of IL-2 has been repositioned to embrace the furanoic acid moiety at one end of the small molecule (▶Figure 17.2). Alanine-scanning mutational studies show that the small molecule and the IL-2Rα bind to the same hotspot residues on IL-2 (Thanos et al. 2006). Although the structures of the small molecule and IL-2Rα differ markedly, the electrostatic potential of the surfaces presented is similar and probably reflects a need to establish the electrostatic complementary principle with IL-2. Electrostatic and surface-shape complementary principle, as well as specific hydrogen-bonding interactions, probably account for the high selectivity of these interactions.

These studies show that the binding surface on IL-2 is adaptive and can bind to a small molecule with high affinity (K_i = 60 nM) using the same main hot-spot residues used for its natural receptor, but again without the benefit of an "O-ring." It is notable that the design of this series of IL-2-binding small molecules did not require knowledge of the structure of the bound receptor complex. Instead, the design was informed by fragment-binding data and by structures of compounds bound to IL-2, coupled with medicinal chemistry and structure–activity relationships (SAR). The small-molecule ligand is not an accurate atomic mimic of the receptor, and it would not have been discovered if it had been assumed that the precise structure of the receptor-bound form of IL-2 needed to be captured. Indeed, the adaptive nature of this protein-protein interface underscores again the challenge of applying structure-based strategies that cannot accurately predict the dynamic nature of the protein surface (Arkin et al. 2003).

17.4 Protein epitope mimetic approaches

Where 3D structural information is available for PPIs, the design of conformation constrained epitope mimetics offers one rational approach to PPI inhibitors. Molecular recognition involving proteins is typically mediated by surface exposed secondary structure elements such as ß turns, ß strands, ß hairpins and α helices. New protein-binders may be found by devising ways to mimic these important secondary structure elements in smaller "semirigid" molecules. Once interesting hits have been identified, their properties can be optimized by exploiting the fact that epitope mimetics tend to

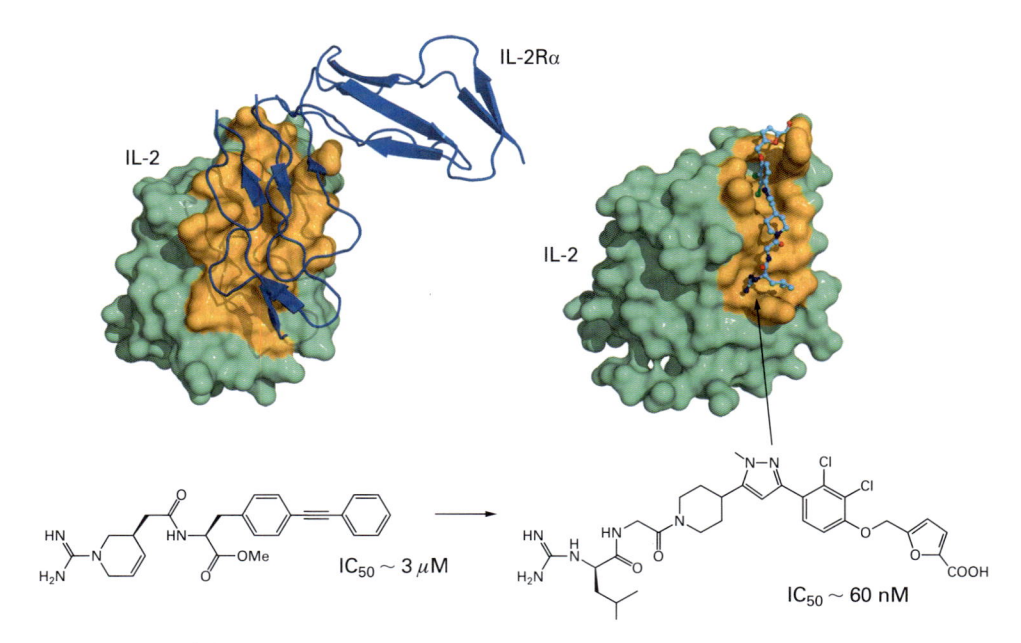

Figure 17.2: Inhibitors of the IL-2–IL-2Rα interaction. The complex of IL-2 (surface representation) and the IL-2Rα chain (ribbon) is shown left, with hot-spot residues on IL-2 shown yellow. The acylphenylalanine (left) was optimized to give the high affinity inhibitor (right) shown bound to IL-2. The small molecule traps a conformation of IL-2 in which a binding-groove is present and in which a loop of IL-2 has been repositioned to embrace the furanoic acid moiety at one end of the small molecule. Alanine-scanning mutagenesis shows that the small molecule and the receptor bind to the same hotspot residues at an adaptive binding surface on IL-2.

be modular in structure. They are typically constructed from building blocks, such as α- or ß-amino acids or peptoids, which can be linked together using robust and efficient methods. By exchanging building blocks, the mimetic structures can then be varied and properties can be optimized, in a combinatorial fashion, using parallel synthetic chemistry. One key property, essential for targeting intracellular PPIs, is permeability across the cell membrane. Progress has also been made recently in identifying cell permeable peptidomimetic PPI inhibitors, suggesting that it may be possible to engineer this property into appropriate peptidomimetic scaffolds.

17.4.1 Helix mimetics

The design of conformationally restrained mimetics of α-helical epitopes continues to attract much attention (Garner and Harding 2007; Henchey et al. 2008). Some of the approaches used so far to stabilize helical conformations in peptides include the use of intramolecular hydrogen-bond surrogates, such as hydrazone or alkenyl links (Cabezas and Satterthwait 1999; Chapman et al. 2004; Henchey et al. 2010; Patgiri et al. 2008; Vernall et al. 2009; Wang et al. 2005, 2006, 2008), and helical end-capping groups (▶Figure 17.3) (Austin et al. 1997; Kemp and Curran 1988a, 1988b; Kemp and

17.4 Protein epitope mimetic approaches

Rothman 1995a, 1995b; Kemp et al. 1995a, 1995b, 1996; Lewis et al. 1998; Maison et al. 2001; Obrecht et al. 1999; Schneider and DeGrado 1998). Other methods to enhance helicity in peptides include incorporating unnatural amino acids (Andrews and Tabor 1999), in particular, α-disubstituted amino acids such as Aib (Venkatraman et al. 2001) and other α-alkylated-α-amino acids.

Another approach for the stabilization of α-helical conformations in peptides is by side-chain cross-linking, also called "stapling,," of *i* and *i*+4 or *i* and *i*+7 side chains lying on the same face of the helix. The earliest reports of cross-linking involved lactam (Felix et al. 1988; Geistlinger and Guy 2003; Judice et al. 1997; Mills et al. 2006; Osapay and Taylor 1992; Phelan et al. 1997; Shepherd et al. 2006; Sia et al. 2002; Taylor 2002), disulfide (Harrison et al. 2010; Jackson et al. 1991; Leduc et al. 2003) and thioether (Brunel and Dawson 2005) bridges, as well as metal-mediated chelation (Beyer et al. 2004; Ghadiri and Choi 1990; Kelso et al. 2003, 2004; Ruan et al. 1990). Hydrocarbon bridges made using olefin metathesis chemistry have also been reported (Blackwell and Grubbs 1998). This concept has been taken a step further by combining within the same amino acids the side-chain cross-link and backbone α-methylation (Schafmeister et al. 2000). In this approach, (*R*)- and (*S*)-α-methyl-α-alkenyl amino acids are incorporated at *i* and *i*+7 positions, respectively, and then linked by ruthenium-catalyzed ring-closing metathesis (▶Figure 17.3). The cross-link results in a stabilized helical conformation in the peptide and improved stability of the peptide toward proteolysis. Several applications

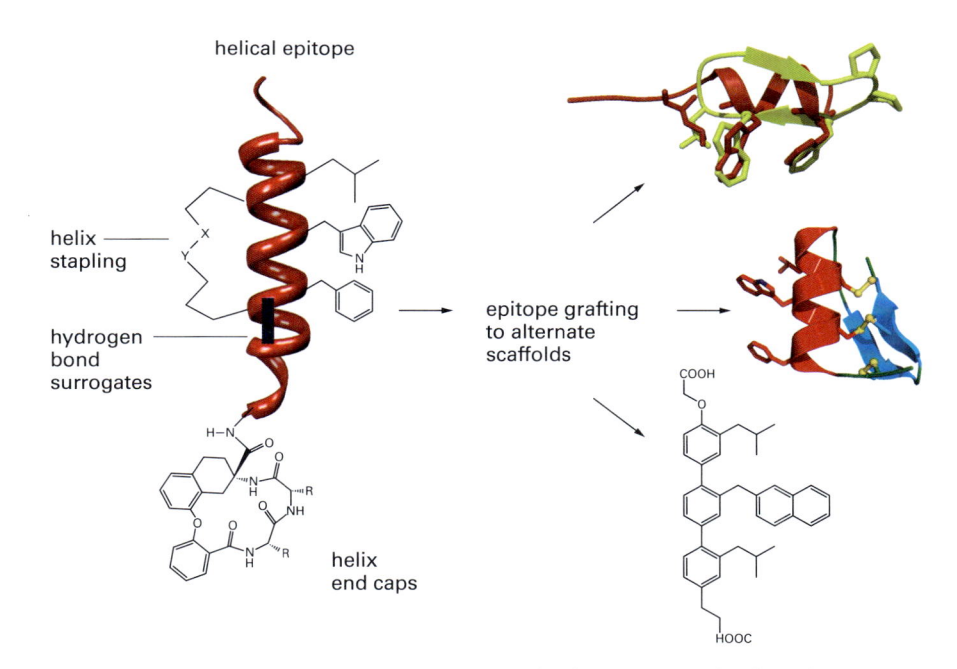

Figure 17.3: Strategies to stabilize or mimic α-helical epitopes. Helical conformations can be stabilized in linear peptides, for example, by helix-stapling, by incorporating hydrogen-bond surrogates, or by adding end-capping groups. Helical epitopes can also be grafted onto new scaffolds, including peptide-based and nonpeptidic scaffolds.

of this technology for helix mimicry have been reported in drug discovery. The conformational constraints imposed by this type of helix cross-linking have favorable effects not only on target binding affinity and proteolytic stability, but surprisingly, also on cell membrane permeability. In one example, a hydrocarbon-stapled peptide was designed to mimic an amphipathic helix in the pro-apoptotic BH3 domain of the BCL-2 family member called BID. The resulting stapled peptide showed not only a much higher helical content in water, compared to a linear control peptide, but also a higher protease stability and a 6-fold higher binding affinity to BCL-2 (39 vs. 269 nM). Moreover, the stapled peptide was apparently taken up by cells in an energy-dependent pinocytotic pathway, where it localized with its target at the outer mitochondrial membrane. The unmodified control peptide was completely impermeable to the intact cells. The stapled peptide mimic of the BH3 domain also specifically activated the apoptotic pathway to kill leukemia cells, and effectively inhibited the growth of human leukemia xenografts in vivo (Walensky et al. 2004). In other examples, a stapled BID BH3 helix was shown to bind and activate BAX, a multidomain BCL-2-family protein that resides in the cytoplasm (Walensky et al. 2006). And a stapled peptide derived from p53 was shown to bind to HDM2, penetrate cells by an endocytic transport mechanism, and reactivate apoptosis in HDM2-overexpressing cells (Bernal et al. 2007). Transcription factors have also been rather difficult to target using small drug-like molecules. However, a hydrocarbon-stapled peptide was shown to mimic MAML1 and bind directly to the oncogenic transcription factor NOTCH1. Treatment of leukaemic cells with the stapled peptide resulted in genome-wide suppression of NOTCH-activated genes (Moellering et al. 2009). These promising results, targeting a range of difficult protein-protein interactions, suggest that such stapled α-helical peptide domains may be widely useful in chemical biology, as tools for biological studies, and potentially also as drug candidates. Further work will be required to assess their pharmacological properties and toxicity, as well as their target affinities and selectivities.

Apart from peptide stapling, α-helical epitopes have been mimicked using other scaffolds, including peptoids (Vaz and Brunsveld 2008), ß hairpins (Fasan et al. 2006; Moehle et al. 2007; Seitz et al. 2010), ß-peptides and various cyclic scaffolds (Oguri et al. 2005, 2006), bi- and ter-phenyls (Ernst et al. 2002; Jacoby 2002; Kutzki et al. 2002; Orner et al. 2001) and related heterocycles (Campbell et al. 2010; Cummings and Hamilton 2010; Lee et al. 2011; Saraogi et al. 2010) and templates (Schneider and Kelly 1995).

Stable miniprotein domains also provide a source of scaffolds for epitope grafting and mimetic design (Vita 1997; Vita et al. 1998). One interesting example is the family of small naturally occurring disulfide cross-linked peptides belonging to the short-chain toxins from scorpion venom, such as charybdotoxin and scyllatoxin (Drakopoulou et al. 1996; Vita et al. 1995, 1999; Zinn-Justin et al. 1996). These toxins typically contain a short α-helical segment at the N-terminal end, cross-linked by three disulfide bridges to a ß-hairpin motif (▶Figure 17.4). Both the α-helical segment and the ß hairpin provide conformation stable segments onto which foreign epitopes can be grafted. For example, the ß-hairpin motif has been exploited to generate mimics of an epitope on CD4, the primary cellular receptor for HIV-1. HIV-1 entry into cells is initiated by the binding of the viral glycoprotein gp120 to CD4 on host cells. Crystallographic studies have shown that a key component of the epitope on CD4 that interacts with gp120 is a surface ß-hairpin loop (the CDR2-like loop) (Kwong et al. 1998). Transplanting this hairpin epitope from CD4 onto the scorpion toxin scaffold afforded, after optimization, mimetics that bind tightly to gp120 and inhibit HIV-1 entry into cells (Huang et al. 2005; Martin et al. 2003; Stricher et al. 2008; Vita et al. 1999).

The same scorpion scaffold was also used in a different way, to generate a mimetic of the α-helical segment of p53 that interacts with its cellular inhibitor HDM2 (Li et al. 2008). This study exploited the α-helical segment of the scorpion toxin scaffold. Thus, the residues along one face of the scorpion toxin helix were replaced with the topologically equivalent residues in p53 (F19, L22, W23 and L26), such that the key hydrophobic side chains required for interaction with HDM2 are displayed in the correct relative orientation (▶Figure 17.4). With some additional optimization, a p53 mimetic based on this toxin scaffold was obtained that bound to HDM2 with submicromolar affinity. The question of cellular membrane permeability was also addressed in this study (Li et al. 2008). In order to enhance cell permeability, the mimetic was reengineered by replacing five residues near the C terminus with arginines, to create a cluster of eight cationic residues. This gave a molecule that, when added to a p53+ cancer cell line, was able to kill cells quantitatively and in a p53-dependent manner, apparently due to its ability to traverse the cell membrane and inhibit the p53-HDM2 interaction.

Another small folded peptide that has been used for engineering experiments is the avian pancreatic polypeptide (aPP) (Hodges and Schepartz 2007). aPP contains 5 turns of α helix in its C-terminal half, linked to a 10-residue extended (ß structure) N-terminal segment, which is back-folded onto the α helix. The p53 epitope has also been successfully grafted onto this aPP scaffold, to create miniprotein inhibitors of the p53-MDM2 interaction (Kritzer et al. 2006). The cyclotides and conotoxins represent two other families of macrocyclic cross-linked peptides, each with great potential as scaffolds in protein epitope mimetic design (Cascales and Craik 2010; Clark et al. 2010; Craik et al. 2006a, 2006b; Green et al. 2007; Henriques and Craik 2010).

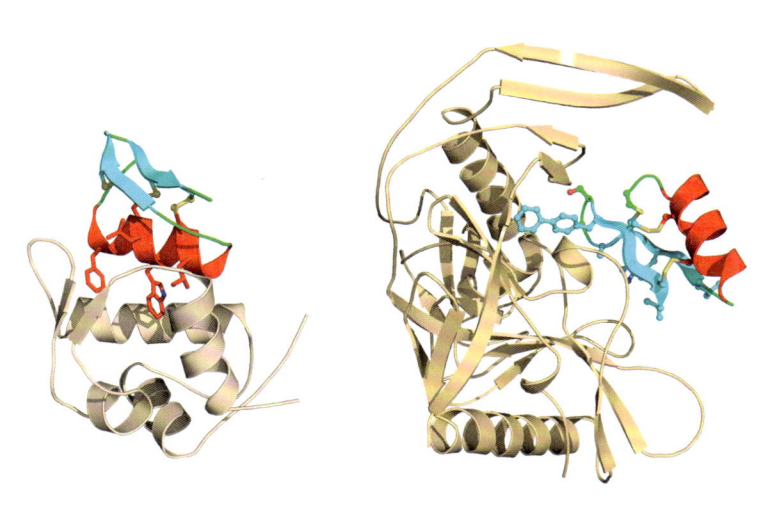

Figure 17.4: Both α-helical and ß-hairpin epitopes have been grafted onto the scorpion toxin scaffold, resulting in mimetics of p53 for binding to HDM2 (shown left) and of CD4 binding to gp120 from HIV-1 (shown right).

A more recent trend is toward the design of novel folding architectures, or "foldamers," that adopt regular structural elements akin to α helices and ß structure, but built using other types of building blocks (i.e. not α-amino acids) (Goodman et al. 2007; Guichard and Huc 2011; Hill et al. 2001; Smith et al. 2011). Stable helix-like conformations have been characterized in oligomers prepared from N-alkylglycines (peptoids) (Fowler and Blackwell 2009) and from ß- and γ-amino acids (Horne and Gellman 2008; Seebach et al. 2004). The ß-peptides have been the most studied family of peptidomimetic foldamers, but whereas many short ß-peptides adopt well-defined conformations in organic solvents, additional helix-stabilizing elements (e.g. ring constraints or helix stapling) are required for stable folds to occur in aqueous solution (Appella et al. 1999; Vaz et al. 2008). Aromatic and aliphatic N,N'-linked oligoureas have also been designed to fold and/or self-assemble in a controlled manner (Fischer and Guichard 2010). The development of foldamers with heterogeneous backbones formed by combining multiple residue types (mixed α/ß-peptides, oligo-urea/amides) has also been reported (Claudon et al. 2010; Horne and Gellman 2008).

Some of these novel foldamer scaffolds have also been applied successfully to PPI inhibitor design. For example, ß-amino acids have been incorporated into a mimic of the helical coiled-coil heptad repeat region 2 (HR2) of gp41 from HIV-1. This HR2 region must interact with the HR1 heptad repeat region before fusion of the viral and target cell membranes can occur. The chimeric α/β-peptides were shown to mimic structural and functional properties of the critical α-helical HR2 epitope in gp41. Biophysical and crystallographic studies, and results from cell-fusion and virus-infectivity assays, collectively indicate that the gp41-mimetic α/β-peptides effectively block HIV-cell fusion via a mechanism comparable to that of gp41-derived α-peptides such as T20 (Fuzeon) (Horne et al. 2009). In a related study, a series of ß3-decapeptides, also based on gp41, were designed with a ß-peptide 14-helical conformation stabilized by macro dipole neutralization and side chain-side chain salt bridges (Stephens et al. 2005). The peptide mimetics were shown to inhibit syncytia formation in cell culture, again by blocking PPIs important for infection by HIV-1. A similar viral-host cell fusion mechanism is also exploited by other viruses (Eckert and Kim 2001). Thus, helical ß-peptides have been designed that target in a similar way the coiled-coil PPIs that occur during infection by human cytomegalovirus (English et al. 2006).

ß-Peptides have also been designed to mimic helical epitopes in the tumor suppressor protein p53 and in antiapoptotic BCL-2 family members such as BH3. For example, salt bridge-stabilized 14-helical ß-peptides have been described, bearing on one face of the helix residues required for binding to MDM2 (Kritzer et al. 2004). The most potent ß-peptides were reported to bind MDM2 and MDMX, and inhibit the p53-MDM2/MDMX PPI, with nanomolar affinities in a direct fluorescence polarization assay (Michel et al. 2009). In later studies, ß-peptides with hydrocarbon and diether staples (bridges) were prepared that not only potently inhibit the p53-MDM2 interaction, but also are taken-up by mammalian cells far more efficiently than unbridged analogues (Bautista et al. 2010). In the case of the BH3/Bcl-x_L interaction, sequence rather than structural information on the interaction of BH3 domains with Bcl-x_L has been exploited to design inhibitors based on chimeric α/ß-peptides (Horne et al. 2008). The approach involved replacing subsets of regularly spaced α residues with β3 residues bearing the original side chains. Each α/β-peptide contains an ααβαααβ backbone repeat, which is derived from the heptad pattern common among α-peptide sequences that form α

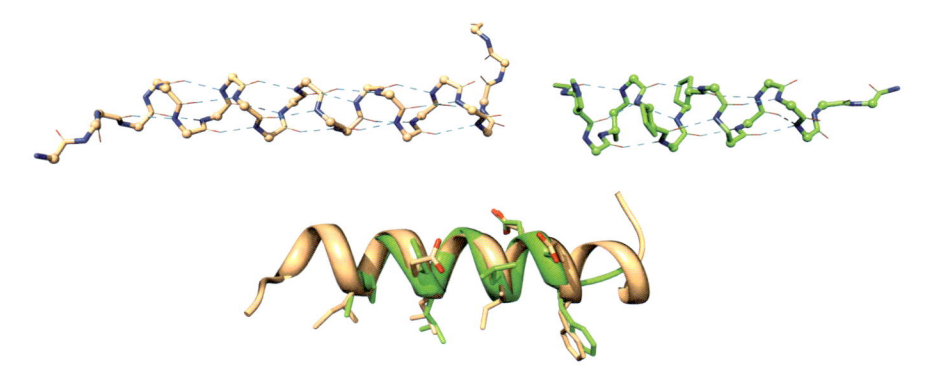

Figure 17.5: Chimeric α/ß-peptide (right) designed to mimic an α-helical epitope (left) in the BH3 domain for binding to Bcl-x$_L$.

helices with a well-developed "stripe" of hydrophobic side chains running along one side. This approach led to α/ß-peptides with comparable binding affinity for the protein target and substantially improved proteolytic stability. A crystal structure of a complex between one α/ß-peptide foldamer and the protein partner Bcl-x$_L$ shows that the foldamer adopts a helical conformation and is oriented within the BH3 recognition groove in a similar fashion to natural BH3 domains (Lee et al. 2009) (▶Figure 17.5). The C-terminal α-peptide portion is α helical, whereas the α/ß segment adopts a characteristic 14/15-helical structure with some cyclic ß-amino acid residues (*trans*-2-aminocyclopentane carboxylic acid) making contacts with the protein surface. The helix formed by the Bcl-x$_L$-bound foldamer features a network of backbone C=O(*i*)···H-N(*i*+4) H-bonds over nearly the entire length of the oligomer. The crystal data suggest that the foldamer achieves high affinity in part by mimicking the three-dimensional display of the canonical side chains projected by natural BH3 domains, however, the β-residue contacts may also contribute significantly to foldamer affinity.

17.4.2 ß-Hairpin mimetics

ß structure and in particular ß hairpins also occur frequently in protein-protein recognition epitopes. Not surprisingly, therefore, the design of hairpin mimetics has also attracted great interest. ß-Hairpin loops are found in the antigen-binding sites of antibodies and the ligand binding sites of many cytokine receptors and polypeptide growth factors, as well as in many integrins and viral proteins, a wide variety of smaller host-defence peptides such as defensins, venom toxins such as ω-conotoxin and three-finger snake toxins, in cyclotides, and in Bowman-Birk and related proteinase inhibitors, to name but a few. Naturally occurring ß-hairpin motifs possess a remarkable degree of structural diversity, due to variations in loop size, variations in the hairpin register, due to the occurrence of ß-bulges within the ß strands, and of course due to variations in sequence. The register of a ß hairpin defines which pairs of cross-strand amino acids occupy hydrogen-bonding (HB), rather than non-hydrogen-bonding (NHB), positions. This in turn determines which amino acid side chains are displayed on the same face of the hairpin (▶Figure 17.6).

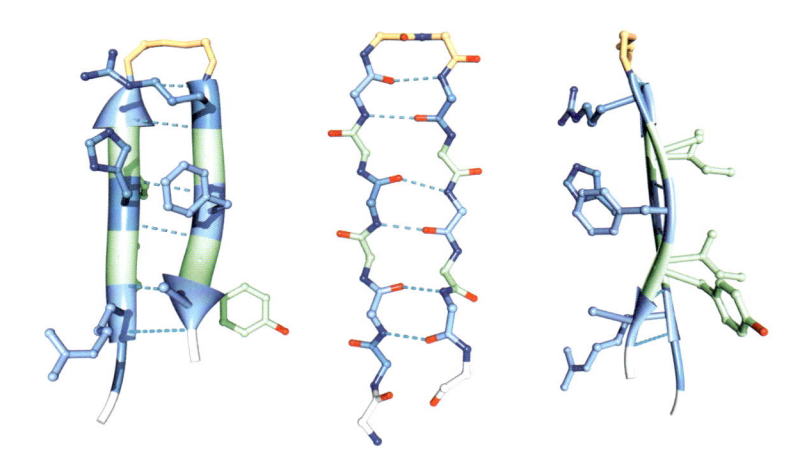

Figure 17.6: The register of a ß-hairpin loop defines which pairs of cross-strand amino acids occupy hydrogen-bonding (HB) (indicated with blue dashed lines) and NHB.

ß-Hairpin loop structures can be stabilized in linear peptides using disulfide bridges. This approach has been particularly valuable in phage-display technology to select peptides that bind to specific protein targets (Sidhu et al. 2000). Spectacular examples have been reported of disulfide-bridged hairpin-loop peptides selected by phage display that can mediate (agonist) or inhibit PPIs. For example, by screening a random phage library against the immobilized erythropoietin receptor (EPO-R), a disulfide-bridged peptide [GGTYSCHFGPLTWVCKPQGG] was found, that binds to EPO-R in a ß-hairpin conformation (▶Figure 17.7) (Livnah et al. 1996; Wrighton et al. 1996). This hairpin peptide mimics the helical cytokine EPO by interacting with the EPO-binding site on the receptor, causing dimerization of the extracellular domain. As a result, both EPO and the peptide induce a similar signalling cascade of phosphorylation events and cell cycle progression in EPO responsive cells. In another example, a phage-library of cyclic peptides was screened for binding to the constant fragment (Fc) of immunoglobulin G (IgG) (DeLano et al. 2000). A disulfide bridged hairpin peptide was isolated that interacts with the Fc at a highly accessible adaptive and hydrophobic site that also is used by other Fc-binding proteins, including protein A, protein G, rheumatoid factor and the neonatal Fc-receptor. Phage display is now a well-established and powerful approach for selecting peptides and proteins with novel binding functions from large combinatorial libraries (Sidhu et al. 2003).

A disulfide bridge, however, has disadvantages, because it can be cleaved in vivo by reaction with a free thiol group, and the disulfide link has many degrees of rotational freedom, which in the absence of other stabilizing interactions (Russell et al. 2002), can make it difficult to stabilize discrete hairpin registers or conformations. ß-Hairpin conformations have been successfully stabilized in linear peptides by exploiting tryptophan zipper motifs (π-π-stacking interactions between cross-ß-strand tryptophans at NHB positions), and related interactions (Eidenschink et al. 2009; Russell et al. 2002; Wu et al. 2010b). The residues in the turn region have a strong influence upon ß-hairpin

17.4 Protein epitope mimetic approaches

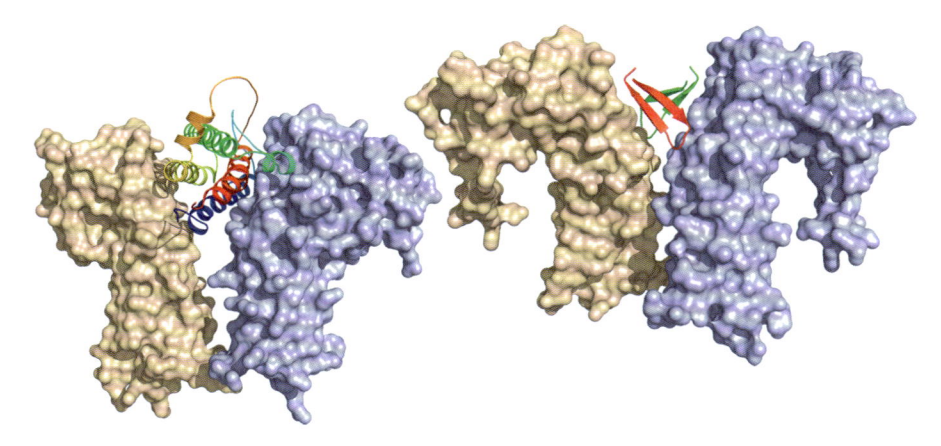

Figure 17.7: A disulfide bridged ß-hairpin peptide selected by phage display (right) mimics the helical cytokine EPO (left) in binding to and activating the EPO-receptor.

stability in linear peptides. Turns formed by Asn-Gly, D-Pro-Gly, Aib-Gly and Aib-D-Pro are strong promoters of ß-hairpin formation due to their propensity to form type-I′ or -II′ turns (Raghavender et al. 2010; Rai et al. 2007), which match the preferred right-handed twist of a hairpin. Other building blocks have been incorporated into turn regions and into the ß-strand regions, including, for example, the 1,2-dihydro-3(6H)-pyridinyl unit that favors extended conformations (Phillips et al. 2005). A novel hairpin-capping motif was described recently, which appears to overcome the problem of fraying at the ends of the ß strands in hairpin motifs in linear peptides (Kier et al. 2010). The capping motif comprises synergistic stacking and hydrogen-bonding interactions between an N-terminal alkanoyl-Trp and a C-terminal -TrpThrGly- motif, which effectively tie together the ß strands as long as the Trp residues are at terminal NHB positions.

Stable ß-hairpin scaffolds can also be achieved by linking together the N and C termini, to produce macrocyclic hairpin structures. Nature has also exploited this strategy, as evidenced by macrocyclic natural products such as gramicidin S, which has a hairpin structure with two D-Phe-L-Pro units marking the turn regions. Other approaches to macrocycles with stable parallel ß-sheet scaffolds have also been described recently (Freire and Gellman 2009; Woods et al. 2007).

Alternatively, ß-hairpin structures in naturally occurring peptides and proteins can be mimicked by transplanting the hairpin loop onto an appropriate (semi-)rigid template; the so-called protein epitope mimetic (PEM) technology (Robinson 2008; Robinson et al. 2008). The dipeptide D-Pro-L-Pro has proved to be an extremely useful template in ß-hairpin mimetic design (▶Figure 17.8). This dipeptide adopts a rigid type-II′ ß turn, which is ideal to initiate antiparallel strands held together by registered cross-strand hydrogen bonds. Upon transplanting the loop from the protein of interest, the cross-strand residue pair directly attached to the D-Pro-L-Pro template then occupay a hydrogen-bonding position. In this way, accurate structural mimetics have been produced of CDR loops found in antibodies (Favre et al. 1999), of a ß-hairpin loop in Tat bound to HIV-1 TAR-RNA (Athanassiou et al. 2004, 2007; Davidson et al. 2009; Lalonde et al. 2011; Leeper et al. 2005), of protease inhibitors related to the Bowman-Birk family (Descours

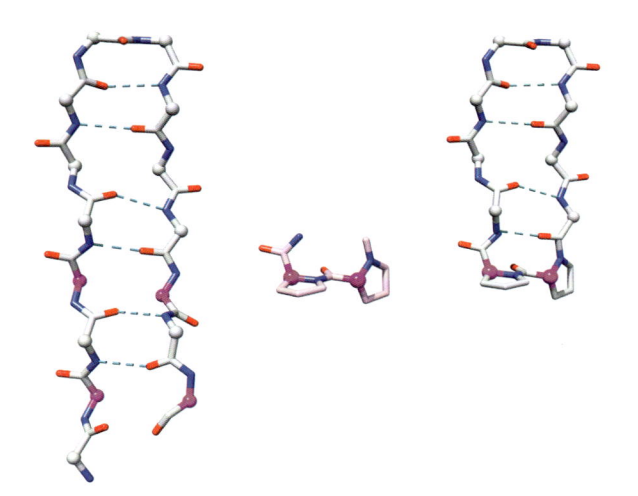

Figure 17.8: ß-Hairpin mimetics can be produced by transplanting the hairpin loop from a protein of interest (left), onto a hairpin-stabilizing template, such as the dipeptide D-Pro-L-Pro (center), to give a cyclic ß-hairpin PEM (right).

et al. 2002), and of cationic host-defence peptides related to protegrin-I (Robinson et al. 2005; Shankaramma et al. 2002, 2003; Srinivas et al. 2010) and polyphemusin (De-Marco et al. 2006). In another example, a 13-residue disulfide cross-linked loop taken from a phage display peptide that binds a human antibody Fc fragment (discussed previously in this section) (DeLano et al. 2000) was transplanted onto a D-Pro-L-Pro template (Dias et al. 2006). The resulting 13-mer loop was shown to adopt a well-defined ß-hairpin structure, which includes a bulge in the second strand close to the turn region, with the unusual result that side-chains of two adjacent residues point to the same side of the hairpin.

The interest in ß-hairpin mimetics is enhanced further by the realization that a hairpin scaffold can also be exploited to mimic α-helical epitopes. Thus, the distance between the Cα atoms of two residues *i* and *i*+2 along one strand of a hairpin is very close to that between the Cα atoms of two residues *i* and *i*+4 on one face of an α helix. Using this information it was possible to design a ß hairpin to mimic the pharmacophore within a helical peptide derived from p53 (▶Figure 17.9), which is involved in binding to its interacting protein HDM2 (Fasan et al. 2004, 2006). Template-bound ß-hairpin mimetics were found that bind with nanomolar affinity to HDM2 (Grässlin et al. 2009). A crystal structure of one mimetic bound to HDM2 confirmed that the hairpin scaffold presents three side chains along the first strand of the hairpin for binding to HDM2, just like the corresponding three residues situated on one face of the helical p53 (▶Figure 17.9). In addition, residues in the second ß strand of the mimetic make additional favorable contacts with HDM2, which are not seen in the p53-HDM2 complex. Other examples of hairpin mimicry of helical epitopes have been reported more recently, based on the interactions of Rev protein to RRE-RNA, and of a helical epitope in the chemokine receptor CCR5 that binds to the HIV-1 glycoprotein gp120 (Moehle et al. 2007; Seitz et

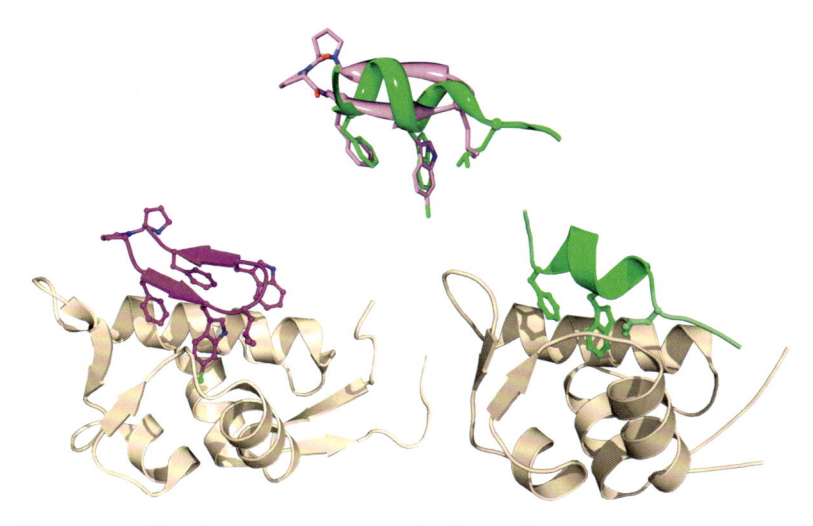

Figure 17.9: ß-Hairpin mimetics can also be used to mimic the α-helical epitope in p53. A crystal structure of one hairpin mimetic bound to HDM2 is shown (left). The cyclic peptide scaffold is used to display hot-spot residues lying along one face of an α-helical epitope in p53, which are important for binding to HDM2 (right).

al. 2010). These examples suggest that the hairpin scaffold might be widely applicable in the design of helical epitope mimetics.

The macrocyclic ß-hairpin mimetics described previously in this section are constructed from building blocks that can be linked together using robust and efficient solid-phase synthesis methods. By exchanging building blocks, the mimetic structures can be quickly varied and their properties optimized, in an efficient combinatorial fashion, using parallel synthetic chemistry (Jiang et al. 2000). The ease with which analogues can be produced in this way, contrasts with the difficulties that often exist in synthesizing analogues of complex natural products of similar size. The variables for creating combinatorial libraries include the size of the ß-hairpin loop, the types of amino acid building blocks, as well as the template (▶Figure 17.10). The proteinogenic α-amino acids often make convenient starting points for mimetic design, but may be exchanged for any of a large number of known nonproteinogenic amino acids, or related building bocks. This provides great scope for optimizing drug-like properties (ADMET), as well as target affinity and specificity, within a lead series. This point is illustrated by recent examples that have led to clinical drug candidates (Obrecht et al. 2011; Robinson et al. 2008). In one case, a family of PEM molecules modeled on the naturally occurring cationic antimicrobial peptide protegrin I (PG-I) was discovered using the PEM technology. The lead compound **L27–11** is a potent antibiotic with a novel mechanism of action, which is active in the nanomolar range against Pseudomonas species and, in particular, the important human pathogen *Pseudomonas aeruginosa* (Srinivas et al. 2010). This family of PEM molecules does not lyse cells, but rather, they are able to target a bacterial ß-barrel outer membrane (OM) protein called LptD, which is essential for the assembly of the outer leaflet of the OM in many gram-negative bacteria (Chng et al. 2010; Ruiz et al. 2009; Sperandeo et al. 2009). When the function of LptD is blocked,

OM biogenesis is disrupted. The discovery of **L27–11** involved an iterative process of library synthesis and screening, in which an initial hit was identified by PEM library synthesis and screening for antimicrobial activity. The optimal hit from each library was used as a starting point for the synthesis and testing of variations in a subsequent library. The same procedure was also followed to optimize the drug-like properties of **L27–11**, including the plasma stability, target selectivity, and toxicology. This process resulted in the clinical lead candidate **POL7080** (Srinivas et al. 2010). The safety of mimetic **POL7080** is now being tested in healthy humans in a phase I clinical study, but it is clear already that this mimetic represents the first in a new class of antibiotics active against gram-negative bacteria.

A second example is provided by a family of PEM molecules that were based upon the naturally occurring peptide polyphemusin II, an 18-amino-acid peptide isolated from the American horseshoe crab (*Limulus polyphemus*), which inhibits the chemokine and G-coupled protein receptor CXCR4. Based on the solution structure of polyphemusin II, several PEM molecules were designed and optimized in biological assays. This led to highly potent and selective CXCR4 antagonist such as **POL3026**, **POL5551** and **POL6326** (DeMarco et al. 2006; Robinson et al. 2008). **POL6326** has now successfully moved into a phase II clinical trial for autologous stem cell transplantation in newly diagnosed multiple myeloma patients. Interim results from this study show that **POL6326** is safe, well tolerated and efficient in mobilizing hematopoetic stem cells. Recently, an

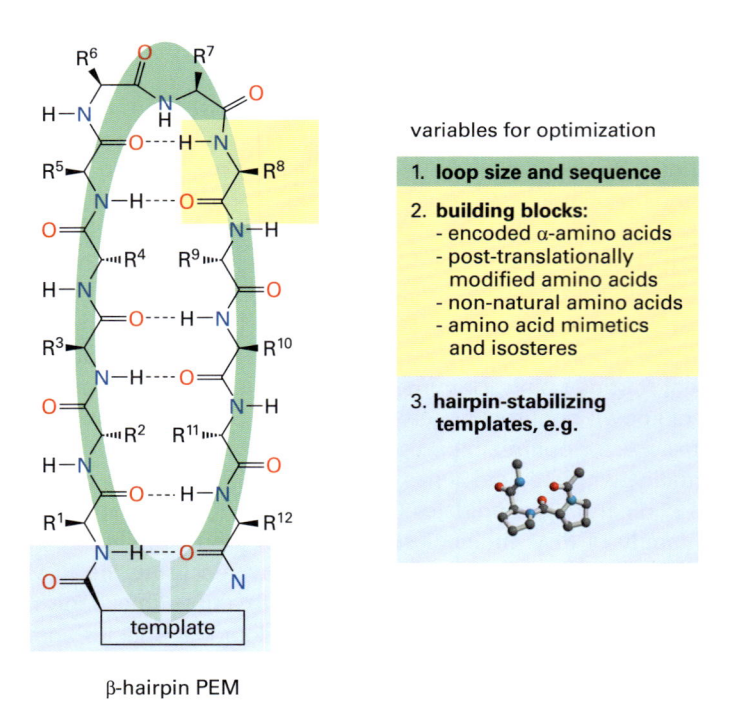

Figure 17.10: The structures and properties of ß-hairpin protein epitope mimetics (PEMs) can be varied and optimized through library synthesis and screening. Some of the variables for library design are indicated.

X-ray structure revealed how a very closely related ß-hairpin-shaped peptide binds to an engineered form of CXCR4 (Wu et al. 2010a). The ß hairpin occupies a binding site on CXCR4 formed by residues in the inward-facing protruding walls of the seven transmembrane helical bundle, several extracellular loops, and the N-terminal segment. A network of polar, hydrogen-bonding and hydrophobic contacts between the ligand and the receptor are responsible for the specific high affinity interaction. The close structural similarity between this bound ligand and **POL3026**, suggests that the macrocyclic ß-hairpin mimetic should interact with CXCR4 in a similar manner.

The genomic and proteomic revolutions continue to provide us with an ever-increasing number of mechanistic insights into biological signalling pathways, and potential targets for inhibition using small molecules. Unfortunately, many of the genes and gene products that are often most attractive from a biological perspective for targeting using small molecules are also often the least tractable from the perspective of small drug-like molecules. Such targets typically include proteins that participate in PPIs or protein-nucleic acid interactions, which are often deemed to be "undruggable" within major pharmaceutical companies (Arkin and Wells 2004; Hopkins and Groom 2002). The difficulty in addressing "undruggable" targets using small drug-like molecules (Wells and McClendon 2007) provides a powerful motivation to explore new scaffolds and new regions of molecular space to overcome this problem. The protein epitope mimetics described in this article are positioned within a huge and relatively unexplored region of molecular space, having a size and complexity between that of traditional small drug-like molecules and much larger 'biologics', such as antibodies. It is clear already that protein epitope mimetics can provide a rich source of interesting leads for use in chemical biology, as well as drug and vaccine research. And the scope for further innovation and discovery in this field appears to be immense.

References

Albeck, S., Unger, R., and Schreiber, G. (2000). Evaluation of direct and cooperative contributions towards the strength of buried hydrogen bonds and salt bridges. J Mol Biol *298*, 503–520.

Andrews, M.J.I., and Tabor, A.B. (1999). Forming stable helical peptides using natural and artificial amino acids. Tetrahedron *55*, 11711–11743.

Appella, D.H., Barchi, J.J., Durell, S.R., and Gellman, S.H. (1999). Formation of short, stable helices in aqueous solution by ß-amino acid hexamers. J Am Chem Soc *121*, 2309–2310.

Arkin, M.R., Randal, M., DeLano, W.L., Hyde, J., Luong, T.N., Oslob, J.D., et al. (2003). Binding of small molecules to an adaptive protein-protein interface. Proc Nat Acad Sci U S A *100*, 1603–1608.

Arkin, M.R., and Wells, J.A. (2004). Small-molecule inhibitors of protein-protein interactions: progressing towards the dream. Nat Rev Drug Discov *3*, 301–317.

Athanassiou, Z., Dias, R.L.A., Moehle, K., Dobson, N., Varani, G., and Robinson, J.A. (2004). Structural mimicry of retroviral Tat proteins by constrained, beta-hairpin peptidomimetics: ligands with high affinity and selectivity for viral TAR RNA regulatory elements. J Am Chem Soc *126*, 6906–6913.

Athanassiou, Z., Patora, K., Dias, R.L.A., Moehle, K., Robinson, J.A., and Varani, G. (2007). Structure-guided peptidomimetic design leads to nanomolar ß-hairpin inhibitors of the Tat-TAR interaction of bovine immunodeficiency virus. Biochemistry *46*, 741–751.

Austin, R.E., Maplestone, R.A., Sefler, A.M., Liu, K., Hruzewicz, W.N., Liu, C.W., et al. (1997). A template for stabilization of a peptide α-helix: synthesis and evaluation of conformational effects by circular dichroism and NMR. J Am Chem Soc *119*, 6461–6472.

Baron, R., Setny, P., and McCammon, J.A. (2010). Water in cavity-ligand recognition. J Am Chem Soc *132*, 12091–12097.

Bautista, A.D., Appelbaum, J.S., Craig, C.J., Michel, J., and Schepartz, A. (2010). Bridged ß3-peptide inhibitors of p53-hDM2 complexation: correlation between affinity and cell permeability. J Am Chem Soc *132*, 2904–2906.

Berg, O.G., and von Hippel, P.H. (1985). Diffusion-controlled macromolecular interactions. Annu Rev Biophys Chem *14*, 131–160.

Bernal, F., Tyler, A.F., Korsmeyer, S.J., Walensky, L.D., and Verdine, G.L. (2007). Reactivation of the p53 tumor suppressor pathway by a stapled p53 peptide. J Am Chem Soc *129*, 2456–2457.

Beyer, R.L., Hoang, H.N., Appleton, T.G., and Fairlie, D.P. (2004). Metal clips induce folding of a short unstructured peptide into an α-helix via turn conformations in water. Kinetic versus thermodynamic products. J Am Chem Soc *126*, 15096–15105.

Blackwell, H.E., and Grubbs, R.H. (1998). Highly efficient synthesis of covalently cross-linked peptide helices by ring-closing metathesis. Angew Chem Int Ed *37*, 3281–3284.

Bogan, A.A., and Thorn, K.S. (1998). Anatomy of hot spots in protein interfaces. J Mol Biol *280*, 1–9.

Braisted, A.C., Oslob, J.D., Delano, W.L., Hyde, J., McDowell, R.S., Waal, N., et al. (2003). Discovery of a potent small molecule IL-2 inhibitor through fragment assembly. J Am Chem Soc *125*, 3714–3715.

Bruncko, M., Oost, T.K., Belli, B.A., Ding, H., Joseph, M.K., Kunzer, A., et al. (2007). Studies leading to potent, dual inhibitors of Bcl-2 and Bcl-xL. J Med Chem *50*, 641–662.

Brunel, F.M., and Dawson, P.E. (2005). Synthesis of constrained helical peptides by thioether ligation: application to analogs of gp41. Chem Comm, 2552–2554.

Cabezas, E., and Satterthwait, A.C. (1999). The hydrogen bond mimic approach: solid-phase synthesis of a peptide stabilized as an alpha-helix with a hydrazone link. J Am Chem Soc *121*, 3862–3875.

Campbell, F., Plante, J.P., Edwards, T.A., Warriner, S.L., and Wilson, A.J. (2010). N-alkylated oligoamide alpha-helical proteomimetics. Org Biomol Chem *8*, 2344–2351.

Cascales, L., and Craik, D.J. (2010). Naturally occurring circular proteins: distribution, biosynthesis and evolution. Org Biomol Chem *8*, 5035–5047.

Chandler, D. (2005). Interfaces and the driving force of hydrophobic assembly. Nature *437*, 640–647.

Chapman, R.N., Dimartino, G., and Arora, P.S. (2004). A highly stable short alpha-helix constrained by a main-chain hydrogen-bond surrogate. J Am Chem Soc *126*, 12252–12253.

Chng, S.-S., Ruiz, N., Chimalakonda, G., Silhavy, T.J., and Kahne, D. (2010). Characterization of the two-protein complex in *Escherichia coli* responsible for lipopolysaccharide assembly at the outer membrane. Proc Natl Acad Sci U S A *107*, 5363–5368.

Clackson, T., Ultsch, M.H., Wells, J.A., and de Vos, A.M. (1998). Structural and functional analysis of the 1:1 growth hormone:receptor complex reveals the molecular basis for receptor affinity. J Mol Biol *277*, 1111–1128.

Clackson, T., and Wells, J.A. (1995). A hot spot of binding energy in a hormone-receptor interface. Science *267*, 383–386.

Clark, R., Jensen, J., Nevin, S., Callaghan, B., Adams, D., and Craik, D. (2010). The engineering of an orally active conotoxin for the treatment of neuropathic pain. Angew Chem Int Ed *49*, 6545–6548.

Claudon, P., Violette, A., Lamour, K., Decossas, M., Fournel, S., Heurtault, B., et al. (2010). Consequences of isostructural main-chain modifications for the design of antimicrobial foldamers: helical mimics of host-defense peptides based on a heterogeneous amide/urea backbone. Angew Chem Int Ed *49*, 333–336.

Clemons, P.A., Bodycombe, N.E., Carrinski, H.A., Wilson, J.A., Shamji, A.F., Wagner, B.K., et al. (2010). Small molecules of different origins have distinct distributions of structural complexity that correlate with protein-binding profiles. Proc Nat Acad Sci U S A *107*, 18787–18792.

Cohen, M., Reichmann, D., Neuvirth, H., and Schreiber, G. (2008). Similar chemistry, but diffeent bond preferences in inter versus intra-protein interactions. Proteins *72*, 741–753.

Conte, L.L., Chothia, C., and Janin, J. (1999). The atomic structure of protein-protein recognition sites. J Mol Biol *285*, 2177–2198.

Coyne, A.G., Scott, D.E., and Abell, C. (2010). Drugging chellenging targets using fragment-based approaches. Curr Opin Chem Biol *14*, 299–307.

Craik, D.J., Cemazar, M., and Daly, N.L. (2006a). The cyclotides and related macrocyclic peptides as scaffolds in drug design. Curr Opin Drug Discov Dev *9*, 251–260.

Craik, D.J., Cemazar, M., Wang, C.K.L., and Daly, N.L. (2006b). The cyclotide family of circular miniproteins: nature's combinatorial peptide template. Biopolymers *84*, 250–266.

Cummings, C.G., and Hamilton, A.D. (2010). Disrupting protein-protein interactions with non-peptidic, small molecule [alpha]-helix mimetics. Curr Opin Chem Biol *14*, 341–346.

Davidson, A., Leeper, T.C., Athanassiou, Z., Patora, K., Karn, J., Robinson, J.A., et al. (2009). Simultaneous recognition of HIV-1 TAR RNA bulge and loop sequences by cyclic peptide mimics of Tat protein. Proc Natl Acad Sci U S A *106*, 11931–11936.

DeLano, W.L. (2002). Unraveling hot spots in binding interfaces: progress and challenges. Curr Opin Struct Biol *12*, 14–20.

DeLano, W.L., Ultsch, M.H., de Vos, A.M., and Wells, J.A. (2000). Convergent solutions to binding at a protein-protein interface. Science *287*, 1279–1283.

DeMarco, S.J., Henze, H., Lederer, A., Moehle, K., Mukherjee, R., Romagnoli, B., et al. (2006). Discovery of novel, highly potent and selective beta-hairpin mimetic CXCR4 inhibitors with excellent anti-HIV activity and pharmacokinetic profiles. Bioorgan Med Chem *14*, 8396–8404.

Descours, A., Moehle, K., Renard, A., and Robinson, J.A. (2002). A new family of ß-hairpin mimetics based on a trypsin inhibitor from sunflower seeds. ChemBioChem *3*, 318–323.

Dias, R.L.A., Fasan, R., Moehle, K., Renard, A., Obrecht, D., and Robinson, J.A. (2006). Protein ligand design: from phage display to synthetic protein epitope mimetics in human antibody Fc-binding peptidomimetics. J Am Chem Soc *128*, 2726–2732.

Drakopoulou, E., Zinn-Justin, S., Guenneugues, M., Gilquin, B., Ménez, A., and Vita, C. (1996). Changing the structural context of a functional ß-hairpin. J Biol Chem *271*, 11979–11987.

Dunitz, J.D. (1994). The entropic cost of bound water in crystals and biomolecules. Science *264*, 670.

Eckert, D.M., and Kim, P.S. (2001). Mechanisms of viral membrane fusion and its inhibition. Annu Rev Biochem *70*, 777–810.

Edwards, A.A., Tipton, J.D., Brenowitz, M.D., Emmett, M.R., Marshall, A.G., Evans, G.B., et al. (2010). Conformational states of human purine nucleoside phosphorylase at rest, at work, and with transition state analogues. Biochemistry *49*, 2058–2067.

Eidenschink, L., Kier, B.L., Huggins, K.N.L., and Andersen, N.H. (2009). Very short peptides with stable folds: building on the interrelationship of Trp/Trp, Trp/cation, and Trp/backbone–amide interaction geometries. Proteins *75*, 308–322.

English, E.P., Chumanov, R.S., Gellman, S.H., and Compton, T. (2006). Rational development of ß-peptide inhibitors of human cytomegalovirus entry. J Biol Chem *281*, 2661–2667.

Erlanson, D.A. (2006). Fragment-based lead discovery: a chemical update. Curr Opin Biotechnol 17, 643–652.

Erlanson, D.A., Braisted, A.C., Raphael, D.R., Randal, M., Stroud, R.M., Gordon, E.M., et al. (2000). Site-directed ligand discovery. Proc Nat Acad Sci U S A 97, 9367–9372.

Ernst, J.T., Kutzki, O., Debnath, A.K., Jiang, S., Lu, H., and Hamilton, A.D. (2002). Design of a protein surface antagonist based on α-helix mimicry: inhibition of gp41 assembly and viral fusion. Angew Chem Int Ed 41, 278–281.

Fasan, R., Dias, R.L.A., Moehle, K., Zerbe, O., Obrecht, D., Mittl, P.R.E., et al. (2006). Structure-activity studies in a family of beta-hairpin protein epitope mimetic inhibitors of the p53-HDM2 protein-protein interaction. ChemBioChem 7, 515–526.

Fasan, R., Dias, R.L.A., Moehle, K., Zerbe, O., Vrijbloed, J.W., Obrecht, D., et al. (2004). Using a beta-hairpin to mimic an alpha-helix: cyclic peptidomimetic inhibitors of the p53-HDM2 protein-protein interaction. Angew Chem Int Ed 43, 2109–2112.

Favre, M., Moehle, K., Jiang, L., Pfeiffer, B., and Robinson, J.A. (1999). Structural mimicry of canonical conformations in antibody hypervariable loops using cyclic peptides containing a heterochiral diproline template. J Am Chem Soc 121, 2679–2685.

Felix, A.M., Heimer, E.P., Wang, C.-T., Lambros, T.J., Fournier, A., Mowles, T.F., et al. (1988). Synthesis, biological activity and conformational analysis of cyclic GRF analogs. Int J Pept Prot Res 32, 441–454.

Fischer, L., and Guichard, G. (2010). Folding and self-assembly of aromatic and aliphatic urea oligomers: towards connecting structure and function. Org Biomol Chem 8, 3101–3117.

Foote, J., and Eisen, H.N. (1995). Kinetic and affinity limits on antibodies produced during immune responses. Proc Natl Acad Sci U S A 92, 1254–1256.

Fowler, S.A., and Blackwell, H.E. (2009). Structure-function relationships in peptoids: recent advances toward deciphering the structural requirements for biological function. Org Biomol Chem 7, 1508–1524.

Freire, F., and Gellman, S.H. (2009). Macrocyclic design strategies for small, stable parallel ß-sheet scaffolds. J Am Chem Soc 131, 7970–7972.

Fry, D.C. (2006). Protein-protein interactions as targets for small molecule drug discovery. Biopolymers (Pept Sci) 84, 535–552.

Gabdoulline, R.R., and Wade, R.C. (2002). Biomolecular diffusional association. Curr Opin Struct Biol 12, 204–213.

Ganesan, A. (2008). The impact of natural products upon modern drug discovery. Curr Opin Chem Biol 12, 306–317.

Garner, J., and Harding, M.M. (2007). Design and synthesis of [small alpha]-helical peptides and mimetics. Org Biomol Chem 5, 3577–3585.

Geistlinger, T.R., and Guy, R.K. (2003). Novel selective inhibitors of the interaction of individual nuclear hormone receptors with a mutually shared steroid receptor coactivator 2. J Am Chem Soc 125, 6852–6853.

Ghadiri, M.R., and Choi, C. (1990). Secondary structure nucleation in peptides. Transition metal ion stabilized .alpha.-helices. J Am Chem Soc 112, 1630–1632.

Gilbreth, R.N., Esaki, K., Koide, A., Sidhu, S.S., and Koide, S. (2008). A dominant conformational role for amino acid diversity in minimalist protein-protein interfaces. J Mol Biol 381, 407–418.

Goodman, C.M., Choi, S., Shandler, S., and DeGrado, W.F. (2007). Foldamers as versatile frameworks for the design and evolution of function. Nat Chem Biol 3, 252–262.

Grässlin, A., Amoreira, C., Baldridge, K.K., and Robinson, J.A. (2009). Thermodynamic and computational studies on the binding of p53-derived peptides and peptidomimetic inhibitors to HDM2. ChemBioChem 10, 1360–1368.

Green, B.R., Catlin, P., Zhang, M.M., Fiedler, B., Bayudan, W., Morrison, A., et al. (2007). Conotoxins containing nonnatural backbone spacers: cladistic-based design, chemical synthesis, and improved analgesic activity. Chem Biol 14, 399–407.

Guichard, G., and Huc, I. (2011). Synthetic foldamers. Chem Comm 47, 5933–5941.

Halperin, I., Wolfson, H., and Nussinov, R. (2004). Protein-protein interactions: coupling of structurally conserved residues and of hot spots across interfaces. Implications for docking. Structure 12, 1027–1038.

Harel, M., Cohen, M., and Schreiber, G. (2007). On the dynamic nature of the transition state for protein-protein association as determined by double-mutant cycle analysis and simulation. J Mol Biol 371, 180–196.

Harrison, R.S., Shepherd, N.E., Hoang, H.N., Ruiz-Gómez, G., Hill, T.A., Driver, R.W., et al. (2010). Downsizing human, bacterial, and viral proteins to short water-stable alpha helices that maintain biological potency. Proc Natl Acad Sci U S A 107, 11686–11691.

Harvey, A.L. (2008). Natural products in drug discovery. Drug Discov Today 13, 894–901.

Henchey, L.K., Jochim, A.L., and Arora, P.S. (2008). Contemporary strategies for the stabilization of peptides in the [alpha]-helical conformation. Curr Opin Chem Biol 12, 692–697.

Henchey, L.K., Kushal, S., Dubey, R., Chapman, R.N., Olenyuk, B.Z., and Arora, P.S. (2010). Inhibition of hypoxia inducible factor 1-transcription coactivator interaction by a hydrogen bond surrogate α-helix. J Am Chem Soc 132, 941–943.

Henriques, S.T., and Craik, D.J. (2010). Cyclotides as templates in drug design. Drug Discov Today 15, 57–64.

Hill, D.J., Mio, M.J., Prince, R.B., Hughes, T.S., and Moore, J.S. (2001). A field guide to foldamers. Chem Rev 101, 3893–4012.

Hodges, A.M., and Schepartz, A. (2007). Engineering a monomeric miniature protein. J Am Chem Soc 129, 11024–11025.

Homans, S.W. (2007). Water, water everywhere – except where it matters. Drug Discov Today 12, 534–539.

Hopkins, A.L., and Groom, C.R. (2002). The druggable genome. Nat Rev Drug Discov 1, 727–730.

Horne, W.S., Boersma, M.D., Windsor, M.A., and Gellman, S.H. (2008). Sequence-based design of α/β-peptide foldamers that mimic BH3 domains. Angew Chem Int Ed 47, 2853–2856.

Horne, W.S., and Gellman, S.H. (2008). Foldamers with heterogeneous backbones. Acc Chem Res 41, 1399–1408.

Horne, W.S., Johnson, L.M., Ketas, T.J., Klasse, P.J., Lu, M., Moore, J.P., et al. (2009). Structural and biological mimicry of protein surface recognition by α/ß-peptide foldamers. Proc Natl Acad Sci U S A 106, 14751–14756.

Hu, Z., Ma, B., Wolfson, H., and Nussinov, R. (2000). Conservation of polar residues as hot spots at protein interfaces. Protein Struct Funct Genet 39, 331–342.

Huang, C.C., Stricher, F., Martin, L., Decker, J.M., Majeed, S., Barthe, P., et al. (2005). Scorpion-toxin mimics of CD4 in complex with human immunodeficiency virus gp120: crystal structures, molecular mimicry, and neutralization breadth. Structure 13, 755–768.

Ikura, T., Urakubo, Y., and Ito, N. (2004). Water-mediated interaction at a protein-protein interface. Chem Phys 307, 111–119.

Jackson, D.Y., King, D.S., Chmielewski, J., Singh, S., and Schultz, P.G. (1991). General approach to the synthesis of short .alpha.-helical peptides. J Am Chem Soc 113, 9391–9392.

Jacoby, E. (2002). Biphenyls as potential mimetics of protein [alpha]-helix. Bioorg Med Chem Lett 12, 891–893.

Janin, J. (1999). Wet and dry interfaces: the role of solvent in protein-protein and protein-DNA recognition. Structure 7, R277–R279.

Jiang, L., Moehle, K., Dhanapal, B., Obrecht, D., and Robinson, J.A. (2000). Combinatorial biomimetic chemistry: parallel synthesis of a small library of ß-hairpin mimetics based on loop III from human platelet-derived growth factor B. Helv Chim Acta 83, 3097–3112.

Jones, S., and Thornton, J.M. (1996). Principles of protein-protein interactions. Proc Natl Acad Sci U S A 93, 13–20.

Judice, J.K., Tom, J.Y.K., Huang, W., Wrin, T., Vennari, J., Petropoulos, C.J., et al. (1997). Inhibition of HIV type 1 infectivity by constrained α-helical peptides: implications for the viral fusion mechanism. Proc Natl Acad Sci U S A 94, 13426–13430.

Karlsson, R., Michaelsson, A., and Mattsson, L. (1991). Kinetic analysis of monoclonal antibody-antigen interactions with a new biosensor analytical system. J Immunol Methods 145, 229–240.

Kelso, M.J., Beyer, R.L., Hoang, H.N., Lakdawala, A.S., Snyder, J.P., Oliver, W.V., et al. (2004). α-Turn mimetics: short peptide α-helices composed of cyclic metallopentapeptide modules. J Am Chem Soc 126, 4828–4842.

Kelso, M.J., Hoang, H.N., Oliver, W., Sokolenko, N., March, D.R., Appleton, T.G., et al. (2003). A cyclic metallopeptide induces α helicity in short peptide fragments of thermolysin. Angew Chem Int Ed 42, 421–424.

Kemp, D.S., Allen, T.J., and Oslick, S.L. (1995a). The energetics of helix formation by short templated peptides in aqueous solution. 1. Characterization of the reporting helical template Ac-Hel1. J Am Chem Soc 117, 6641–6657.

Kemp, D.S., Allen, T.J., Oslick, S.L., and Boyd, J.G. (1996). The structure and energetics of helix formation by short templated peptides in aqueous solution. 2. Characterization of the helical structure of Ac-Hel1-Ala6-OH. J Am Chem Soc 118, 4240–4248.

Kemp, D.S., and Curran, T.P. (1988a). (2S,5S,8S,11S)-1-Acetyl-1,4-diaza-3-keto-5-carboxy-10-thia-tricyclo – [2.8.04,8]-tridecane, 1 the preferred conformation of 1 (1 [reverse not equivalent] [alpha]Temp-OH) and its peptide conjugates [alpha]Temp-L-(Ala)n-OR (n=1 to 4) and [alpha]-Temp-L-Ala-L-Phe-L-Lys([epsilon]Boc)-L-Lys([epsilon]-Boc)-NHMe studies of templates for [alpha]-helix formation. Tetrahedron Lett 29, 4935–4938.

Kemp, D.S., and Curran, T.P. (1988b). (2S,5S,8S,11S)-1-Acetyl-1,4-diaza-3-keto-5-carboxy-10-thia-tricyclo-[2.8.04,8]-tridecane, 1 synthesis of prolyl-proline-derived, peptide-functionalized templates for [alpha]-helix formation. Tetrahedron Lett 29, 4931–4934.

Kemp, D.S., and Rothman, J.H. (1995a). Efficient helix nucleation by a macrocyclic triproline-derived template. Tetrahedron Lett 36, 4023–4026.

Kemp, D.S., and Rothman, J.H. (1995b). Synthesis and analysis of a macrocyclic triproline-derived template containing a local conformational constraint. Tetrahedron Lett 36, 4019–4022.

Kemp, D.S., Rothman, J.H., Curran, T.C., and Blanchard, D.E. (1995b). A macrocyclic triproline-derived template for helix nucleation. Tetrahedron Lett 36, 3809–3812.

Keskin, O., Ma, B., and Nussinov, R. (2005). Hot regions in protein-protein interactions: the organization and contribution of structurally conserved hot spot residues. J Mol Biol 345, 1281–1294.

Kier, B.L., Shu, I., Eidenschink, L.A., and Andersen, N.H. (2010). Stabilizing capping motif for ß-hairpins and sheets. Proc Natl Acad Sci U S A 107, 10466–10471.

Koide, A., Tereshko, V., Uysal, S., Margalef, K., Kossiakoff, A.A., and Koide, S. (2007). Exploring the capacity of minimalist protein interfaces: interface energetics and affinity maturation to picomolar K-D of a single-domain antibody with a flat paratope. J Mol Biol 373, 941–953.

Kortemme, T., and Baker, D. (2002). A simple physical model for binding energy hot spots in protein-protein complexes. Proc Natl Acad Sci U S A 99, 14116–14121.

Kossiakoff, A.A., and Koide, S. (2008). Understanding mechanisms governing protein-protein interactions from synthetic binding interfaces. Curr Opin Struct Biol 18, 499–506.

Kritzer, J.A., Lear, J.D., Hodsdon, M.E., and Schepartz, A. (2004). Helical ß-peptide inhibitors of the p53-hDM2 interaction. J Am Chem Soc 126, 9468–9469.

Kritzer, J.A., Zutshi, R., Cheah, M., Ran, F.A., Webman, R., Wongjirad, T.M., et al. (2006). Miniature protein inhibitors of the p53-hDM2 interaction. ChemBioChem 7, 29–31.

Kutzki, O., Park, H.S., Ernst, J.T., Orner, B.P., Yin, H., and Hamilton, A.D. (2002). Development of a potent Bcl-xL antagonist based on α-helix mimicry. J Am Chem Soc 124, 11838–11839.

Kwong, P.D., Wyatt, R., Robinson, J., Sweet, R.W., Sodroski, J., and Hendrickson, W.A. (1998). Structure of an HIV gp120 envelope glycoprotein in complex with the CD4 receptor and a neutralizing human antibody. Nature 393, 648–659.

Lalonde, M.S., Lobritz, M.A., Ratcliff, A., Chamanian, M., Athanassiou, Z., Tyagi, M., et al. (2011). Inhibition of both HIV-1 reverse transcription and gene expression by a cyclic peptide that binds the Tat-transactivating response element (TAR) RNA. PLoS Pathogens 7, 17.

Leduc, A.-M., Trent, J.O., Wittliff, J.L., Bramlett, K.S., Briggs, S.L., Chirgadze, N.Y., et al. (2003). Helix-stabilized cyclic peptides as selective inhibitors of steroid receptor-coactivator interactions. Proc Natl Acad Sci U S A 100, 11273–11278.

Lee, E.F., Sadowsky, J.D., Smith, B.J., Czabotar, P.E., Peterson-Kaufman, K.J., Colman, P.M., et al. (2009). High-resolution structural characterization of a helical α/β-peptide foldamer bound to the anti-apoptotic protein Bcl-xL. Angew Chem Int Ed 48, 4318–4322.

Lee, J.H., Zhang, Q., Jo, S., Chai, S.C., Oh, M., Im, W., et al. (2011). Novel pyrrolopyrimidine-based alpha-helix mimetics: cell-permeable inhibitors of protein-protein interactions. J Am Chem Soc 133, 676–679.

Leeper, T.C., Athanassiou, Z., Dias, R.L.A., Robinson, J.A., and Varani, G. (2005). TAR RNA recognition by a cyclic peptidomimetic of Tat protein. Biochemistry 44, 12362–12372.

Lehne, B., and Schlitt, T. (2009). Protein-protein interaction databases: keeping up with growing interactomes. Hum Genomics 3, 291–297.

Lewis, A., Rutherford, T.J., Wilkie, J., Jenn, T., and Gani, D. (1998). Design, construction and properties of peptide N-terminal cap templates devised to initiate [small alpha]-helices. Part 3. Caps derived from N-[(2S)-2-chloropropionyl]-(2S)-Pro-(2R)-Ala-(2S,4S)-4-thioPro-OMe. J Chem Soc Perkin I Trans , 3795–3806.

Li, C., Liu, M., Monbo, J., Zou, G., Li, C., Yuan, W., et al. (2008). Turning a scorpion toxin into an antitumor miniprotein. J Am Chem Soc 130, 13546–13548.

Li, J.W.H., and Vederas, J.C. (2009). Drug discovery and natural products: end of an era or an endless frontier? Science 325, 161–165.

Livnah, O., Stura, E.A., Johnson, D.L., Middleton, S.A., Mulcahy, L.S., Wrighton, N.C., et al. (1996). Functional mimicry of a protein hormone by a peptide agonist: the EPO receptor complex at 2.8 Å. Science 273, 464–471.

Ma, B., Elkayam, T., Wolfson, H., and Nussinov, R. (2003). Protein-protein interactions: structurally conserved residues distinguish between binding sites and exposed protein surfaces. Proc Natl Acad Sci U S A 100, 5772–5777.

Maison, W., Arce, E., Renold, P., Kennedy, R.J., and Kemp, D.S. (2001). Optimal N-caps for N-terminal helical templates: effects of changes in H-bonding efficiency and charge. J Am Chem Soc 123, 10245–10254.

Malham, R., Johnstone, S., Bingham, R.J., Barratt, E., Phillips, S.E.V., Laughton, C.A., et al. (2005). Strong solute-solute dispersive interactions in a protein-ligand complex. J Am Chem Soc 127, 17061–17067.

Martin, L., Stricher, F., Missé, D., Sironi, F., Pugniére, M., Barthe, P., et al. (2003). Rational design of a CD4 mimic that inhibits HIV-1 entry and exposes cryptic neutralization epitopes. Nat Biotech 21, 71–76.

Michel, J., Harker, E.A., Tirado-Rives, J., Jorgensen, W.L., and Schepartz, A. (2009). In silico improvement of ß3-peptide inhibitors of p53-hDM2 and p53-hDMX. J Am Chem Soc 131, 6356–6357.

Mills, N.L., Daugherty, M.D., Frankel, A.D., and Guy, R.K. (2006). An α-helical peptidomimetic inhibitor of the HIV-1 Rev-RRE interaction. J Am Chem Soc *128*, 3496–3497.

Moehle, K., Athanassiou, Z., Patora, K., Davidson, A., Varani, G., and Robinson, J.A. (2007). Design of ß-hairpin peptidomimetics that inhibit binding of alpha-helical HIV-1 Rev protein to the Rev response element RNA. Angew Chem Int Ed *46*, 9101–9104.

Moellering, R.E., Cornejo, M., Davis, T.N., Bianco, C.D., Aster, J.C., Blacklow, S.C., et al. (2009). Direct inhibition of the NOTCH transcription factor complex. Nature *462*, 182–188.

Moreira, I.S., Fernandes, P.A., and Ramos, M.J. (2007). Hot spots-a review of the protein-protein interface determinant amino-acid residues. Proteins *68*, 803–812.

Murray, C.W., and Blundell, T.L. (2010). Structural biology in fragment-based drug design. Curr Opin Struct Biol *20*, 497–507.

Northrup, S.H., Boles, J.O., and Reynolds, J.C. (1988). Brownian dynamics of cytochrome c and cytochrome c peroxidase association. Science *241*, 67–70.

Northrup, S.H., and Erickson, H.P. (1992). Kinetics or protein-protein association explained by Brownian dynamics computer simulation. Proc Natl Acad Sci U S A *89*, 3338–3342.

Obrecht, D., Altorfer, M., and Robinson, J.A. (1999). Novel peptide mimetic building blocks and strategies for efficient lead finding. Adv Med Chem *4*, 1–68.

Obrecht, D., Chevalier, E., Moehle, K., and Robinson, J.A. (2011). ß-Hairpin protein epitope mimetic technology in drug discovery. Drug Discov Today Technol *8*, in press.

Oguri, H., Oomura, A., Tanabe, S., and Hirama, M. (2005). Design and synthesis of a trans-fused polycyclic ether skeleton as an [alpha]-helix mimetic scaffold. Tetrahedron Lett *46*, 2179–2183.

Oguri, H., Tanabe, S., Oomura, A., Umetsu, M., and Hirama, M. (2006). Synthesis and evaluation of [alpha]-helix mimetics based on a trans-fused polycyclic ether: sequence-selective binding to aspartate pairs in [alpha]-helical peptides. Tetrahedron Lett *47*, 5801–5805.

Olsson, T.S.G., Williams, M.A., Pitt, W.R., and Ladbury, J.E. (2008). The thermodynamics of protein-ligand interaction and solvation: insight for ligand design. J Mol Biol *384*, 1002–1017.

Oltersdorf, T., Elmore, S.W., Shoemaker, A.R., Armstrong, R.C., Augeri, D.J., Belli, B.A., et al. (2005). An inhibitor of Bcl-2 family proteins induces regression of solid tumours. Nature *435*, 677–681.

Orner, B.P., Ernst, J.T., and Hamilton, A.D. (2001). Toward proteomimetics: terphenyl derivatives as structural and functional mimics of extended regions of an α-helix. J Am Chem Soc *123*, 5382–5383.

Osapay, G., and Taylor, J.W. (1992). Multicyclic polypeptide model compounds. 2. Synthesis and conformational properties of a highly .alpha.-helical uncosapeptide constrained by three side-chain to side-chain lactam bridges. J Am Chem Soc *114*, 6966–6973.

Padlan, E.A. (1990). On the nature of antibody combining sites: unusual structural features that may confer on these sites an enhanced capacity for binding ligands. Proteins Struct Funct Genet *7*, 112–124.

Park, C.-M., Bruncko, M., Adickes, J., Bauch, J., Ding, H., Kunzer, A., et al. (2008). Discovery of an orally bioavailable small molecule inhibitor of prosurvival B-cell lymphoma 2 proteins. J Med Chem *51*, 6902–6915.

Patgiri, A., Jochim, A.L., and Arora, P.S. (2008). A hydrogen bond surrogate approach for stabilization of short peptide sequences in α-helical conformation. Acc Chem Res *41*, 1289–1300.

Phelan, J.C., Skelton, N.J., Braisted, A.C., and McDowell, R.S. (1997). A general method for constraining short peptides to an α-helical conformation. J Am Chem Soc *119*, 455–460.

Phillips, S.T., Piersanti, G., and Bartlett, P.A. (2005). Quantifying amino acid conformational preferences and side-chain-side-chain interactions in ß-hairpins. Proc Natl Acad Sci U S A *102*, 13737–13742.

Potapov, V., Reichmann, D., Abramovich, R., Filchtinski, D., Zohar, N., Ben Halevy, D., et al. (2008). Computational redesign of a protein-protein interface for high afinity and binding specificity using modular architecture and naturally occurring template fragments. J Mol Biol *384*, 109–119.

Raghavender, U.S., Aravinda, S., Rai, R., Shamala, N., and Balaram, P. (2010). Peptide hairpin nucleation with the obligatory Type I' ß-turn Aib-DPro segment. Org Biomol Chem *8*, 3133–3135.

Rai, R., Raghothama, S., Sridharan, R., and Balaram, P. (2007). Tuning the β-turn segment in designed peptide β-hairpins: construction of a stable type I' β-turn nucleus and hairpin–helix transition promoting segments. Pept Sci *88*, 350–361.

Raimundo, B.C., Oslob, J.D., Braisted, A.C., Hyde, J., McDowell, R.S., Randal, M., et al. (2004). Integrating fragment assembly and biophysical methods in the chemical advancement of small-molecule antagonists of IL-2. An approach for inhibiting protein-protein interactions. J Med Chem *47*, 3111–3130.

Raman, C.S., Jemmerson, R., Nall, B.T., and Allen, M.J. (1992). Diffusion-limited rates for monoclonal antibody binding to cytochrome c. Biochemistry *31*, 10370–10379.

Reichmann, D., Rahat, O., Albeck, S., Meged, R., Dym, O., and Schreiber, G. (2005). The modular architecture of protein-protein binding interfaces. Proc Natl Acad Sci U S A *102*, 57–62.

Reichmann, D., Rahat, O., Cohen, M., Neuvirth, H., and Schreiber, G. (2007). The molecular architecture of protein-protein binding sites. Curr Opin Struct Biol *17*, 67–76.

Rickert, M., Wang, X., Boulanger, M.J., Goriatcheva, N., and Garcia, K.C. (2005). The structure of interleukin-2 complexed with its alpha receptor. Science *308*, 1477–1480.

Robinson, J.A. (2008). beta-Hairpin peptidomimetics: design, structures and biological activities. Acc Chem Res *41*, 1278–1288.

Robinson, J.A., DeMarco, S., Gombert, F., Moehle, K., and Obrecht, D. (2008). The design, structures and therapeutic potential of protein epitope mimetics. Drug Discov Today *13*, 944–951.

Robinson, J.A., Shankaramma, S.C., Jettera, P., Kienzl, U., Schwendener, R.A., Vrijbloed, J.W., et al. (2005). Properties and structure-activity studies of cyclic beta-hairpin peptidomimetics based on the cationic antimicrobial peptide protegrin I. Bioorg Med Chem *13*, 2055–2064.

Rodier, F., Bahadur, R.P., Chakrabarti, P., and Janin, J. (2005). Hydration of protein–protein interfaces. Proteins Struct Funct Bioinformatics *60*, 36–45.

Ruan, F., Chen, Y., and Hopkins, P.B. (1990). Metal ion-enhanced helicity in synthetic peptides containing unnatural, metal-ligating residues. J Am Chem Soc *112*, 9403–9404.

Ruiz, N., Kahne, D., and Silhavy, T.J. (2009). Transport of lipopolysaccharide across the cell envelope: the long road of discovery. Nat Rev Microbiol *7*, 677–683.

Russell, S.J., Blandl, T., Skelton, N.J., and Cochran, A.G. (2002). Stability of cyclic ß-hairpins: asymmetric contributions from side chains of a hydrogen-bonded cross-strand residue pair. J Am Chem Soc *125*, 388–395.

Saraogi, I., Hebda, J.A., Becerril, J., Estroff, L.A., Miranker, A.D., and Hamilton, A.D. (2010). Synthetic α-helix mimetics as agonists and antagonists of islet amyloid polypeptide aggregation. Angew Chem Int Ed *49*, 736–739.

Sattler, M., Liang, H., Nettesheim, D., Meadows, R.P., Harlan, J.E., Eberstadt, M., et al. (1997). Structure of Bcl-xL-Bak peptide complex: recognition between regulators of apoptosis. Science *275*, 983–986.

Schafmeister, C.E., Po, J., and Verdine, G.L. (2000). An all-hydrocarbon cross-linking system for enhancing the helicity and metabolic stability of peptides. J Am Chem Soc *122*, 5891–5892.

Schneider, J.P., and DeGrado, W.F. (1998). The design of efficient α-helical C-capping auxiliaries. J Am Chem Soc *120*, 2764–2767.

Schneider, J.P., and Kelly, J.W. (1995). Templates that induce .alpha.-helical, .beta.-sheet, and loop conformations. Chem Rev *95*, 2169–2187.

Schreiber, G. (2002). Kinetic studies of protein-protein interactions. Curr Opin Struct Biol *12*, 41–47.

Schreiber, G., and Fersht, A. (1993). Interaction of barnase with its polypeptide inhibitor barstar studied by protein engineering. Biochemistry *32*, 5145–5150.

Schreiber, G., and Fersht, A. (1995). Energetics of protein-protein interactions: analysis of the Barnase-Barstar interface by single mutations and double mutant cycles. J Mol Biol *248*, 478–486.

Schreiber, G., and Fersht, A. (1996). Rapid, electrostatically assisted association or proteins. Nat Struct Biol *3*, 427–431.

Seebach, D., Beck, A.K., and Bierbaum, D.J. (2004). The world of β- and γ-peptides comprised of homologated proteinogenic amino acids and other components. Chem Biodivers *1*, 1111–1239.

Seitz, M., Rusert, P., Moehle, K., Trkola, A., and Robinson, J.A. (2010). Peptidomimetic inhibitors targeting the CCR5-binding site on the human immunodeficiency virus type-1 gp120 glycoprotein complexed to CD4. Chem Comm *46*, 7754–7756.

Selzer, T., and Schreiber, G. (1999). Predicting the rate enhancement of protein complex formation from the electrostatic energy of interaction. J Mol Biol *287*, 409–419.

Selzer, T., and Schreiber, G. (2001). New insights into the mechanism of protein-protein association. Proteins *45*, 190–198.

Setny, P., Baron, R., and McCammon, J.A. (2010). How can hydrophobic association be enthalpy driven? J Chem Theory Comput *6*, 2866–2871.

Shankaramma, S.C., Athanassiou, Z., Zerbe, O., Moehle, K., Mouton, C., Bernardini, F., et al. (2002). Macrocyclic hairpin mimetics of the cationic antimicrobial peptide protegrin I: a new family of broad-spectrum antibiotics. ChemBioChem *3*, 1126–1133.

Shankaramma, S.C., Moehle, K., James, S., Vrijbloed, J.W., Obrecht, D., and Robinson, J.A. (2003). A family of macrocyclic antibiotics with a mixed peptide-peptoid ß-hairpin backbone conformation. Chem Comm, 1842–1843.

Shepherd, N.E., Hoang, H.N., Desai, V.S., Letouze, E., Young, P.R., and Fairlie, D.P. (2006). Modular α-helical mimetics with antiviral activity against respiratory syncitial virus. J Am Chem Soc *128*, 13284–13289.

Shuker, S.B., Hajduk, P.J., Meadows, R.P., and Fesik, S.W. (1996). Discovering high-affinity ligands for proteins: SAR by NMR. Science *274*, 1531–1534.

Sia, S.K., Carr, P.A., Cochran, A.G., Malashkevich, V.N., and Kim, P.S. (2002). Short constrained peptides that inhibit HIV-1 entry. Proc Natl Acad Sci U S A *99*, 14664–14669.

Sidhu, S.S., Fairbrother, W.J., and Deshayes, K. (2003). Exploring protein-protein interactions with phage display. ChemBioChem *4*, 14–25.

Sidhu, S.S., Lowman, H.B., Cunningham, B.C., and Wells, J.A. (2000). Phage display for selection of novel binding peptides. In: Methods in Enzymology, Academic Press, San Diego, USA, 2000, *328*, 333–363.

Smith, A.B., Charnley, A.K., and Hirschmann, R. (2011). Pyrrolinone-based peptidomimetics. "Let the enzyme or receptor be the judge." Acc Chem Res *44*, 180–193.

Smith, A.J.T., Zhang, X.Y., Leach, A.G., and Houk, K.N. (2009). Beyond picomolar affinities: quantitative aspects of noncovalent and covalent binding of drugs to proteins. J Med Chem *52*, 225–233.

Sperandeo, P., Dehó, G., and Polissi, A. (2009). The lipopolysaccharide transport system of gram-negative bacteria. Biochim Biophys Acta *1791*, 594–602.

Srinivas, N., Jetter, P., Ueberbacher, B.J., Werneburg, M., Zerbe, K., Steinmann, J., et al. (2010). Peptidomimetic antibiotics target outer-membrane biogenesis in *Pseudomonas aeruginosa*. Science *327*, 1010–1013.

Stephens, O.M., Kim, S., Welch, B.D., Hodsdon, M.E., Kay, M.S., and Schepartz, A. (2005). Inhibiting HIV fusion with a ß-peptide foldamer. J Am Chem Soc *127*, 13126–13127.

Stricher, F., Huang, C.C., Descours, A., Duquesnoy, S., Combes, O., Decker, J.M., et al. (2008). Combinatorial optimization of a CD4-mimetic miniprotein and cocrystal structures with HIV-1 gp120 envelope glycoprotein. J Mol Biol *382*, 510–524.

Stumpf, M.P.H., Thorne, T., de Silva, E., Stewart, R., An, H.J., Lappe, M., et al. (2008). Estimating the size of the human interactome. Proc Natl Acad Sci U S A *105*, 6959–6964.

Tang, C., Iwahara, J., and Clore, G.M. (2006). Visualization of transient encounter complexes in protein-protein association. Nature *444*, 383–386.

Taylor, J.W. (2002). The synthesis and study of side-chain lactam-bridged peptides. Pept Sci *66*, 49–75.

Thanos, C.D., DeLano, W.L., and Wells, J.A. (2006). Hot-spot mimicry of a cytokine receptor by a small molecule. Proc Natl Acad Sci U S A *103*, 15422–15427.

Tilley, J.W., Chen, L., Fry, D.C., Emerson, S.D., Powers, G.D., Biondi, D., et al. (1997). Identification of a small molecule inhibitor of the IL-2/IL-2R receptor interaction which binds to IL-2. J Am Chem Soc *119*, 7589–7590.

Toogood, P.L. (2002). Inhibition of protein-protein association by small molecules: approaches and progress. J Med Chem *45*, 1543–1558.

Tsai, C.-J., Lin, S.L., Wolfson, H., and Nussinov, R. (1996). Protein-protein interfaces: architectures and interactions in protein-protein interfaces and in protein cores. Their similarities and differences. Crit Rev Biochem Mol Biol *31*, 127–152.

Tsai, C.-J., Lin, S.L., Wolfson, H.J., and Nussinov, R. (1997). Studies of protein-protein interfaces: a statistical analysis of the hydrophobic effect. Protein Sci *6*, 53–64.

Vaz, B., and Brunsveld, L. (2008). Stable helical peptoids via covalent side chain to side chain cyclization. Org Biomol Chem *6*, 2988–2994.

Vaz, E., Pomerantz, W.C., Geyer, M., Gellman, S.H., and Brunsveld, L. (2008). Comparison of design strategies for promotion of β-peptide 14-helix stability in water. ChemBioChem *9*, 2254–2259.

Venkatraman, J., Shankaramma, S.C., and Balaram, P. (2001). Design of folded peptides. Chem Revs *101*, 3131–3152.

Vernall, A.J., Cassidy, P., and Alewood, P.F. (2009). A single α-helical turn stabilized by replacement of an internal hydrogen bond with a covalent ethylene bridge. Angew Chem Int Ed *48*, 5675–5678.

Vita, C. (1997). Engineering novel proteins by transfer of active sites to natural scaffolds. Curr Opin Biotech *8*, 429–434.

Vita, C., Drakopoulou, E., Vizzavona, J., Rochette, S., Martin, L., Ménez, A., et al. (1999). Rational engineering of a miniprotein that reproduces the core of the CD4 site interacting with HIV-1 envelope glycoprotein. Proc Natl Acad Sci U S A *96*, 13091–13096.

Vita, C., Roumestand, C., Toma, F., and Ménez, A. (1995). Scorpion toxins as natural scaffolds for protein engineering. Proc Natl Acad Sci U S A *92*, 6404–6408.

Vita, C., Vizzavona, J., Drakopoulou, E., Zinn-Justin, S., Gilquin, B., and Ménez, A. (1998). Novel miniproteins engineered by the transfer of active sites to small natural scaffolds. Pept Sci *47*, 93–100.

Walensky, L.D., Kung, A.L., Escher, I., Malia, T.J., Barbuto, S., Wright, R.D., et al. (2004). Activation of apoptosis in vivo by a hydrocarbon-stapled BH3 helix. Science *305*, 1466–1470.

Walensky, L.D., Pitter, K., Morash, J., Oh, K.J., Barbuto, S., Fisher, J., et al. (2006). A stapled BID BH3 helix directly binds and activates BAX. Mol Cell *24*, 199–210.

Wang, D., Chen, K., Kulp, J.L., III, and Arora, P.S. (2006). Evaluation of biologically relevant short α-helices stabilized by a main-chain hydrigen-bond surrogate. J Am Chem Soc *128*, 9248–9256.

Wang, D., Liao, W., and Arora, P.S. (2005). Enhanced metabolic stability and protein-binding properties of artificial α helices derived from a hydrogen-bond surrogate: application to Bcl-xL. Angew Chem Int Ed *117*, 6683–6687.

Wang, D., Lu, M., and Arora, P.S. (2008). Inhibition of HIV-1 fusion by hydrogen-bond-surrogate-based α helices. Angew Chem Int Ed *47*, 1879–1882.

Wang, L., Berne, B.J., and Friesner, R.A. (2011). Ligand binding to protein-binding pockets with wet and dry regions. Proc Natl Acad Sci U S A *108*, 1326–1330.

Wells, J.A. (1996). Binding in the growth hormone receptor complex. Proc Natl Acad Sci U S A *93*, 1–6.

Wells, J.A., and McClendon, C.L. (2007). Reaching for high-hanging fruit in drug discovery at protein-protein interfaces. Nature *450*, 1001–1009.

Williams, D.H. (2010). Enzyme catalysis from improved packing in their transition-state structures. Curr Opin Chem Biol *14*, 666–670.

Woods, R.J., Brower, J.O., Castellanos, E., Hashemzadeh, M., Khakshoor, O., Russu, W.A., et al. (2007). Cyclic modular ß-sheets. J Am Chem Soc *129*, 2548–2558.

Wrighton, N.C., Farrell, F.X., Chang, R., Kashyap, A.K., Barbone, F.P., Mulcahy, L.S., et al. (1996). Small peptides as potent mimetics of the protein hormone erythropoietin. Science *273*, 458–463.

Wu, B.L., Chien, E.Y.T., Mol, C.D., Fenalti, G., Liu, W., Katritch, V., et al. (2010a). Structures of the CXCR4 chemokine GPCR with small-molecule and cyclic peptide antagonists. Science *330*, 1066–1071.

Wu, L., McElheny, D., Takekiyo, T., and Keiderling, T.A. (2010b). Geometry and efficacy of cross-strand Trp/Trp, Trp/Tyr, and Tyr/Tyr aromatic interaction in a ß-hairpin peptide. Biochemistry *49*, 4705–4714.

Zhang, X., and Houk, K.N. (2005). Why enzymes are proficient catalysts: beyond the pauling paradigm. Acc Chem Res *38*, 379–385.

Zinn-Justin, S., Guenneugues, M., Drakopoulou, E., Gilquin, B., Vita, C., and Ménez, A. (1996). Transfer of a ß-hairpin from the functional site of snake curaremimetic toxins to the scaffold of scorpion toxins: three-dimensional solution structure of the chimeric protein. Biochemistry *35*, 8535–8543.

18 The structural biology of α_1-antitrypsin deficiency and the serpinopathies

Bibek Gooptu and David A. Lomas

The serpinopathies result from naturally occurring point mutations in members of the serine protease inhibitor or serpin superfamily. These mutations cause a delay in folding with unstable intermediates being cleared by endoplasmic-reticulum-associated degradation (ERAD). The remaining protein is either fully folded and secreted or retained as ordered polymers within the endoplasmic reticulum of the cell of synthesis. We describe here the use of a range of different techniques to characterize the misfolding of mutants of the serpins in order to define the structure of the pathogenic polymer.

18.1 Clinical phenotypes of the serpinopathies

18.1.1 α_1-Antitrypin deficiency

α_1-Antitrypsin is the most abundant circulating protease inhibitor. It is synthesized by the liver and is present in the plasma at a concentration of 1.5–3.5 g/L. It functions primarily as an inhibitor of the enzyme neutrophil elastase. Most individuals are homozygous for the M allele with the commonest deficiency alleles being the severe Z (Glu342Lys) and the mild S (Glu264Val) variants. The Z mutation causes the retention of α_1-antitrypsin within hepatocytes as polymers that form periodic acid Schiff (PAS)–positive, diastase-resistant inclusions. These inclusions are associated with neonatal hepatitis, cirrhosis and hepatocellular carcinoma (Eriksson et al. 1986; Sveger 1976). Only 10%–15% of Z α_1-antitrypsin is folded and released into the circulation. This leaves the lungs exposed to enzymatic damage by neutrophil elastase and so predisposes the Z homozygote to early onset panlobular emphysema. The S allele results in less retention of protein within hepatocytes with plasma levels being 60% of those of the M allele. This does not result in any clinical sequelae. Other rare alleles have been described that also cause hepatic inclusions with severe plasma deficiency: the Siiyama-Ser53Phe (Lomas et al. 1993b), Mmalton-Δ52Phe (Lomas et al. 1995) and King's-His334Asp (Miranda et al. 2010) alleles. There are also null alleles that result in a complete absence of protein.

18.1.2 Familial encephalopathy with neuroserpin inclusion bodies (FENIB)

The process of disease related polymerization is most striking for mutations in the neurone-specific serpin neuroserpin. This protein is expressed during the late stage of development in neurons of the central and peripheral nervous system and in the adult brain (Hastings et al. 1997). The target protease of neuroserpin is tissue plasminogen activator (tPA), and thus it is likely to be important in the control of synaptic plasticity and in learning and memory (Yepes and Lawrence 2004). Mutations in neuroserpin result in the autosomal dominant inclusion body dementia, FENIB (Bradshaw et al. 2001; Davis et al. 1999a, 1999b; Miranda and Lomas 2006). This is characterized by eosinophilic

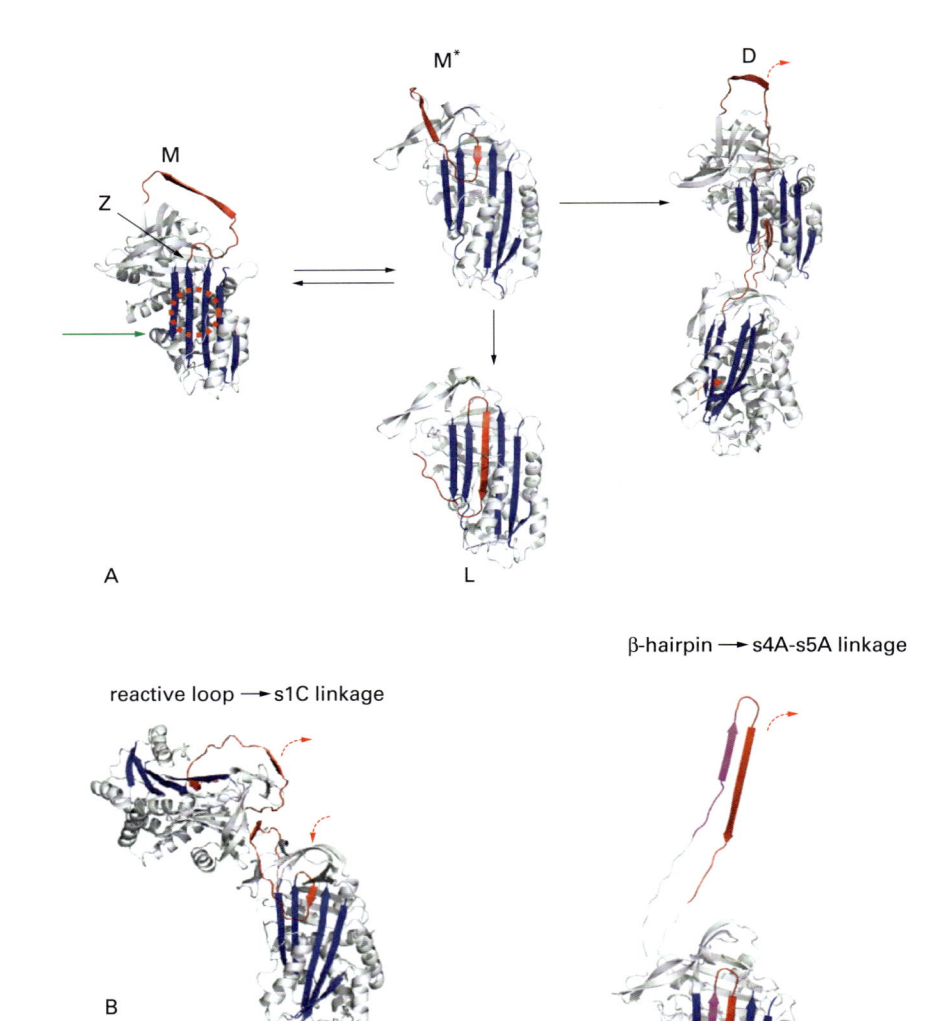

Figure 18.1: (A) The classic pathway of serpin polymerization (left). This is best described for Z α$_1$-antitrypsin and mutants of neuroserpin. The structure of a typical serpin such as

(*Continued*)

Figure 18.1: (*Continued*)

α1-antitrypsin (M) is centered on β-sheet A (blue) and the mobile reactive center loop (red) (Elliott et al. 2000). The Z mutation at the base of the reactive loop (arrowed) and mutations in the shutter domain (red circle) result in the formation of an unstable intermediate (M*) that has an open β-sheet A (Dafforn et al. 1999; Gooptu et al. 2000; Mahadeva et al. 2002). This patent β-sheet A can either accept the loop of another molecule to form a dimer (D), which then extends into polymers, or its own loop to form the latent conformation (L) (Im et al. 2002). Alternative linkages have been proposed between the reactive center loop and s1C (B) (Carrell et al. 1994; Zhang et al. 2008), s7A (C) (McGowan et al. 2006; Sharp et al. 1999) and recently between a β hairpin of the reactive center loop (red) and s5A (magenta) and β-sheet A (D) (Yamasaki et al. 2008). The red arrows show the linkage of the reactive center loop and the green arrow the I helix.

neuronal inclusions of neuroserpin (Collins bodies) in the deeper layers of the cerebral cortex and the substantia nigra. The inclusions are PAS positive and diastase resistant and bear a striking resemblance to those of Z α_1-antitrypsin that form within the liver.

There is a direct relationship between the magnitude of the intracellular accumulation of neuroserpin and the severity of FENIB (Davis et al. 2002). Ser49Pro neuroserpin (neuroserpin Syracuse) is associated with diffuse small intraneuronal inclusions of neuroserpin with an onset of dementia between the ages of 45 and 60 years (Bradshaw et al. 2001; Davis et al. 1999a, 1999b). The conformationally more severe Ser52Arg neuroserpin (neuroserpin Portland) results in larger inclusions and an onset of dementia in early adulthood, while His338Arg neuroserpin is associated with even more inclusions and an onset of dementia in adolescence. The most striking examples are in the most severe mutations of neuroserpin, Gly392Glu and Gly392Arg. These are associated with an even earlier onset of disease with the Gly392Arg mutation causing a profound intellectual decline in an 8-year-old girl, seizures and electrical brain activity in keeping with "epilepsy of slow-wave sleep (ESES)" (Coutelier et al. 2008; Gooptu and Lomas 2009).

18.1.3 Other serpinopathies

The phenomenon of serpin polymerization has been reported in mutants of the plasma proteins C1-inhibitor (Aulak et al. 1993; Eldering et al. 1995), antithrombin (Bruce et al. 1994; Picard et al. 2003) and α_1-antichymotrypsin (Faber et al. 1993; Gooptu et al. 2000; Poller et al. 1993). These mutants are retained in the liver with the lack of circulating protein resulting in uncontrolled activity of proteolytic cascades and hence angio-oedema, thrombosis and chronic obstructive pulmonary disease, respectively (Gooptu and Lomas 2009; Stein and Carrell 1995). A mutation in heparin cofactor II (Glu428Lys) has also been associated with plasma deficiency but as yet this has not been shown to cause disease (Corral et al. 2004). This mutation is of particular interest as it is the same as the Z allele that causes polymerization and deficiency of α_1-antitrypsin.

18.2 The serpin mechanism of protease inhibition

Members of the serine protease inhibitors or serpin superfamily are found in all branches of life and play an important role in the regulation of enzymes involved in proteolytic cascades. The superfamily is characterized by more than 25% identity with the archetypal serpin α_1-antitrypsin, and conservation of tertiary structure. Serpins adopt a metastable conformation composed in most cases of 9 α helices, three β sheets (A to C) and an exposed mobile reactive center loop (RCL). This flexible RCL typically contains 20 residues that act as a pseudosubstrate for the target protease (Elliott et al. 1996; Ryu et al. 1996). After formation of a Michaelis complex, the enzyme cleaves the P1-P1' bond of the serpin, releasing the P1' residue and almost forming an ester bond between the protease and the serpin. This is then followed by a dramatic conformational transition from a "stressed to relaxed" (S→R) conformation that flips the enzyme from the upper to the lower pole of the serpin as the reactive loop inserts as an extra strand in β-sheet A (Dementiev et al. 2006; Huntington et al. 2000). The inactivated enzyme is then cleared from the circulation or site of inflammation. It is this mechanism that is subverted by mutations of the serpins to cause polymerization and disease.

18.3 Folding, misfolding and polymerization

The observation that FENIB was associated with mutations Ser49Pro and His338Arg in the neuroserpin gene that are homologous to those in α_1-antitrypsin that cause liver disease (Ser53Phe and His334Asp, respectively) (Lomas et al. 1993b; Miranda et al. 2010) strongly indicated a common molecular mechanism. This was confirmed by the finding that the neuronal inclusion bodies of FENIB were formed by entangled polymers of neuroserpin with identical morphology to the polymers of mutant α_1-antitrypsin present in hepatocytes from a child with α_1-antitrypsin deficiency related cirrhosis (Davis et al. 1999b). Moreover mutants of neuroserpin that cause FENIB have greatly accelerated rates of polymerization when compared to the wild-type protein when assessed at the protein level (Belorgey et al. 2002) or within cell and *Drosophila* models of disease (Belorgey et al. 2004; Miranda et al. 2004, 2008). It is clearly important to establish the mechanism by which the monomeric protein is linked to form the chain of polymers.

18.3.1 Polymer linkage via β-sheet A

Biochemical, biophysical and crystallographic studies have been used to dissect the molecular basis by which monomeric Z α_1-antitrypsin forms polymers. The Glu342Lys mutation associated with the Z allele is located at the head of strand 5 of β-sheet A and the base of the mobile reactive loop (▶Figure 18.1). This mutation causes a conformational transition and the formation of an unstable intermediate termed M* (Dafforn et al. 1999; James and Bottomley 1998) in which β-sheet A opens (Dafforn et al. 1999; James and Bottomley 1998; Lomas et al. 1992; Purkayastha et al. 2005; Sivasothy et al. 2000) and the upper part of helix F unwinds (Baek et al. 2007; Cabrita et al. 2004; Gooptu et al. 2000, 2009). The patent β-sheet A can then accept the loop of another molecule to form a loop-sheet dimer, which extends to form longer chains of loop-sheet polymers (Dafforn et al. 1999; James and Bottomley 1998; Lomas et al. 1992; Purkayastha et al. 2005; Sivasothy et al. 2000) (▶Figure 18.2). There is evidence that it is the dimer that initiates and propagates polymerization of the serpins (Chiou et al. 2009; Zhou and Carrell 2008).

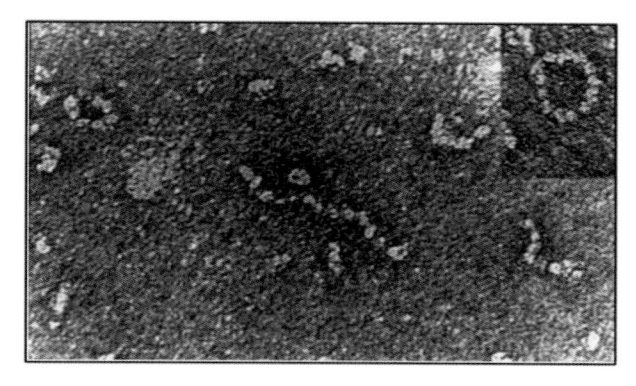

Figure 18.2: The polymers have a "beads on a string appearance" on electron microscopy. Reproduced from Lomas et al. (1993b).

Although many α_1-antitrypsin deficiency variants have been described, only three other mutants of α_1-antitrypsin have similarly been associated with profound plasma deficiency and hepatic inclusions: α_1-antitrypsin Siiyama (Ser53Phe) (Lomas et al. 1993b), Mmalton (ΔPhe52) (Lomas et al. 1995) and King's (His334Asp) (Miranda et al. 2010). All of these mutations are within the "shutter" region that underlies the center of β-sheet A and regulates its expansion. They therefore directly destabilise the 5 stranded β-sheet A that characterizes the native conformer (▶Figure 18.1). This predisposes to the accumulation of intermediates that have an expanded β-sheet A, both during and subsequent to folding, and hence favors polymerization (Kang et al. 1997; Kim et al. 1995; Yu et al. 1995) (Miranda et al. 2010).

18.3.2 Alternative pathways to polymerization

Polymers can be induced to form in vitro under a variety of conditions, for example, by heating or incubation with denaturants (urea or guanidine) or at low pH (Dafforn et al. 1999; Devlin et al. 2002; Ekeowa et al. 2010). Moreover, a number of intermolecular linkage mechanisms have been described crystallographically, for example, between the reactive center loop and strand 1C (Bottomley et al. 1998; Bruce et al. 1994; Carrell et al. 1994; Chang et al. 1994; Eldering et al. 1995; Lomas et al. 1995; Zhang et al. 2008) and between the reactive center loop and strand 7A (McGowan et al. 2006; Sharp et al. 1999; Zhou et al. 2001). However, such linkages between native monomers are not stable in aqueous solution and so are unlikely to be the mechanism by which pathological polymers assemble. A more recent suggestion is that polymers are formed by a larger domain swap involving exchange of the reactive center loop and strand 5A. This is based on the crystal structure of a dimer of antithrombin (Yamasaki et al. 2008). The authors argue that polymers form while the protein is folding rather than from protein with a near-native structure. Indeed there is strong evidence that pathological mutations in both α_1-antitrypsin and neuroserpin are associated with a significant delay in folding (Baek et al. 2007; Takehara et al. 2010).

18.4 Serpin folding

The folding of serpins can be summarized as follows:

$U \Leftrightarrow I \Leftrightarrow N$

where U is the unfolded ensemble; I represents a transiently stable, partially folded ensemble and N represents the natively folded protein (Whisstock and Bottomley 2006).

The current data suggest that I is a single intermediate ensemble through which all serpins pass to attain a native conformation. This species has approximately 80% of the native secondary structure with partially formed β-sheets A and C, a well-formed β-sheet B and a nonnative helix F (Whisstock and Bottomley 2006). The significance of this work is underscored by the findings of Wu and colleagues who showed that the Z (Glu342Lys) and Siiyama (Ser53Phe) mutations of $α_1$-antitrypsin result in a long lived folding intermediate (Kang et al. 1997; Kim et al. 1995; Yu et al. 1995). Thus, the polymers may be the result of aggregation of misfolded intermediates and therefore have a different structure from that proposed in the classical pathway. However, mutants that favor polymer formation also predispose to the latent conformer (Onda et al. 2005) suggesting that latency and polymerization compete from a common unstable intermediate (▶ Figure 18.1).

Similar data were observed in studies that assessed the refolding of neuroserpin (Takehara et al. 2010). The protein was first denatured in 5 M guanidine-HCL and then refolded by 30-fold dilution in a buffer without guanidine-HCL. In this study, an initial refolding intermediate I_{IN} was reported for wild type, Ser49Pro and the severe His338Arg mutant of neuroserpin, followed by a late intermediate I_R. This intermediate could then form either the native protein or polymers.

18.5 Dissecting the pathways of polymerization

The two different pathways to polymerization can be simplified by the question of whether mutants of $α_1$-antitrypsin and other serpins form polymers while folding or from a near-native conformation. There are theoretical and experimental inconsistencies with both pathways. The β-sheet A linkage has proved difficult to model (Huntington and Whisstock 2010) while polymer formation while the protein is folding is difficult to reconcile with the spontaneous polymerization of folded mutant serpins under physiological conditions (Lomas et al. 1992), the presence of polymers in tissues such as the skin from which they are not secreted (Gross et al. 2009) and in vivo polymer formation from wild-type protein (Inagi et al. 2005). It is clearly critical to define the intermolecular linkage in order to develop strategies to block serpin polymerization and so treat the serpinopathies. This has been addressed using a range of strategies.

18.5.1 Limited proteolysis and hydrogen/deuterium exchange

The linkage in the serpin polymer has been assessed by limited proteolysis aimed at assessing the availability of cleavage sites in helix I. Helix I must unravel if linear and flexible polymers are formed by a reactive center loop/strand 5A-β-sheet A linkage (Yamasaki et al. 2008), while helix I plays no role in polymers formed by a reactive center loop-β-sheet A interaction (▶Figure 18.1). Previous data had suggested the exposure of cryptic sites in the I helix in polymers of wild-type M α_1-antitrypsin in support for the β-hairpin model of polymerization (Yamasaki et al. 2008). However endopeptidase cleavage data with the enzymes Lys-C and Asp-N showed that the I helix remained entirely protected from digestion in monomeric Z α_1-antitrypsin, in polymers of Z α_1-antitrypsin formed under physiological conditions in vitro and in polymers isolated from the endoplasmic reticulum of hepatocytes of a Z α_1-antitrypsin homozygote (Ekeowa et al. 2010). These data are consistent with hydrogen/deuterium exchange studies that showed no change in the I helix of α_1-antitrypsin upon formation of the polymerogenic intermediate or chains of polymers (Tsutsui et al. 2008).

18.5.2 Monoclonal antibodies

The conformation of the pathological polymer was also addressed using our monoclonal antibody that specifically recognises polymers of α_1-antitrypsin (Ekeowa et al. 2010). Polymers formed from folded protein can be created by heating the monomeric serpin. Polymers formed while folding can be prepared by treating the folded protein with urea or guanidine. In the case of α_1-antitrypsin, the polymers that form as a consequence of treatment with denaturants or at low pH are not recognized by the 2C1 monoclonal antibody that recognizes the pathological polymers that form in vivo (Ekeowa et al. 2010) (▶Figure 18.3). This implies that the folding intermediate does not give rise to the pathological polymer. An explanation is that this intermediate is efficiently degraded by the proteasome via the pathway of ERAD (Belorgey et al. 2011) (▶Figure 18.4). Moreover, experiments performed in vitro with guanidine or urea produce conformers that are unlikely to be relevant to disease (Yamasaki et al. 2008).

It is clear that the quality-control pathway is able to fold some of the material to a near native conformation. Indeed, for many serpin mutants, the protein is trafficked through the Golgi apparatus and then secreted. However, a proportion of the secreted material forms polymers. This is more in keeping with a reactive center loop/β-sheet A linkage (▶Figure 18.1) (Belorgey et al. 2002; Gooptu and Lomas 2009) with the pathologically relevant conformational transitions being recapitulated by heating the folded monomeric protein (Ekeowa et al. 2010). The resulting polymers of α_1-antitrypsin are recognized by the monoclonal antibody that recognizes the pathological polymers that form in vivo.

18.5.3 Mass spectrometry

Ion mobility–mass spectrometry (IM-MS) provides multidimensional separation of gas phase protein ions and has only recently been applied to large noncovalent protein complexes (Ruotolo et al. 2008). The drift times of protein ions in a gas-filled drift tube under the influence of a weak electric field can be converted into collision cross-sections that depend directly on their size and geometry (Smith et al. 2009). More compact ions transverse the drift tube faster than elongated ions for a given mass and charge state, thus allowing a measure of the overall topology and conformation of a polymer ion in the gas phase and from these values the intermolecular distances for the components of the polymer chain can be estimated (Smith et al. 2009). We used this method successfully to assess different species of α_1-antitrypsin formed during polymerization (Ekeowa et al. 2010).

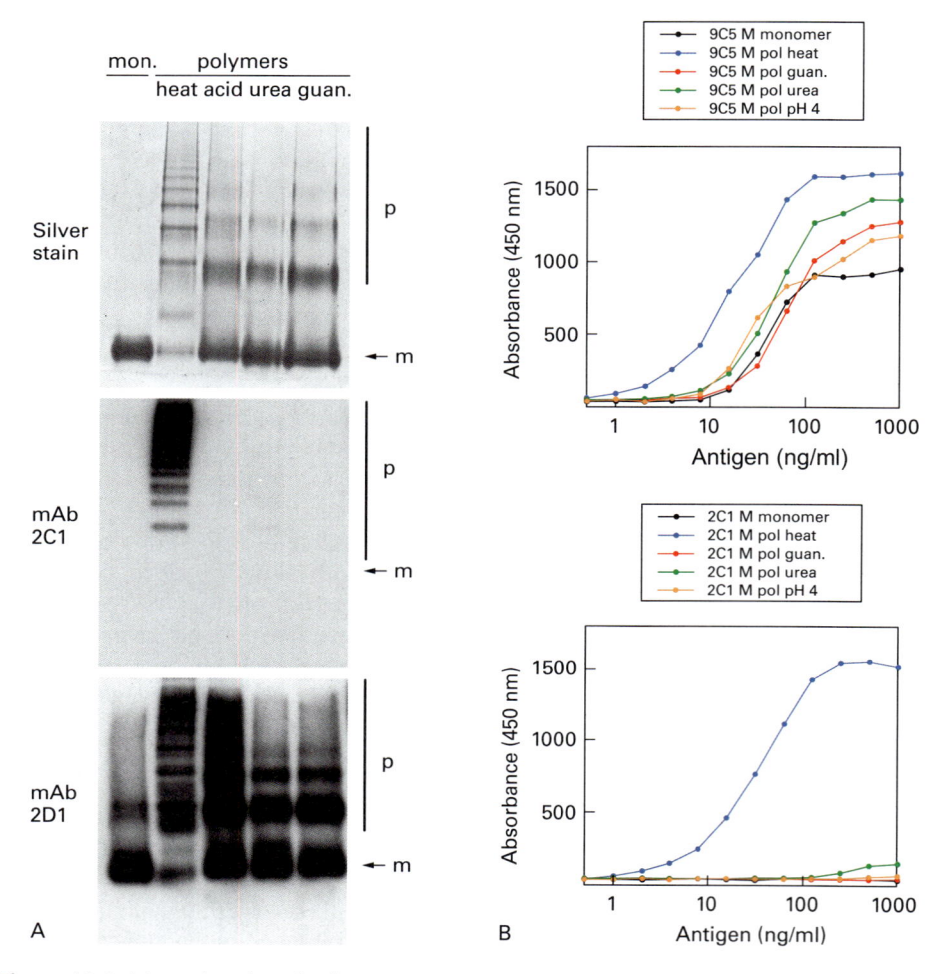

Figure 18.3: Monoclonal antibody 2C1 recognizes polymers of α_1-antitrypsin prepared by heating but not by other denaturing conditions. (A) 7.5% (w/v) nondenaturing PAGE to show polymer analysed by silver stain (upper panel) or Western blot, first with mAb 2C1 (central panel) and then with mAb 2D1 that recognizes all conformers of α_1-antitrypsin (lower panel). mAb 2C1 recognized heat-induced polymers but gave no signal for polymers prepared by treating α_1-antitrypsin at low pH, or with 1–3 M guanidine or 1–4 M urea. The monomer (m) and polymer (p) are indicated. (B) The same polymers assayed in sandwich ELISA, using either mAb 9C5 (upper graph) or 2C1 (lower graph) as the detecting antibodies. mAb 9C5 detected all species with high affinity but mAb 2C1 detected only heat-induced polymers. Reproduced from Ekeowa et al. (2010).

A distribution of peaks from 12+ to 15+ was observed for monomeric M α_1-antitrypsin (▶Figure 18.5). The splitting of the peaks observed for each charge state is due to the microheterogeneity at the N terminus of α_1-antitrypsin (Lomas et al. 1993a). There is no crystal structure of the pathological α_1-antitrypsin polymer; however, there are structures of polymers linked by a cleaved RCL (Dunstone et al. 2000; Huntington et al.

1999) (▶Figure 18.6). This provides a structural anchor for the IM-MS data. The reactive loop cleaved spectrum contains four charge state distributions with series of peaks at 2,250–3,000 m/z centring on the 9+ ion corresponding to the enzyme glycylendopeptidase (23 kDa) that was used to cleave the reactive center loop. Residual unpolymerised cleaved monomer has a similar charge distribution to the untreated monomer centerd on the 13+ ion. The distribution of peaks around 5,250 *m/z* corresponds to the cleaved polymer dimer (100 kDa) and that around 6,250 *m/z* to the trimer (150 kDa).

Polymers of M or Z α_1-antitrypsin were formed by heating the monomeric protein at 60°C in 200 mM ammonium acetate pH 6.5. The heat-induced polymer spectrum has five distinct charge state distributions corresponding to five different species. The charge state distributions were comparable to those found for the cleaved polymer ladder except for the unique series of peaks at 2,775–3,250 *m/z*. These low intensity charge states (15+ to 18+) correspond to the mass of the monomer but have increased charge relative to the monomer distribution that is observed for all three samples. These peaks are consistent with the intermediate species. Similar data were obtained by real time mass spectral analysis of M and Z α_1-antitrypsin polymerization within temperature regulated electrospray capillaries. The polymerogenic intermediate of Z α_1-antitrypsin predictably formed at an earlier time point than that of M α_1-antitrypsin and its appearance was coincident with the formation of the higher order polymers.

18.5.3.1 Structural analysis of the polymer by ion mobility spectrometry

▶Figure 18.5B shows the ion mobility spectra and corresponding collision cross-sections for the monomer and both reactive loop cleaved and heat-induced polymers. The monomeric intermediate, whose collision cross-section is significantly larger than those observed for the other monomers, was again observed for the heat-induced polymers. The collision cross-section plot (▶Figure 18.5C) showed a striking resemblance to the nondenaturing PAGE of heat-induced α1-antitrypsin polymers, with the intermediate species eluting between the monomer and the dimer.

The theoretical collision cross-sections for the α_1-antitrypsin monomer 1QLP (Elliott et al. 2000) and the cleaved dimer1QMB (Huntington et al. 1999) were calculated for comparison with the experimental data. The model for the β-hairpin dimer of α_1-antitrypsin was generated using PyMOL and the domain swapped antithrombin dimer 2ZNH (Yamasaki et al. 2008) as a template. The disparity between the experimental collision cross-sections (▶Figure 18.6A), and the calculated collision cross-sections (▶Figure 18.6B) is due to the three N-linked oligosaccharides that are not seen in the X-ray structures of α_1-antitrypsin. However, a measure of the relative intermolecular arrangement and distance was obtained by the calculation of the percentage increase in collision cross-section in the dimer relative to the monomer. The experimental data show that both the heat-induced and glycylendopeptidase cleaved dimers display a similar increase in collision cross-section when compared with the monomer (▶Figure 18.6C). This is indicative of comparable intermolecular separation. The percentage increase in the theoretical collision cross-section for the cleaved dimer crystal structure is in good agreement with the experimental percentage increase in collision cross-section for the glycylendopeptidase cleaved dimer. Significantly, the loop-sheet dimer model rather than the β-hairpin dimer model gave values consistent with the experimental increase in collision cross-section.

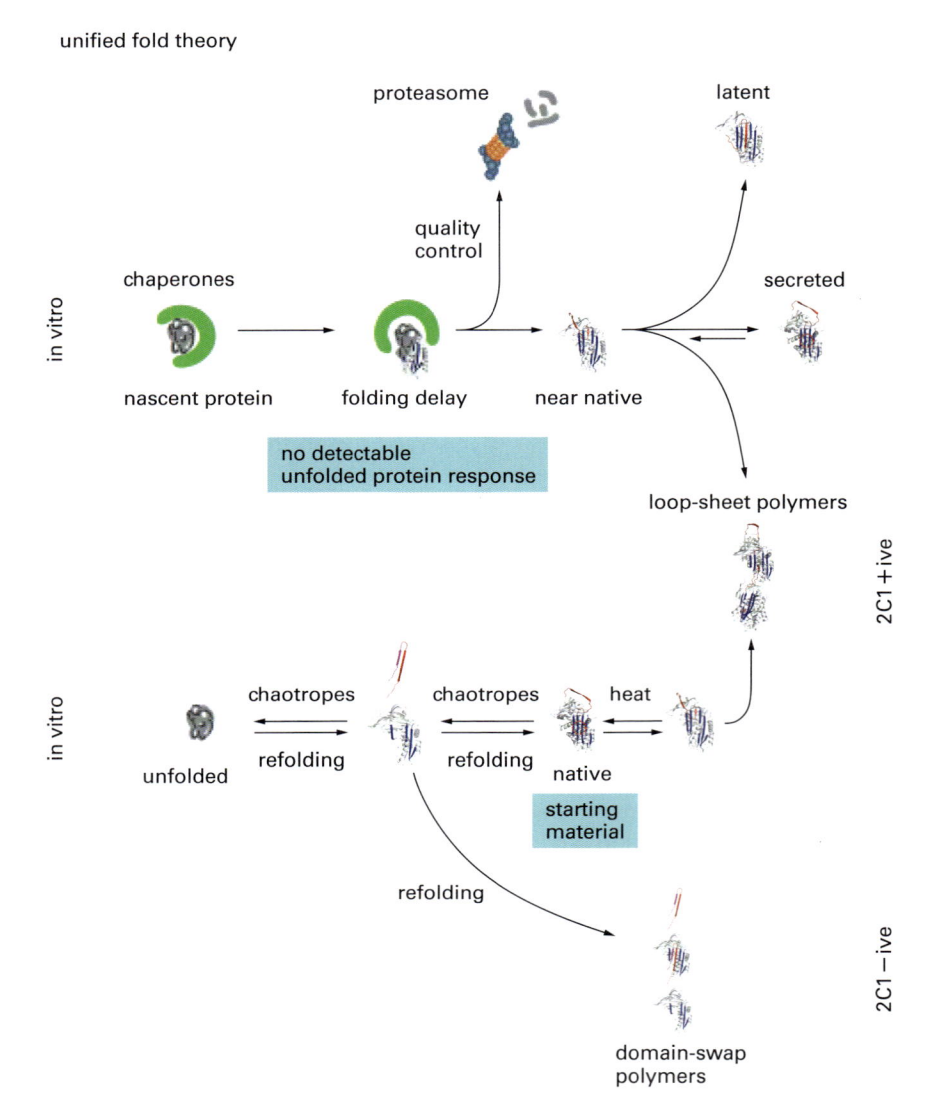

334 | 18 The structural biology of α1-antitrypsin deficiency and the serpinopathies

Figure 18.4: The "unified fold pathway," a unifying hypothesis to explain the cellular handling of mutants of neuroserpin and α_1-antitrypsin. The nascent protein binds to chaperones while in the folding pathway. The mutants cause a significant delay to folding (Baek et al. 2007; Takehara et al. 2010) causing much of the newly synthesized protein to be "timed out" (Wu et al. 2003) and targeted for ERAD via the proteasome (Kröger et al. 2009). Efficient chaperone function and ERAD prevent accumulation of incompletely folded protein and a failure of ER homeostasis; thus no unfolded protein response is detected (Davies et al. 2009). A proportion of the protein is folded by the chaperone machinery to a near-native conformation. Some of this will fold fully and be trafficked through the Golgi apparatus for secretion. However the near native structure can form either the monomeric latent conformer or loop-sheet polymers that are linked by the reactive loop of one molecule inserting

(Continued)

Figure 18.4: (*Continued*)

into β-sheet A of another. These polymers are retained within the endoplasmic reticulum and, when formed by mutants of α_1-antitrypsin, can be stained with the 2C1 monoclonal antibody (Miranda et al. 2010). The in vivo pathway can be recapitulated by in vitro experiments. Monomeric proteins can be unfolded in urea and guanidine and then refolding monitored over time (Takehara et al. 2010). The folding delay is followed by the formation of polymers that are not recognized by the 2C1 antibody (Ekeowa et al. 2010). Thus these polymers are likely to be folding artefacts that are not found in vivo. Heating the monomeric protein produces polymers that bind the 2C1 antibody and so recapitulate the epitope associated with disease. Reproduced from Belorgey et al. (2011).

18.5.4 Two-color coincidence detection (TCCD)

Other biophysical techniques can be used to assess the different conformers of a serpin and the pathways by which they are formed. TCCD, a single molecule fluorescence method, has been used to study the early stages of neuroserpinpolymerization (Chiou et al. 2009). In this method, two distinct fluorophores are excited with two lasers, which permit the detection and identification of associated molecules and oligomers. These data show that neuroserpin polymerization proceeds first by the unimolecular formation of an active monomer, followed by competing processes of polymerization and formation of a latent monomer from the activated species. This can be described using Equation A in Scheme 18.1. Equation B shows a mechanism in which the formation of dimers requires that the two monomers are activated to form M* and that the addition of monomer to the oligomer results exclusively by the addition of M*. This mechanism provided a good global fit with the experimental data. However, because the formation of dimers requires two molecules of activated monomer, this reaction needs to be very fast to explain the appearance of dimers within a few minutes of the start of the reaction. Indeed, the fit gave a value for k_1 (dimer formation) of 2×10^5 M^{-1} s^{-1}, which is close to the diffusion limited reaction rate constant for a 45 kDa protein. Moreover, as the orientation of the molecules is an important factor in dimer formation (because the loop of one monomer is inserted into β-sheet A of a second monomer), it seems very unlikely that such a rapid reaction takes place.

In the β-hairpin domain swap model, polymers form during protein folding and the intermediate is characterized by an exposed β hairpin composed of strands s4A and s5A (Yamasaki et al. 2008). As polymers also form from the folded protein, then in this case each monomer must be activated before addition to the polymer chain. Once again, this is unlikely given the rate of the reaction. Both strands s4A and s5A must be incorporated into β-sheet A in the latent species (▶Figure 18.1A), and thus the latent protein is more likely to be formed by a competing pathway from the unactivated monomer (Equation C in Scheme 18.1). The TCCD findings demonstrate that polymerization proceeds by the unimolecular formation of an active monomer, followed by competing processes of polymerization and formation of a latent monomer from the activated species. This is incompatible with Equation C and the proposed domain swap model of polymer formation.

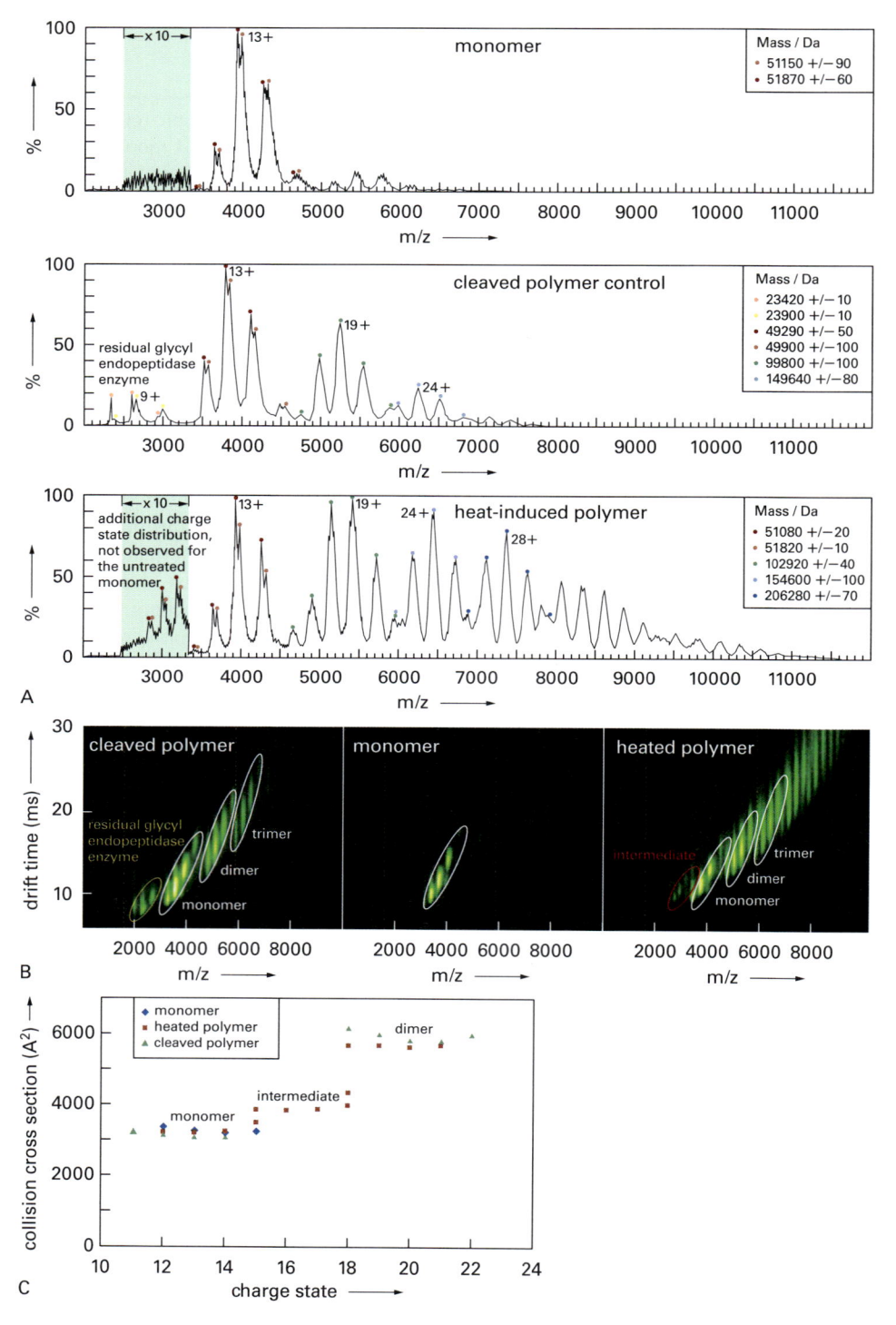

Figure 18.5: α_1-Antitrypsin polymers studied by IM-MS. (A) Mass spectra of M α1-antitrypsin monomer, the heat-induced polymer and P9-P10 reactive loop cleaved polymer. The mass

(Continued)

Figure 18.5: (*Continued*)

spectrum of the monomer has a charge state distribution centered on the 13+ ion (upper spectrum). The mass spectrum of the reactive loop cleaved polymer has four charge state distributions for the reactive loop cleaved polymer (middle spectrum). The peaks at low m/z are attributed to residual enzyme and the series of peaks at 4,000 and 5,250 and 6,250 m/z represent the cleaved monomer, dimer and trimer, respectively. Similar spectra were observed for the heat-induced sample (lower spectrum). However, this spectrum exhibits a second monomer population with higher charge states (15+ to 18+), which are consistent with the polymerogenic intermediate. (B) Ion mobility drift times observed for the M α_1-antitrypsin monomer, the heat-induced polymer and reactive loop cleaved polymer (corresponding mass spectra shown in A). The monomer and polymer species are labelled and highlighted with white ellipses, the intermediate with a red ellipse and the cleavage enzyme with a yellow ellipse. (C) Drift times from (B) converted into a collision cross-sectional plot. The monomer collision cross-sections for all the samples are almost identical, and similarly, the collision cross-sections for the cleaved polymer and heat-induced polymer dimers also overlap. The intermediate in the heat-induced polymer sample shows an expanded collision cross-section relative to the monomer. Reproduced from Ekeowa et al. (2010).

18.6 Cellular processing of polymers

The cellular handling of mutant serpins has been assessed by developing cell lines that conditionally express the wild-type and mutant proteins. Typically the accumulation of proteins within the endoplasmic reticulum of cells activates the PKR-like ER kinase (PERK), the inositol-requiring kinase 1 (IRE1) and the activating transcription factor 6 (ATF6) limbs of the unfolded protein response (UPR). The UPR serves to attenuate protein translation, increase levels of ER chaperones and to enhance the degradation of misfolded ER proteins (a process termed ERAD). Surprisingly, the polymerogenic mutants both α_1-antitrypsin and neuroserpin are predominantly degraded by ERAD but without activating the UPR (Davies et al. 2009; Graham et al. 1990; Hidvegi et al. 2005; Kröger et al. 2009). This striking property results from the ordered structure of serpin polymers and is likely to reflect the near-native characteristics of the intermediate. This contrasts with the exposed hydrophobic β hairpin of the intermediate for the model in ▶Figure 18.1D which would predictably activate the unfolded protein response.

The development of monoclonal antibodies to polymers of neuroserpin allow this species to be detected in cell models of disease. Polymers of neuroserpin were identified by specific immunoprecipitation following metabolic labelling for 30 minutes. They were detected following 3 hours of chase; this suggests that the full-length polypeptide chains are first synthesized and only then are polymers assembled from preexisting neuroserpin molecules. This still does not preclude the formation of polymers from incompletely folded subunits, but it does indicate that monomers require at least 30 minutes of ER residence before being incorporated into polymers (Kröger et al. 2009).

Finally, it is possible that a mixture of different polymer types form in vivo and that antipolymer monoclonal antibodies only detect a portion of them in ELISA, Western blot analysis and immunocytochemistry. This possibility was assessed by using the anti-α_1-antitrypsin monoclonal antibody (2C1) to immunodeplete polymers formed by His334Asp α_1-antitrypsin. COS-7 cells were transiently transfected with either M or

A

polymer species	average experimental CCS (Å²)	% dimer / monomer
M monomer	3252 ± 228	–
Z monomer	3348 ± 234	–
M intermediate	3857 ± 270	118 ± 17
heat-induced dimer	5646 ± 395	174 ± 24
cleaved dimer	5943 ± 416	183 ± 26

B

model structure	calculated CCS (Å²)	% dimer / monomer
1QLP monomer	2675	–
linear loop-sheet dimer	4705	176
flexed loop-sheet dimer	4705	173
β-hairpin dimer	6014	225
1QMB cleaved dimer	4650	174

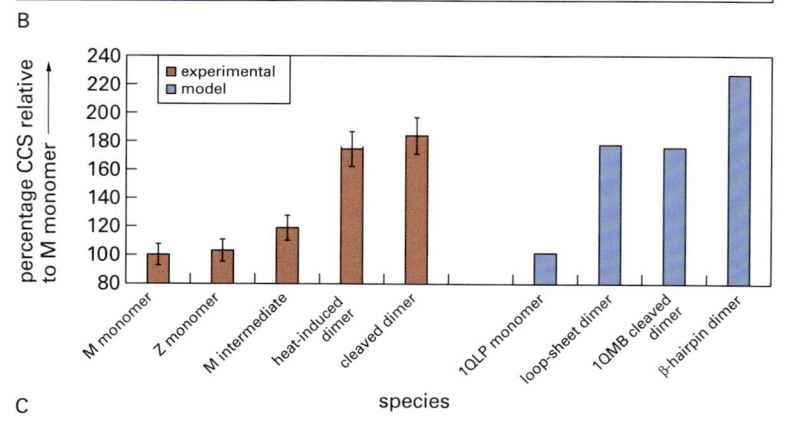

C

Figure 18.6: Comparison of experimental and calculated collision cross-sections for models of α₁antitrypsin polymers. (A) Experimental collision cross-sections for α₁antitrypsin

(*Continued*)

Figure 18.6: (*Continued*)

monomers, the heat-induced and reactive loop cleaved M α_1-antitrypsin dimers. The dimer collision cross-sections are indistinguishable. (B) Table of calculated collision cross-section measurements for the X-ray crystallographic structures of monomeric α_1-antitrypsin (1QLP), the cleaved dimer (1QMB) and the 3D dimer models for the loop-sheet and β-hairpin pathways. The flexed loop-sheet, linear loop-sheet dimers and 1QMB cleaved dimers collision cross-sections are very similar, whereas the β-hairpin dimer has a significantly larger collision cross-section. (C) Plot comparing the percentage increases in collision cross-section upon dimer formation for experimental (red) and calculated (blue) data. The relationship between the experimental heat-induced and reactive loop cleaved dimer collision cross-sections are consistent with a loop-sheet linkage and the crystallographic structure of cleaved dimers (1QMB). The calculated collision cross-section of the β-hairpin dimer model is significantly larger than that observed experimentally for dimers formed by heating α_1-antitrypsin. Reproduced from Ekeowa et al. (2010).

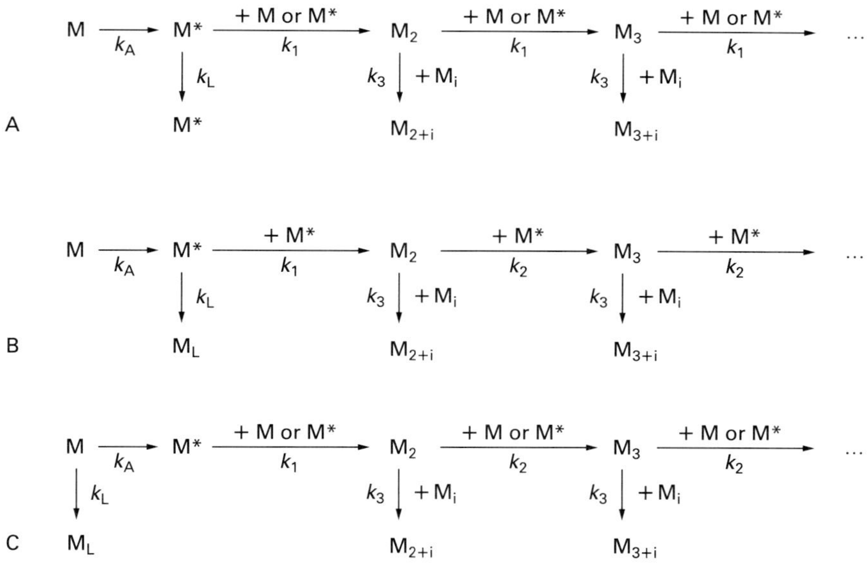

Scheme 18.1 Neuroserpin polymerization kinetic schemes. (A) Neuroserpin monomer M is activated to M* with an activation rate k_A; depending on the model, either a normal (M) or an active monomer (M*) can undergo a transformation into the latent form (M_L) or form a dimer M_2 by adding either M or M* that we assume to occur at the same rate k_1; the polymers can grow by addition of a monomer (M or M*) with a rate k_2, or by addition of an oligomer of size i (M_i) with any other oligomer of size j to form an oligomer M_{i+j} at a rate k_3 that we assume to be the same for whatever value of i and j. In (B) the dimer formation and monomer addition only occur through activated monomers M*; whereas in (C) the transformation to nonpolymerizing latent form is a parallel pathway of the monomer M that may transform either to the active M* or to the latent monomer (M_L) (Chiou et al. 2009).

His334Asp α_1-antitrypsin and cell lysates were collected after 24 h expression. The polyclonal antibody removed all the α_1-antitrypsin from the supernatants of both M and His334Asp α_1-antitrypsin expressing cells after the second round of immunoprecipitation. In contrast, the 2C1 monoclonal antibody immunoprecipitated minimal amounts of α_1-antitrypsin from cells expressing the M variant but large amounts of His334Asp α_1-antitrypsin, virtually depleting the sample after three rounds of immunoprecipitation. Similar results were obtained when the experiment was repeated using cells expressing Z α_1-antitrypsin. These data show that only one type of polymer, recognized by the 2C1 mAb, is detectable in the lysates of cells expressing His334Asp and Z α_1-antitrypsin (Miranda et al. 2010).

18.7 Stem cell technology to generate models of disease

One of the issues of issues with cell models of disease is that they massively overexpress the protein of interest. While this is useful to see phenotypes it can produce unwanted artefacts. Human induced pluripotent stem cells (hIPSCs) promise great opportunities for modeling of human disease as they allow the expression of proteins of interest from endogenous promoters. Dermal fibroblasts have been isolated from individuals with α_1-antitrypsin deficiency and used to generate patient specific hIPSC lines. Each of the hIPSC lines was differentiated into "hepatocyte-like cells" using a novel and simple three step differentiation protocol in chemically defined conditions. The patient specific hIPSC derived hepatocytes show the key pathological features of α_1-antitrypsin deficiency: protein misfolding, the formation of pathological polymers and the retention of polymers in the ER (Rashid et al. 2010). These cells provide an invaluable resource for modeling α_1-antitrypsin deficiency and other serpinopathies.

18.8 Conclusions

The serpinopathies are characterized by delays in protein folding and the retention of ordered polymers of the mutant serpin within the cell of synthesis. The clinical phenotype results from either a toxic gain of function from the inclusions or a loss of function, as there is insufficient protease inhibitor to regulate important proteolytic cascades. It is important to recognize that the conditions in which experiments are performed will have a major effect on the findings. For example, incubation of monomeric serpins with guanidine or urea will produce polymers that are not found in vivo. The single strand reactive loop-β-sheet A interaction best explains the data from limited proteolysis, hydrogen-deuterium exchange, monoclonal antibodies, mass spectrometry, TCCD and from cell models of disease. This linkage should be used to develop strategies for blocking polymerization to treat the serpinopathies.

References

Aulak, K.S., Eldering, E., Hack, C.E., Lubbers, Y.P.T., Harrison, R.A., Mast, A., et al. (1993). A hinge region mutation in C1-inhibitor (Ala[436]→Thr) results in nonsubstrate-like behavior and in polymerization of the molecule. J Biol Chem *268*, 18088–18094.

Baek, J.H., Im, H., Kang, U.B., Seong, K.M., Lee, C., Kim, J., et al. (2007). Probing the local conformational change of a$_1$-antitrypsin. Protein Sci *16*, 1842–1850.

Belorgey, D., Crowther, D.C., Mahadeva, R., and Lomas, D.A. (2002). Mutant neuroserpin (Ser49Pro) that causes the familial dementia FENIB is a poor proteinase inhibitor and readily forms polymers *in vitro*. J Biol Chem *277*, 17367–17373.

Belorgey, D., Irving, J.A., Ekeowa, U.I., Freeke, J., Roussel, B.D., Miranda, E., et al. (2011). Characterisation of serpin polymers in vitro and in vivo. Methods *53 (3)*, 255–266.

Belorgey, D., Sharp, L.K., Crowther, D.C., Onda, M., Johansson, J., and Lomas, D.A. (2004). Neuroserpin Portland (Ser52Arg) is trapped as an inactive intermediate that rapidly forms polymers: implications for the epilepsy seen in the dementia FENIB. Eur J Biochem *271*, 3360–3367.

Bottomley, S.P., Hopkins, P.C., and Whisstock, J.C. (1998). Alpha 1-antitrypsin polymerisation can occur by both loop A and C sheet mechanisms. Biochem Biophys Res Commun *251*, 1–5.

Bradshaw, C.B., Davis, R.L., Shrimpton, A.E., Holohan, P.D., Rea, C.B., Fieglin, D., et al. (2001). Cognitive deficits associated with a recently reported familial neurodegenerative disease. Arch Neurol *58*, 1429–1434.

Bruce, D., Perry, D.J., Borg, J.-Y., Carrell, R.W., and Wardell, M.R. (1994). Thromboembolic disease due to thermolabile conformational changes of antithrombin Rouen VI (187 Asn→Asp). J Clin Invest *94*, 2265–2274.

Cabrita, L.D., Dai, W., and Bottomley, S.P. (2004). Different conformational changes within the F-helix occur during serpin folding, polymerization and proteinase inhibition. Biochemistry *43*, 9834–9839.

Carrell, R.W., Stein, P.E., Fermi, G., and Wardell, M.R. (1994). Biological implications of a 3Å structure of dimeric antithrombin. Structure *2*, 257–270.

Chang, W.-S.W., Whisstock, J., Carrell, R.W., and Wardell, M.R. (1994). The mechanism of antithrombin polymerization: a pathological process. Blood *84 (suppl. 1)*, 391a.

Chiou, A., Hägglöf, P., Orte, A., Yuyin Chen, A., Dunne, P., Belorgey, D., et al. (2009). Probing neuroserpin polymerization and interaction with amyloid-b peptides using single molecule fluorescence. Biophys J *97*, 2306–2315.

Corral, J., Aznar, J., Gonzalez-Conejero, R., Villa, P., Minano, A., Vaya, A., et al. (2004). Homozygous deficiency of heparin cofactor II: relevance of P17 glutamate residue in serpins, relationship with conformational diseases, and role in thrombosis. Circulation *110*, 1303–1307.

Coutelier, M., Andries, S., Ghariani, S., Dan, B., Duyckaerts, C., van Rijckevorsel, K., et al. (2008). Neuroserpin mutation causes electrical status epilepticus of slow-wave sleep. Neurology *71*, 64–66.

Dafforn, T.R., Mahadeva, R., Elliott, P.R., Sivasothy, P., and Lomas, D.A. (1999). A kinetic mechanism for the polymerisation of α_1-antitrypsin. J Biol Chem *274*, 9548–9555.

Davies, M.J., Miranda, E., Roussel, B.D., Kaufman, R.J., Marciniak, S.J., and Lomas, D.A. (2009). Neuroserpin polymers activate NF-kB by a calcium signalling pathway that is independent of the unfolded protein response. J Biol Chem *284*, 18202–18209.

Davis, R.L., Holohan, P.D., Shrimpton, A.E., Tatum, A.H., Daucher, J., Collins, G.H., et al. (1999a). Familial encephalopathy with neuroserpin inclusion bodies (FENIB). Am J Pathol *155*, 1901–1913.

Davis, R.L., Shrimpton, A.E., Carrell, R.W., Lomas, D.A., Gerhard, L., Baumann, B., et al. (2002). Association between conformational mutations in neuroserpin and onset and severity of dementia. Lancet *359*, 2242–2247.

Davis, R.L., Shrimpton, A.E., Holohan, P.D., Bradshaw, C., Feiglin, D., Sonderegger, P., et al. (1999b). Familial dementia caused by polymerisation of mutant neuroserpin. Nature *401*, 376–379.

Dementiev, A., Dobó, J., and Gettins, P.G. (2006). Active site distortion is sufficient for proteinase inhibition by serpins: structure of the covalent complex of alpha1-proteinase inhibitor with porcine pancreatic elastase. J Biol Chem *281*, 3452–3457.

Devlin, G.L., Chow, M.K.M., Howlett, G.J., and Bottomley, S.P. (2002). Acid denaturation of α_1-antitrypsin: characterization of a novel mechanism of serpin polymerization. J Mol Biol *324*, 859–870.

Dunstone, M.A., Dai, W., Whisstock, J.C., Rossjohn, J., Pike, R.N., Feil, S.C., et al. (2000). Cleaved antitrypsin polymers at atomic resolution. Protein Sci *9*, 417–420.

Ekeowa, U.I., Freekeb, J., Miranda, E., Gooptu, B., Bush, M.F., Pérez, J., et al. (2010). Defining the mechanism of polymerization in the serpinopathies. Proc Natl Acad Sci U S A *107*, 17146–17151.

Eldering, E., Verpy, E., Roem, D., Meo, T., and Tosi, M. (1995). COOH-terminal substitutions in the serpin C1 inhibitor that cause loop overinsertion and subsequent multimerization. J Biol Chem *270*, 2579–2587.

Elliott, P.R., Lomas, D.A., Carrell, R.W., and Abrahams, J.-P. (1996). Inhibitory conformation of the reactive loop of a_1-antitrypsin. Nat Struct Biol *3*, 676–681.

Elliott, P.R., Pei, X.Y., Dafforn, T.R., and Lomas, D.A. (2000). Topography of a 2.0Å structure of α_1-antitrypsin reveals targets for rational drug design to prevent conformational disease. Protein Sci *9*, 1274–1281.

Eriksson, S., Carlson, J., and Velez, R. (1986). Risk of cirrhosis and primary liver cancer in alpha$_1$-antitrypsin deficiency. N Engl J Med *314*, 736–739.

Faber, J.-P., Poller, W., Olek, K., Baumann, U., Carlson, J., Lindmark, B., et al. (1993). The molecular basis of α_1-antichymotrypsin deficiency in a heterozygote with liver and lung disease. J Hepatol *18*, 313–321.

Gooptu, B., Hazes, B., Chang, W.-S.W., Dafforn, T.R., Carrell, R.W., Read, R., et al. (2000). Inactive conformation of the serpin α_1-antichymotrypsin indicates two stage insertion of the reactive loop; implications for inhibitory function and conformational disease. Proc Natl Acad Sci U S A *97*, 67–72.

Gooptu, B., and Lomas, D.A. (2009). Conformational pathology of the serpins – themes, variations and therapeutic strategies. Annu Rev Biochem *78*, 147–176.

Gooptu, B., Miranda, E., Nobeli, I., Mallya, M., Purkiss, A., Leigh Brown, S.C., et al. (2009). Crystallographic and cellular characterisation of two mechanisms stabilising the native fold of alpha-1-antitrypsin: implications for disease and drug design. J Mol Biol *387*, 857–868.

Graham, K.S., Le, A., and Sifers, R.N. (1990). Accumulation of the insoluble PiZ variant of human α_1-antitrypsin within the hepatic endoplasmic reticulum does not elevate the steady-state level of grp78/BiP. J Biol Chem *265*, 20463–20468.

Gross, B., Grebe, M., Wencker, M., Stoller, J.K., Bjursten, L.M., and Janciauskiene, S. (2009). New findings in PiZZ alpha(1)-antitrypsin deficiency-related panniculitis. Demonstration of skin polymers and high dosing requirements of intravenous augmentation therapy. Dermatology *218 (4)*, 370–375.

Hastings, G.A., Coleman, T.A., Haudenschild, C.C., Stefansson, S., Smith, E.P., Barthlow, R., et al. (1997). Neuroserpin, a brain-associated inhibitor of tissue plasminogen activator is localized primarily in neurones. J Biol Chem *272*, 33062–33067.

Hidvegi, T., Schmidt, B.Z., Hale, P., and Perlmutter, D.H. (2005). Accumulation of mutant alpha$_1$-antitrypsin Z in the endoplasmic reticulum activates caspases-4 and –12, NFkappaB, and BAP31 but not the unfolded protein response. J Biol Chem *280*, 39002–39015.

Huntington, J.A., Pannu, N.S., Hazes, B., Read, R., Lomas, D.A., and Carrell, R.W. (1999). A 2.6Å structure of a serpin polymer and implications for conformational disease. J Mol Biol *293*, 449–455.

Huntington, J.A., Read, R.J., and Carrell, R.W. (2000). Structure of a serpin-protease complex shows inhibition by deformation. Nature *407*, 923–926.

Huntington, J.A., and Whisstock, J.C. (2010). Molecular contortionism-on the physical limits of serpin 'loop-sheet' polymers. Biol Chem *973–982*.

Im, H., Woo, M.-S., Hwang, K.Y., and Yu, M.-H. (2002). Interactions causing the kinetic trap in serpin protein folding. J Biol Chem *277*, 46347–46354.

Inagi, R., Nangaku, M., Usuda, N., Shimizu, A., Onogi, H., Izuhara, Y., et al. (2005). Novel serpinopathy in rat kidney and pancreas induced by overexpression of megsin. J Am Soc Nephrol *16*, 1339-1349.

James, E.L., and Bottomley, S.P. (1998). The mechanism of α_1-antitrypsin polymerization probed by fluorescence spectroscopy. Arch Biochem Biophys *356*, 296–300.

Kang, H.A., Lee, K.N., and Yu, M.-H. (1997). Folding and stability of the Z and Siiyama genetic variants of human α_1-antitrypsin. J Biol Chem *272*, 510–516.

Kim, J., Lee, K.N., Yi, G.-S., and Yu, M.-H. (1995). A thermostable mutation located at the hydrophobic core of α_1-antitrypsin suppresses the folding defect of the Z-type variant. J Biol Chem *270*, 8597–8601.

Kröger, H., Miranda, E., MacLeod, I., Pérez, J., Crowther, D.C., Marciniak, S.J., et al. (2009). Endoplasmic reticulum-associated degradation (ERAD) and autophagy cooperate to degrade polymerogenic mutant serpins. J Biol Chem *284*, 22793–22802.

Lomas, D.A., Elliott, P.R., Sidhar, S.K., Foreman, R.C., Finch, J.T., Cox, D.W., et al. (1995). Alpha$_1$-antitrypsin Mmalton (^{52}Phe deleted) forms loop-sheet polymers *in vivo*: evidence for the C sheet mechanism of polymerisation. J Biol Chem *270*, 16864–16870.

Lomas, D.A., Evans, D.L., Finch, J.T., and Carrell, R.W. (1992). The mechanism of Z α_1-antitrypsin accumulation in the liver. Nature *357*, 605–607.

Lomas, D.A., Evans, D.L., Stone, S.R., Chang, W.-S.W., and Carrell, R.W. (1993a). Effect of the Z mutation on the physical and inhibitory properties of α_1-antitrypsin. Biochemistry *32*, 500–508.

Lomas, D.A., Finch, J.T., Seyama, K., Nukiwa, T., and Carrell, R.W. (1993b). α_1-antitrypsin S$_{iiyama}$ (Ser^{53}→Phe); further evidence for intracellular loop-sheet polymerisation. J Biol Chem *268*, 15333–15335.

Mahadeva, R., Dafforn, T.R., Carrell, R.W., and Lomas, D.A. (2002). Six-mer peptide selectively anneals to a pathogenic serpin conformation and blocks polymerisation: implications for the prevention of Z α_1-antitrypsin related cirrhosis. J Biol Chem *277*, 6771–6774.

McGowan, S., Buckle, A.M., Irving, J.A., Ong, P.C., Bashtannyk-Puhalovich, T.A., Kan, W.T., et al. (2006). X-ray crystal structure of MENT: evidence for functional loop-sheet polymers in chromatin condensation. EMBO J *25*, 3144–3155.

Miranda, E., and Lomas, D.A. (2006). Neuroserpin: a serpin to think about. Cell Mol Life Sci *17*, 1527–1539.

Miranda, E., McLeod, I., Davies, M.J., Pérez, J., Römisch, K., Crowther, D.C., et al. (2008). The intracellular accumulation of polymeric neuroserpin explains the severity of the dementia FENIB. Hum Mol Genet *17*, 1527–1539.

Miranda, E., Pérez, J., Ekeowa, U.I., Hadzic, N., Kalsheker, N., Gooptu, B., et al. (2010). A novel monoclonal antibody to characterise pathogenic polymers in liver disease associated with α_1-antitrypsin deficiency. Hepatology *52*, 1078–1088.

Miranda, E., Römisch, K., and Lomas, D.A. (2004). Mutants of neuroserpin that cause dementia accumulate as polymers within the endoplasmic reticulum. J Biol Chem *279*, 28283–28291.

Onda, M., Belorgey, D., Sharp, L.K., and Lomas, D.A. (2005). Latent S49P neuroserpin spontaneously forms polymers: identification of a novel pathway of polymerization and implications for the dementia FENIB. J Biol Chem *280*, 13735–13741.

Picard, V., Dautzenberg, M.-D., Villoutreix, B.O., Orliaguet, G., Alhenc-Gelas, M., and Aiach, M. (2003). Antithrombin Phe229Leu: a new homozygous variant leading to spontaneous antithrombin polymerisation in vivo associated with severe childhood thrombosis. Blood *102*, 919–925.

Poller, W., Faber, J.-P., Weidinger, S., Tief, K., Scholz, S., Fischer, M., et al. (1993). A leucine-to-proline substitution causes a defective α_1-antichymotrypsin allele associated with familial obstructive lung disease. Genomics *17*, 740–743.

Purkayastha, P., Klemke, J.W., Lavender, S., Oyola, R., Cooperman, B.S., and Gai, F. (2005). α_1-antitrypsin polymerisation: a fluorescence correlation spectroscopic study. Biochemistry *44*, 2642–2649.

Rashid, S.T., Corbineau, S., Hannan, N., Marciniak, S.J., Miranda, E., Alexander, G., et al. (2010). Modeling inherited metabolic disorders of the liver using human induced pluripotent stem cells. J Clin Invest *120*, 3127–3136.

Ruotolo, B.T., Benesch, J.L., Sandercock, A.M., Hyung, S.J., and Robinson, C.V. (2008). Ion mobility-mass spectrometry analysis of large protein complexes. Nat Protoc *3*, 1139–1152.

Ryu, S.-E., Choi, H.-J., Kwon, K.-S., Lee, K.N., and Yu, M.-H. (1996). The native strains in the hydrophobic core and flexible reactive loop of a serine protease inhibitor: crystal structure of an uncleaved α_1-antitrypsin at 2.7Å. Structure *4*, 1181–1192.

Sharp, A.M., Stein, P.E., Pannu, N.S., Carrell, R.W., Berkenpas, M.B., Ginsburg, D., et al. (1999). The active conformation of plasminogen activator inhibitor 1, a target for drugs to control fibrinolysis and cell adhesion. Structure *7*, 111–118.

Sivasothy, P., Dafforn, T.R., Gettins, P.G.W., and Lomas, D.A. (2000). Pathogenic a_1-antitrypsin polymers are formed by reactive loop-β-sheet A linkage. J Biol Chem *275*, 33663–33668.

Smith, D.P., Knapman, T.W., Campuzano, I., Malham, R.W., Berryman, J.T., Radford, S.E., et al. (2009). Deciphering drift time measurements from travelling wave ion mobility spectrometry-mass spectrometry studies. Eur J Mass Spectrom *15*, 113–130.

Stein, P.E., and Carrell, R.W. (1995). What do dysfunctional serpins tell us about molecular mobility and disease? Nat Struct Biol *2*, 96–113.

Sveger, T. (1976). Liver disease in alpha$_1$-antitrypsin deficiency detected by screening of 200,000 infants. N Engl J Med *294*, 1316–1321.

Takehara, S., Zhang, J., Yang, X., Takahashi, N., Mikami, B., and Onda, M. (2010). Refolding and polymerization pathways of neuroserpin. J Mol Biol *403*, 751–762.

Tsutsui, Y., Kuri, B., Sengupta, T., and Wintrode, P.L. (2008). The structural basis of serpin polymerization studied by hydrogen/deuterium exchange and mass spectrometry. J Biol Chem *283*, 30804–30811.

Whisstock, J.C., and Bottomley, S.P. (2006). Molecular gymnastics, serpin structure, folding and misfolding. Curr Opin Struc Biol *16*, 761–768.

Wu, Y., Swulius, M.T., Moremen, K.W., and Sifers, R.N. (2003). Elucidation of the molecular logic by which misfolded α_1-antitrypsin is preferentially selected for degradation. Proc Natl Acad Sci U S A *100*, 8229–8234.

Yamasaki, M., Li, W., Johnson, D.J., and Huntington, J.A. (2008). Crystal structure of a stable dimer reveals the molecular basis of serpin polymerization. Nature *455*, 1255–1258.

Yepes, M., and Lawrence, D.A. (2004). New functions for an old enzyme: nonhemostatic roles for tissue-type plasminogen activator in the central nervous system. Exp Biol Med *229*, 1097–1104.

Yu, M.-H., Lee, K.N., and Kim, J. (1995). The Z type variation of human α_1-antitrypsin causes a protein folding defect. Nat Struct Biol *2*, 363–367.

Zhang, Q., Law, R.H., Bottomley, S.P., Whisstock, J.C., and Buckle, A.M. (2008). A structural basis for loop C-sheet polymerization in serpins. J Mol Biol *376*, 1348–1359.

Zhou, A., and Carrell, R.W. (2008). Dimers initiate and propagate serine protease inhibitor polymerisation. J Mol Biol *375*, 36–42.

Zhou, A., Faint, R., Charlton, P., Dafforn, T.R., Carrell, R.W., and Lomas, D.A. (2001). Polymerisation of plasminogen activator inhibitor-1. J Biol Chem *276*, 9115–9122.

Index

activation energy 78
acyl carrier protein (ACP) 274, 275, 276, 277, 278, 279, 280, 281, 282, 284, 285, 286, 287
— AcpS 199, 203, 205, 208, 218
— ACP-tag 198, 203, 205, 207
aggregation 187, 190, 194
Aib 309
Alanine-scanning mutagenesis 297
alginate 6, 8
alkaline phosphatase 6
α-amylases 6
alpha-crystallin B 244, 245, 247
α-cyano-4-hydroxycinnamic acid 117
α-galactosidase 7
α-helices 87
alpha helix 172, 173, 177, 178, 181
Alzheimer's disease 187, 188
amplitudes 86
amyloid fibrils 187
analytical ultracentrifugation (AUC) 30, 33, 37
anilinonaphtalene sulfonic acid 51
anion exchange chromatography 280
antibody-antigen interaction 296
anti-chaotropic effects 3
antimicrobial peptide 311
apoptosis 96
association rates 296
A-TMP-tag 202
ATP-hydrolysis 15
ATP sepharose 17
autophagy 19
avian pancreatic polypeptide 305

bacterial periplasm 95, 96
BAG-domain co-chaperone 15
bakers yeast 95, 98
BAL 220, 221, 222, 223, 224, 225, 226, 227, 228, 229, 230, 231
barnase-barstar complex 297
BCL-2 304

Bcl-XL 299
benzylcytosine 199, 201
benzylguanine 199, 200
beta-barrel membrane protein 23
ß-hairpin 84, 307
ß-lactamase 297
ß-peptide 304
β-sheets 172, 173, 177, 181
ß-strand 309
BH3 domain 304
bimolecular rate constant 76
bioconjugation 197, 198
biomarker 219, 220, 221, 222, 223, 225, 233, 236, 247, 248
biosensor 8, 210, 215
biotin 197, 198, 199, 201, 203, 214, 216, 218
biotinbiotin ligase (BirA) 199, 203, 205, 208
bis(sulfosuccinimidyl) glutarate (BS2G) 128
bis(sulfosuccinimidyl) suberate (BS3) 128
BL21 17
Bowman-Birk 309
Bradford 236, 238, 242
brassinosteroid receptor 51

CAL 253, 256, 259, 260, 262, 264, 265, 269
CALI 210
calcium 50, 235, 245, 246, 247, 248
calnexin 47
calreticulin 47
cancer 21
CAP 19
carbonic anhydrase I and II 6
CASA 19
cathepsin B 6
cathepsin D 5, 6
cathepsin L 3, 4, 6
cation exchange 4

CCR5   310
CD4   304
C8E4   28, 34
cell culture   236, 237, 240, 241, 248
cell-free transcription system   100
cell surface   201, 202, 203, 212, 213, 214, 215, 218
cellulase complex of Trichoderma reesei   1, 6
CF   265
CFTR   259, 264, 265, 266, 267, 268, 269
chaperone   13
— chaperone-assisted degradation   22
CHAPS   28
charge state distribution (CSD)   284
charybdotoxin   304
chemical cross-linking   127
— data analysis   134
— experimental parameters   132
— mass spectrometric analysis   133
— protocol for   131
— reagents for   128
— sample requirements for   131
— software   134
— spacer length of reagents   129
chemical probes   197, 198, 201, 202, 204, 206, 208
chemiluminescence   277
chitosan   6, 8
chloroplasts   95
choline oxidase   155, 156
chymotrypsin inhibitor 2   51
circular dichroism   271, 281, 282, 283, 284
— spectroscopy   30, 41
CLIP-tag   199, 201, 207, 208, 210
cluster   297
CMA   19
CoA   199, 203, 205
coarse-grained approach   180
complementation assay   285, 287
Complete protease inhibitor   18
conductive double-sided adhesive tape   115, 120

conformational change   271, 272, 275, 288
conformational dynamics   73
conformational equilibrium   85
conformational state   271
conotoxin   305
constant fragment (Fc)   308
contact formation   73
Coomassie Blue   238, 242
COPD   220, 221, 222, 225, 226, 227, 228, 231, 232
cosolvents differentiating   77
cosolvents leveling   2
cristae   95
cross-linking probes   210
cross-linking reagents
— affinity-tagged   129
— cleavable   130
— conventional   128
— isotope-coded   129
cruzipain   51
crystallization   98
CXCR4   312
cyanocysteine   273
cyclic polypeptide   272, 273, 274
cyclotides   272, 273, 305
CyDye   222, 225, 227
cysteine peptidases   7
cysteine proteinases   6
cystic fibrosis   21, 252, 260
— CF-cells   263
cystic fibrosis transmembrane regulator   39, 252, 260
— CFTR modulator   253
— CTFR trafficking   253, 264
cytochrome c   97, 103, 105
cytochrome c oxidase   98

DEAE sepharose   17
Debye formula   171, 179
degradation   13
denaturant   91
detergent
— critical micelle concentration (CMC)   25, 29, 30
— micelle   24, 29, 31, 35, 37

Dexter mechanism 74
diamond-like carbon-coated stainless steel plates 124
differentially expressed proteins 223, 229, 230, 231
diffusion coefficients 77
diffusion controlled reactions 76
DIGE 219, 221, 222, 223, 224, 225, 227, 229, 231, 233
dimerisation of serum albumin 3
dipole-dipole 74
dissociation constants 295
disuccinimidyl glutarate (DSG) 128
disuccinimidyl suberate (DSS) 128
disulfide bonds 96
dithiothreitol (DTT) 277, 279, 280
DNA 6, 7, 8
dodecyl phosphocholine (DPC) 25, 27, 35
double mutant cycle 297
Duchenne musclar dystrophy (DMD) 235, 240, 242, 248
— dehydrogenase in DMD 247
— diagnosis of DMD 235, 236
— DMD myoblasts 240, 242
— DMD progression 240, 244
— DMD samples 242
— proteins in DMD 245, 246
— proteome analysis of DMD 247
dystrophin 235, 236, 237, 240, 247, 248

eDHFR 202, 205
Einstein-Smoluchowski 296
electroblotting 117
electroblotting efficiency 120
electronic absorption spectroscopy 149, 156, 158
electrophoresis 238, 241, 242
— blue native electrophoresis 119
— gel electrophoresis 277, 281
— 1DE-IEF-electrophoresis 238
— phos-tag gel electrophoresis 118
— polyacrylamide gel electrophoresis 39, 41

— 2-DE electrophoresis 113, 117, 118
— 2-DE gel electrophoresis 225
— 2D-PAGE electrophoresis 236
emulsion stability 4
end-capping groups 302
endo-β-N-acetylglucosaminidase H 59
endoplasmic reticulum (ER) 47, 95
— associated degradation 48, 325
enrichment of cross-linked peptides 131
enthalpy 298
entropic 298
entropy 298
enzyme 295
enzymeenzyme-mediated labeling 199, 202
enzymeerythrocyte enzyme 7
epitope 297
equilibrium constant 85
ER 96
ERp57 48, 106
Erv1 96, 97, 98, 101, 103, 104, 105, 108
erythropoietin 308
ESI-LIT/TOF MS 118
ESI-Q/TOF MS 118
EX1-limit 86
EX2-limit 86
excreted-secreted proteins of Corynebacterium pseudotuberculosis 6
exo-polygalacturonase 7

fixation 197, 212, 213
fluctuation 166, 167, 179, 180, 181
fluorescence spectroscopy 271, 282, 283, 284
fluorescent proteins 96
fluorophores 197, 198, 201, 202, 204, 209, 216
Fluorotrans 113
foldamers 306
förster resonance energy transfer 32, 39, 272

Index

fourier transform infrared spectroscopy   30
FP   197, 201, 210, 212, 213
fragment-based approach   299
FRET   74, 201, 210, 214, 215, 217, 218

Gal4 transcription factor   15
gene replacement   285
glucoamylase   6
glucosidase II   47
glutathione   96, 101
— glutathione-S-transferase   280
glycophorin A   37
gp41   306
gp120   304
gramicidin S   309
growth complementation   286, 287, 288
growth hormone   296

hairpin   307
haloalkane   199, 201, 202
HaloTag   198, 199, 201, 202, 205, 207, 208, 217, 218
HDM2   304
heat shock protein (Hsp)   244, 245, 248
— Hsc70   13, 15
— Hsp40   15
— Hsp70   13
— Hsp72   13
— Hsp73   13
— HspA1   13
— HspA8   13
— HspBP1   15
helical epitopes   302
Helix mimetics   302
hemoglobin   166, 170, 174, 175, 176, 177
high-throughput screening   299
Hip   15
HIV-1   304
homogenization artifacts   3
homopolymer   83
hot-residues   297
hot spot   296
HP35   89
hydration   179

hydrogen-bond surrogates   302
hydrogen/deuterium (H/D) exchange   85
hydrogen peroxide   96, 103
hydrophobic interaction   296

IEF   221, 222, 225, 227, 238
IL2   301
Immobilon
— Immobilon CD   115
— Immobilon P   113
— Immobilon PSQ   113
immunofluorescence   213
immunoglobulin G   308
inhibitors   295
internal standard   225
intracellular labeling   200, 204, 206
intrachain diffusion   73
invertase   7
Ionizing radiation   146
IPG   238, 241, 242
Ising model   88
isopropyl β-D-1-thiogalactopyranoside (IPTG)   276, 277, 280, 285, 287, 288

lactoglobulin   177, 178
lactose permease   36
lasso peptides   272, 273
lauryl dimethylamine oxide   28, 34
LC-MS/MS   223, 229, 230, 231
ligand
— binding   165, 175, 177
— protein-ligand interaction   182
lipids   24, 32, 33, 35
Lipoic acid ligase   204
liquid chromatography   113, 279, 284
loop formation   79
loop size   80
low molecular weight pigments   2
LplA   204, 205
LptD   311
LTQ-Orbitrap MS   118
lysophospholipids   28, 30
lysosomal proteinases   1
lysosomes   4

macrocyclic 309
major histocompatibility complex 52
MALDI 239, 242
    — IR-MALDI-TOF MS 115
    — MALDI MS 113, 115
    — MALDI-TOF MS 117
    — MALDI-TOF/TOF MS 118
manganese superoxide dismutase 6
Man 6-P Receptor Homologous 52
MARS 224, 225
MASCOT 118, 120, 239, 242
mass fingerprint 115, 124
    — peptide mass fingerprinting 117
mass spectrometry (MS) 21, 113, 127, 271, 279, 280, 283, 284
    — MS efficiency 120
    — MS/MS 113
    — MS/MS analysis 118
MCP-tag 203, 205
MDM2 306
membrane
    — lipids 2
    — protein 5
metabolism 240, 247, 248
metathesis 303
Mia40 97, 98, 99, 101, 102, 103, 104, 105, 108
micelle 24
microinjection 212
microscopic rate constants 85
mitochondria 47, 95
modeling
    — use of cross-linking for 137
molecular crowding or exclusion 4
molecular exclusion chromatography 4
myoblast 235, 236, 237, 240, 241
myoglobin 172, 174
myopathy 97

naphthalene 74
native chemical ligation (NCL) 273, 274
native PAGE 279, 280, 281
natural product 299
n-dodecyl-β-D-maltoside (DDM) 26, 28, 34

needle complex
    — analysis by cross-linking 137
N-glycosylation 47
NHERF 256, 259, 260, 264
NMR 98
NMR spectroscopy 73
n-octyl-β-D-gluocoside (OG) 26, 28, 34
nomenclature of cross-linking products 128
non-fluorescent substrate 209, 213
NOTCH1 304
nuclear magnetic resonance (NMR) 25, 27, 28, 29, 30
    — in-cell NMR 271
nucleated polymerization 187, 188

oligoureas 306
On-membrane digestion 117
O-ring 297
oxidative stress 244
oxidoreductases 96

pair distribution function 171, 172, 178, 180
PALM 209
parvalbumin 83
P-domain 55
PDZ domain 250, 251, 253, 260, 262, 264, 265, 266, 267, 268, 269
peptides 73
    — array 249, 250, 253, 254, 255, 256, 260
    — chemically synthesized peptide 190
    — library 274
    — for MALDI 239
    — self-labeling peptide 198
    — sequencing 279
    — synthesis 39, 190
peptoid 304
permeability 302
peroxidase 6, 7
peroxisomes 95
persistence length 81
p53 304
phage-display 308

phage transduction 285
phospholipase D 6
Photobleaching 214
photo-crosslinker 204
photoelectric effect 146
photoswitchable probes 209
piezoelectric chemical inkjet printer 115
polyacrylamide gel electrophoresis 34, 37
polyethlylenglycol 77
polyphemusin 310
polyvinylidene difluoride (PVDF) 113, 117, 119
PPTases 199, 203
PRIME 204, 205, 208
principal components analysis 172, 176
prion 187, 188
programmed cell death 96
pro-MMP-9 6
Pro peptide bond 83
prosaposin 51
protease 7
proteasome
 — analysis by cross-linking 136
protegrin-I 310
protein
 — binding protein (BiP) 51
 — corn protein 6
 — cyclic protein 273, 277, 285
 — differentially expressed protein 219, 221
 — disordered protein 165, 184
 — fluorescent protein 197, 214, 218
 — green fluorescent protein 5
 — intrinsically disordered protein 271
 — LAMP-2 integral membrane proteins 4
 — natively unfolded protein 275
 — protein aggregation disease 21
 — protein complex 127
 — protein cyclization 273, 274, 275, 276, 277, 279, 288
 — protein degradation 14, 16, 19, 22
 — protein domain 172, 173
 — protein dynamics 271, 272, 274, 288
 — protein engineering 272
 — protein extraction 236, 237
 — protein flexibility 271
 — protein function 210, 217
 — protein interaction 127
 — protein interaction domain (PID) 249
 — protein labeling 199, 204, 205, 208, 217, 218
 — protein labeling system 198, 200, 201, 202
 — protein microarray 203
 — protein misfolding 187
 — protein-protein interaction 200, 201, 202, 203, 204, 210, 214
 — protein-protein interfaces 295
 — protein purification 201, 277, 280
 — protein structure 165, 170, 271, 274, 281
 — protein structure-function relationships 274
 — protein trafficking 197, 206
 — protein trans-splicing 274, 275, 277, 280
 — protein turnover 213
 — unfolded protein response 50
protein disulfide isomerases 96
Protein Epitope Mimetic (PEM) technology 309
protein folding 84
proteome 219, 220, 221, 222, 223, 224
 — analysis 235, 240, 247, 248
proteostasis 13
pseudo first-order conditions 77
Pseudomonas 311
pullulanase 6
pulse-chase 198, 201, 202, 206, 209, 213
purine nucleoside phosphorylase 295
Pyroglutamyl peptidase type I 6

quantum dots 202, 203, 216
quencher 76

radiation damage 147
radical formation 78
Raf-1 21
Ramachandran map 81
Raman scattering 150
Raman spectroscopy 149, 154
rate constants 86
reactive boundary 77
resonance Raman 150, 157, 158
respiratory chain 95, 97, 98
Rev protein 310
Rg 165, 171, 172
Rieske cluster 158
— Rieske [2Fe-2S] cluster 157
RNA polymerase II
— analysis by cross-linking 136
RRE-RNA 310

Saccharomyces cerevisiae 96
SAR by NMR 299
scorpion 304
scyllatoxin 304
SDS-PAGE 113, 117, 221, 222, 227, 228, 277, 279, 280
self-labeling 201
— approach 203
Sephadex G-75 4
Sfp 199, 203, 205, 208, 218
Sf21 16
side-chain cross-linking 303
signal transduction 236, 244, 248
single-crystal spectroscopy 150, 151, 152, 153, 154
size exclusion chromatography (SEC) 31
skeletal muscle cells 236
small multidrug resistance transporters 39
Smoluchowski equation 77
SNAP-tag 198, 199, 200, 201, 205, 206, 207, 208, 210, 211, 212, 213, 214, 215, 216, 217
SOD (superoxide dismutase) 244, 245, 247

sodium dodecyl sulfate (SDS) 25, 26, 35, 37
sodium perfluoroctanoate (SPFO) 29, 34, 39
solid-phase peptide synthesis (SPPS) 273
solvated electrons 147
Sortase A 199, 204, 205, 273
soybean agglutinin 51
soybean trypsin inhibitor 6, 8
speed-limit for protein folding 84
split intein 274, 275, 276, 277, 278, 279, 288
SPOT synthesis 250, 256, 258, 261, 264, 265, 266, 268
SrtA 204
S-Sepharose 5
stachydrinedemethylase 157, 158
stapling 303
starch 6, 8
STED 209
stefin B 3, 4
Stern-Volmer equation 77
Stokes-Einstein equation 77
STORM 209
structural ensemble 165, 166, 180, 181, 182
structural fluctuation 165
subtiligase 273
sulfhydryl oxidases 96
superoxide dismutase 6
super-resolution microscopy 198, 202, 206, 209
SWISSPROT 118, 262
synchrotron x-ray sources 165

targeting signal 95, 98
TAR-RNA 309
T-cell receptor 51
temperature factor (B-factor) 7
tetracysteine 198, 199, 200, 206, 214, 216, 217
tetraserine 199, 200, 215
therapies 235
thia-zip reaction 273
thioester bond 3
thiol-disulfide interchange 33

three-phase partitioning (TPP)   1, 2
— activity after TPP   5
— macroaffinity ligand-facilitated three-phase partitioning   8
— mechanism of TPP   3
— other uses   8
— TPP-modified proteinase K   5
three-state model   87
three-state reaction   75
thylakoid membrane proteome   6
time-resolved study   182
TMP-tag   198, 199, 201, 202, 205, 207, 208, 215
Transblot   113
transglutaminase   273
translation   100
transmembrane protein
— bilayer insertion   24, 35, 38
— denaturation   25, 27, 29, 30, 31, 35
— detergent solubilization   25, 27, 28
— folding   24, 37, 41
— helix-helix association   24, 27, 29, 32, 34, 35, 37, 39
— secondary structure   30
— transporter   24
trimethoprim   199, 202, 205, 215, 217, 218
triplet-triplet energy transfer   73
trypsin   115, 117, 120, 236, 237, 239, 242, 279
— inhibitor   7
tryptophan
— fluorescence   31
— zipper   308
tunicamycin   50

2D-PAGE   239, 240
two-photon imaging   202, 215

UbcH5b   20
ubiquitin   172, 179, 181
ubiquitylation   21
UDP-[14C]Glc   58
UDP-Glc:glycoprotein glucosyltransferase   48
undruggable   313
unfolded state   85
unstructured polypeptide chains   79

villin headpiece subdomain   89
viscosity-dependence   77

water molecules   297
Western blot   277, 279
Westran   113
wheat germ
— agglutinin   6, 8
— bifunctional protease/amylase inhibitor   6
— lipase   8
Wide-angle X-ray Solution Scattering (WAXS)   165

xanthone   74
x-ray crystallography   28, 32, 34
xylanase   6

yeast
— proteins   120
— yeast-2-hybrid   15
Yos9p   52

ZIP-TIP   239, 242